AF895731

Clinico-Pathological Atlas of Cardiovascular Diseases

Joaquín S. Lucena
Pablo García-Pavía
M. Paz Suárez-Mier
Luis A. Alonso-Pulpón
Editors

Clinico-Pathological Atlas of Cardiovascular Diseases

Springer

Editors
Joaquín S. Lucena
Department of Forensic Pathology
Institute of Legal Medicine
University of Cadiz
Seville
Spain

M. Paz Suárez-Mier
Histopathology Department
National Institute Toxicology
and Forensic Sciences
Madrid
Spain

Pablo García-Pavía
Cardiology Department
Hospital Universitario Puerta de Hierro
Majadahonda
Madrid
Spain

Luis A. Alonso-Pulpón
Cardiology Department
Hospital Universitario Puerta de Hierro
Majadahonda
Madrid
Spain

ISBN 978-3-319-11145-2 ISBN 978-3-319-11146-9 (eBook)
DOI 10.1007/978-3-319-11146-9
Springer Cham Heidelberg New York Dordrecht London

Library of Congress Control Number: 2014956504

© Springer International Publishing Switzerland 2015
This work is subject to copyright. All rights are reserved by the Publisher, whether the whole or part of the material is concerned, specifically the rights of translation, reprinting, reuse of illustrations, recitation, broadcasting, reproduction on microfilms or in any other physical way, and transmission or information storage and retrieval, electronic adaptation, computer software, or by similar or dissimilar methodology now known or hereafter developed. Exempted from this legal reservation are brief excerpts in connection with reviews or scholarly analysis or material supplied specifically for the purpose of being entered and executed on a computer system, for exclusive use by the purchaser of the work. Duplication of this publication or parts thereof is permitted only under the provisions of the Copyright Law of the Publisher's location, in its current version, and permission for use must always be obtained from Springer. Permissions for use may be obtained through RightsLink at the Copyright Clearance Center. Violations are liable to prosecution under the respective Copyright Law.
The use of general descriptive names, registered names, trademarks, service marks, etc. in this publication does not imply, even in the absence of a specific statement, that such names are exempt from the relevant protective laws and regulations and therefore free for general use.
While the advice and information in this book are believed to be true and accurate at the date of publication, neither the authors nor the editors nor the publisher can accept any legal responsibility for any errors or omissions that may be made. The publisher makes no warranty, express or implied, with respect to the material contained herein.

Printed on acid-free paper

Springer is part of Springer Science+Business Media (www.springer.com)

Foreword I

It is a pleasure and an honour to write these introductory lines for the 'Clinicopathological Atlas of cardiovascular disease'. The main purpose of this book is to contribute to the diffusion of the morphological findings in cardiovascular pathology among the several specialists who deal with the care of patients affected by cardiovascular diseases: cardiologists, cardiac surgeons, pathologists, forensic doctors, sports doctors, etc.

Cardiovascular diseases represent the main cause of death in the world, but paradoxically, the specialists who care for the patients affected by these diseases know little about the disease after the death of their patients. The significant explosion of imaging techniques and minimally invasive tests in clinical practice has perhaps given physicians a false feeling of already knowing the morphological aspect of the diseases they care for. Furthermore, the progressive decline in the number of autopsies performed at hospitals has undoubtedly helped little to change this tendency.

The present work manages to fill this gap through the coordinated endeavour of forensic pathologists throughout the Spanish territory and of cardiovascular specialists, the vast majority of whom come from the 'Hospital Universitario Puerta de Hierro'. The directors of this work, Dr. Lucena ('Instituto de Medicina Legal' in Sevilla) and Dr. Suarez-Mier ('Instituto Nacional de Toxicología y Ciencias Forenses' in Madrid) on the side of the forensic pathologists, together with Dr. Garcia-Pavia and Prof. Alonso-Pulpón (Hospital Universitario Puerta de Hierro'. Madrid) on the clinicians' side, have succeeded in involving numerous experts in diverse fields to assemble the best of both worlds in each chapter.

I would like to congratulate the editors and each of the authors for the excellent work done and the great enthusiasm that they have shown. I am convinced that this Atlas will contribute to the expansion of the morphological knowledge of cardiovascular pathology and will become a very important work for future generations of specialists that tackle these diseases both from the clinical and from the forensic side.

Valentin Fuster, MD, PhD
Mount Sinai Heart Institute, Medicine Mount Sinai
School of Medicine, Miami Beach, FL, USA
Centro Nacional de Investigaciones Cardiovasculares
(CNIC, Madrid), Madrid, Spain

Foreword II

Cardiovascular diseases represent the main cause of death for people of developed countries, and frequently they may account for premature fatal outcomes even in the apparently healthy young. The morbid entities are mostly structural, affecting the major components of the heart (aorta, pulmonary artery, pericardium, coronary arteries, myocardium, endocardium and conduction system). The mean age of humankind has increased by 6 years in the span of the last 30 years, and two-thirds of this gain in life expectancy are due to improvement in the diagnosis, therapy and prevention of cardiovascular diseases. Clearly, clinical diagnosis, surgical or interventional therapy and prevention are based on the knowledge of anatomy and pathology so that a post-mortem examination of fatal cases still plays a pivotal role.

This beautiful atlas of cardiovascular diseases is the result of a collaborative Spanish network of scholars in cardiovascular pathology. It includes gross and histological findings illustrated in a very professional manner. Acquired diseases are mostly considered even though relevant congenital anomalies are also covered.

The changing spectrum of cardiovascular pathology with surgical and device pathology is a significant part of this atlas. More than 600 colour illustrations with detailed legends render the atlas impressive.

The Editors of the book and the authors of the chapters should be congratulated. The interest in cardiovascular pathology, which has been scientifically neglected for years, is rising throughout the world, revitalizing a discipline which, in tight collaboration with clinicians, is giving major contributions to the fight against cardiac diseases and early death. The great advances in cardiovascular medicine, following the introduction of cardiac surgery and interventional cardiology, could have not been achieved without the precise knowledge of the structural basis of cardiac diseases in the light of the development of new imaging tools, like three-dimensional echocardiography, computed tomography, positron emission tomography and magnetic resonance, not only for a morpho-functional assessment of the heart and great vessels but also for in vivo tissue characterization.

Certainly, the obligation, in Spain, of forensic autopsy in any case of sudden death has fostered the interest in cardiovascular pathology, and the Spanish colleagues should be congratulated for their enthusiasm and the quality of their commitment.

Gaetano Thiene
Cristina Basso
Cardiovascular Pathology, University of Padua
Padua, Italy

Preface

The paradigm of anatomo-clinical thinking in medical practice, introduced by Bichat (1771–1802, considered the father of modern pathological anatomy), presented the lesions in a tissue as a fundamental element in the diagnosis and physiology of a disease. Through the knowledge of the lesion, the physician should establish a relationship with the patient's signs and symptoms. This mainstream thinking remained in force throughout the first half of the twentieth century.

The understanding of the signs, symptoms and physiopathology of a disease could only be achieved by an accurate diagnosis of the originating histological lesion. It was logical, then, that the practice of necropsies in medical schools and hospitals was routine and mandatory. Diagnosis tools were emerging virtually in parallel at this time with the objective to obtain indirect signs of the causal lesion. Initially, these tools were very rudimentary such as the stethoscopes used, but their complexity progressively increased: X-rays (1895), computed tomography (1972) and magnetic resonance (1973), which, as known by everyone, had completely revolutionised medicine.

The degree of clinico-radiological correlation, including the fact that it is also possible to obtain tissue samples by guided puncture, is so high that the practice of clinical necropsies has been reduced to a minimum and has even been discarded by the hospital pathologist. Young medical doctors and current medical students are scarcely familiar with the macroscopic vision of different pathologies.

Today, we are witnessing a renaissance in cardiovascular pathology thanks to forensic medicine, given that sudden deaths (whose origin is cardiovascular in 80 % of cases) demand, according to Spanish laws and European Union guidelines, a legal autopsy. In Spain, it is estimated that 12.5 % of naturally occurring deaths are sudden deaths, which implies between 10,000 and 15,000 cases per year. Also, more than one half of legal autopsies performed in the Departments of Forensic Medicine are cases of natural deaths, and among them, sudden deaths constitute the most numerous group.

But as in other areas of Medicine or yet more, pathology has to walk alongside the clinic (cardiology) as it is almost impossible to establish an accurate diagnosis based only on morphology. The main purpose of this Atlas is to assemble the knowledge and points of view of two medical specialties, cardiology and forensic pathology, which, although apparently very remote from each other, must, however, converge in order to gain a better understanding of the disease.

This Atlas compiles, in 660 images distributed in 12 chapters, the spectrum of pathologies affecting the structures of the cardiovascular system (aorta, pulmonary vessels, coronary arteries, heart valves, myocardium, pericardium, cardiac conduction system, etc.), most frequently encountered in sudden cardiac deaths. It has been designed as a very practical book that includes a summary of clinical characteristics, many high-quality images and clear tables containing the main morphological characteristics of each disease. It is conceived as a first step in cardiovascular pathology diagnoses, and more comprehensive scientific references are offered in addition at the end of each chapter.

The coordinated work of forensic pathologists and histopathologists from one side and of cardiologists from the other has allowed the summation of these two separate worlds in each chapter to produce a work that, we hope, will be enjoyed by a wide range of specialists including cardiologists, forensic doctors, pathologists, sports doctors, etc.

We anticipate that the knowledge obtained from this Atlas will contribute not only to a better understanding of cardiovascular pathology but also to a better treatment of patients.

Seville, Spain — Joaquín S. Lucena
Madrid, Spain — Pablo García-Pavía
Madrid, Spain — M. Paz Suárez-Mier
Madrid, Spain — Luis A. Alonso-Pulpón

Acknowledgements

We would like to thank the various professionals who have participated in different chapters of this Atlas, especially our colleagues at the ‘University Hospital Puerta de Hierro’ from Majadahonda (Madrid), at the ‘Institutes of Legal Medicine’ in Bilbao, Valencia, Huelva and Seville, and at the ‘National Institute of Toxicology and Forensic Sciences’ from Madrid. Our special thanks also to the pathologists and forensic doctors from Seville (Spain), Brazil, Chile, Italy and Panama for providing us with interesting cases.

We would like to give a special thank-you to the numerous forensic doctors who, understanding the importance of cardiovascular studies for legal autopsies, have granted us access to thousands of hearts, which has allowed us to compile a great deal of the material presented in this Atlas. Without their contribution, this book would have never seen the light.

And finally, a thank-you to our families for the constant support and infinite patience that they have endured during the time that we have dedicated to this work.

Contents

Contributors

Beatriz Aguilera Histopathology Service, National Institute of Toxicology and Forensic Sciences, Las Rozas (Madrid), Spain

Luis A. Alonso-Pulpón Cardiology Department, Hospital Universitario Puerta de Hierro, Majadahonda (Madrid), Spain

Universidad Autónoma de Madrid, Madrid, Spain

Roberto Barriales-Villa Inherited Cardiovascular Diseases Unit, Cardiology Department, Complejo Hospitalario Universitario A Coruña, Coruña, Spain

Evaristo Castedo Cardiac Surgery Department, Hospital Universitario Puerta de Hierro, Majadahonda (Madrid), Spain

Marta Cobo Heart Failure and Inherited Cardiac Diseases Unit, Cardiology Department, Hospital Universitario Puerta de Hierro, Majadahonda (Madrid), Spain

Juan Manuel Escudier Coronary Care Unit, Cardiology Department, Hospital Universitario Puerta de Hierro, Majadahonda (Madrid), Spain

Ignacio Fernández-Lozano Arrhythmia Section, Cardiology Department, Hospital Universitario Puerta de Hierro, Majadahonda (Madrid), Spain

Universidad Autónoma de Madrid, Madrid, Spain

Pedro M. Garamendi Department of Forensic Pathology, Institute of Legal Medicine, Huelva, Spain

María Dolores García-Cosío Carmena Heart Failure and Transplant Unit, Department of Cardiology, Hospital Santa Creu i Sant Pau, Barcelona, Spain

Pablo García-Pavía Heart Failure and Inherited Cardiac Diseases Unit, Cardiology Department, Hospital Universitario Puerta de Hierro, Majadahonda (Madrid), Spain

Manuel Gómez-Bueno Advanced Heart Failure and Transplant Unit, Cardiology Department, Hospital Universitario Puerta de Hierro, Majadahonda (Madrid), Spain

Gonzalo Guzzo-Merello Heart Failure Unit, Cardiology Department, Hospital Universitario de Getafe, Getafe (Madrid), Spain

Joaquín S. Lucena Department of Forensic Pathology, Institute of Legal Medicine, Seville, Spain

Forensic Medicine, University of Cadiz, Cadiz, Spain

Pilar Molina Histopathology Section, Department of Forensic Pathology, Institute of Legal Medicine, Valencia, Spain

University of Valencia, Valencia, Spain

Benito Morentin Department of Forensic Pathology, Institute of Legal Medicine, Bizkaia, Spain

Forensic Medicine, Basque Country University, Bizkaia, Spain

Luís Nombela-Franco Cardiovascular Institute, Hospital Clínico San Carlos, Universidad Complutense de Madrid, Madrid, Spain

Javier Segovia Advanced Heart Failure and Transplant Unit, Cardiology Department, Hospital Universitario Puerta de Hierro, Majadahonda (Madrid), Spain

Universidad Autónoma de Madrid, Madrid, Spain

M. Paz Suárez-Mier Histopathology Service, National Institute of Toxicology and Forensic Sciences, Las Rozas (Madrid), Spain

Methodological Approach to Cardiac Autopsy

1

M. Paz Suárez-Mier and Beatriz Aguilera

Contents

M.P. Suárez-Mier (✉) • B. Aguilera
Histopathology Service, National Institute of Toxicology and Forensic Sciences, Las Rozas (Madrid), Spain
e-mail: mariapaz.suarez@justicia.es

1.1 Introduction

Pathological study of the heart can be performed either after the death of a subject (during an autopsy) or during the lifetime of a patient (biopsy and cardiac transplantation). The purpose of each of these studies is summarised in Table 1.1.

Presently, the largest number of histopathological studies on the whole heart is carried out in forensic investigation due to the following reasons:

1. Cardiac pathology is responsible for 80 % of the sudden deaths that require a forensic autopsy to discard a violent cause of death.
2. Cardiac pathology can be a coadjuvant factor for deaths that occur during extracardiac surgical interventions or during hospital stays; situations where medical malpractice claims are frequent.
3. Cardiovascular surgical procedures such as correction of congenital defects, catheterizations, ablations, pacemaker implantations, prostheses, coronary by-pass grafts, etc., can result in fatal outcomes that also require legal investigation.
4. Cardiac pathology can be responsible for traffic-related accidents, or work or home-related accidents, which cause death due to traumatic lesions.

In the context of sudden death, it is also essential to study the heart in order to identify potential hereditary diseases, and through this offer

J.S. Lucena et al. (eds.), *Clinico-Pathological Atlas of Cardiovascular Diseases*,
DOI 10.1007/978-3-319-11146-9_1, © Springer International Publishing Switzerland 2015

Table 1.1 Purpose of the histopathological study of the heart

Biopsy	Explanted heart	Clinical autopsy	Medico-legal autopsy
Diagnosis of myocarditis and cardiomyopathies	Diagnosis	Correlation with imaging techniques	Cause of death
Diagnosis of rejection, infection, or lymphoproliferative disorder in a transplanted heart		Effect of a treatment	Contribution to the death
		Extension of disease	Detection of hereditary diseases
			Malpractice claims

counselling and protect other family members. Logically, before performing an autopsy on the heart it is necessary to have as much information as possible regarding: family and personal history, previous cardiovascular events, surgical interventions and circumstances of the death (whether witnessed, related to physical activity, duration and type of resuscitation manoeuvres attempted).

1.2 Study of the Heart

The study of the heart begins with the removal of the sternum and the inspection for possible intrapericardial effusion (quantity and characteristics: haematic, serous, purulent); or mediastinal haemorrhage due to rupture of the extrapericardial aorta. If gas embolism is suspected, the pericardial sac must be opened under water to check for leakage of gas bubbles through an incision on the pericardial surface. Once the pericardial sac is opened, and to discard thromboembolism, the trunk of the pulmonary artery can be sectioned 2 cm above the pulmonary valve, although many authors recommend for this purpose the opening of the pulmonary arteries from the posterior side. In cases of aortic dissection, the aorta should be examined along its entire length to look for a tear in the intima, and also in the external wall in cases of cardiac tamponade or mediastinal or retroperitoneal haemorrhage.

The heart must be removed as a whole by cutting the inferior vena cava close to and above the diaphragm; the pulmonary veins after checking their correct drainage in the left atrium; and the aorta and pulmonary trunk 2 cm above the semilunar valves.

Once the heart has been removed it should be washed to eliminate blood clots, and weighed. The heart weight must be compared with the expected weight for the gender and body weight. It is estimated that there is an increase of 5 % in the weight of a fixed heart in formalin. It can also be useful to measure the ventricular axes in the posterior surface of the heart, taking as a reference the *crux cordis* (indicative of the presence of dilation and/or ventricular hypertrophy). A transversal axis of 10–11 cm, and a longitudinal axis of 9–10 cm, is considered normal. Next, a macroscopic study is initiated, which is summarised in Table 1.2. The Annex at the end of this chapter provides the normal reference values for cardiac dimensions and weight. It is recommended that photographs of any lesions are taken, or even the absence of alterations, in specific legal cases.

The study of the heart begins with an external examination to assess its shape, appearance of the pericardium, absolute mass of epicardial adipose tissue, etc. (Fig. 1.1).

There are several methods to open and dissect the heart depending on the pathology to be studied but, taking into account that in the majority of

Table 1.2 Gross examination of the heart

External examination	**Shape**: Normal, globoid, ventricular aneurysms
	Pericardium: Roughened or nodular appearance, whitish plaques
	Epicardium: Quantity and distribution of epicardial adipose tissue
Ventricles	**Wall thickness:** concentric or asymmetric hypertrophy
	Subaortic septum: Sigmoid, whitish endocardium
	Ventricular lumen: Diameter, presence of mural thrombi
	Myocardial lesions: Scars, recent necrosis, haemorrhage, disarray, mottled appearance, adipose infiltration, location (basal mid-ventricular, apical) and distribution (subendocardial, transmural, subepicardial)
Valves	**Tricuspid**: Perimeter, vegetations, myxoid changes
	Pulmonary: Number of leaflets, vegetations, dysplastic leaflets
	Mitral: Perimeter, commissure fusion, prolapse, annular calcification, vegetations, shortening and fibrosis or rupture of the tendinous cords
	Aortic: Perimeter, number of leaflets, calcifications in the sinuses or in the sinotubular junction, commissure fusion, stenosis, vegetations, aneurysms, dysplastic leaflets
Coronary arteries	*Ostia coronaria* position
	Course
	Permeability (degree of luminal reduction)
	Dominance
Atria	**Foramen ovale**: patency, aneurysms
	Interatrial septum thickness
	Mural or appendage thombi
Ascending	Atheromas
	Dilatations, aneurysms
Aorta	Intimal tears in dissections
	Orifices of coronary bypass grafts

forensic cases there is no medical antecedents or clinical records of the deceased, the method of choice to study the heart must be the one that allows the diagnosis of the most frequent diseases. The classical method of dissection following the direction of the blood flow can be used in cases of congenital heart disease in children, but not for the study of adult heart pathology (myocardial infarction, cardiomyopathies, infiltrating diseases, etc.) in which several parameters such as the presence of myocardial lesions, wall thickness and ventricular size need to be evaluated. For this reason the most convenient method of dissection of the adult heart involves sectioning the ventricular cone into slices of 1–2 cm in thickness, parallel to the posterior atrioventricular groove, starting from the apex to the base of the papillary muscles. Using this method, about four or five biventricular slices are obtained and the cardiac base is left intact (Fig. 1.2). This sectioning is best performed on a fixed heart and with long-bladed knives.

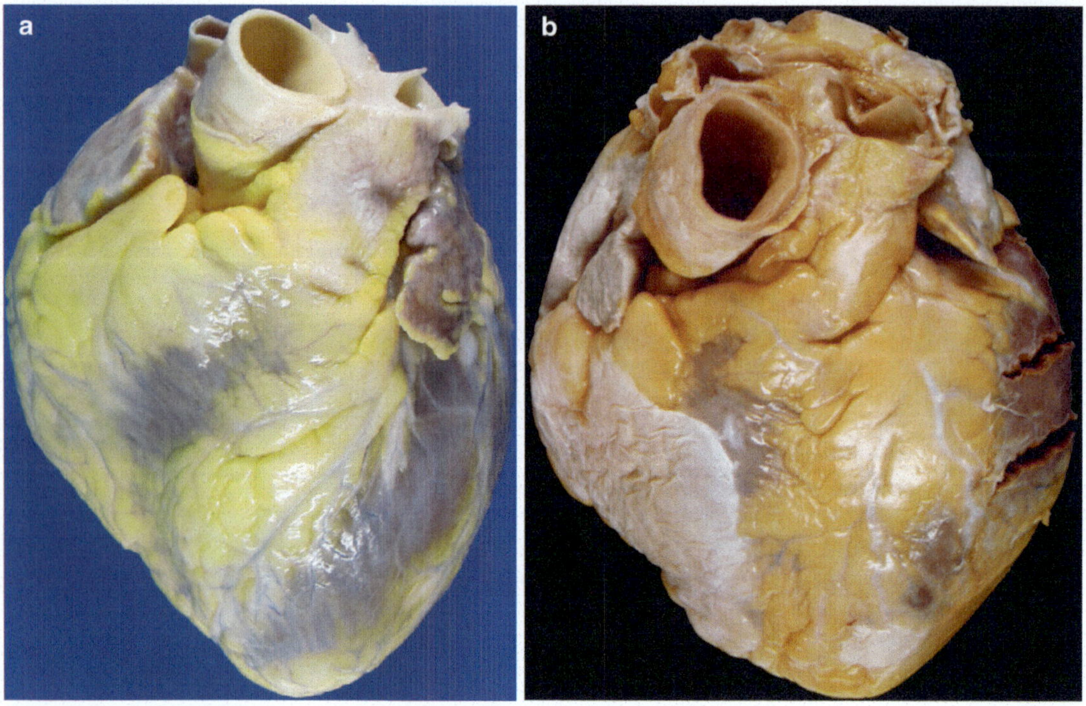

Fig. 1.1 (**a**) Normal heart with epicardial adipose tissue distributed along the AV groove and in the course of the anterior descending coronary artery. (**b**) Whitish patches (*Soldier's patches*) in visceral pericardium classically attributed to mechanical trauma or healed pericarditis

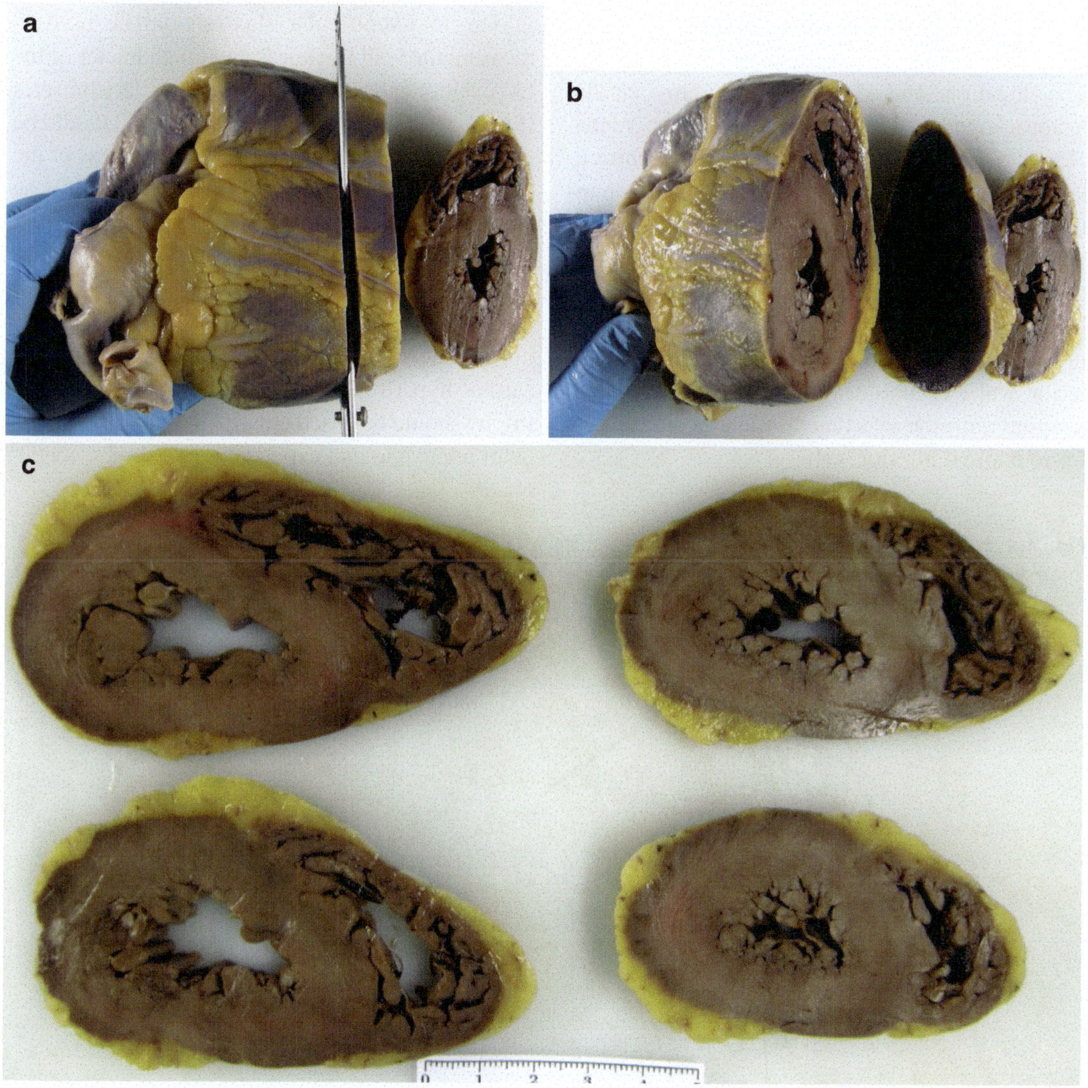

Fig. 1.2 (**a**–**c**). Protocol to study the heart by transversal sectioning from the apex to the papillary muscles, obtaining a closed base with intact valves and four to five biventricular slices

1.2.1 Study of the Myocardium

The location and distribution of myocardial lesions, together with the diameter of the ventricles, should be evaluated in the sliced biventricular sections. The anterior wall of the LV spans from the anterolateral papillary muscle to the septum; the lateral wall is situated between the two papillary muscles; and the posterior/inferior LV wall is adjacent to the posterior papillary muscle (Fig. 1.3). Pathologists maintain the posterior wall terminology, whereas clinicians, in keeping with the heart anatomical position, prefer to name it inferior wall.

The mid-ventricular slice should be used to measure the wall thickness and the diameter of the LV, excluding the trabeculae and papillary muscles (Fig. 1.4). A possible dilation of the RV should be evaluated in the cardiac apex. The thickness of the anatomical sections is not equivalent to those obtained by echocardiography since this is performed with the heart in telediastole. Values considered normal are 10–15 mm for the thickness of the LV and the septum, and 3–4 mm for the RV, during necropsy. Larger values are indicative of cardiac hypertrophy; however, diagnosis of hypertrophy should be based chiefly on the weight of the heart relative to that expected for the body weight.

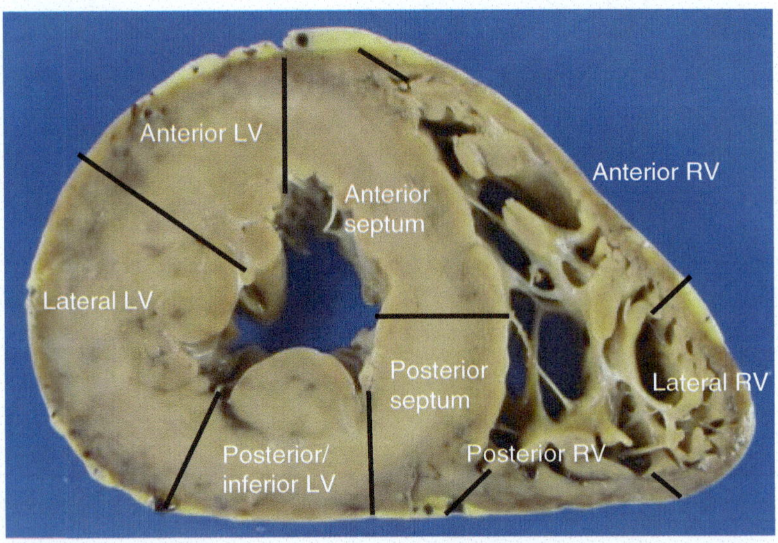

Fig. 1.3 Nomenclature of the ventricle walls

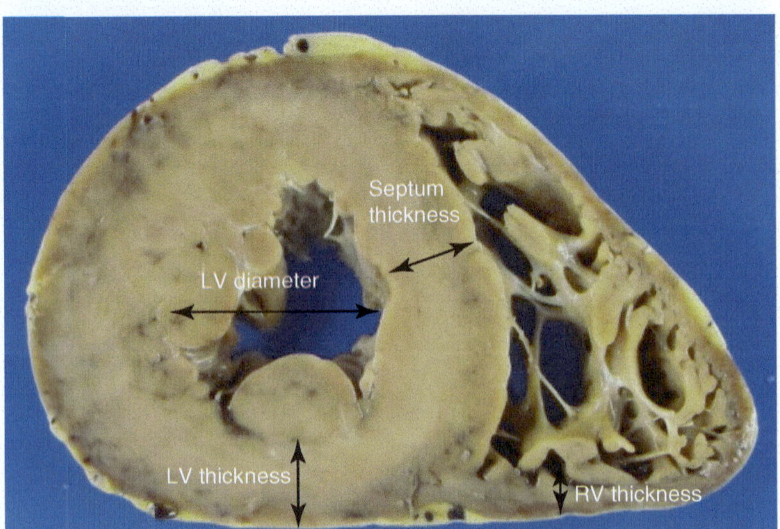

Fig. 1.4 Measurement of the ventricular thickness and the diameter of the LV, excluding the papillary muscles and trabeculae. The thickness of the RV is measured at its posterior wall since the anterior wall presents more epicardial adipose tissue and adipose infiltration, especially in elderly people

1.2.2 Study of the Cardiac Base and Valves

The cardiac base is opened following the direction of blood flow, starting in the right atrium that is opened from the inferior vena cava to the apex of the right atrial appendage, to preserve the sinoatrial node of the conduction system (Fig. 1.5) (see Chap. 10). Then examine the tricuspid valve through which, in normal conditions, three average-sized fingers can be placed.

Then, open the lateral border of the RV (Fig. 1.6a) and inspect the opened tricuspid valve, the right atrium and the basal portion of the RV. The perimeter of the valve is measured at the level of the annulus (Fig. 1.6b).

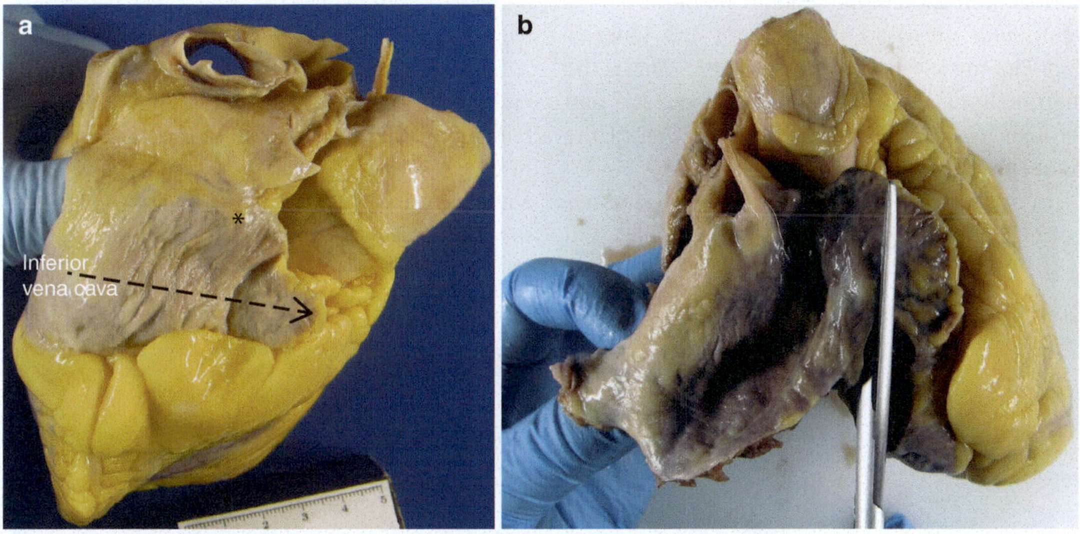

Fig. 1.5 (**a**) The discontinous *arrow* points to the direction of the cut to open the right atrium. The *asterisk* indicates the position of the sinoatrial node. (**b**) Opening with scissors

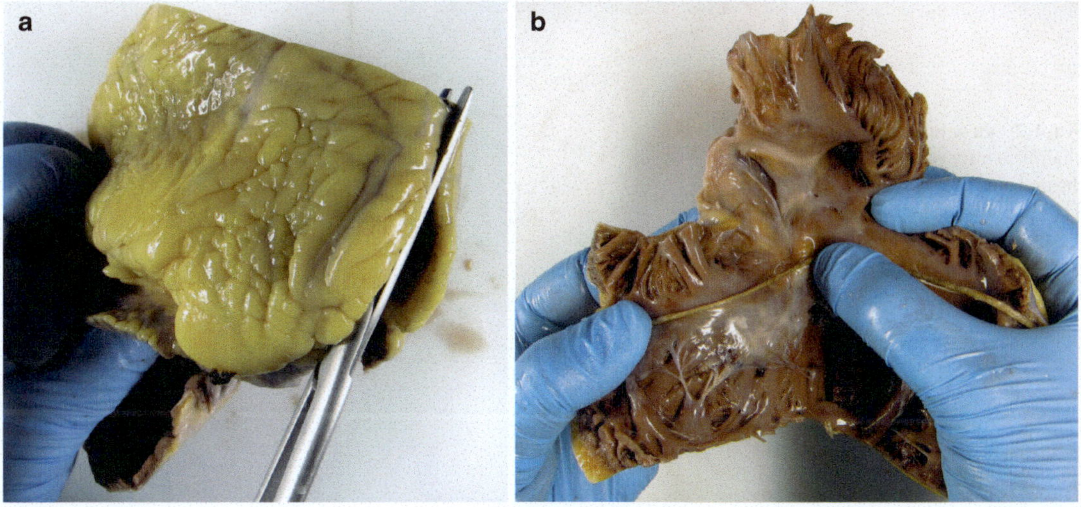

Fig. 1.6 (**a**) Opening of the lateral border of the RV. (**b**) Measurement of the tricuspid valve perimeter with the aid of a length of string

The most important structures of the right side of the heart are compiled in Fig. 1.7. The AV node is positioned in front of the mouth of the coronary sinus (see Chap. 10). The tricuspid valve has one large anterior papillary muscle, a septal muscle (*Lancisi's muscle*) which can be double or multiple, and a posterior papillary muscle formed by several smaller ones. The septomarginal trabecula (also known as moderator band) extends from the inferior part of the septum (continuation of the septal band of the supraventricular crest) to the base of the anterior papillary muscle at the tricuspid valve. This is the inferior limit of the inflow tract of the RV and through its interior runs the continuation of the right bundle of the conduction system.

In one third of adults, the foramen ovale does not close correctly (Patent foramen ovale) and functions as a competent flap valve, ensuring that the orifice remains closed since greater pressure exists in the left atrium (Fig. 1.15).

After, following in the direction of the blood flow, open the outflow tract of the RV and the pulmonary artery (Fig. 1.8). The annulus of the pulmonary artery is located 1.5 cm above the annulus of the aortic valve and has three leaflets: the right leaflet is located near to the aortic valve right leaflet, and the left leaflet is found near the aortic valve left leaflet (Fig. 1.9). The valve perimeter is measured at the sinotubular junction.

The opening of the LA is more arbitrary, generally between the left and right pulmonary veins, but attention must be paid to the atrial appendage in case possible thrombus is present (Fig. 1.10). View the mitral valve from the left atrium searching for prolapse, fusion of commissures, etc.

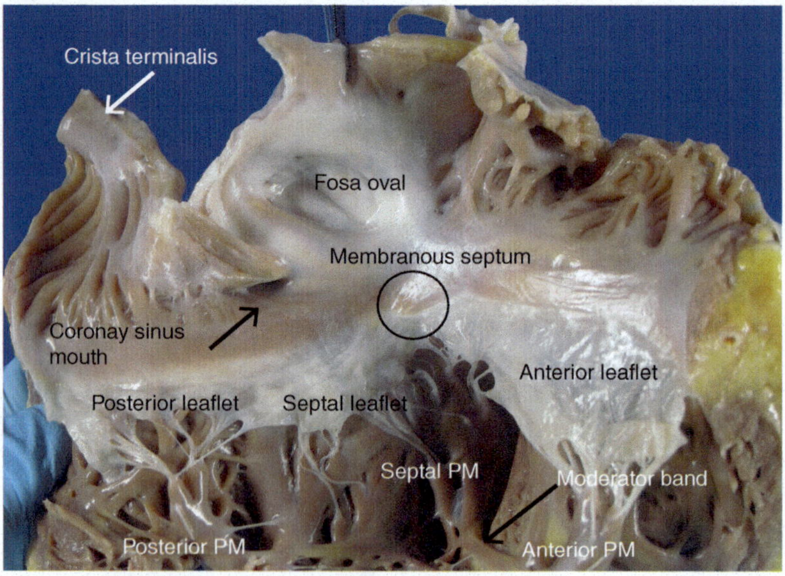

Fig. 1.7 View of the right chambers after opening through the lateral border. The most relevant structures are indicated

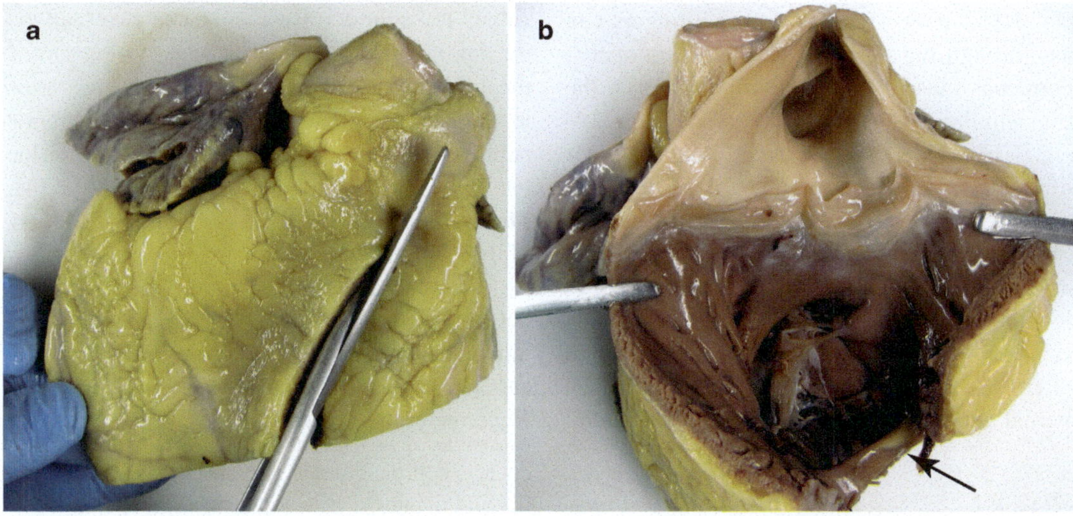

Fig. 1.8 (**a**) Sectioning through the RV outflow tract. (**b**) View of the opened pulmonary valve and the origin of the pulmonary arteries. In the bottom, the moderator band can be observed (*arrow*). In some cases the pulmonary valve can have four leaflets (see Chap. 7)

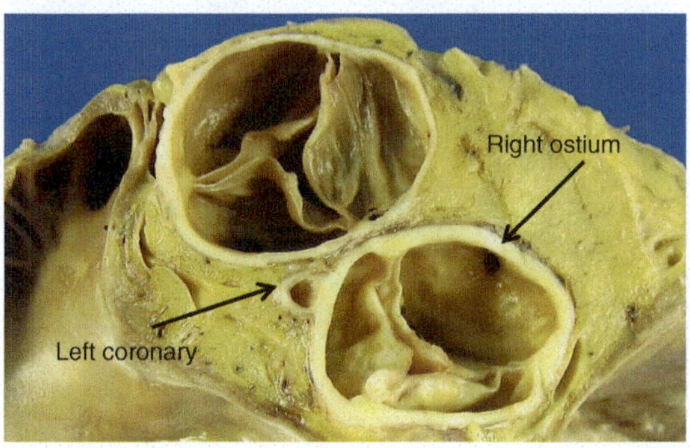

Fig. 1.9 Transversal section at the level of the semilunar valves. At the top of the image, the pulmonary artery and, at the bottom, the aortic valve where the *ostium* of the right coronary artery and the origin of the left coronary artery trunk can be visualised

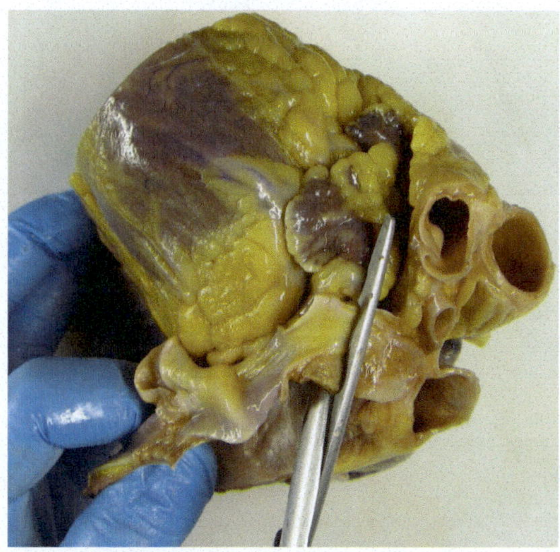

Fig. 1.10 Opening of the left atrium and appendage

(Fig. 1.11). It is estimated that two fingers can pass through the mitral annulus. Exceptionally, anomalous muscular bands can be found in the atria, similar to the false tendons of the LV (Fig. 1.12), which are associated in some cases with patent foramen ovale and to *Chiari's network*, and can cause mitral valve insufficiency.

Next, open the LV laterally with a longitudinal cut between the papillary muscles (Fig. 1.13) to view the left atrium, the opened mitral valve (its perimeter is measured at the annulus level) and the left ventricle, paying special attention to the internal structure of the chambers and the mitral valve. The mitral valve consists of two leaflets: the anterior one is semicircular and forms part of the left ventricular outflow tract; the posterior leaflet has a scalloped appearance. The anterior papillary muscle is single whereas the posterior papillary muscle is usually double (Fig. 1.14). The area facing the LV is very trabeculated (1/4 of the total thickness).

Visual inspection of the aortic valve should be performed from above before its opening to determine the number of cusps, the presence of calcifications, commissural fusion, etc. (Fig. 1.16). Next, the outflow tract of the LV is opened by an

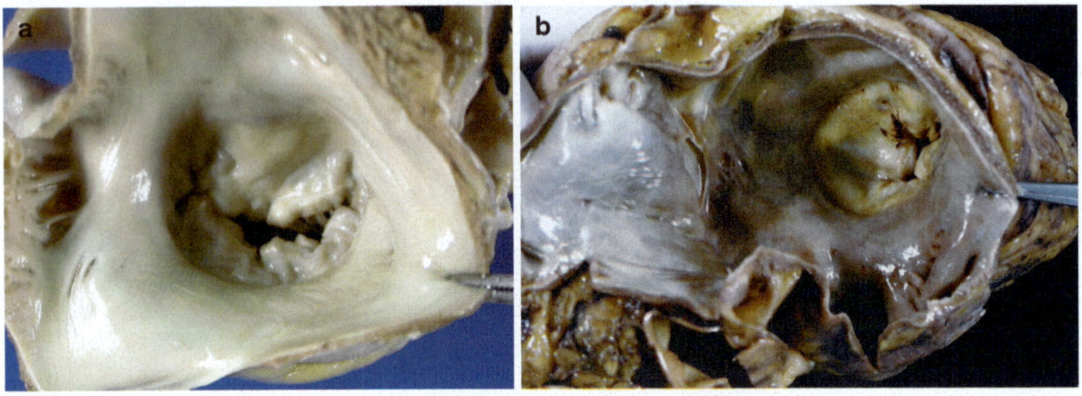

Fig. 1.11 (**a**) A normal mitral valve as seen from the left atrium. (**b**) A rheumatic mitral valve: the atrium is dilated, with valvular fibrosis in the leaflets and commissure fusion

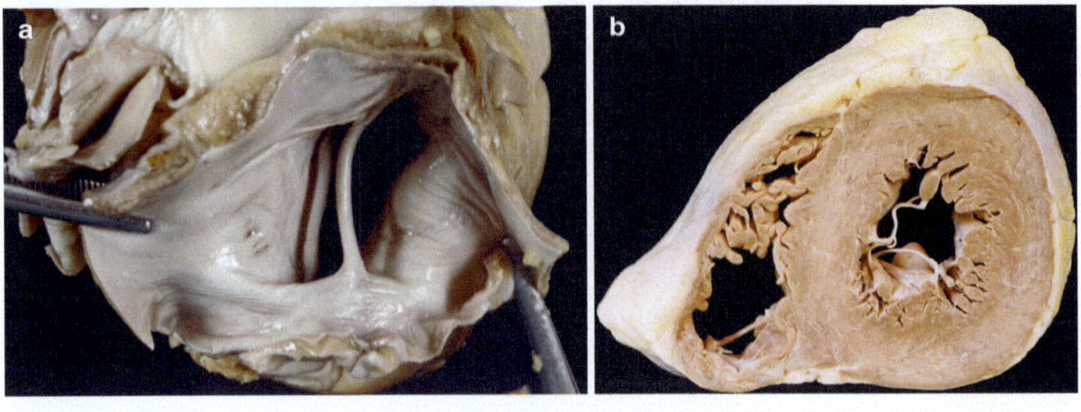

Fig. 1.12 (**a**) An anomalous band between the anterior and posterior wall of the left atrium. (**b**) False tendons between the papillary muscles and the septum, and in the RV

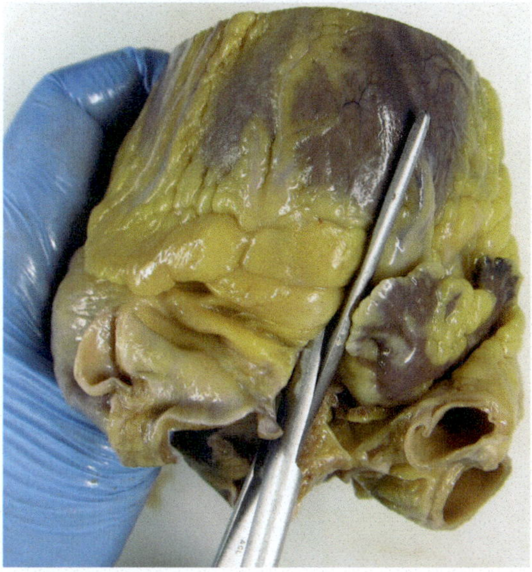

Fig. 1.13 Opening of the lateral border of the LV

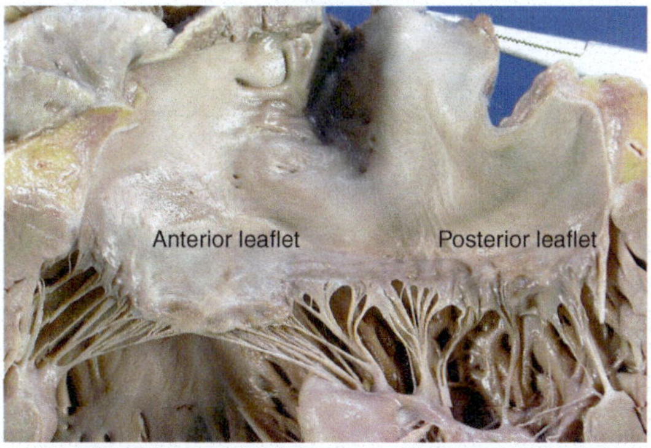

Fig. 1.14 View of the opened mitral valve

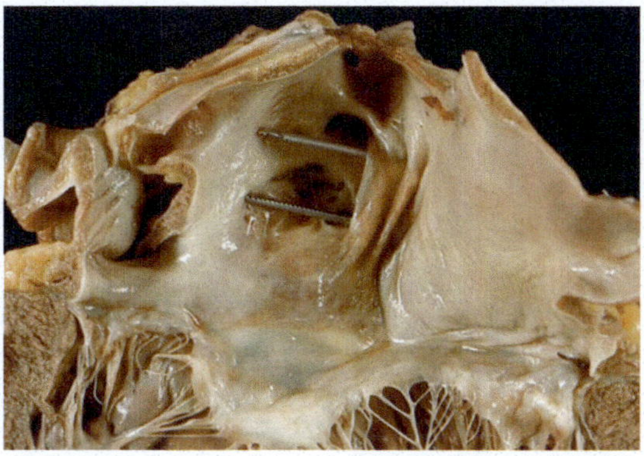

Fig. 1.15 *Patent foramen ovale* as seen from the left side

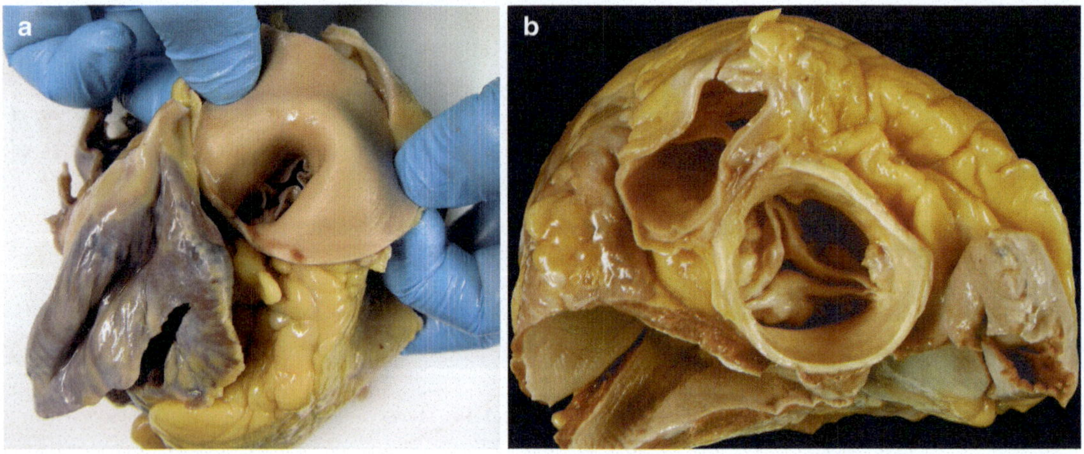

Fig. 1.16 (**a**) Inspection of the aortic valve before its opening. (**b**) Aortic valve with calcification in the three leaflets, in an 82-year-old woman with senile aortic stenosis

oblique cut in the anterior wall proximal to the septum (Fig. 1.17) to inspect the ascending aorta (check for atheromas, aneurysms, intimal tears in aortic dissections, intramural haematomas, origin of coronary bypass, etc.) and the aortic root (consisting of the fibrous annulus, the sinuses of Valsalva and the sinotubular junction).

The aortic and mitral annuli are fused through the intervalvular fibrosa (fibrous continuity between the anterior leaflet of the mitral valve and the left and posterior sigmoids) (Fig. 1.18). The anterior Valsalva sinuses give rise to the coronary arteries (Fig. 1.9). Fenestrations are frequently found between the valve closure line and its free border (lunula) (Fig. 1.21a), and also fibrocartilaginous nodules (*Arancio-Morgagni's nodules*) and papillary excrescences (*Lambl's excrescences*) in the valve closure line (Fig. 1.19).

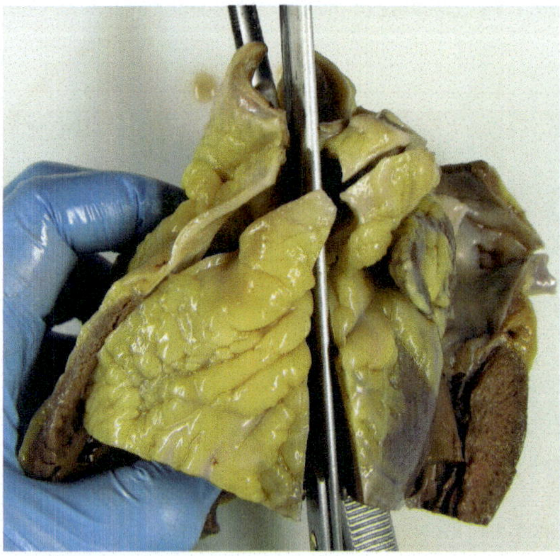

Fig. 1.17 Opening of the LV outflow tract. Long forceps serve as a guide

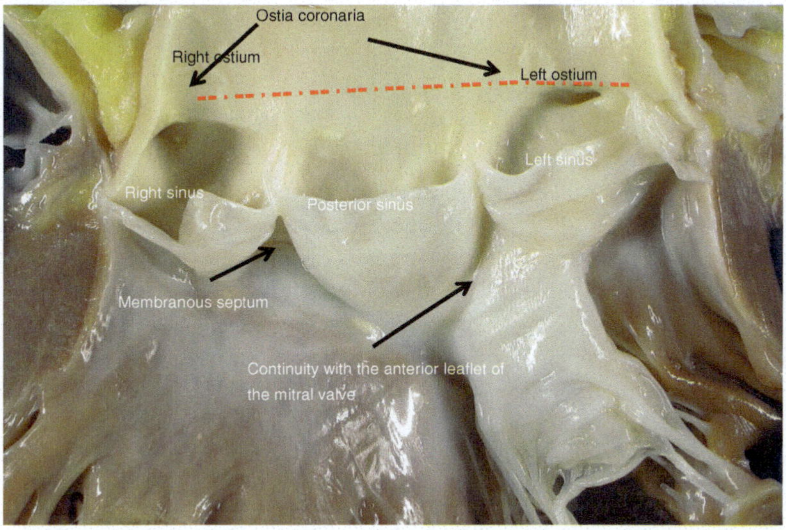

Fig. 1.18 Normal opened aortic valve where the Valsalva sinuses can be seen, with the origin of the coronary arteries (right (RC) and left (LC)). Discontinuos line = dashed line

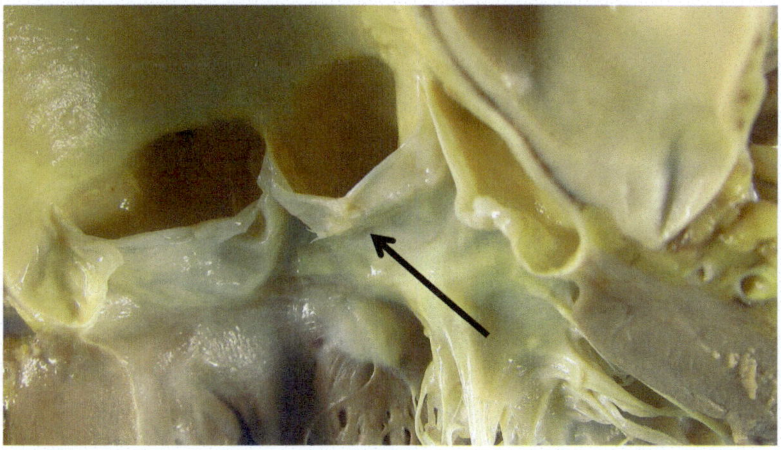

Fig. 1.19 Fibrous nodules and *Lambl's excrescences* in the valve closure line (*arrow*)

1.2.3 Study of the Coronary Arteries

The left and right coronary arteries arise from the corresponding aortic Valsalva sinuses, and their branches irrigate the entire heart. The most frequent coronary artery tree is depicted in Fig. 1.20. According to this tree, the posterior descending coronary artery originates from the right coronary artery (right dominance), although there are three possible types of dominance whose characteristics and frequency are described in Table 1.3.

In one third of the population, the left main coronary artery is divided in three branches: the left anterior descending coronary artery (LAD), the circumflex coronary artery (Cx) and in between the intermediate branch (not shown in the figure). In these cases, the intermediate coronary artery replaces the first diagonal.

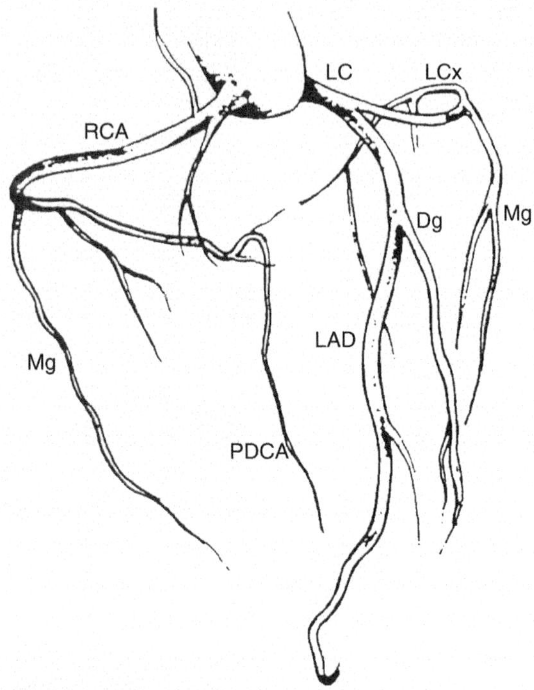

Fig. 1.20 Scheme of the coronary artery tree. *RCA* right coronary artery, *PDCA* posterior descending coronary artery, *LC* left coronary artery trunk, *LAD* left anterior descending coronary artery, *Dg* diagonal, *LCx* left circumflex, *Mg* marginal

1.2.3.1 Study of the Origin of the Coronary Arteries

The study of the coronary arteries begins by examining their origin in the corresponding Valsalva sinuses. The diameter of the right coronary artery at its *ostium* is commonly slightly smaller than that of the left coronary artery *ostium*. There is usually one *ostium* in the sinus, in a central position, although in some cases variations occur that are considered normal: two or more *ostia* in the right sinus or two *ostia* in the left sinus (Fig. 1.21). When both coronary arteries arise from the same sinus, sudden death can occur (see Chap. 6). In up to 30 % of human hearts, the left anterior descending coronary artery penetrates the myocardial septum to 2–3 mm depth and for a length of 10 mm in the basal third portion (Fig. 1.22). When the depth is more than 4 mm, the permeability of the artery during contraction of the ventricle can be compromised, provoking myocardial ischaemia (see Chap. 6).

Table 1.3 Types of coronary dominance (Baroldi 1967)

Type I (77 %) (right)	The right coronary artery gives rise to the posterior descending coronary artery and ends: In the posterior interventricular groove without significant branches on the LV (5 %) At distinct levels on the LV (55 %) In the left cardiac edge with the inferior LV wall totally irrigated by the right coronary artery
Type II (8 %) (left)	The circumflex coronary artery gives rise to the posterior descending artery
Type III (15 %) (balanced or codominance)	There are two posterior descending branches, one from each artery. In this type, the apex and the inferior ventricular septum rely on the left coronary artery since the LAD re-ascends through the posterior interventricular groove in one quarter or more of its length

1.2.3.2 Study of the Distribution and Permeability of the Coronary Arteries

In cases of ischaemic heart disease, the ideal method is to perform radiography on the whole heart before sectioning, in order to identify

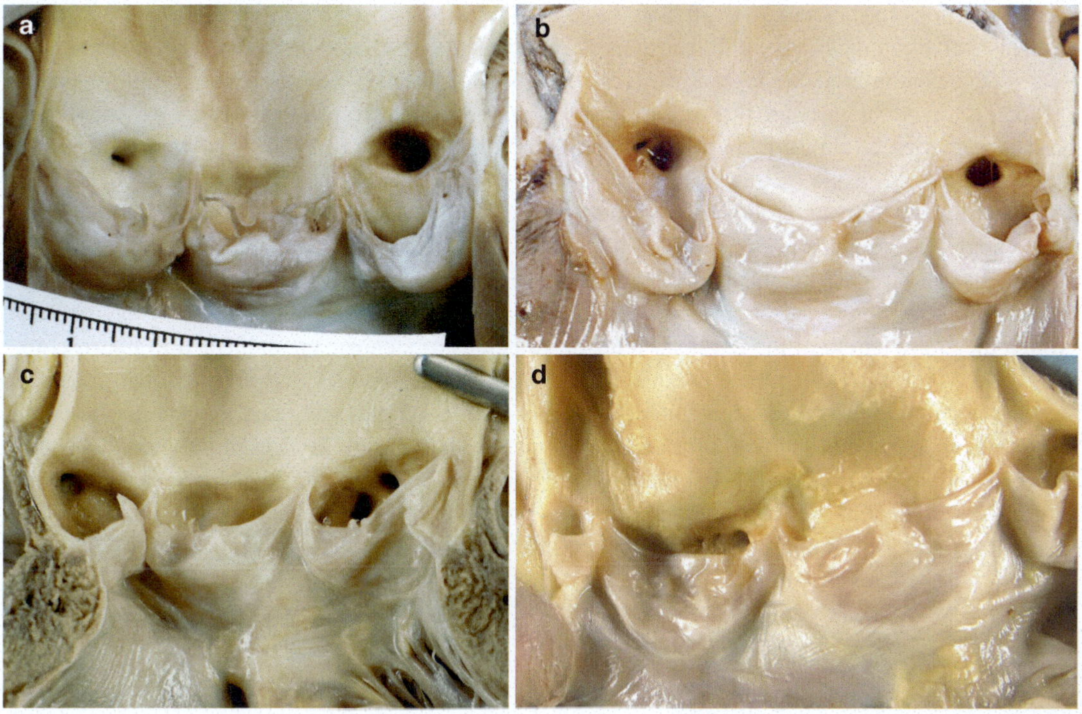

Fig. 1.21 Normal variations in the origin of the coronary arteries. (**a**) A unique *ostium* in each coronary sinus. The posterior leaflet is fenestrated. (**b**) Two orifices in the right sinus; the smallest one is the origin of the conus artery. This occurs in more than half of the hearts examined. (**c**) Independent origins of the left anterior descending and circumflex arteries in the left sinus (very infrequent). (**d**) In the right sinus three *ostia* can be recognized, corresponding to the right coronary artery, the conus artery and the right superior septal artery

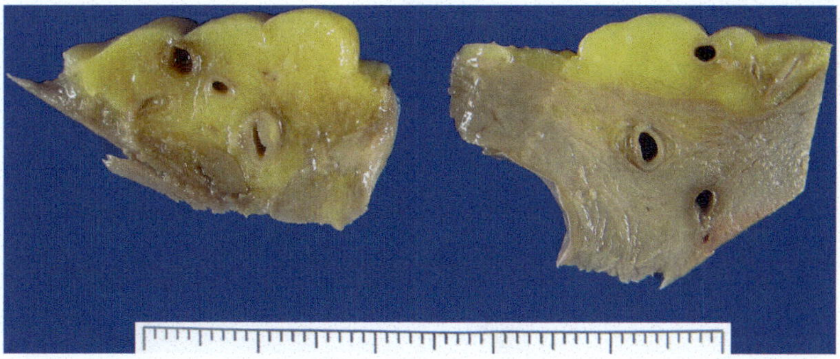

Fig. 1.22 Intramural course of the left anterior descending coronary artery

calcified portions in the coronary arteries, possible stents, valve calcification, etc. It is also useful to carry out a post-mortem angiography by inserting contrast material through the origin of the coronary arteries, followed by radiography (Fig. 1.23) (the downside of this method is that the coronary *ostia* cannot be properly examined).

The study of coronary permeability, with or without previous radiological study, should be performed with transversal 3–4 mm sections through the entire length of the coronary arteries (Fig. 1.24a). This study offers better results if the heart has been previously fixed in formaldehyde. To examine the proximal segment of the circumflex coronary artery, it is useful to lift, or even better, to resect, the left atrial appendage. If there is calcification in the coronary arteries, they have to be dissected (the entire artery or the affected segments only), and decalcified in 7 % nitric acid (or an alternative decalcifying solution) before performing the transversal sections, as mentioned before (Fig. 1.24b). Classically, the degree of coronary stenosis can fall within four grades: light (25–50 % stenosis); moderate (50–75 %); severe (>75 %); and occlusive (100 %) (Fig. 1.24c).

Afterwards, the portions with the greatest degrees of stenosis are subjected to microscopic analysis, noting its location (proximal, medial or distal). The right, left main, left anterior descending and circumflex arteries must always be examined and, optionally, the marginal, diagonal, posterior descending and intermediate coronary arteries also.

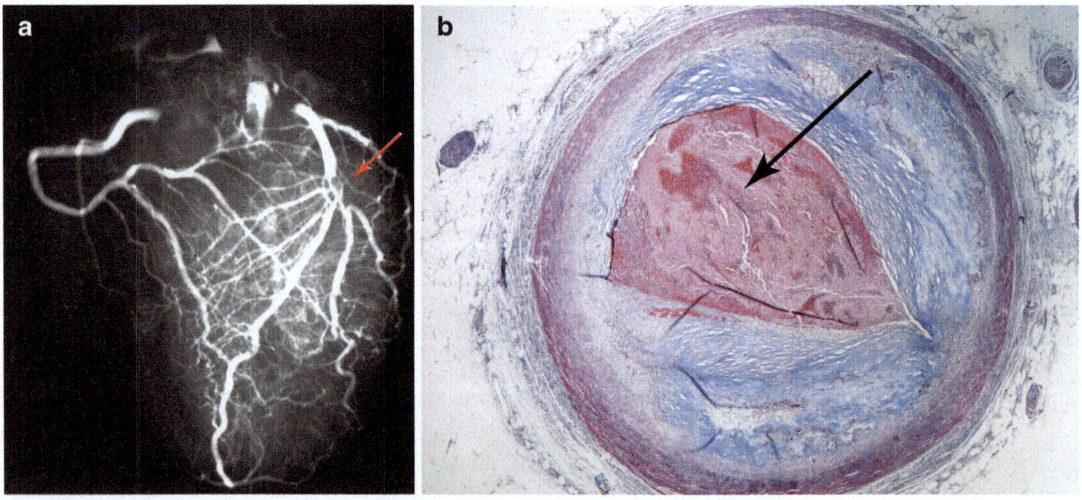

Fig. 1.23 (**a**) Post-mortem coronary angiography, where an absence of filling in the proximal segment of the left anterior descending coronary artery (*arrow*) can be appreciated. (**b**) In the histology, an occlusive platelet thrombus adherent to a fibrous atheroma plaque (*arrow*) was found. Results are from a 33-year-old man that had been examined in a casualty department with precordial pain one hour before his death. The ECG was negative

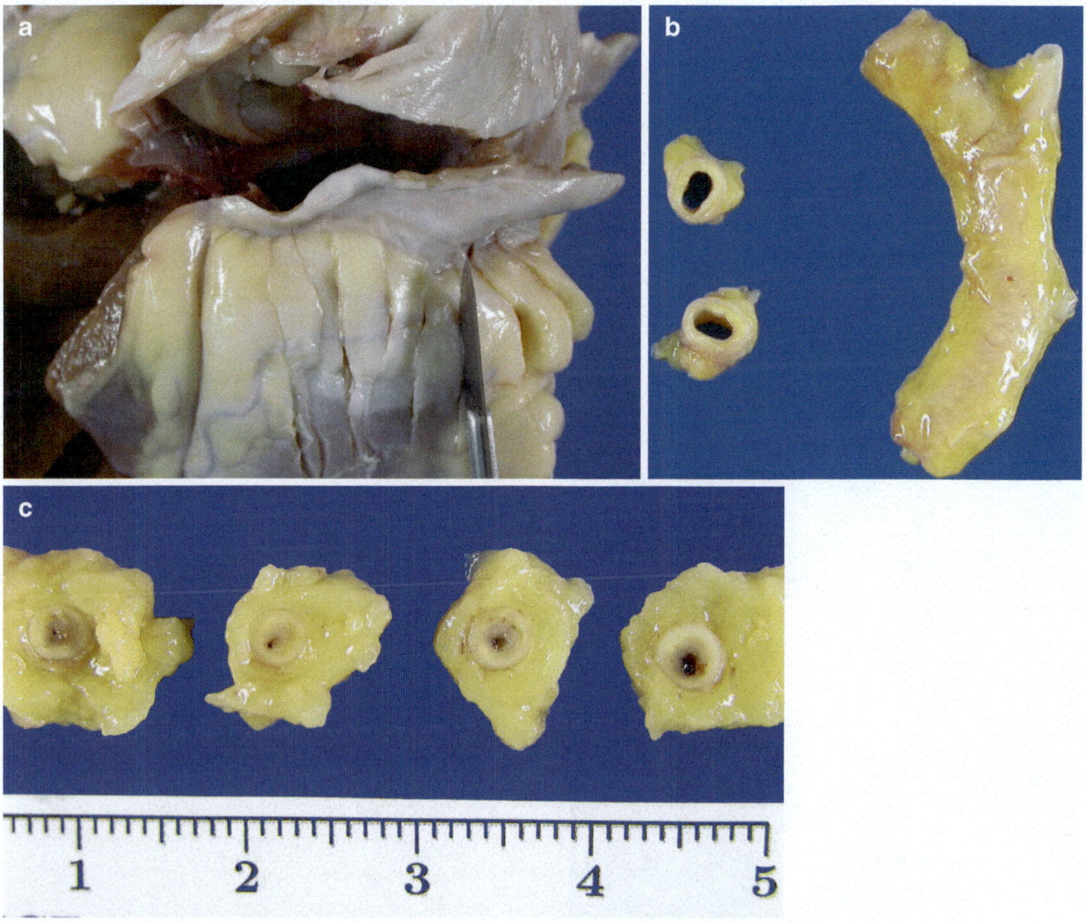

Fig. 1.24 (**a**) Transversal sections along the right coronary artery. (**b**) Dissected coronary segment ready for decalcification and serial sectioning. (**c**) Serial study of the coronary segment after decalcification, where an obstruction of >75 % of the lumen can be appreciated

1.2.4 Study of Stent-carrying Coronary Arteries

In hearts with coronary artery stents, as mentioned previously, it is convenient to take an X-ray of the heart before its manipulation. An additional option is to dissect the coronary artery and then perform the radiological study (Figs. 1.25 and 1.26). The X-ray result allows the determination of the type of stent used, whether it has been implanted in the appropriate vessel, and if the stent has expanded adequately.

Ideally, a histological study of intravascular stents requires methacrylate-embedding of stents in situ followed by sectioning at 4–5 μm. This is the method of choice if stents have been implanted for less than 1 month and are scarcely covered by thrombuses and newly-formed intimal tissue (Fig. 1.26). For longer-term stents, the coronary artery can be sectioned with scissors every 3–4 mm to carefully extract the metallic wire for routine processing (Fig. 1.27).

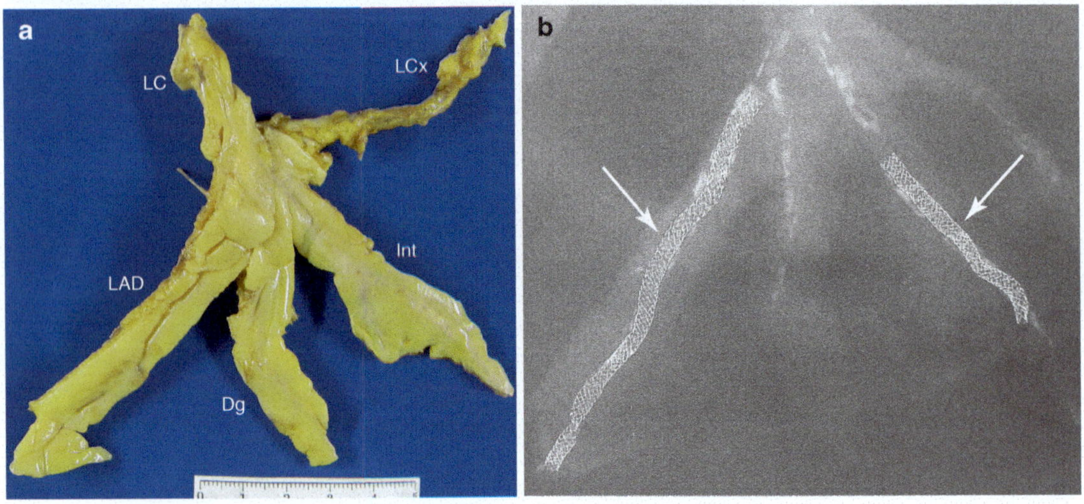

Fig. 1.25 (**a**) Dissection of the left coronary arterial tree. *LC* left coronary artery trunk, *LAD* left anterior descending coronary artery, *Dg* diagonal, *Int* intermediate coronary artery, *LCx* left circumflex coronary artery. (**b**) Radiography of the segment, where the calcified atheroma plaques and the stents implanted in the anterior descending and intermediate coronary arteries (*arrows*) can be appreciated

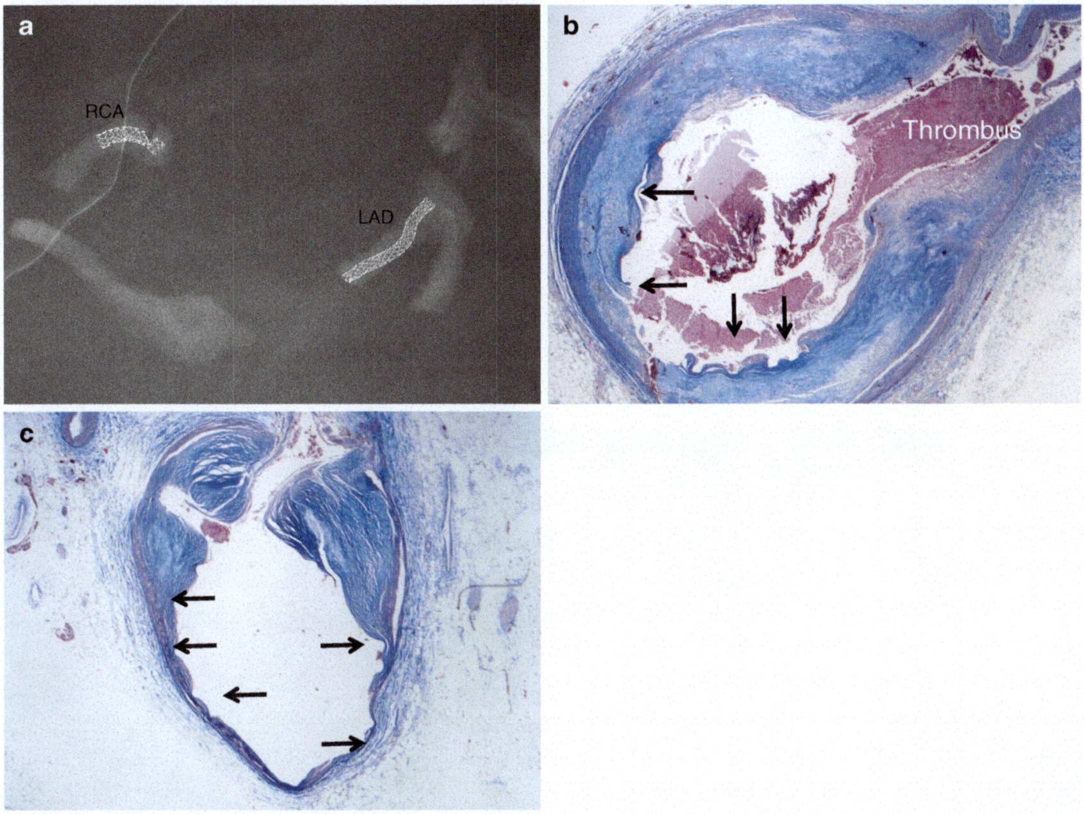

Fig. 1.26 A 68-year-old male was admitted to hospital with precordial pain. After catheterization, stents were implanted in the right and left anterior descending coronary arteries. The patient died at home on the day of his release. (**a**) Radiography of the coronary arteries after dissection, where a stent of 15 mm in the proximal segment of the right coronary artery (RCA) and another of 25 mm in the left anterior descending coronary artery (LAD) can be identified. (**b**) After removing the stent wires, an occlusive thrombus can be observed in the right coronary artery. (**c**) Image of the left anterior descending coronary artery after removal of the stent. In both cases, the metallic imprint of the stent can be recognised (*arrows*). The posterior/inferior wall of the LV showed an acute infarct of >2 weeks and the posterior wall of the RV revealed an infarct of 1–3 days evolution (Masson's trichrome)

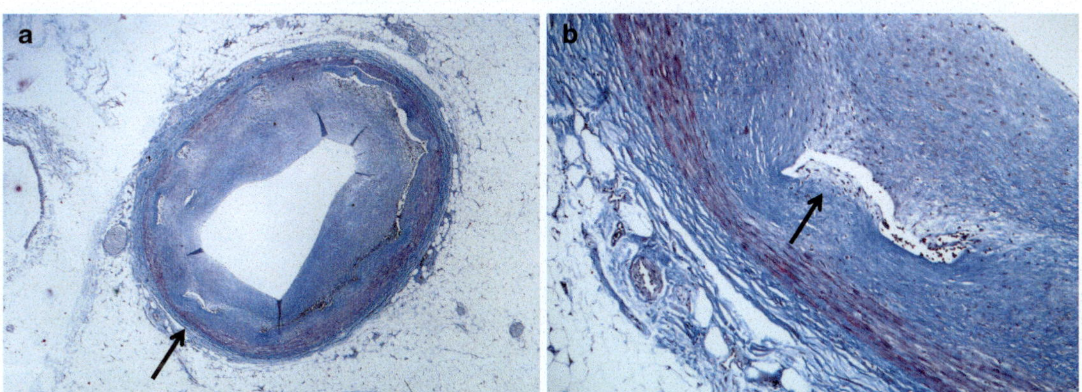

Fig. 1.27 A 44-year-old male who died in a motorbike accident. During the examination of the coronary arteries, a stent of 15 mm could be identified in the left anterior descending coronary artery. (**a**) Image of the coronary artery after removal of the stent. (**b**) Detail of the previous figure. In the arterial wall, restenosis of the coronary segment due to concentric intimal fibrosis can be observed. The orifices left behind after stent removal can be seen near the medial layer (*arrows*) (Masson's trichrome)

1.2.5 Study of Coronary By-pass Grafts

The classical aortocoronary bypass involves the connection of the aorta or some of its branches, to portions of the coronary arteries distal to the obstructive lesions through a vascular graft. The vascular graft can be a vein, an artery, or prosthetic. The anastomosis of the venous graft to the ascending aorta and coronary artery is of the terminolateral type (Fig. 1.28). The most commonly-used vein is the internal saphenous vein, while the external saphenous and cephalic veins are less used. The arterial grafts can be pediculated or free. In pediculated arterial grafts, one of the ends is kept in its anatomical location while the other end is directed to the coronary artery that requires re-vascularization. The arterial graft most widely used is the pediculated internal mammary artery, which remains

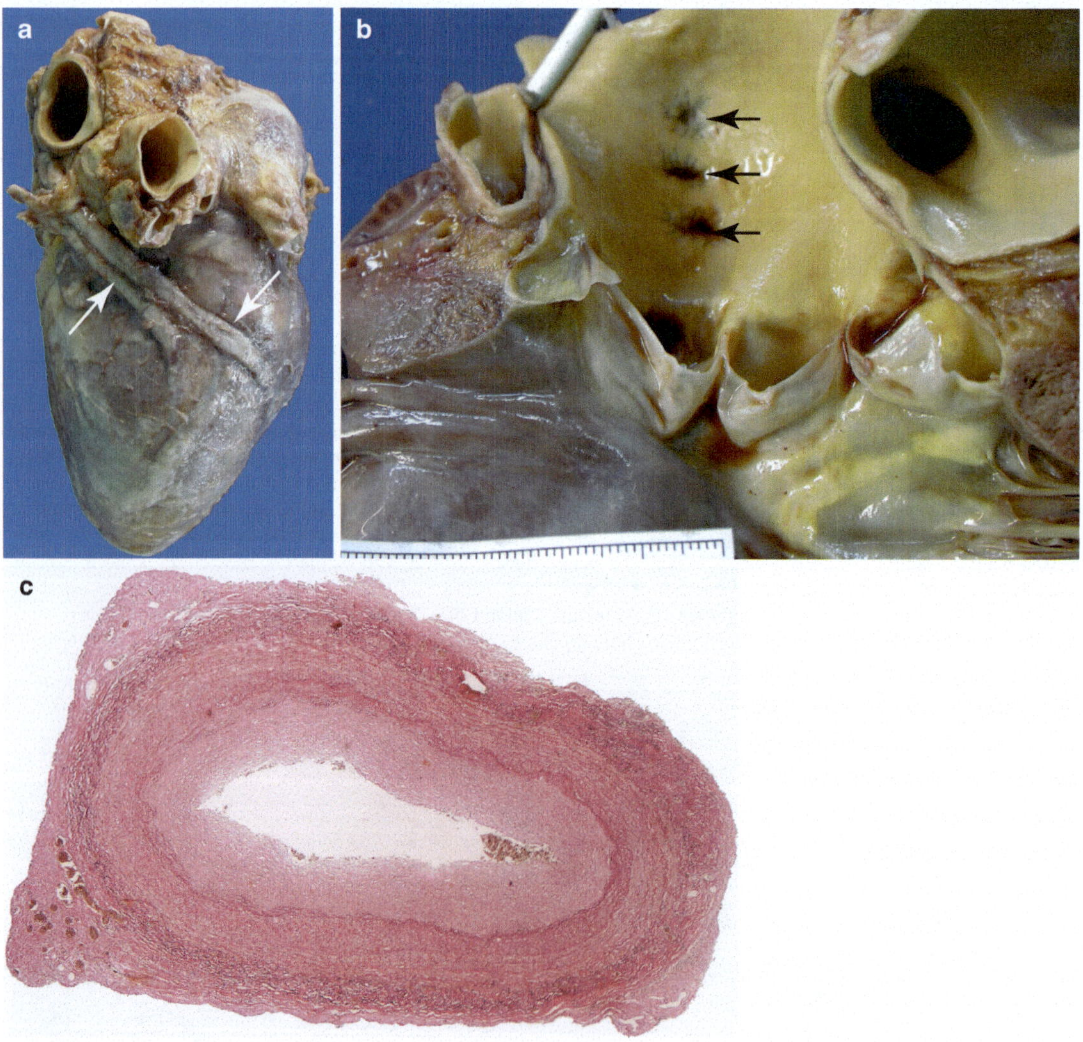

Fig. 1.28 (**a**) Two venous grafts at the left oblique posterior and left marginal coronary arteries (*arrows*). (**b**) Image of the ascending aorta from the same case showing the origin of a triple by-pass (*arrows*) (the third graft is anastomosed in the posterior descending coronary artery and is not visible in the photographs). (**c**) Microscopic image of a venous graft at the left marginal artery. Hyperplasia of the intima can be observed (Weigert)

connected to its natural site of origin, the subclavian artery. The internal mammary graft is the best choice to re-vascularize the left anterior descending artery (Fig. 1.29). In the majority of cases, pathological examination of coronary bypass is difficult due to post-surgical pericardial adherences. The inspection for possible tightness and twisted areas is necessary. Macroscopic study, as in the native coronary arteries, is done by transversal sectioning of the vessel every 3–4 mm and, for the microscopic study, the section containing the area of anastomosis with the native artery must always be examined (Fig. 1.30). Generally, with very recent grafts (the first 3 days) it is necessary to look for thrombus, compression in the insertion site and severe atherosclerosis in the native coronary artery distal to the insertion; in older grafts, stenosis due to fibrointimal proliferation and atherosclerosis should be explored.

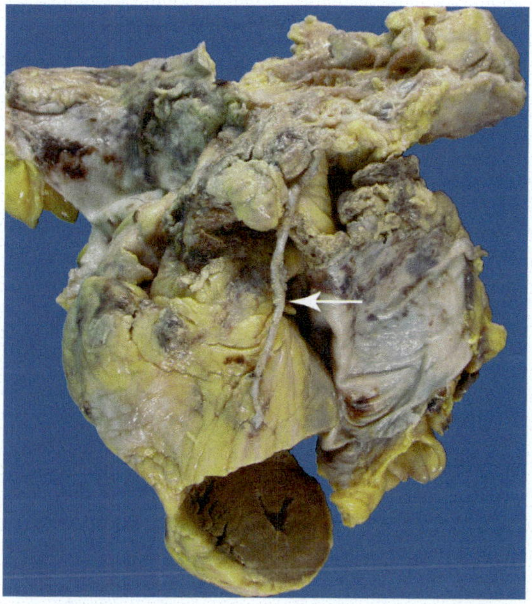

Fig. 1.29 Pediculated arterial graft from the internal mammary artery (*arrow*)

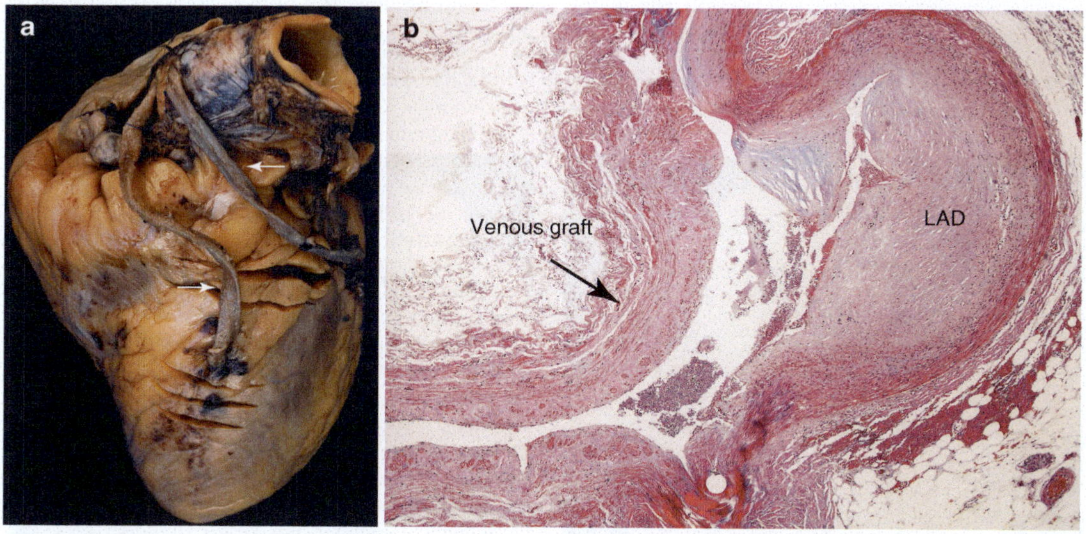

Fig. 1.30 (**a**) Double by-pass to the left anterior descending and intermediate coronary arteries (*arrows*). The image shows transversal cuts in the area of anastomoses to the left anterior descending coronary artery and in the native coronary artery distal to it. (**b**) Microscopic image of the area of the venous graft insertion to the native anterior descending coronary artery (LAD) showing significant atheromatosis (Masson's trichrome)

1.2.6 Study of Prosthetic Valves

The purpose of the pathological study of heart valve prostheses is to identify the type of prosthesis (biological or mechanical); determine the size and position in relation to the annulus and cardiac chamber; evaluate the valve movement; search for thrombi, abscesses, vegetations or paravalvular leaks; and assess degenerative alterations (see Chap. 7).

Initially, the prosthesis needs to be examined in situ, to identify paravalvular leaks, tissue overgrowth (*pannus*) surrounding the prosthesis, and to evaluate its correct orientation (Figs. 1.31 and 1.32). Next, the prosthesis is removed and the native annulus is examined to discard paravalvular abscesses. Later, the prosthesis is carefully examined on both sides (Fig. 1.33) and, if possible, radiography is performed (Fig. 1.34). Histological study can be done from the annulus, possible thrombus, vegetations or from the leaflets when the prosthesis is biological. Table 1.4 summarises the alterations to search for in the macroscopic, microscopic and radiological studies. For more specialised studies, techniques such as scanning electron microscopy, transmission electron microscopy, calcium/phosphorus determination, lipid biochemistry, etc. are used.

Table 1.4 Lesions to evaluate in a prosthetic valve study

Macroscopic	***Radiological***
Anatomical orientation of the valve	Valve identification
Tissue overgrowth (*pannus*)	Calcification
Thrombi	Microfracture
Vegetations	***Histological***
Paravalvular leaks	Leaflet configuration
Cuspal stiffness	Inflammation
Cuspal haematomas	Vegetations
Calcifications	Thrombi
Fenestrations/tears	Prosthesis/native annulus relationship
Cuspal abrasions	Degenerative lesions
Cuspal stretching	Abrasions
Mechanical dysfunction	Calcifications
Extrinsic interference or damage	Endothelialization
Dehiscence	Haematoma of the valves
	Foreign material

From Schoen et al. (1992)

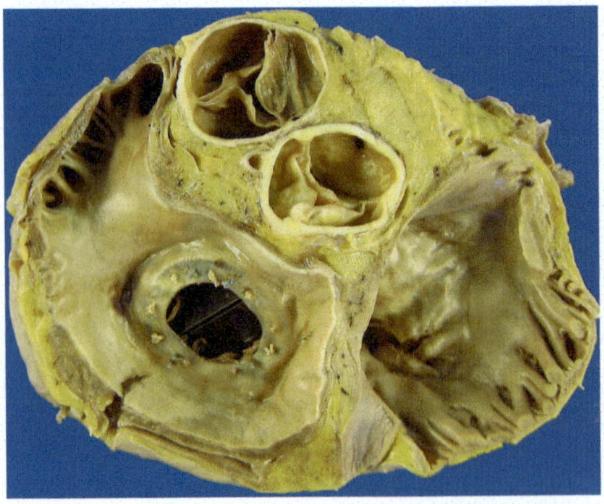

Fig. 1.31 View of a mitral prosthesis after sectioning the base near the atrioventricular junction. Circumferential tissue growth over the prosthesis can be appreciated

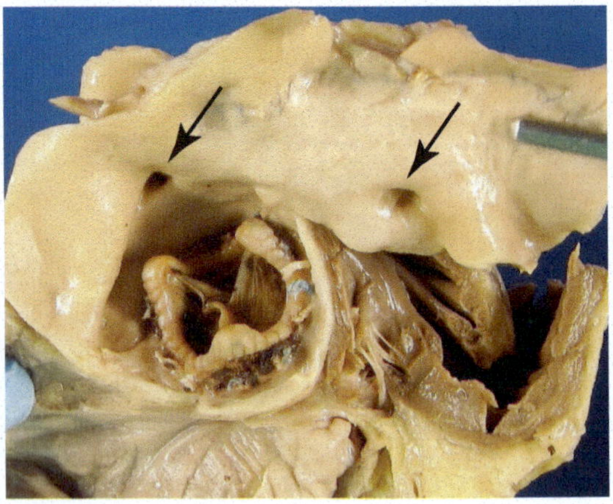

Fig. 1.32 Biological aortic prosthesis and orifices of the origin of two aortocoronary bypass grafts (*arrows*)

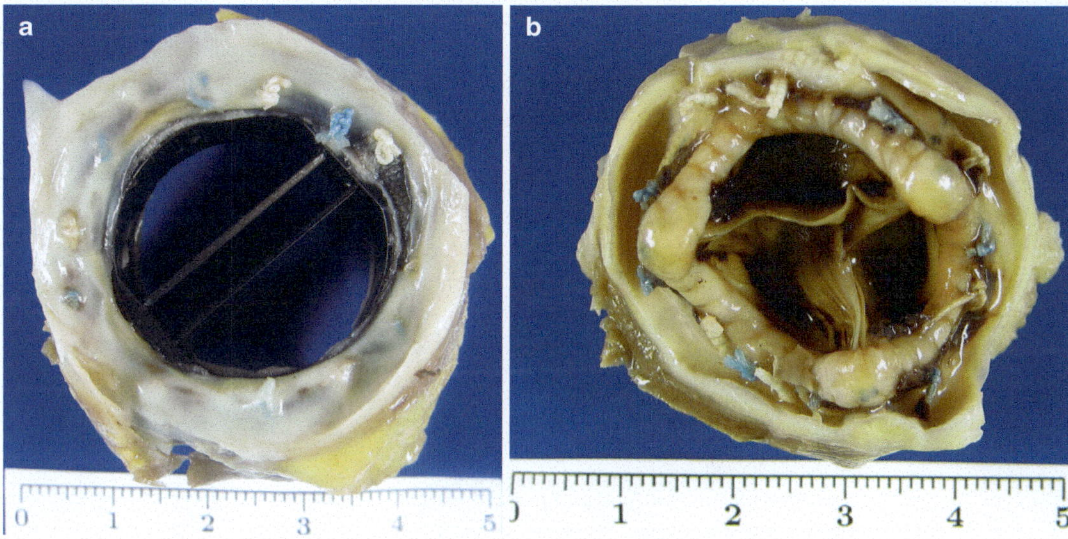

Fig. 1.33 (**a**) Mechanical prosthetic bi-leaflet mitral valve (*St. Jude Medical*) viewed from the atrial side. (**b**) Biological aortic prosthesis viewed from the aortic side

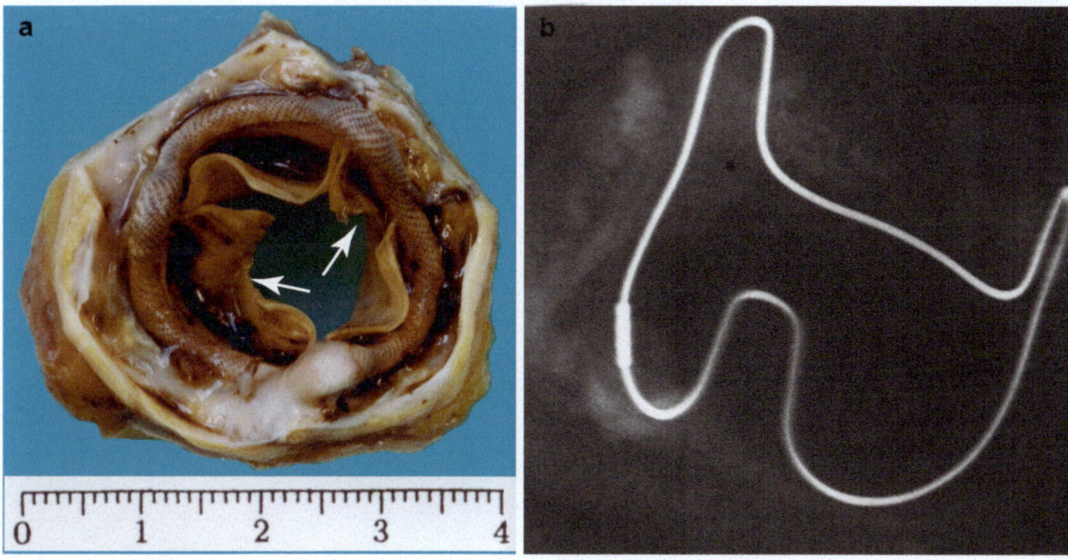

Fig. 1.34 (**a**) Biological valve with tearing of cusps. (**b**) The type of scaffolding observed in the radiological study allows the identification of the prosthesis as *Ionescu-Shiley* type

1.2.7 Selection of Samples for Microscopic Study

The selection of samples for microscopic study is summarised in Table 1.5. There are some areas of the heart that must always be studied, whereas other areas are examined only under certain circumstances. The study of the conduction system can be performed by a simplified method described in Chap. 10. The stain routinely used is haematoxylin-eosin and, depending on the pathological findings, other stains such as Masson's trichrome or Weigert's stain for connective tissue, Congo red stain for amyloid, PAS-Alcian blue staining for mucopolysaccharides, etc. are used. In cases of myocarditis, immunohistochemistry for CD3, CD20, CD68, etc. is recommended.

Table 1.5 Sampling of the heart for microscopic study

Routine	Anterior LV wall with anterior papillary muscle Lateral LV wall Posterior LV wall with posterior papillary muscle Interventricular septum (one or two sections) Anterior and posterior RV wall Longitudinal section with right atrium, right coronary artery, tricuspid valve and RV wall (lateral border or posterior side) Longitudinal section with left atrium, coronary sinus with circumflex artery, mitral valve and LV wall (lateral border or posterior side)
Depending on pathological findings	Extended sampling of myocardium, atria or valves
If arrhythmogenic cardiomyopathy is suspected	Apart from the routine sampling: Lateral RV wall Longitudinal RV wall at the outflow tract
If asymmetric or concentric left ventricular hypertrophy	Apart from the routine sampling: At least three sections of the interventricular septum
In sudden death without gross pathological findings	Apart from the routine sampling: Transversal blocks of the apical ventricles Conduction system (see Chap. 10)
Coronary arteries	When >50 % stenosis or a thrombus, haemorrhage or aneurysm is suspected
Aorta	In cases of aortic pathology (dissection, aneurysm, aortitis, etc.) longitudinal and transversal sections of the pathological and normal areas

1.3 Age-Related Changes

Analogous to other organs, the heart suffers age-related modifications that must be differentiated from possible concomitant cardiac pathologies. The most important modifications are summarised in Table 1.6.

Table 1.6 Age-related cardiac changes (Figs. 1.35, 1.36, 1.37, 1.38, 1.39 and 1.40)

Cardiac weight: increases in women associated with menopause
Pericardium: becomes more opaque
Epicardium: the mass of epicardial adipose tissue increases, especially on the RV and the AV groove, with adipocyte infiltration in the free wall of the RV
Interventricular septum with sigmoid morphology
Shortening of the length base-apex → relative elongation of tendinous cords → mitral valve posterior leaflet pseudo-prolapse
Endocardium: The thickness in the atria increases, especially the left atrium
Left atrium dilation
Mitral valve: thickening, atheromas in anterior leaflet, annular calcification (can cause AV and bundle branch blocks) (see Chaps. 7 and 10)
Aortic valve: Nodules of calcification on the cusps (causes stenosis); dilation of the aortic annulus (causes insufficiency) (see Chap. 7)
Ascending aorta: dilation
Coronary arteries: rigid, tortuous and dilated
Myocardium: slight interstitial fibrosis, lipofuscin deposits, basophilic degeneration
Amyloid deposits in myocardium and endocardium
Conduction system: fibrosis and adipose infiltration (see Chap. 10)

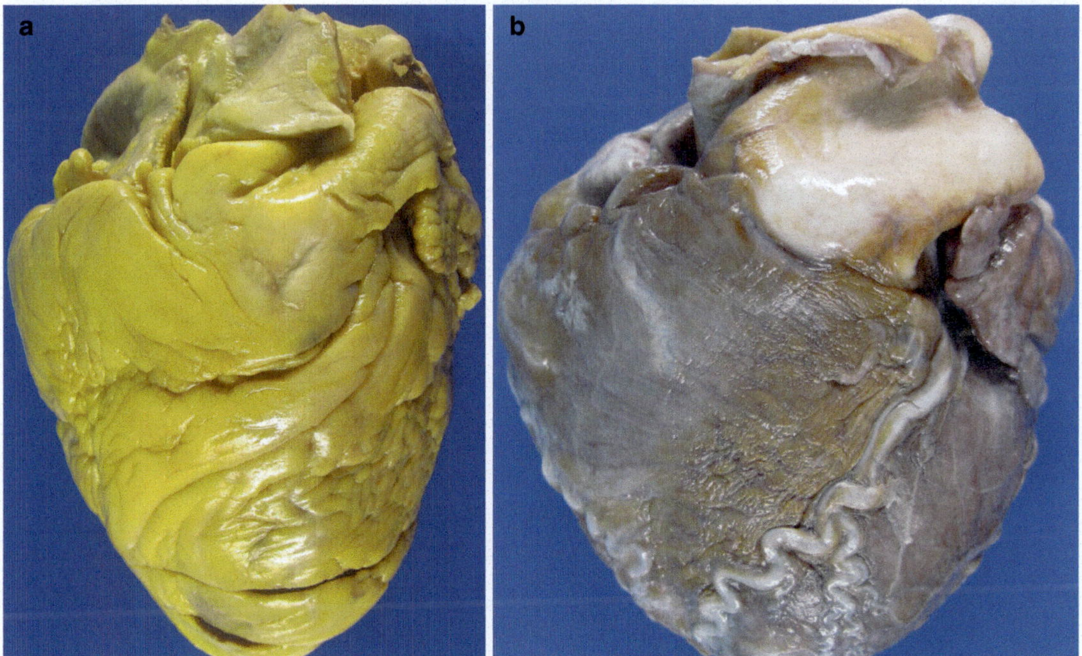

Fig. 1.35 (**a**) Marked increase in the mass of epicardial adipose tissue in a 79-year-old male. (**b**) Rigid coronary arteries with a tortuous course in a woman of advanced age

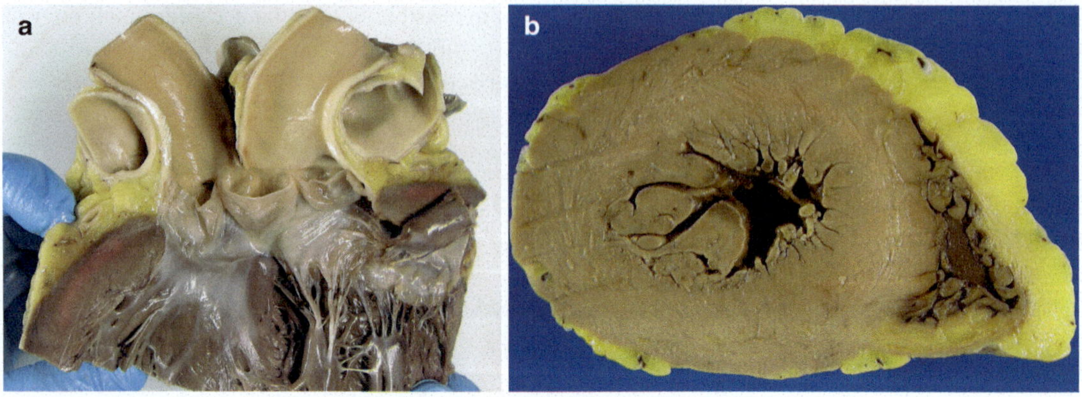

Fig. 1.36 (**a**) Sigmoid-shaped septum with endocardial fibrosis. (**b**) Adipose infiltration in the RV

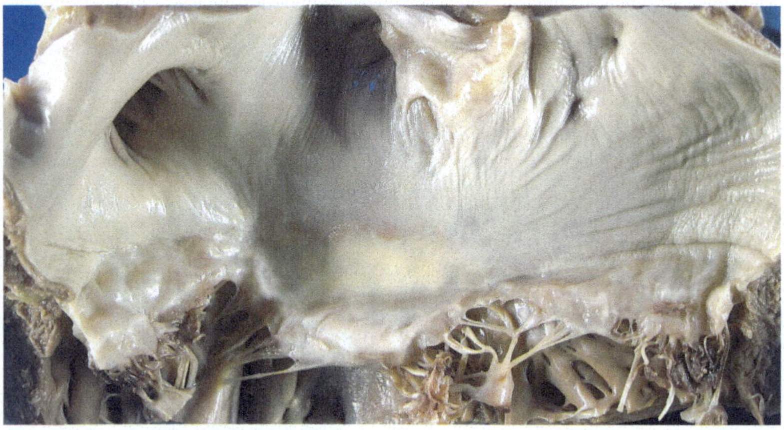

Fig. 1.37 Dilated left atrium with rough endocardium and yellowish plaques in the anterior mitral valve leaflet, in a 77-year-old male

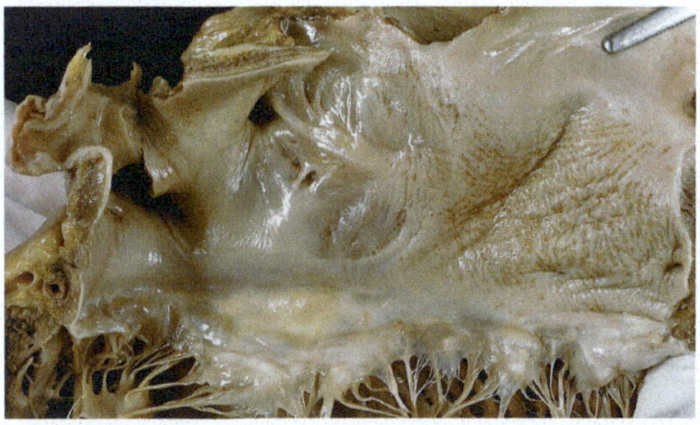

Fig. 1.38 Granular appearance of the endocardium of the left atrial wall characteristic of amyloid, in a 66-year-old male

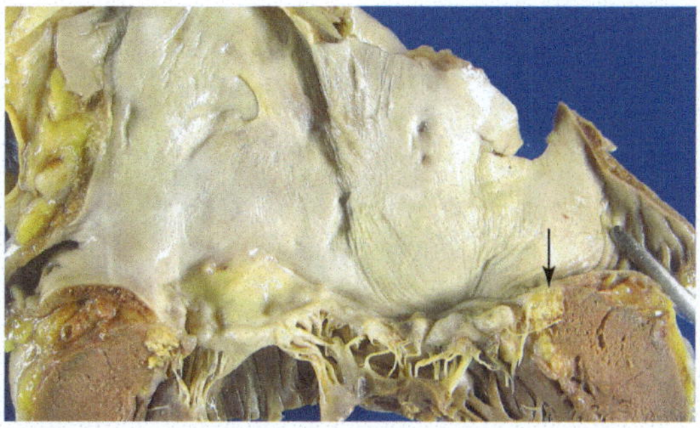

Fig. 1.39 A 90-year-old female. Dilation of the left atrium, atheromas in the anterior leaflet, thickening of the closure line and calcification in the mitral valve annulus (*arrow*)

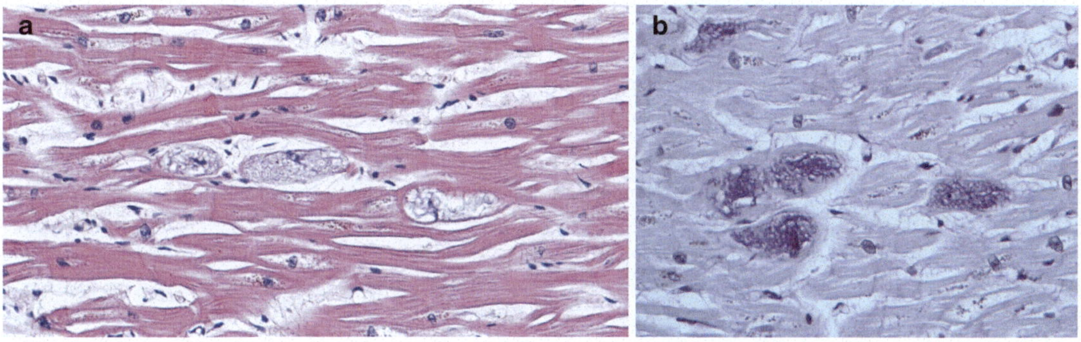

Fig. 1.40 (**a**) Basophilic degeneration of myocytes (Haematoxylin-eosin). (**b**) Positive PAS stain

1.4 Molecular Studies

As previously mentioned in the introduction, cardiac pathology accounts for 80 % of sudden deaths, and thus the anatomopathological study of the heart following defined protocols described in this chapter is essential. However, there are numerous cases, especially among the young, in which sudden death, although cardiac in nature, arises from illnesses where the heart structure is normal but the ionic channels responsible for its electrical activity do not function correctly. These are the so-called *channelopathies*, that include primarily the long QT syndrome, Brugada's syndrome, short QT syndrome and catecholaminergic polymorphic ventricular tachycardia (CPVT). These are hereditary syndromes and, during the last decades, mutations in genes encoding different components of these ionic channels, particularly the sodium and potassium channels, have been discovered. The diagnosis of these channelopathies during an autopsy can be achieved through the genetic study from blood or tissue samples from the deceased, the so-called *molecular autopsy* (term given by Ackerman in 2001).

Furthermore, etiological diagnosis in cases of myocarditis also requires molecular studies by PCR to identify virus or bacterial genetic material. Recently, viruses have been detected in hearts without histological lesions.

It is also very important to carry out a genetic analysis in cases of structural cardiomyopathies or aortic disease to detect mutations present in other family members.

During the last years, special attention has been paid to the study of sudden death, especially among young people, which has led to the publication of several guidelines (Basso et al. 2008; TRAGADY 2008; Oliva et al. 2011), that expressly indicate which samples need to be taken for molecular studies (Table 1.7). It is important to remember the need to carry out toxicological determinations in cases of sudden death, especially among adolescents and young adults, to discard the involvement of alcohol or drug abuse.

Finally, as mentioned in the introduction of this chapter, we cannot forget the fact that in order to determine the cause of death, a histopathological study of the heart and other organs is not sufficient. It is essential to be aware of the clinical antecedents, the circumstances surrounding the death, and the symptoms previous to the death, in order to perform an effective and complete clinico-pathological correlation that allows us to determine the cause and mechanisms of death.

Table 1.7 Samples to be collected at the autopsy for molecular studies

EDTA-blood (10 ml). It must be kept frozen at −80 °C
5 g of fresh heart or spleen
Fresh ventricular myocardium (3 mm^3) in a tube specific for viruses

1.5 Annex: Normal Cardiac Parameters

Normal weight and dimensions (Silver and Freedom 1991)

	Females	Males
Cardiac weight	0.40 % of the body weight	0.45 % of the body weight
Valvular perimeters:	10–11.1 cm	11.2–11.8 cm
Tricuspid	5.7–7.4 cm	6.1–7.7 cm
Pulmonary	8.2–9.1 cm	9.2–9.9 cm
Mitral	5.7–7.9 cm	6.0–8.5 cm
Aortic		
Wall thickness	10–15 mm	
LV	10–15 mm	
Septum	3–4 mm	
RV		

Normal cardiac weights in relation to the body weight in adults (Kitzman et al. 1988)

	Females			Males		
Weight (kg)	L95	P	U95	L95	P	U95
30	133	196	287	162	213	282
32	137	201	295	167	220	291
34	141	206	302	172	227	300
36	144	211	310	177	234	309
38	148	216	317	182	240	317
40	151	221	324	187	247	325
42	154	226	331	191	253	334
44	157	230	337	196	259	341
46	160	234	344	200	265	349
48	163	239	350	205	270	357
50	166	243	356	209	276	364
52	169	247	362	213	281	371
54	171	251	368	217	287	379
56	174	255	374	221	292	386
58	177	259	379	225	297	392
60	179	262	385	229	302	399
62	182	266	390	233	307	406
64	184	270	395	237	312	412
66	187	273	401	240	317	419
68	189	277	406	244	322	425
70	191	280	411	248	327	431
72	194	284	416	251	331	437
74	196	287	420	255	336	444
76	198	290	425	258	341	450
78	200	293	430	261	345	455
80	202	297	435	265	349	461
82	205	300	439	268	354	467
84	207	303	444	271	358	473
86	209	306	448	275	362	478
88	211	309	453	278	367	484
90	213	312	457	281	371	489
92	215	315	461	284	375	495
94	217	318	465	287	379	500
96	219	320	470	290	383	506
98	221	323	474	293	387	511
100	222	326	478	296	391	516
102	224	329	482	299	395	521
104	226	331	486	302	399	526
106	228	334	490	305	403	531
108	230	337	494	308	406	536
110	232	339	497	311	410	541
112	233	342	501	314	414	546
114	235	345	505	316	418	551
116	237	347	509	319	421	556
118	239	350	513	322	425	561
120	240	352	516	325	429	566
122	242	355	520	327	432	570
124	244	357	523	330	436	575
126	245	360	527	333	439	580
128	247	362	531	335	443	584
130	249	364	534	338	446	589
132	250	367	537	341	450	593
134	252	369	541	343	453	598
136	253	371	544	346	456	602
138	255	374	548	348	460	607
140	257	376	551	351	463	611
142	258	378	554	353	466	616
144	260	381	558	356	470	620
146	261	383	561	358	473	624
148	263	385	564	361	476	629
150	264	387	567	363	479	633

Bibliography

Ackerman MJ, Tester DJ, Driscoll DJ. Molecular autopsy of sudden unexplained death in the young. Am J Forensic Med Pathol. 2001;22:105–11.

Angiología. In: William PL and Warwick R, editors. Gray anatomía. vol 1. 35th ed. Barcelona: Salvat editores; 1985. pp. 683–881.

Baroldi G, Scomazzoni G. Coronary circulation in the normal heart and the pathologic heart. Washington, DC: U.S. Government Printing Office: Armed Force Institute of Pathology; 1967. p. 5–90.

Basso C, Burke M, Fornes P, Gallagher PJ, de Gouveia RH, Sheppard M, et al. Guidelines for autopsy investigation of sudden cardiac death. Virchows Arch. 2008;452(1):11–8.

Burke A, Tavora F. Techniques and approach to cardiovascular pathology. In: Practical cardiovascular pathology: an atlas. Philadelphia: Wolters Kluwer Health/Lippincott Williams & Wilkins; 2011. p. 3–11.

Kitzman DW, Scholz DG, Hagen PT, Ilstrup DM, Edwards WD. Age-related changes in normal human hearts during the first 10 decades of life. Part II (maturity): a quantitative anatomic study of 765 specimens from subjects 20 to 99 years old. Mayo Clin Proc. 1988;63(2):137–46.

Kitzman DW, Edwards WD. Age-related changes in the anatomy of the normal human heart. J Gerontol. 1990;45(2):M33–9.

Larsen MK, Nissen PH, Berge KE, Leren TP, Kristensen IB, Jensen HK, et al. Molecular autopsy in young sudden cardiac death victims with suspected cardiomyopathy. Forensic Sci Int. 2012;219(1–3):33–8.

Monteforte N, Napolitano C, Priori SG. Genetics and arrhythmias: diagnostic and prognostic applications. Rev Esp Cardiol (Engl Ed). 2012;65(3):278–86.

Oliva A, Brugada R, D'Aloja E, Boschi I, Partemi S, Brugada J, et al. State of the art in forensic investigation of sudden cardiac death. Am J Forensic Med Pathol. 2011;32(1):1–16.

Pomerance A. Age-related cardiovascular changes and mechanically induced endocardial pathology. In: Silver MD, editor. Cardiovascular pathology. 2nd ed. New York: Churchill Livingstone; 1991. p. 155–94.

The Royal College of Pathologists' Working Party on the autopsy. Guidelines on autopsy practice: Scenario 1: Sudden death with likely cardiac disease. London: Royal College of Pathologists; 2005.

Schoen FJ, Piehler H, Levy RJ. Pathological considerations in replacement cardiac valves. Cardiovasc Pathol. 1992;1:29–52.

Schoen FJ. Approach to the analysis of cardiac valve prostheses as surgical pathology or autopsy specimens. Cardiovasc Pathol. 1995;4:241–55.

Scholz DG, Kitzman DW, Hagen PT, et al. Age-related changes in normal human hearts during the first 10 decades of life. Part I (Growth): a quantitative anatomic study of 200 specimens from subjects from birth to 19 years old. Mayo Clin Proc. 1988;63:126–36.

Sheppard MN. Approach to the cardiac autopsy. J Clin Pathol. 2012;65(6):484–95.

Silver MM, Freedom RM. Gross examination and structure of the heart. In: Silver MD, editor. Cardiovascular pathology. 2nd ed. New York: Churchill Livingstone; 1991. p. 1–42.

Stone JR, Basso C, Baandrup UT, Bruneval P, Butany J, Gallagher PJ, et al. Recommendations for processing cardiovascular surgical pathology specimens: a consensus statement from the Standards and Definitions Committee of the Society for Cardiovascular Pathology and the Association for European Cardiovascular Pathology. Cardiovasc Pathol. 2012;21(1):2–16.

The members of Trans-Tasman Response Against sudden Death in the Young (TRAGADY). Post-mortem in sudden unexpected death in the young: guidelines on autopsy practice. Available in: http://www.cidg.org/webcontent/LinkClick.aspx?fileticket=DO9YIQWqegI%3D&tabid=161. 2008.

Virmani R, Burke A, Farb A, Atkinson JB. Examination of the heart. In: Cardiovascular pathology, Major problems in pathology series, vol. 40. 2nd ed. Philadelphia: Saunders Company; 2001. p. 1–25.

Pulmonary Artery Pathology

2

Pedro M. Garamendi, Joaquín S. Lucena, Manuel Gómez-Bueno, and Javier Segovia

Contents

2.1 Introduction

Pathologies associated with alterations in the vascular system of the pulmonary circuit constitute a separate arena within cardiovascular pathology. Disorders affecting the pulmonary circulation are heterogenous in nature, although all involve the presentation of significant alterations in the pulmonary arterial and/or venous systems, which are identified through macroscopic and microscopic examination (Table 2.1).

For reasons of image economy, this chapter will deal only with the two most significant pathologies encountered by the pathologist during autopsy: pulmonary arterial hypertension, specifically idiopathic and familial forms, and pulmonary thromboembolic disease.

For a detailed analysis and study of additional pulmonary pathologies, please refer to the general pathology textbooks listed in the bibliography section of this chapter.

P.M. Garamendi (✉)
Department of Forensic Pathology,
Institute of Legal Medicine, Huelva, Spain
e-mail: imanolgaramendi@gmail.com

J.S. Lucena
Department of Forensic Pathology,
Institute of Legal Medicine, Seville, Spain

Forensic Medicine,
University of Cadiz, Cadiz, Spain
e-mail: joaquin.lucena@uca.es

M. Gómez-Bueno
Advanced Heart Failure and Transplant Unit,
Cardiology Department, Hospital Universitario
Puerta de Hierro, Majadahonda (Madrid), Spain

J. Segovia
Advanced Heart Failure and Transplant Unit,
Cardiology Department, Hospital Universitario
Puerta de Hierro, Majadahonda (Madrid), Spain

Universidad Autónoma de Madrid, Madrid, Spain

J.S. Lucena et al. (eds.), *Clinico-Pathological Atlas of Cardiovascular Diseases*,
DOI 10.1007/978-3-319-11146-9_2, © Springer International Publishing Switzerland 2015

Table 2.1 Pathologies associated with alterations in the pulmonary vascular system

1. Congenital vascular malformations: Unilateral absence of a pulmonary artery, atresia, pulmonary artery stenosis associated with extreme tetralogy of Fallot or anomalies in the intraparenchymal vessels, due to defects in the media or to misalignment of the pulmonary vessels (Fig. 2.1) Pulmonary artery aneurysm associated with mycosis, infections such as syphilis, vasculitis (Behçet's disease), pulmonary hypertension, trauma or congenital cardiac diseases (principally, patent ductus arteriosus and atrial septal defects)
2. Thrombotic pulmonary embolism
3. Non-thrombotic pulmonary embolisms, caused by fragments of bone marrow, tumours, fat, foreign bodies, air bubbles, parasites or those associated with pregnancy and puerperium
4. Thrombotic angiopathy associated with chronic haemolytic anemia in sickle cell disease
5. Pulmonary veno-occlusive disease
6. Pulmonary capillary haemangiomatosis
7. Pulmonary hypertension originating in arterial or venous vessels (pulmonary edema and chronic venous hypertension)
8. Hypoxic pulmonary hypertension, such as high-altitude cardiopulmonary diseases, sleep apnea syndrome or kyphoscoliosis
9. Structural lung diseases, infiltrative or obstructive, which secondarily affect the pulmonary circulation

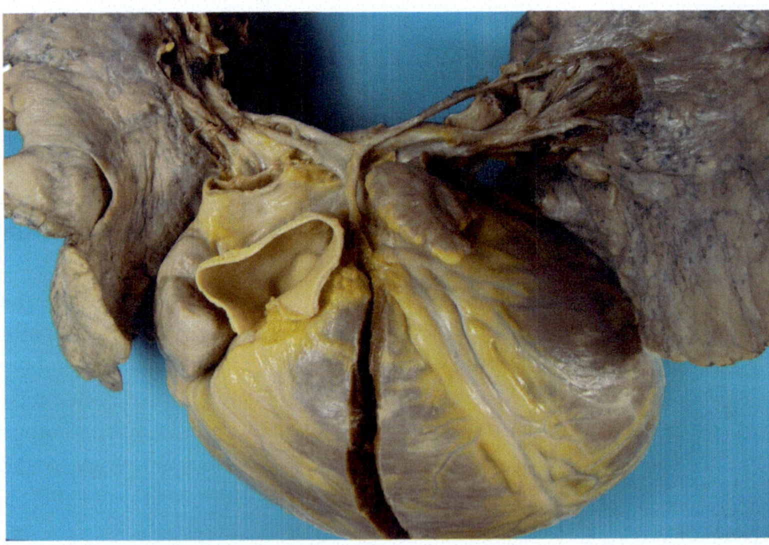

Fig. 2.1 A 33-year-old male with extreme tetralogy of Fallot: pulmonary atresia and severe hypoplasia in the pulmonary arteries, which are reduced to a fibrous cord. The patient presented with severe heart failure. Severe cardiomegalia (>1,000 g) with dilation and hypertrophy of the right cavities

2.2 Pulmonary Thromboembolism

Pulmonary thromboembolism (PTE) and deep venous thrombosis (DVT) encompass different clinical manifestations of a single disease, venous thromboembolism (VTE). Because of its non-specific clinical manifestations, diagnosis is complex and the disease is often identified at autopsy. In autopsy series PTE constitutes the cause of death in approximately 4 % of sudden deaths that occur outside of the hospital.

The so-called *Virchow's triad* of interrelated factors: vascular stasis, endothelial damage and abnormal blood flow, still constitutes the basis for understanding the pathogenesis of thrombus formation. Identified thrombogenic risk factors include: reduced mobility, old age, malignancy, recent medical or surgical disease, trauma, spinal lesion, obesity, smoking, pregnancy and puerperium, polycythemia vera, and several medications such as oral contraceptive drugs. Recently, a correlation has been established between fatal PTE and a history of psychiatric disorder, or treatment with antipsychotic drugs.

Together with these acquired factors, additional hereditary disorders have been related with a state of hypercoagulability, which facilitates the development of venous thrombosis. Included among these hereditary disorders are hereditary thrombophilias such as mutations in Factor V (Factor V Leiden), prothrombin gene mutation (20210A), antithrombin I, II and III deficiencies, Protein S, Protein C or plasminogen deficiencies, etc. Additional syndromes including antiphospholipid syndrome or hyperhomocysteinemia are also associated with hypercoagulability.

In 95 % of PTE cases, the thrombus originates in the deep venous system of the lower extremities (Fig. 2.2a–c). Different pathological conditions result in thrombus formation in the crural, femoral or iliac veins. In exceptional cases, the origin of the embolizing thrombus has been discovered in diverse venous territories such as the mesenteric veins (Fig. 2.3), the right heart (Fig. 2.4a–c) and the upper extremities. Furthermore, the presence of a central venous cathether constitutes a causal factor for thrombosis and subsequent pulmonary embolism, especially in children.

Once the embolizing thrombus has reached the territory of the pulmonary arteries it can provoke two different effects in the pulmonary circulation: on the one hand a hypoxic state can arise since the thrombus obstructs, totally or partially, blood flow in the pulmonary respiratory units, altering the ventilation/perfusion relationship. This can lead to the release of substances that would subsequently produce pulmonary vasoconstriction, aggravating the mechanical ventilation effect. Alternatively, the flow obstruction could mediate a haemodynamic effect, resulting in acute pulmonary artery hypertension (PAH). In turn, this acute state of PAH produces right ventricular pressure overload that ultimately causes right cardiac failure.

The clinical severity of PTE depends on factors such as the size of the embolus (from lodged thombi that affect both pulmonary arteries at the proximal level, to multiple disseminated microthrombi), which determines the number of obstructed vessels and the relative percentage of the affected pulmonary artery bed, the speed of establisment of an acute PAH, and the previous state of the pulmonary circulation. Thus, pre-existence of pulmonary or cardiac pathologies determines whether or not the double circulation through the bronchial arteries can counteract the damage caused by obstruction in the main pulmonary arteries. The existence of this supplementary circulation, and the speed at which the effects of PAH appear, especially in cases of massive PTE, explains the relatively low incidence of pulmonary infarctions found at autopsy series following a PTE episode. This ranges from 6 to 20 % depending on the autopsy series, and can be even lower in cases of sudden death with non-organised thromboemboli.

Clinical diagnosis is complex. Patients with DVT in the lower extremities can be asymptomatic or can present with pain, heat or inflammation. Pulmonary embolism must always be considered in cases of unexplained dyspnea. The majority of pulmonary emboli are multiple and affect the inferior lobes. In the case of small emboli that do not cause significant obstruction

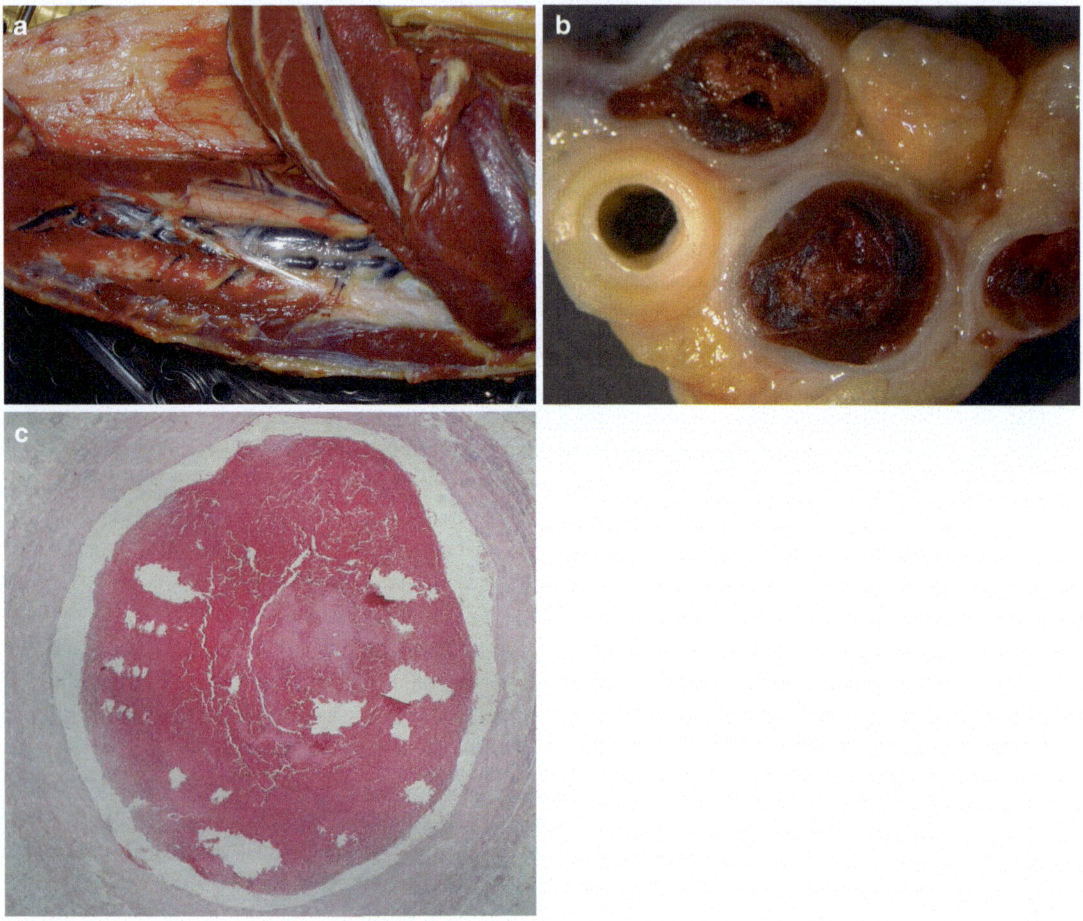

Fig. 2.2 (**a**) Thrombosis in the deep venous system of the leg. (**b**) Stereoscopic image of a transversal section of the venous thrombosis. (**c**) Histological image of a thrombus located in the posterior tibial vein, formed by platelets and red blood cells

in the pulmonary circulation, the clinical symptoms can manifest as acute pleuritic chest pain, or as haemoptysis. Only 10 % of emboli cause pulmonary infarcts, usually in patients with pre-existing cardiopulmonary pathology. In the most severe cases, the clinical picture can be syncope and hypotension. Symptoms can be more subtle in older patients, where it is not unexpected to find that the main clinical manifestation is syncope. This contrasts with younger patients, where the clinical presentation is more florid. In up to 25 % of cases DVT can present as sudden death.

Imaging techniques are fundamental for the diagnosis of PTE. The most commonly-used are ventilation/perfusion gammagraphy and angio-TC. The choice between one technique and the other depends on its availability, and the experience of the centre. Magnetic resonance imaging and pulmonary arteriography are techniques that are not usually available in an emergency setting. Doppler echography is the method of choice to diagnose the presence of DVT. Additional tests including ECG, echocardiogram, dimer-D test, troponin and natriuretic peptide determination, can be helpful in the diagnosis of PTE and have independent prognostic value.

When death occurs, the most significant pathological findings are discovered in the pulmonary arteries, pulmonary tissue, and the heart. With PTE, the autopsy must be exceptionally methodical in order to locate the thrombogenic focus, usually situated in the deep veins of the lower

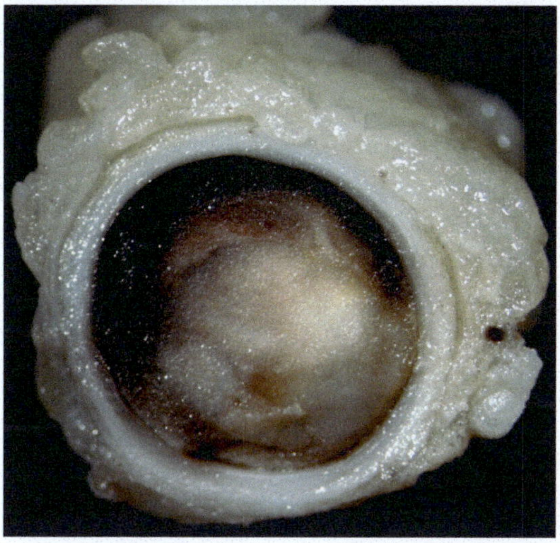

Fig. 2.3 Acute thrombosis in the superior mesenteric vein in a 16-year-old male who died suddenly due to PTE. Analysis for hereditary thrombophilia was negative

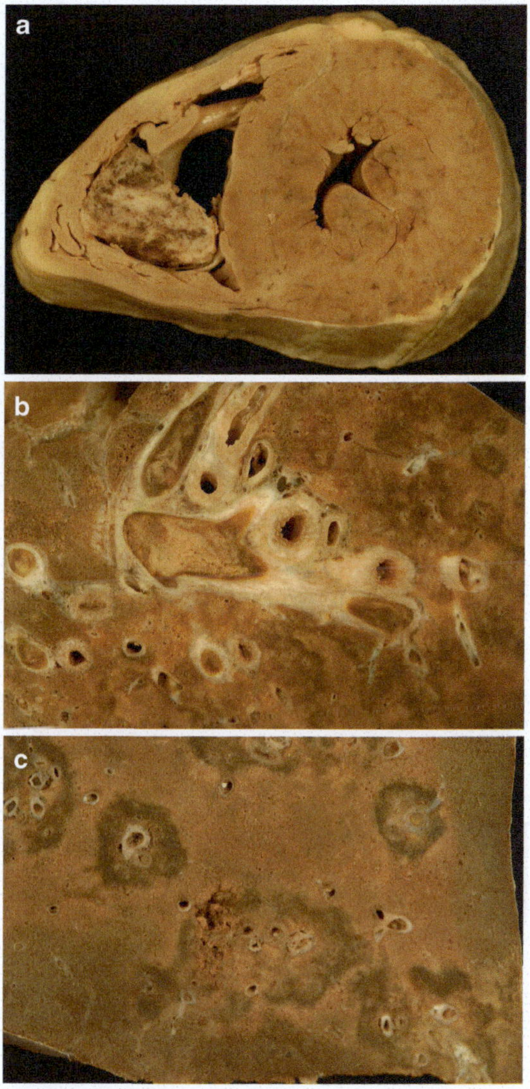

Fig. 2.4 (**a**) Intracavitary thrombus adhered to the right ventricular endocardium, which was the cause of a masive fatal PTE in an 8-year-old boy with corticosteroid-dependent nephrotic syndrome. (**b**) Massive thrombosis in secondary and tertiary pulmonary arteries. (**c**) Necrotizing-haemorrhagic perivascular lesions. Analysis for hereditary thrombophilia indicated a heterozygous mutation in the prothrombin 20210A allele

extremities which must be systematically examined during necropsy.

Thromboemboli causing PTE are found in the pulmonary arteries. Very large thrombi tend to remain located in the main arterial trunk, causing in some cases an obstruction of both pulmonary arteries ("saddle" embolus) (Fig. 2.5a). Smaller thrombi can provoke an obstruction in the main branches of the pulmonary artery (Figs. 2.6 and 2.7), while the smallest thrombi obstruct the intrapulmonary branches.

Thrombi can be divided morphologically into two major types according to the degree of chronological evolution: grey thrombi (Fig. 2.8) and red thrombi (Figs. 2.5 and 2.6). Grey thrombi are compact, whitish in appearance, and are laminated when sectioned. Red thrombi are also compact, but are dark red in colour and display alternate white and red layers when sectioned (*Zahn's lines*). Both thrombi occlude, but do not perfectly mould, the arterial lumen. The contour of the thrombi reflects the shape of the original venous mould, and even the imprinting of the cardiac valves of the right heart, through which they progress towards the pulmonary arteries. In microscopy, the *Zahn's lines* correspond to alternating layers of erythrocyte aggregates and platelets mixed with fibrin (Fig. 2.5b, c). These layers, or lines, reflect the mechanism through which thrombi are produced in situations of venous stasis at the thrombogenic focus and result from the progressive deposition of blood elements on a thrombus originally adhered to a damaged venous endothelium. In situations where many neutrophils can be microscopically

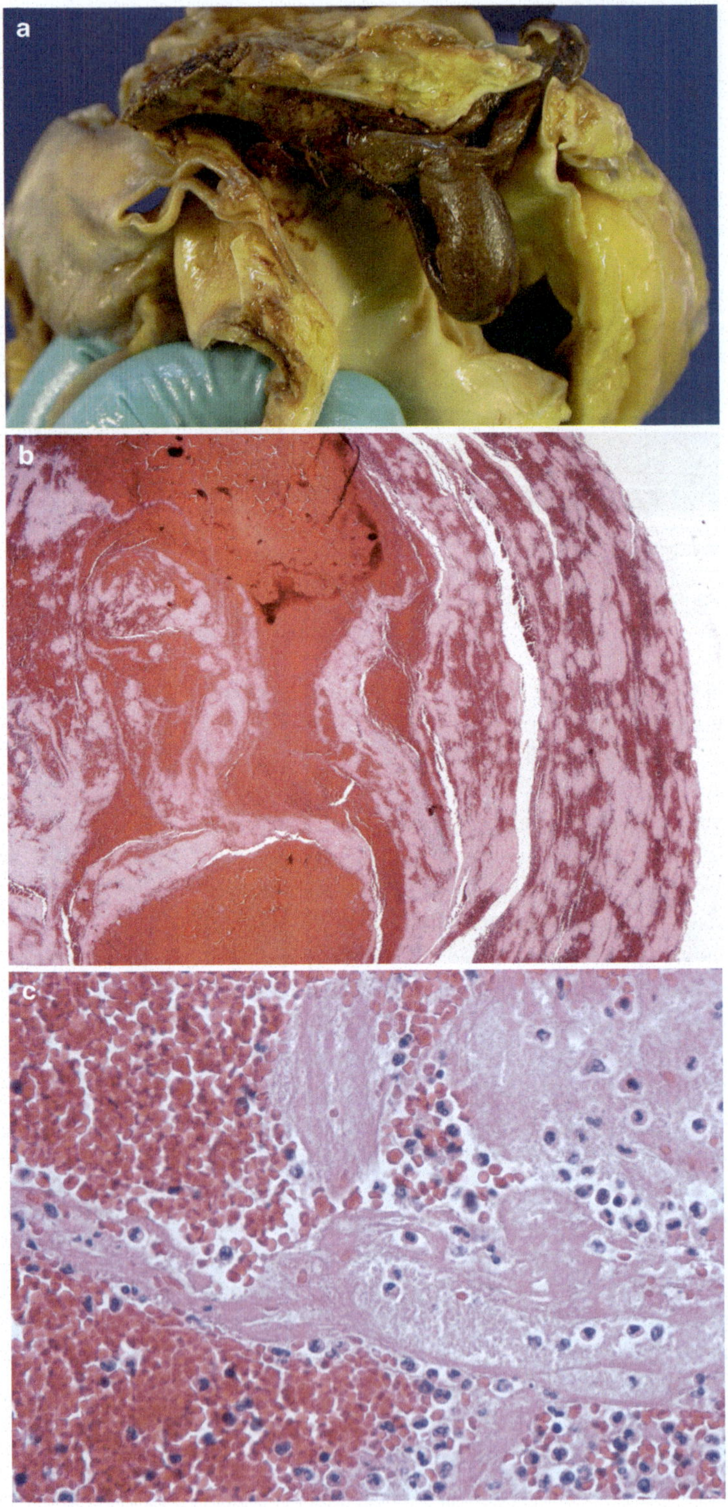

Fig. 2.5 Sudden death in a 39-year-old pregnant woman. (**a**) Thrombus lodged at the bifurcation of the pulmonary artery *("saddle" embolus)*. (**b**) Microscopic image of the thrombus where the *Zahn lines* are evident. (**c**) Microscopic detail of the thrombus with abundant platelets

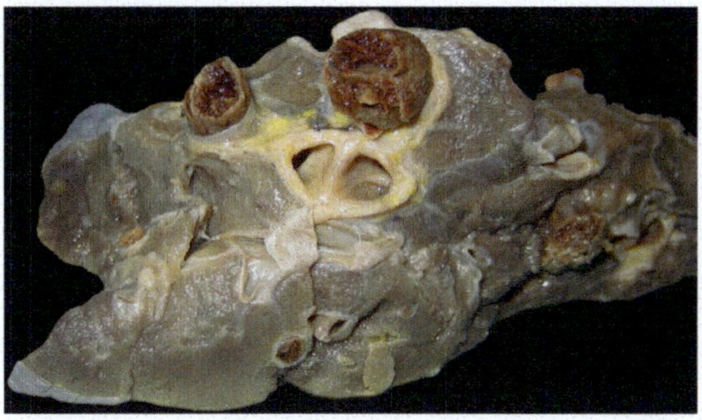

Fig. 2.6 Thromboemboli in the main pulmonary arteries. Image resembling *"toothpaste squeezed from a tube"*

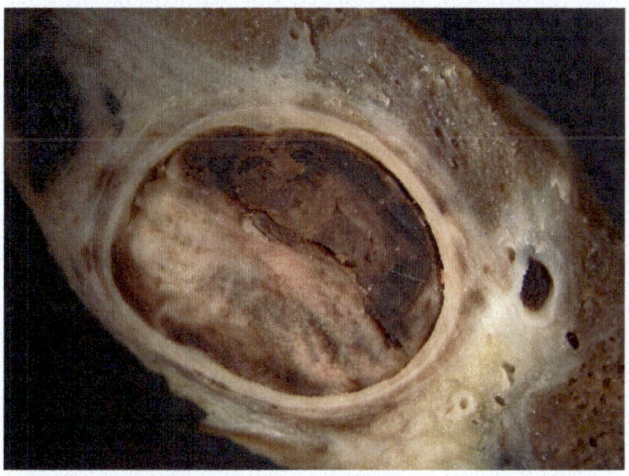

Fig. 2.7 Detailed image of a red thromboembolus occluding a main pulmonary artery

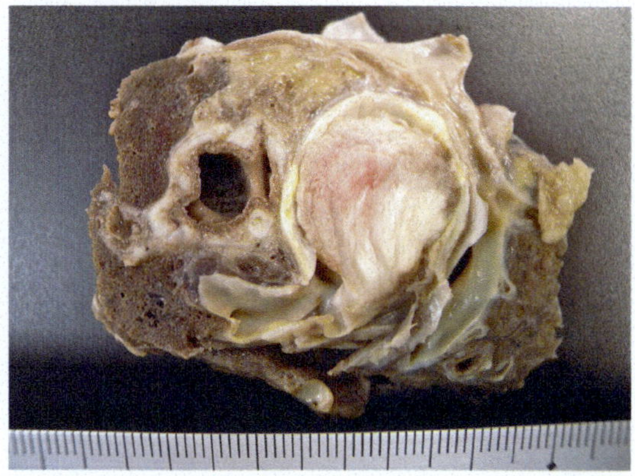

Fig. 2.8 Gray thromboembolus in the main pulmonary artery with a near-complete occlusion in the vascular lumen and a laminated surface on section

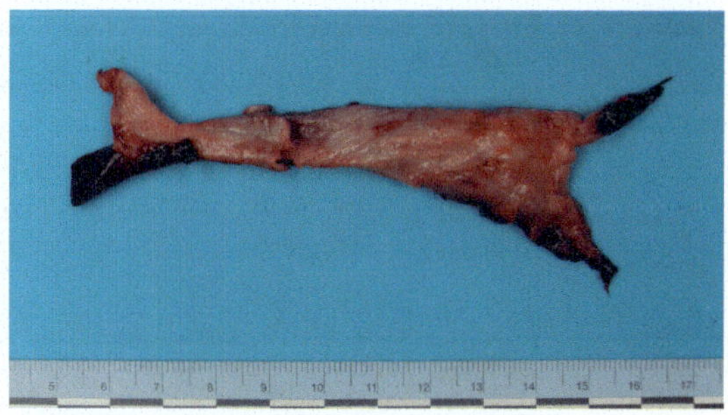

Fig. 2.9 Post-mortem blood clot. Biphasic colour can be observed with light areas clearly separated from dark areas (at the extremes of the blood clot)

identified in the thrombus, the term septic thrombus is given, and suggests the presence of an associated infectious process.

Red thrombi found in the pulmonary artery bed have to be distinguished from postmortem clots (Fig. 2.9). Table 2.2 describes the morphological differences between acute vital thrombi, responsible for PTE, and postmotem clots.

Venous thrombi, once they obstruct the pulmonary arterial lumen, undergo a remodelling process that provokes, in the first instance, the adherence of the thrombus to the arterial endothelium (Fig. 2.10). Next, the thrombus is replaced with connective scar tissue and finally, the thrombus is rechanneled with newly-formed lumen in the intially-occluded arterial lumen (Fig. 2.11). This temporal evolution can be reflected in the findings from a microscopic study of the thrombus. However, this classical scheme, which is applicable to thrombotic lesions found in peripheral veins, could be accelerated in the pulmonary arteries. Recent studies have proposed a classification system based on only three identifiable temporal phases, as summarised in Table 2.3. Several authors recommend to focus the thrombus-dating study at the origin of DVT, rather than at the pulmonary thrombi, given the more unpredictable evolution in this latter location.

Table 2.2 Macroscopic characteristics of the vital thrombi and postmortem blood clots (Groshong et al. 2008)

	Acute vital thrombus	Postmortem blood clot
Colour	Dark red	Biphasic (red/yellow)
Consistency	Firm	Soft and jelly-like
Shape	Coiled	Forms branches
	Mould of the leg veins	Mould of the pulmonary artery
Lineal striations	Present	Absent
Vascular distension	Present	Absent
Adhesion to vessels	Present	Absent
Perivascular haemorrhage	Present	Absent

Macroscopic examination of the pulmonary tissue does not often reveal significant findings, other than the presence of thrombi emerging from the arterial lumen which resemble "toothpaste squeezed from a tube" (Fig. 2.6). Similarly, microscopic examination reveals only vascular congestion, edema and ocasionally pulmonary haemorrhage.

As indicated previously, pulmonary infarcts associated with PTE are seen in less than 20 % of patients with previous pathologies including severe cardiocirculatory alterations or cancer, or in cases of emboli affecting more than two lobules that are not associated with sudden death. The pathologic findings of the pulmonary infarct are described in Table 2.4.

Table 2.3 Phases of the chronologic evolution of thrombi and thromboemboli (Fineschi et al. 2009)

Phase I: Day 1–7	Flowing blood on an eroded endothelium, eliciting a platelet plug and fibrin deposition with a layered growth (Zahn's lines) No reaction between endothelium and thrombus is visible Erythrocytes are preserved and agglomerated Initial white blood cells picnosis Monocytes cells with enlarged nuclei Calcium is observed as precipitates with Von Kossa stain. The thrombus at its initiation is firmly attached to a small portion of the vessel wall and is not easily removed to leave fragments in situ. In contrast, a blood clot maintains the usual blood composition (i.e., prevailing red cells plus leukocytes and platelets and a fine network of fibrin), is not attached to the endothelium, and can be easily removed
Phase II: Week 2–8	Endothelial budding and proliferative changes of the medial ring represented by penetration of fibroblasts. Macrophages containing hemosiderin predominate, red blood cells ghosts and fibrinous transformation. The ribbons of fibrin changing to coalescences trapping white cells. Free surface of thrombus is covered by endothelium. Scattered nuclear debris of white blood cells
Phase III: Older than 2 months	Completely hyalinized thrombus with central sinous cavities and more advanced recanalizing neo-formed larger vessels with fresh flowing blood. Few white cells are visible between compact, fibre-rich and cell-deficient connective tissue

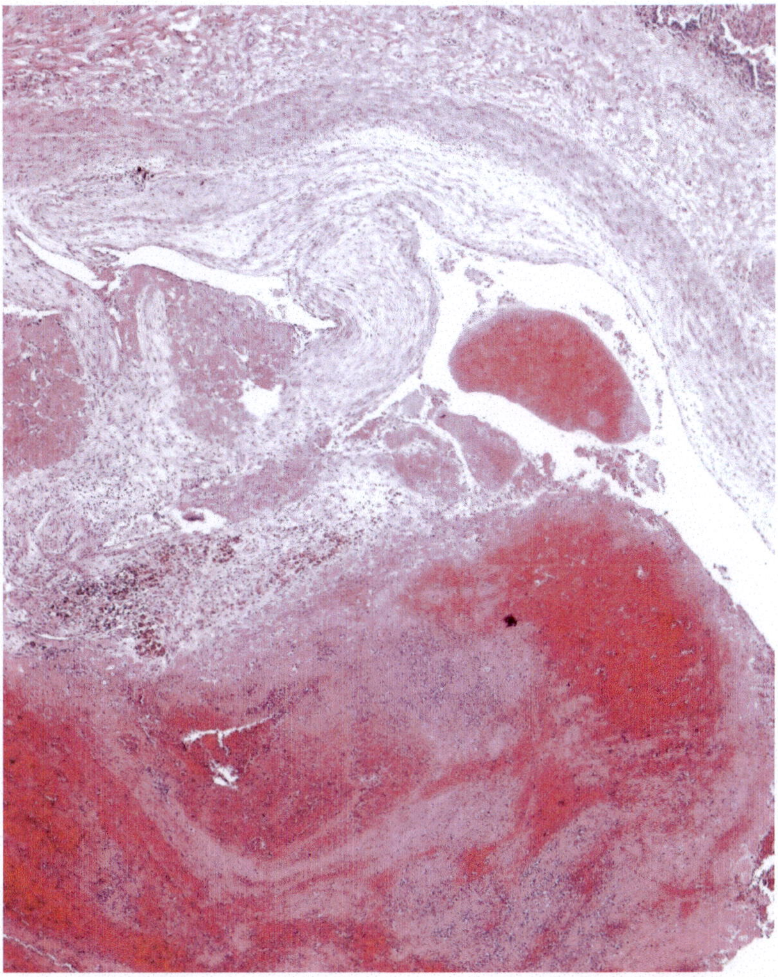

Fig. 2.10 Thrombus adhered to the vascular endothelium of the pulmonary artery

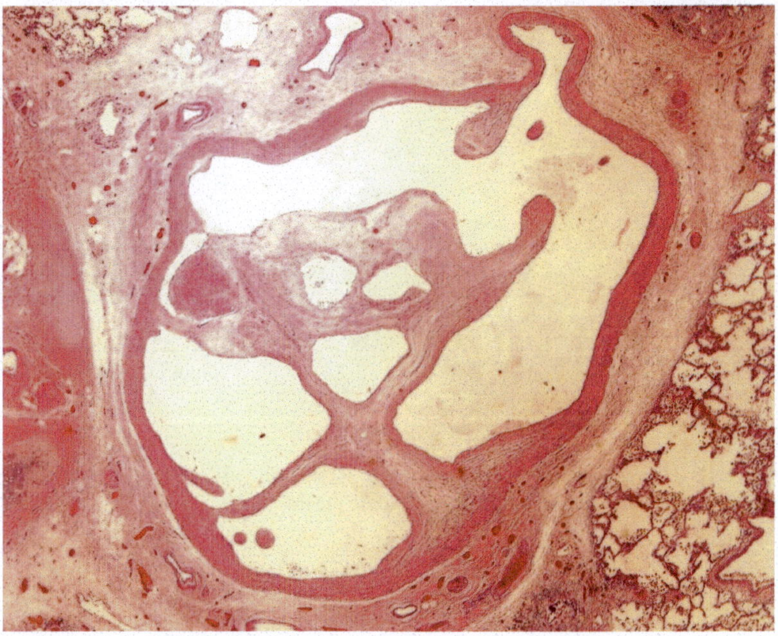

Fig. 2.11 Rechanneled thrombus with newly formed lumen

Table 2.4 Pathology of pulmonary infarction

Macroscopic features
Areas with pulmonary infarction have a wedge-shaped morphology, with the vertex orientated towards the pulmonary hilium
In the initial phase, they are haemorrhagic in appearence, blue-redish in colour with an over-elevated surface (Fig. 2.12) that can be associated with pleural fibrinous exudates (Fig. 2.13)
In the second phase, the colour of the infarcted tissue changes to brown due to haemosiderin lysis (Fig. 2.14)
In the third phase, fibrosis occurs due to the replacement of necrotic tissue with connective tissue, giving it a white-greyish colour that progresses from the periphery to the central area of the infarct. In the final phase, all infarcted tissue is replaced by fibrous tissue, which retracts, forming a scar
Microscopic features
In the initial phases of a pulmonary infarct, only haemorrhage, alveolar edema and arterial congestion can be appreciated, with no macroscopic evidence of necrosis. In the established infarct phase these findings are associated with a cellular inflammatory reaction and nuclear disapearence in the alveolar septi (Fig. 2.15)

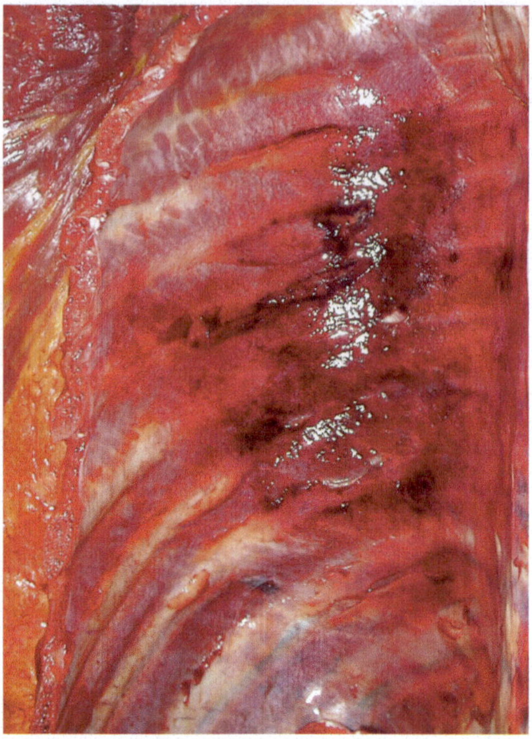

Fig. 2.13 Fibrotic reaction of the parietal pleura juxtaposed to a pulmonary infarct caused by PTE

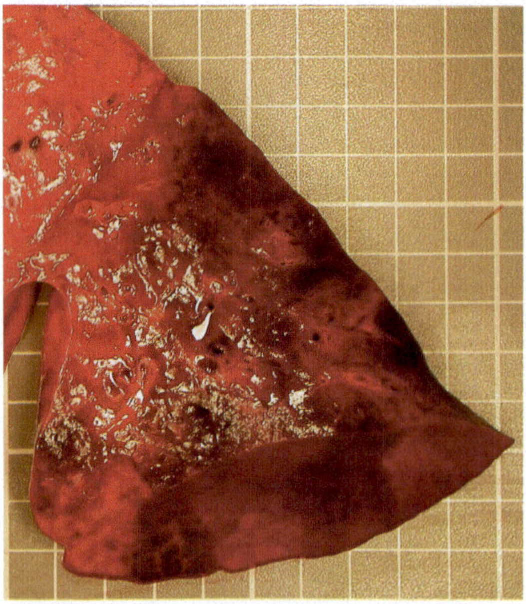

Fig. 2.12 Haemorrhagic infarct secondary to PTE, with the typical triangular shape whose base is at the pleura

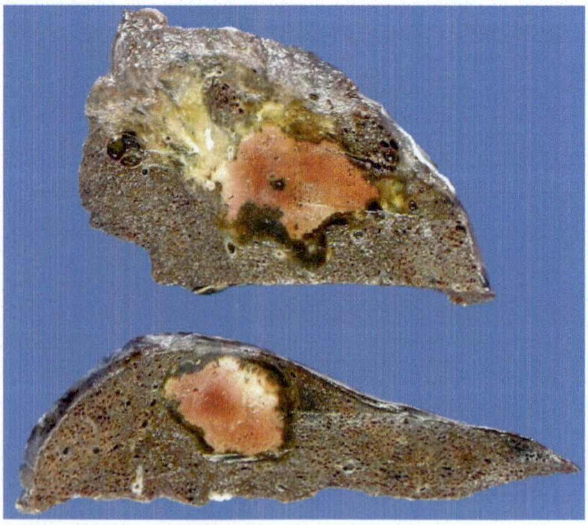

Fig. 2.14 A mature pulmonary infarct, brownish in colour and surrounded by a congestive halo

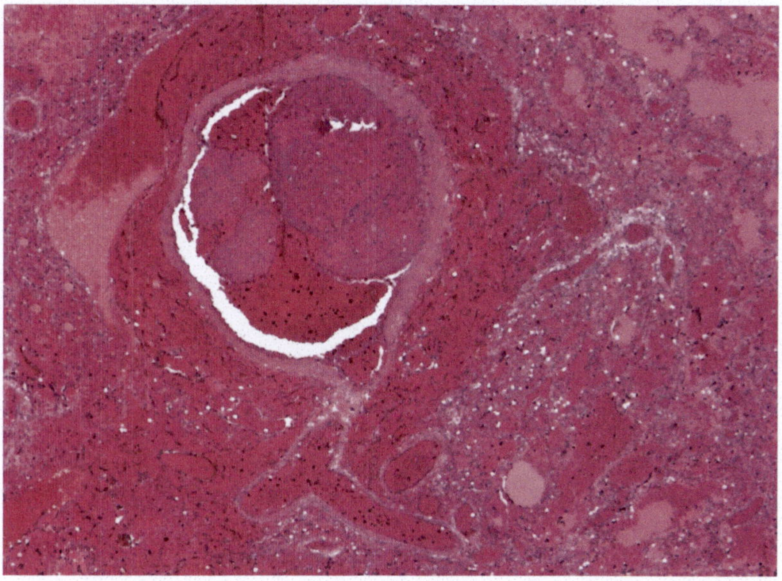

Fig. 2.15 Arterial thrombosis with associated pulmonary infarct, characterised by coagulative necrosis in the pulmonary parenchyma within an area of haemorrhage and edema

2.3 Pulmonary Hypertension

During extrauterine life, the arterial pulmonary circulation is a low pressure circulation. The average pressure in the pulmonary artery is typically less than 20 mmHg in the resting state, but can reach up to 30 mmHg during exercise. Pulmonary arterial hypertension (PAH) can be defined as a clinical situation when the average pulmonary blood pressure is equal to or higher than 25 mmHg at rest. The term PAH embraces several clinical entities from diverse origins that all share an ability to destabilize the pressure within the pulmonary artery by:

(a) Primary involvement of the different layers in the pulmonary arteries which ultimately produce vasoconstriction, cellular proliferation and intravascular thrombosis.
(b) Retrograde mechanical effect due to an increase in blood pressure in the venous pulmonary circuit (cardiac diseases such as mitral stenosis).
(c) Destruction of respiratory units (pulmonary diseases) (Fig. 2.16).
(d) Active obstruction in the pulmonary artery circulation (recurring embolism).
(e) Stenosis affecting the blood flow in the intrapulmonary arterial vessels (autoimmune and congenital diseases) (Fig. 2.17a, b).

Table 2.5 includes the classification of these clinical entities accepted by the World Health Organization (WHO) in 2008 and adopted by the European Society of Cardiology.

In the majority of pathologies associated with PAH, the precipitating factor is endothelial cell dysfunction in the intrapulmonary arteries, due either to mechanical effects (stenosis), hypoxia, or a chemical influence (such as that mediated by fibrin adherence to the endothelium in recurrent PTE cases, or chronic post-thrombotic syndrome by PTE). Apart from these factors, there is also a vasospastic factor that explains the efficiency of certain vasodilator drugs in the clinical treatment of some of these pathologies.

Clinically, PAH in its idiopatic and familial forms tends to be more common in women aged between 30 and 50 years-old. Once established, PAH evolves rapidily and, without treatment, leads to death in 2–3 years. Symptons only appear in the pathological-advanced stages of the disease. The initial symptom is progressive dyspnea on exertion. When the dysfunction in the right

Table 2.5 Updated classification of pulmonary hypertension (Galié et al. 2009; Simonneau et al. 2009)

1. Pulmonary arterial hypertension (PAH)
1.1. Idiopathic
1.2. Familial
1.2.1. BMPR2
1.2.2. ALK1, endoglin (with or without hereditary haemorrhagic telangiectasia)
1.2.3. Unknown
1.3. Drug and toxin induced
1.4. Associated with:
1.4.1. Connective tissue diseases
1.4.2. HIV infection
1.4.3. Portal hypertension
1.4.4. Congenital heart diseases
1.4.5. Schistosomiasis
1.4.6. Chronic haemolytic anemia
1.5. Persistent pulmonary hypertension of the newborn
1.6. Pulmonary veno-occlusive disease (PVOD) and/or pulmonary capillary hemangiomatosis (PCH)
2. Pulmonary hypertension due to left heart disease
2.1. Systolic dysfunction
2.2. Diastolic dysfunction
2.3. Valvular disease
3. Pulmonary hypertension due to lung disease and/or hypoxia
3.1. Chronic obstructive pulmonary disease (COPD)
3.2. Interstitial lung disease
3.3. Other pulmonary diseases with a mixed restrictive and obstructive pattern
3.4. Sleep-disordered breathing (Disnea-apnea)
3.5. Alveolar hypoventilation disorders
3.6. Chronic exposure to high altitude
3.7. Developmental abnormalities
4. Chronic thromboembolic pulmonary hypertension
5. Pulmonary hypertension with unclear multifactorial mechanisms
5.1. Haematologic disorders: myeloproliferative syndrome, splenectomy
5.2. Systemic disorders: sarcoidosis, pulmonary Langerhans'-cell histiocytosis, lymphangioleiomyomatosis, neurofibromatosis, vasculitis
5.3. Metabolic disorders: glycogen storage disease, Gaucher disease, thyroid disorders
5.4. Others: tumoral obstruction, fibrosing mediastinitis, chronic renal failure on dialysis

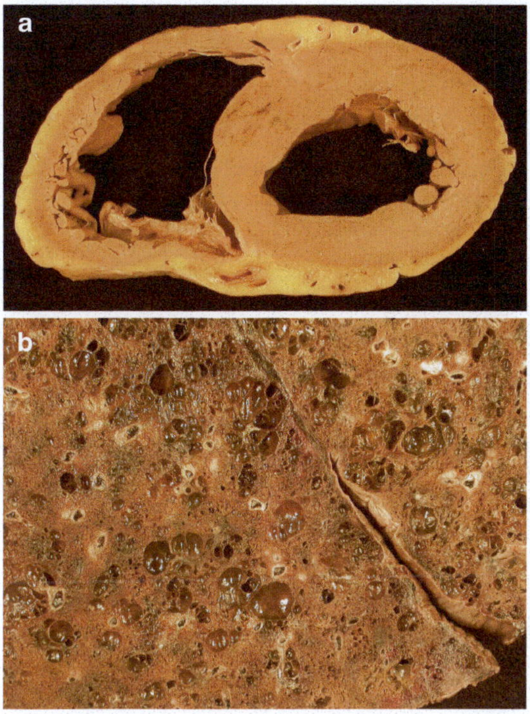

Fig. 2.16 A 56-year-old male with a history of COPD was found dead at home. (**a**) *Cor pulmonale* secondary to (**b**) centroacinar pulmonary emphysema

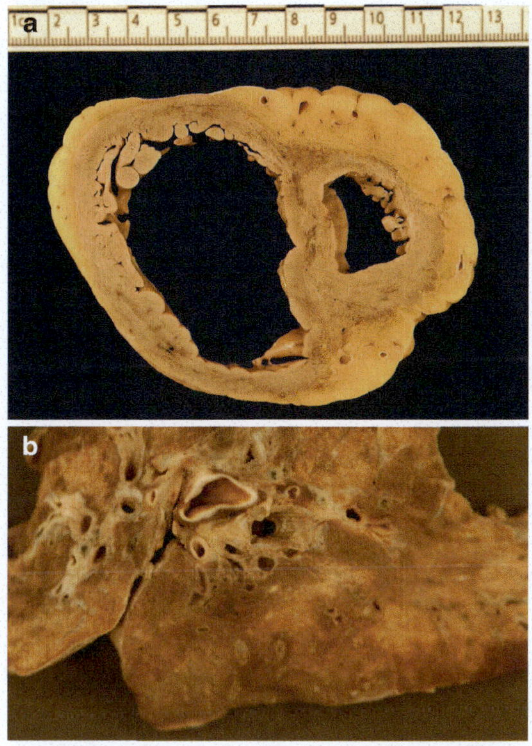

Fig. 2.17 A 26-year-old woman, diagnosed with a connective tissue mixed disease, died suddenly at home. (**a**) Chronic *cor pulmonale* with cardiomegalia (405 g), RV hypertrophy (9 mm) and dilation of the right ventricular chamber (5 cm). (**b**) Wall thickening in the pulmonary arteries

ventricle (RV) progresses, angina or syncope with exercise appear due to the inability of the RV to adapt the cardiac output to exercise. In advaced stages, these symptoms can also present at rest. In 80 % of the cases, death is due to RV failure, although in the remainder, death results from overlapping complications such as pneumonias or pulmonary embolism. Sudden death in these patients is not common but has been recorded associated with cardiac arrhythmia, pulmonary haemorrhage, pulmonary embolism, and acute right ventricular ischaemia.

In cases of death due to PAH, specific pathological findings at necropsy have been described at three differentiated levels: heart, pulmonary artery trunk and intrapulmonary arteries.

In the heart, the most significant finding is the presence of a chronic *cor pulmonale*, characterised morphologically by a remodelling of the RV, manifested by a dilated ventricular chamber, trabeculae hypertrophy, wall thickening and an increase in the total weight of the ventricular wall (Figs. 2.16, 2.17, 2.18, 2.25a and 2.26a). An estimation of hypertrophy can be performed by measuring the free right ventricular wall, which can reach a thickness >1 cm; although, several authors recommend a more sophisticated approach to assess hypertophy, based on dissection and weight of both ventricles independently. In the primary stages of hypertrophy it is possible to observe a thickening in the muscular fibres below the pulmonary valve, and in the moderator band. In more advanced stages, the increased size of the RV can compress and deform the left ventricular chamber and cause tricuspid insufficiency with fibrous thickening of the valve.

In the pulmonary artery trunk and the elastic great branches, the phenomenon of "aortization" can be observed in the walls, with a thickening in the muscular layer, a return to the pre-natal state, and arteriosclerotic events, although less severe than those normally observed in the aorta and systemic arteries (Fig. 2.19).

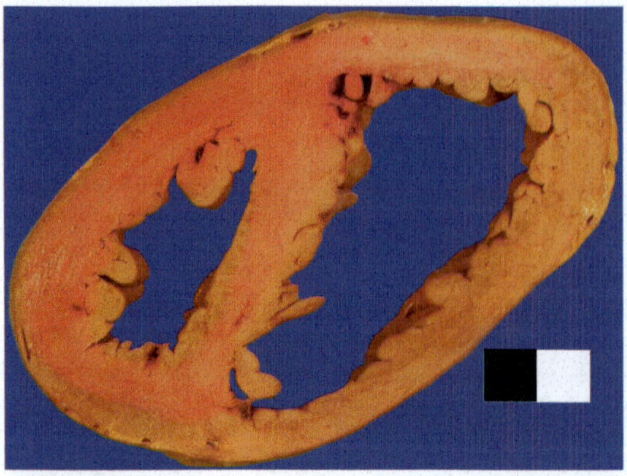

Fig. 2.18 Chronic *cor pulmonale* in a 36-year-old male drug addict with evolved pulmonary enphysema. Heart weight of 475 g with RV hypertrophy (9 mm) and dilation of the right ventricular chamber (4.5 cm)

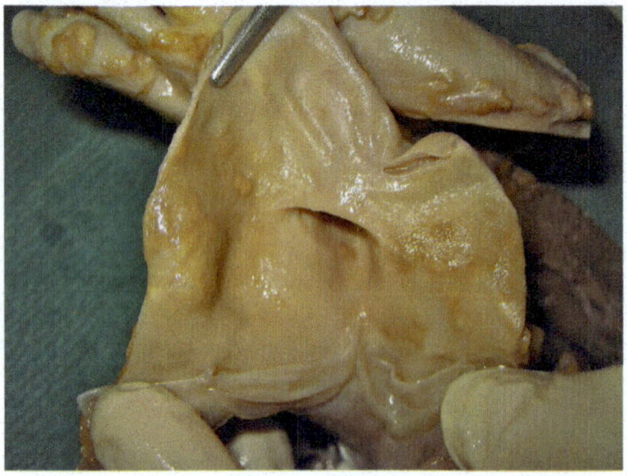

Fig. 2.19 Sudden death in a 43-year-old woman with grade III PAH. Atheroma plaque and fatty streaks in the pulmonary artery

Table 2.6 Normal pulmonary artery vasculature (Travis et al. 2002)

Elastic arteries
(Pulmonary trunk, main pulmonary arteries, intrapulmonary arteries)
Elastic laminae well developed and parallel
Diameter of 0.5–1 mm (depending of whether they are perfused or not)
Muscular arteries
Muscular medial layer lined by an outer and an inner elastic lamina.
Diameter <500 μm
Arterioles
Without muscular layer
Diameter <100 μm
Capillaries
(In the alveolar septum)
Single layer of endothelial cells resting on a basement membrane

Table 2.7 Heath and Edwards grading (1958) for hypertensive pulmonary vascular disease (Travis et al. 2002)

Grade I	Distal extension of muscle into distal arterioles and medial thickenning of muscular arteries (Figs. 2.20 and 2.21)
Grade II	Cellular intimal proliferation limited to small muscular arteries, usually mostly cellular endothelial reaction (Fig. 2.22)
Grade III	Medial hypertrophy and concentric laminar intimal fibrosis (Fig. 2.23)
Grade IV	Progressive generalized arterial dilatation with plexiform lesions
Grade V	Chronic dilatation with medial as well as intimal fibrosis; prominent plexiform lesions, vein-like branches and angiomatoid lesiones; pulmonary hemosiderosis (Figs. 2.24, 2.25 and 2.26)
Grade VI	Necrotizing arteritis

In the intrapulmonary arteries several changes fundamentally affecting the intimal and medial layers are observed, especially in the arteries ranging from 40 to 300 μm in diameter. For a better understanding of the lesions produced during arterial hypertension, Table 2.6 summarises the normal histology of the pulmonary arterial vessels.

Table 2.7 describes the classification of lesion grades found in the intrapulmonary arteries, as defined classically by Heath and Edwards (1958). This classification is based on the authors' initial observations in cases of idiopathic and familial PAH. More recently, some authors have suggested alternative classification systems with fewer grades which adjust better with the severity pattern and evolution in the non-primary forms; although they include similar findings and are not generally used.

At the histological level, hypertrophy is evident in the media of the intrapulmonary arterial vessels, which correlates with the clinical severity. The thickness of the media divided by the total diameter of the vessel, multiplied by 100, provides a percentual value of the thickness of the media with respect to the total diameter of the vessel. Thus, the normal thickness percentage of the media ranges between 3 and 7 %; values from 7 to 9 % indicate slight hypertrophy, 10–14 % moderate, and >15 % severe hypertrophy.

Changes in the intima do not directly correlate with clinical severity. Some intimal hyperplasia can be observed in the initial phases that is nonspecific and tends to be associated with scarce collagen deposits. In idiopathic and familial forms, however, hyperplasia tends to be laminar and concentric, with dense deposits of collagen and elastin. In more advanced presentations, plexiform lesions occur as a consequence of arterial dilation and luminal filling with bizarre phenomena of concentric hyperplasia. These complex phenomena are particular for the advanced idiopathic and heritable forms, but are not pathognomonic.

Thickening of the adventitia can be observed in forms associated with collagen disease, which can be linked with constrictive bronchiolitis and lymphoid aggregates with germinal centres. Together with the classical pathologic findings of PAH, other findings related to the PAH-determining pathologies can be found. For example, it is possible to discover organised or rechanneled thrombi in cases of chronic thromboembolic PAH, or some pulmonary fibrosis, emphysema or chronic bronchitis in PAH due to pulmonary and/or hypoxic diseases.

For an accurate evaluation of the pulmonary vascular findings it is recommended to insufflate the lungs with formaldehyde prior to serial macroscopic examination.

Medial hypertrophy in the muscular arteries in the case described in Fig. 2.19.

Intimal cellular proliferation in the muscular artery in the case described in Fig. 2.1 (extreme tetralogy of Fallot).

Medial thickening and laminar concentric intimal fibrosis (Orcein stain).

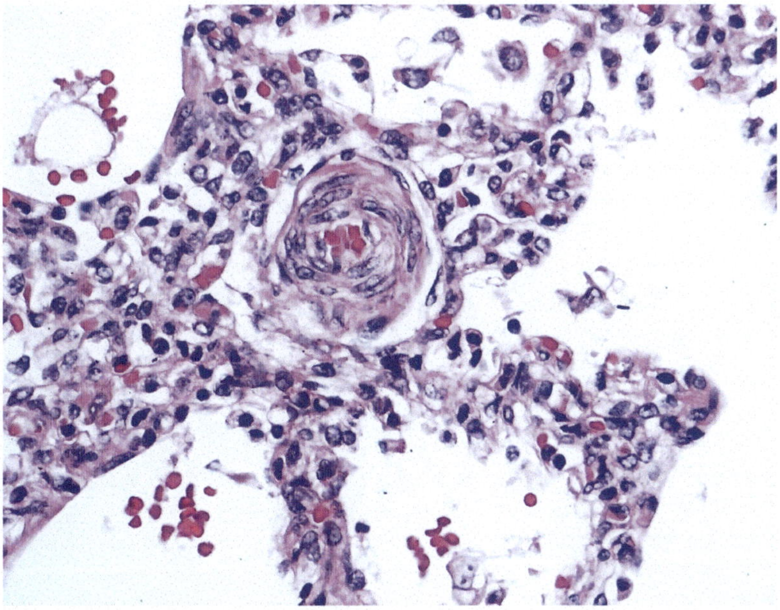

Fig. 2.20 Grade I pulmonary hypertension. Muscularization phenomenon in a pulmonary arteriole

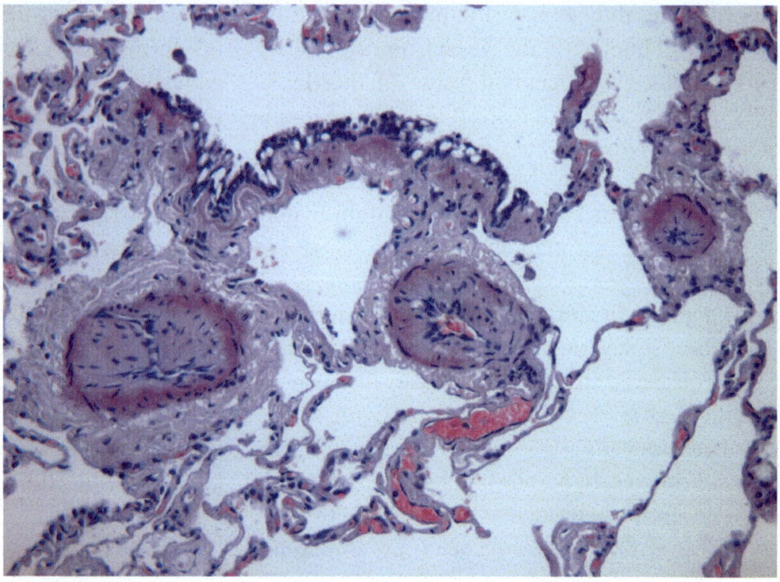

Fig. 2.21 Grade I pulmonary hypertension

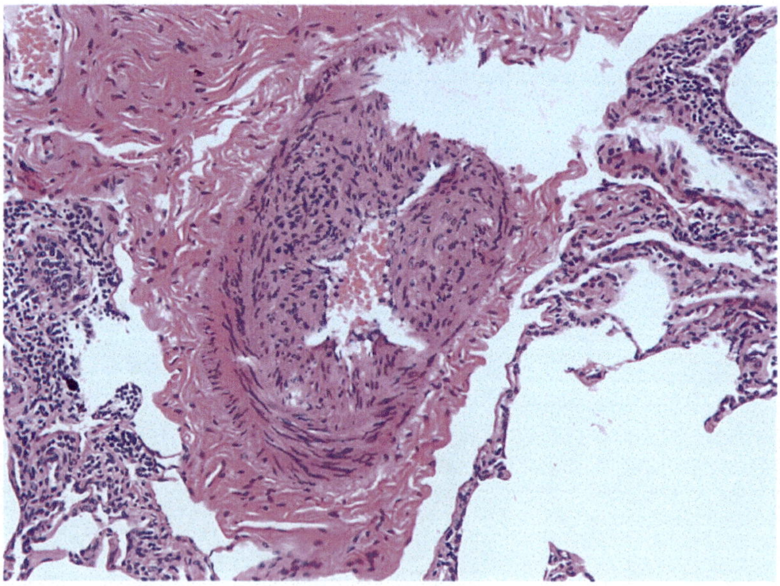

Fig. 2.22 Grade II pulmonary hypertension

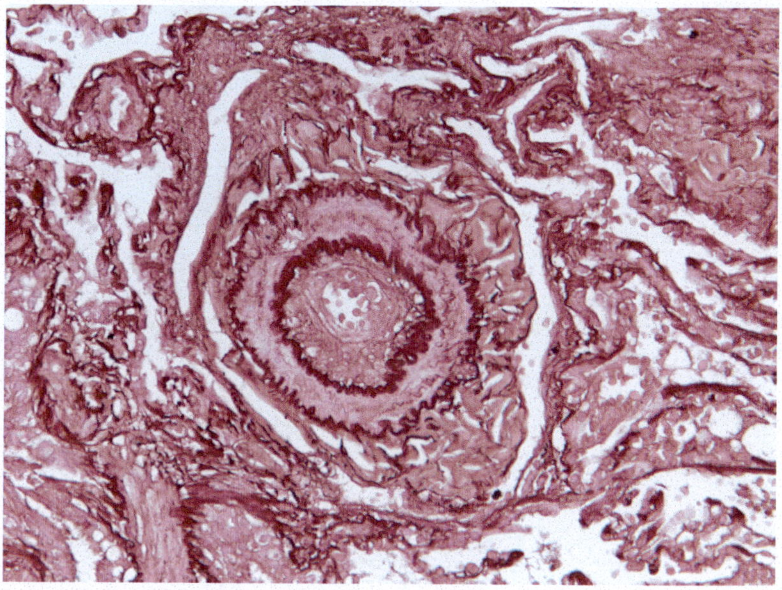

Fig. 2.23 Grade III pulmonary hypertension

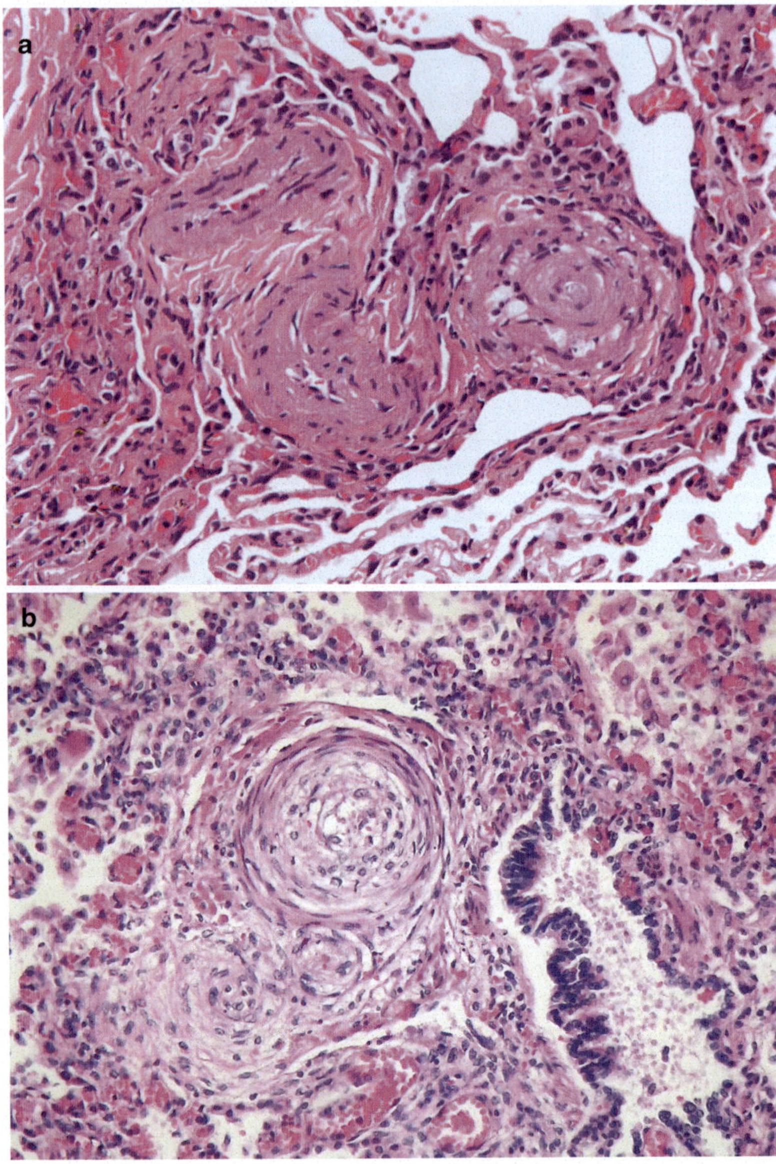

Fig. 2.24 Grade V pulmonary hypertension. (**a**, **b**) Plexiform and angiomatous lesions in the case described in Fig. 2.17

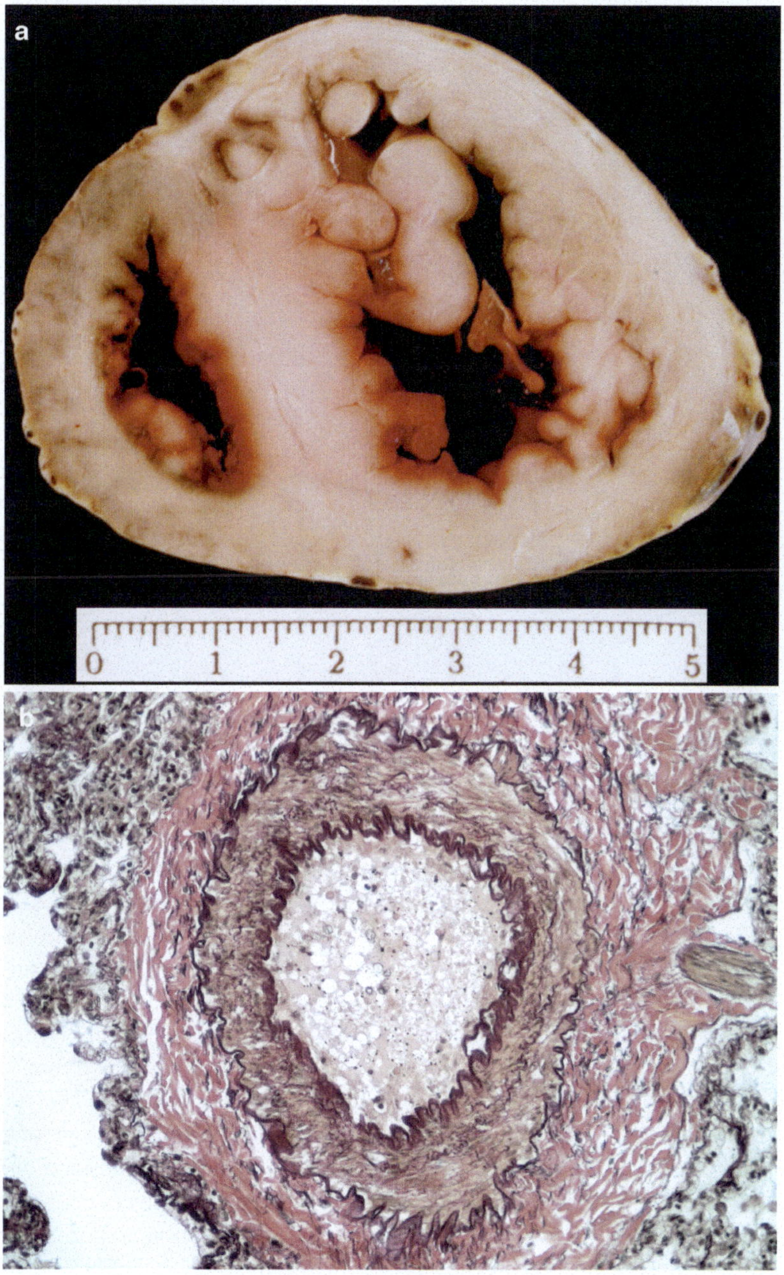

Fig. 2.25 Pulmonary hypertension. A 4-year-old girl without previous medical history was admitted to the casualty department due to general decay that had persisted during the previous 4 days. She presented with cyanosis, paleness, tachycardia and an O_2 saturation of 69 %. After 40 min she died. Cardiac weight was 161 g (normal <85 g) with marked RV hypertrophy. (**a**) In the biventricular transversal section dilation and RV, hypertrophy can be appreciated in both the free wall and in the trabeculae and papillary muscles. Thickness of the LV: 6 mm; thickness of the RV: 8 mm. (**b**) Muscular artery with medial thickening (Grade I) (Weigert stain). (**c**) Arteriole with intimal proliferation (Grade II) (Weigert stain). (**d**) Arteriole with occlusive intimal cell proliferation (Grade II)

Fig. 2.25 (continued)

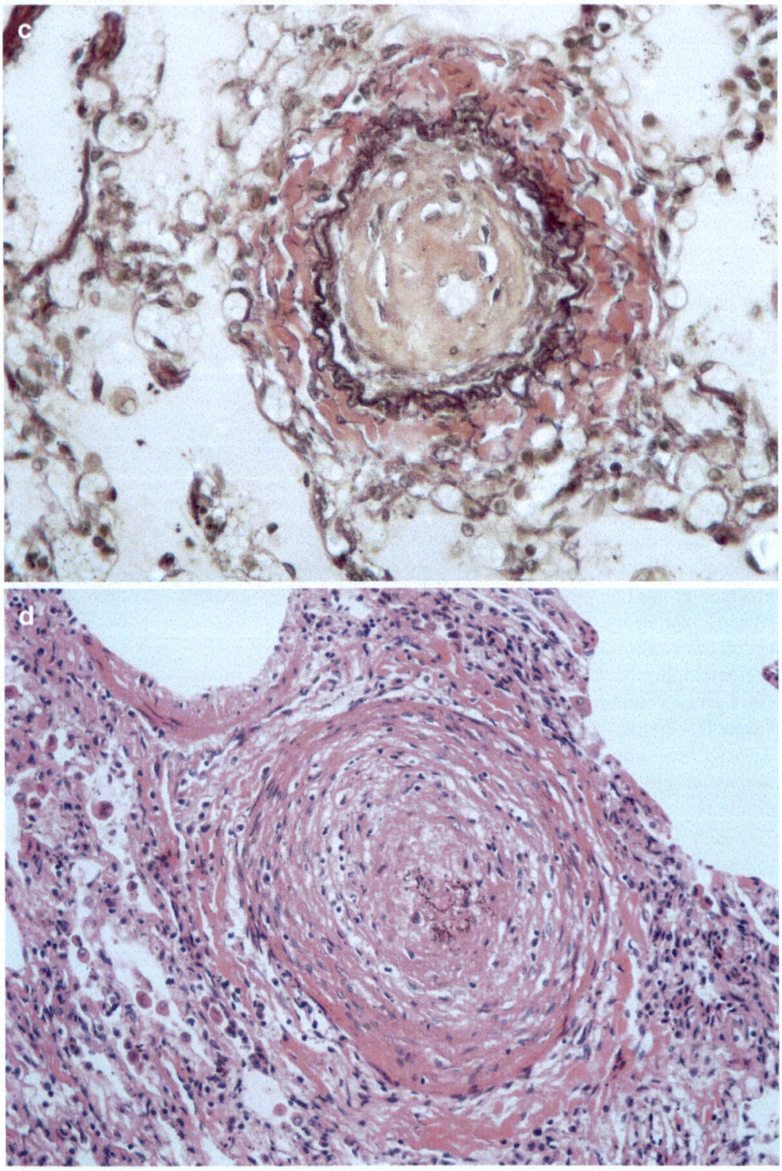

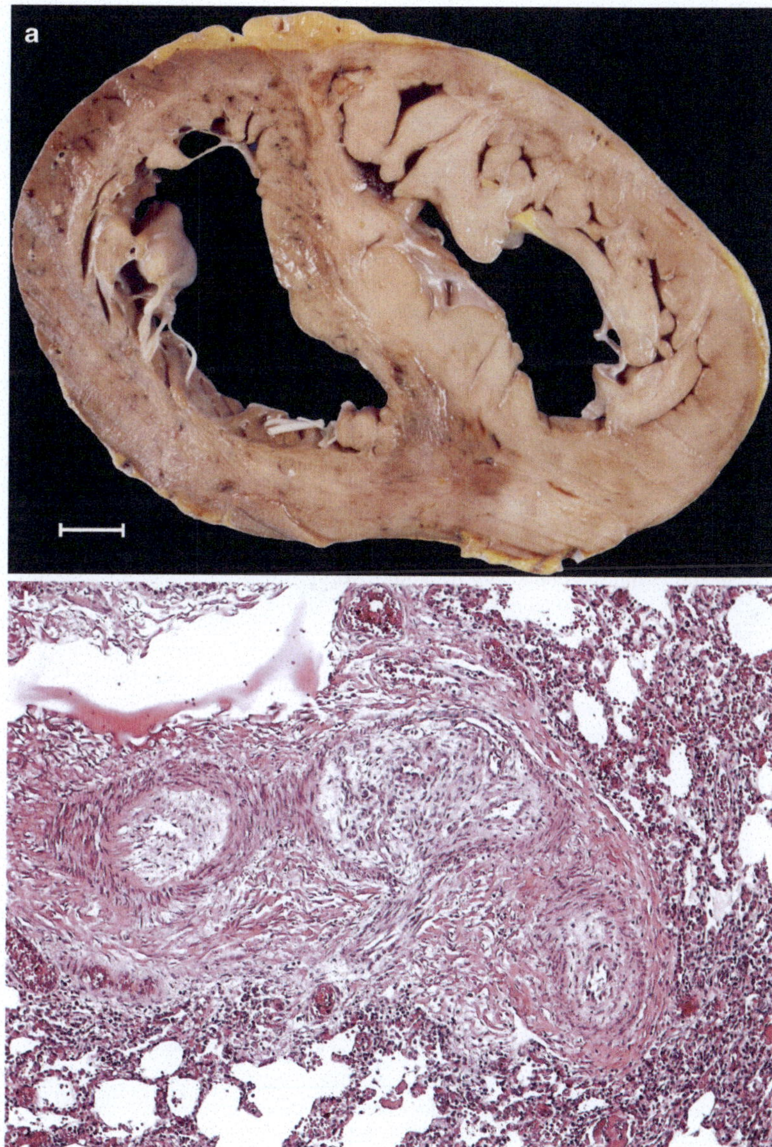

Fig. 2.26 Grade V pulmonary hypertension. A 28-year-old male. Chronic *Cor pulmonale*. (**a**) Heart of 602 g with RV hypertrophy (1 cm). (**b**) Histological image showing plexiform lesions

Bibliography

Dettmeyer RB. Thrombosis and embolism. In: Dettmeyer RB, editor. Forensic histopathology: fundamentals and perspectives. Berlin/Heilderberg: Springer; 2011. p. 173–90.

Fineschi V, Turilazzi E, Neri M, Pomara C, Riezzo I. Histological age determination of venous thrombosis: a neglected forensic task in fatal pulmonary thrombo-embolism. Forensic Sci Int. 2009;186:22–8.

Galié N, Hoeper MM, Humbert M, et al. Guidelines for the diagnosis and treatment of pulmonary hypertension: the Task Force for the Diagnosis and Treatment of Pulmonary Hypertension of the European Society of Cardiology (ESC) and the European Respiratory Society (ERS), endorsed by the International Society of Heart and Lung Transplantation (ISHLT). Eur Heart J. 2009;30:2493–537.

Groshong SD, Tomashefski Jr SD, Cool CD, et al. Pulmonary vascular diseases. In: Tomashefski Jr SD, Cagle PH, Farver CF, editors. Non neoplastic diseases. Dail and Hammar's pulmonary pathology, vol. 1. 3rd ed. New York: Springer-Science+Business Media, LLC; 2008. p. 1032–87.

Hamanaka S, Kamijo Y, Nagai T, Kurihara K, Tanaka K, Soma K, Miyaoka H. Massive pulmonary thromboembolism demonstrated at necropsy in Japanese psychiatric patients treated with neuroleptics including atypical antipsychotics. Circ J. 2004;68:850–2.

Heath D, Edwards JE. The pathology of hypertensive pulmonary vascular disease; a description of six grades of structural changes in the pulmonary arteries with special reference to congenital cardiac septal defects. Circulation. 1958;18:533–47.

Hoendermis ES. Pulmonary artery hypertension: an update. Neth Heart J. 2011;19:514–22.

Houser S, Balis UJ, Mark EJ. Pulmonary vascular disease. In: Houser S, Balis UJ, Mark EJ, editors. Lung pathology: a consultative atlas. Totowa: Humana Press; 2005. p. 299–394.

Husain AN, Kumar V. El Pulmón. In: Kumar V, Abbas AK, Fausto N, editors. Robbins y Cotran Patología Funcional y Estructural. 7th ed. Madrid: MMV Elsevier-España S.A; 2007. p. 715–76.

Kannan R, Molina DK. Olanzapine: a new risk factor for pulmonary embolus? Am J Forensic Med Pathol. 2008;29:368–70.

Kokturk N, Oguzulgen K, Demir N, Demirel K, Ekim N. Differences in clinical presentation of pulmonary embolism in older vs younger patients. Circ J. 2005;69: 981–6.

Lucena J, Rico A, Vázquez R, Marín R, Martínez C, Salguero M, Miguel L. Pulmonary embolism and sudden-unexpected death: prospective study on 2477 forensic autopsies performed at the Institute of Legal Medicine in Seville. J Forensic Leg Med. 2009;16: 196–210.

Nakanishi N. 2009 ESC/ERS pulmonary hypertension guidelines and connective tissue disease. Allergol Int. 2011;60:419–24.

Nieto JA, Ruiz-Ribó MD. Pulmonary thromboembolism. Light and shadows. Rev Esp Cardiol. 2008;61(3): 229–32.

Parasuraman S, Goldhaber SZ. Venous thromboembolism in children. Circulation. 2006;113:e12–6.

Ro A, Kageyama N, Tanifuji T, Fukunaga T. Pulmonary thromboembolism: overview and update from medico legal aspects. Leg Med. 2008;10:57–71.

Rulon JJ, Cho CG, Guerra LL, Bux RC, Gulley ML. Activated protein C resistance is uncommon in sudden death due to pulmonary embolism. J Forensic Sci. 1999;44:1111–3.

Ryan J, Thenappan T, Luo N, et al. The WHO classification of pulmonary hypertension: a case-based imaging compendium. Pulm Circ. 2012;2:107–21.

Saukko P, Knight B. Complications of injury. In: Saukko P, Knight B, editors. Forensic pathology. 3rd ed. London: Arnold Publisher; 2004. p. 339–51.

Simonneau G, Robbins IM, Beghetti M, et al. Updated clinical classification of pulmonary hypertension. J Am Coll Cardiol. 2009;54:S43–54.

Subirana MT, Juan-Babot JO, Puig T, Lucena J, Rico A, Salguero M, Borondo JC, Ordóñez J, Arimany J, Vázquez R, Badimon L, Thiene G, de Luna AB. Specific characteristics of sudden death in a Mediterranean Spanish population. Am J Cardiol. 2011;107:622–7.

Tapson VF. Acute pulmonary embolism. N Engl J Med. 2008;358:1037–52.

Travis WD, Colby TV, Koss MN, Rosado-de-Christenson ML, Müller NL, King TE. Pulmonary hypertension and other vascular disorders. In: Non-neoplastic disorders of the lower respiratory tract. Washington: American Registry of Pathology and the Armed Forces Institute of Pathology; 2002. p. 767–92.

Wagenvoort CA. Classifying pulmonary vascular disease. Chest. 1973;64:503–4.

Pathology of the Aorta

3

Pedro M. Garamendi, Joaquín S. Lucena, M. Paz Suárez-Mier, and Juan Manuel Escudier

Contents

P.M. Garamendi (✉)
Department of Forensic Pathology,
Institute of Legal Medicine, Huelva, Spain
e-mail: imanolgaramendi@gmail.com

J.S. Lucena
Department of Forensic Pathology,
Institute of Legal Medicine, Seville, Spain

Forensic Medicine, University of Cadiz, Cadiz, Spain
e-mail: joaquin.lucena@uca.es

M.P. Suárez-Mier
Histopathology Service, National Institute
of Toxicology and Forensic Sciences,
Las Rozas (Madrid), Spain
e-mail: mariapaz.suarez@justicia.es

J. Manuel Escudier
Coronary Care Unit, Cardiology Department,
Hospital Universitario Puerta de Hierro,
Majadahonda (Madrid), Spain

3.1 Introduction

During the course of a post-mortem examination, aortic pathology can constitute either a mere incidental finding or the main pathology related to the cause of death. In either of these occasions, a detailed post-mortem examination is essential to provide clarifying data on the cause of death and also about the nature of the disorder and its congenital or acquired origin.

The spectrum of aortic pathology is wide and includes different modalities (Table 3.1).

In this chapter we will focus on the five main aortic pathologies with greater frequency and incidence as a cause of death, many times with a sudden and unexpected onset: atherosclerosis, aneurysms, acute aortic syndrome, coartaction and aortitis.

3.2 Atherosclerosis

Atherosclerosis is the most prevalent aortic disease and is morphologically similar to coronary artery disease (see Chap. 5). The main characteristics of atherosclerosis are summarised in Table 3.2.

J.S. Lucena et al. (eds.), *Clinico-Pathological Atlas of Cardiovascular Diseases*,
DOI 10.1007/978-3-319-11146-9_3, © Springer International Publishing Switzerland 2015

Table 3.1 Pathology of the aorta

Congenital and developmental disorders
Transposition of great arteries, hypoplasia of the aorta, double aortic arch, aortic arch interruption (often associated with aortic coartaction) aortic valve malformations (supravalvular stenosis, bicuspid aortic valve and aortic coartaction)
Degenerative and acquired disorders
Atherosclerosis, aortic aneurysms, aortic dissection, intramural haematoma, penetrating atherosclerotic ulcer, aortitis (Takayasu, giant cells, syphilitic, tuberculous, collagenopathies)
Disorders associated with heredofamilial syndromes and connectivopathies
Marfan syndrome, Beals-Hecht syndrome, type IV Ehlers-Danlos syndrome, Loeys-Dietz syndrome
Traumatic pathologies
Caused by thoracic (closed or penetrating), abdominal and pelvic trauma. Iatrogenic lesions due to invasive (aortocoronary bypass) and non-invasive (catheterism) surgical treatments can also be included in this category

Table 3.2 Pathology of aortic atherosclerosis

Fatty streaks (intimal xanthomas): (Fig. 3.1)
Yellowish plaque on the intimal surface
Microscopically: lipid deposits in macrophages and smooth muscle cells (foam cells) and lipid accumulation in the extracellular matrix
Fibroatheromatous plaques: (Fig. 3.2)
Bulging lesions that may calcify and ulcerate
Fibrous cap formed by smooth muscle cells and dense extracellular matrix
Macrophages, smooth muscle cells and T lymphocytes present under the cap and towards the sides (in the plaque shoulders)
Internally, necrotic core containing disorganized lipid areas (cholesterol and cholesterol esters), cholesterol crystals, fragments of dead cells, foam cells, fibrin, thrombosis and other plasma proteins
Neovascularization in the periphery of the lesion
Fibrous plaques are formed by smooth muscle cells and fibrous tissue
Advanced (complicated) lesions: (Figs. 3.3, 3.4, 3.5 and 3.6)
Ulceration, thrombosis and calcification
Can occlude the origin of the intercostal and mesenteric arteries.

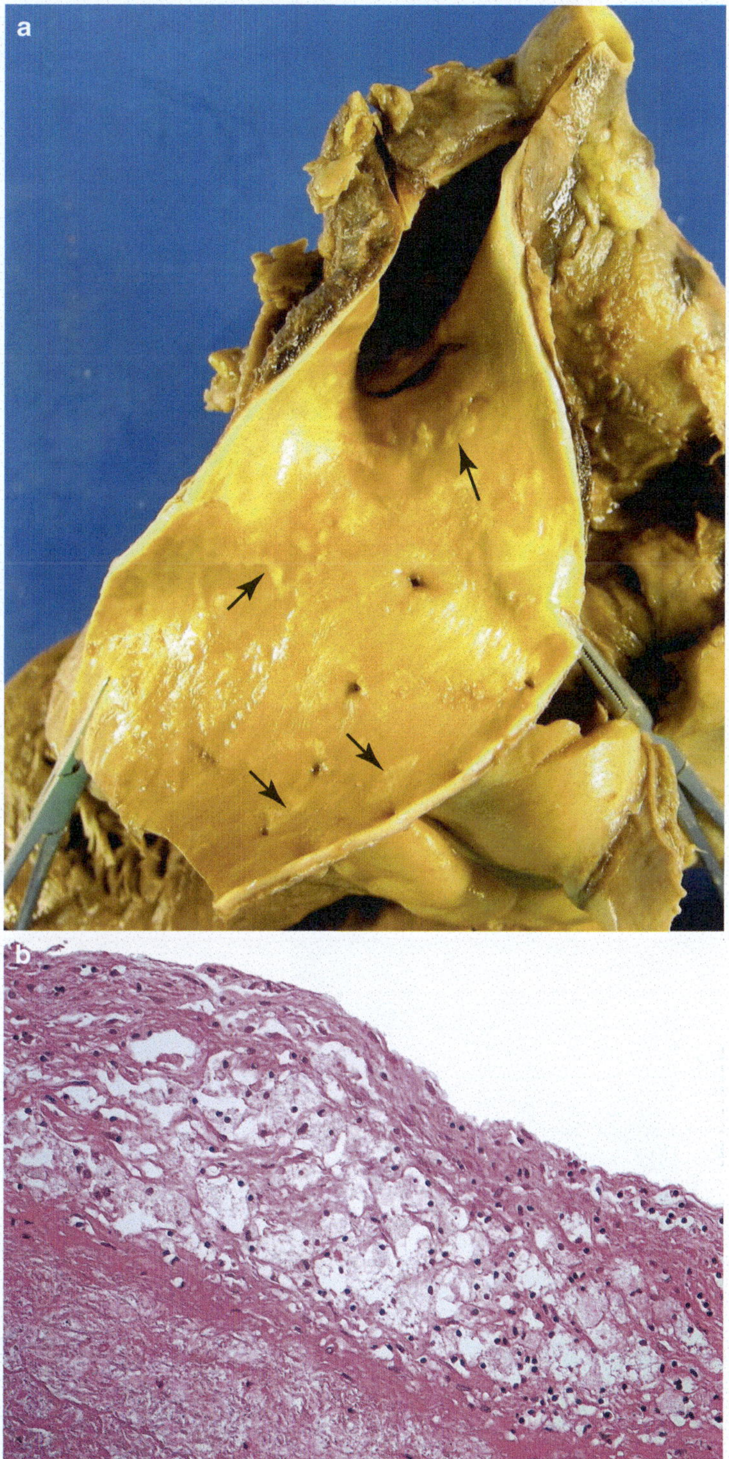

Fig. 3.1 *Fatty streaks.* (**a**) Fatty streaks in descending aorta (*arrows*). (**b**) Microscopic image where foam cells can be observed in the intima

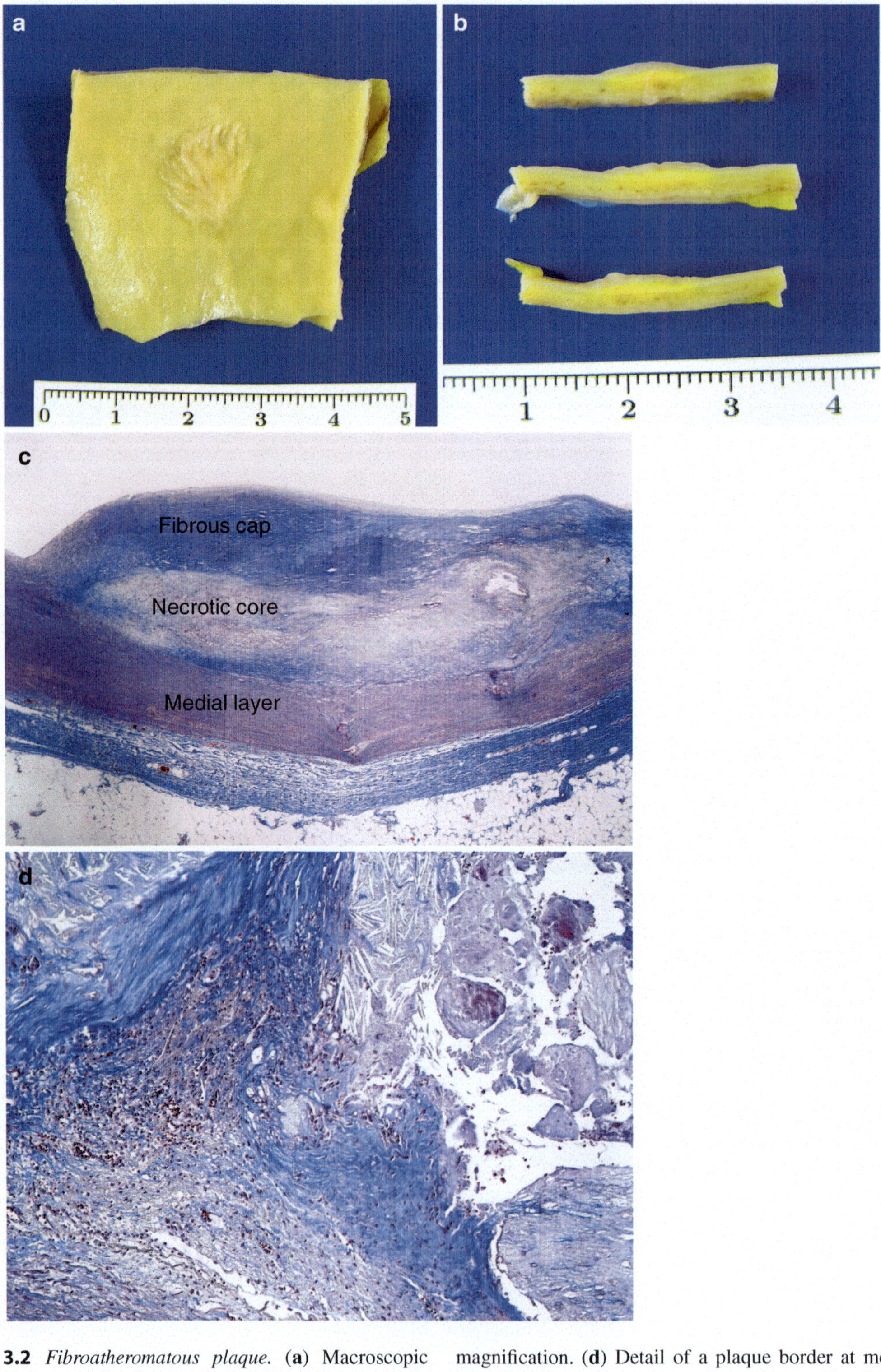

Fig. 3.2 *Fibroatheromatous plaque.* (**a**) Macroscopic image of an aortic plaque. (**b**) Serial sectioning of the lesion where the yellowish lipid content can be appreciated at depth. (**c**) Microphotography of a plaque at low magnification. (**d**) Detail of a plaque border at medium magnification where neovascularization can be observed (Masson's trichrome)

Fig. 3.3 Ulcerated atheromatosis in the thoracic aorta of a 75-year-old male who died suddenly from an acute coronary syndrome

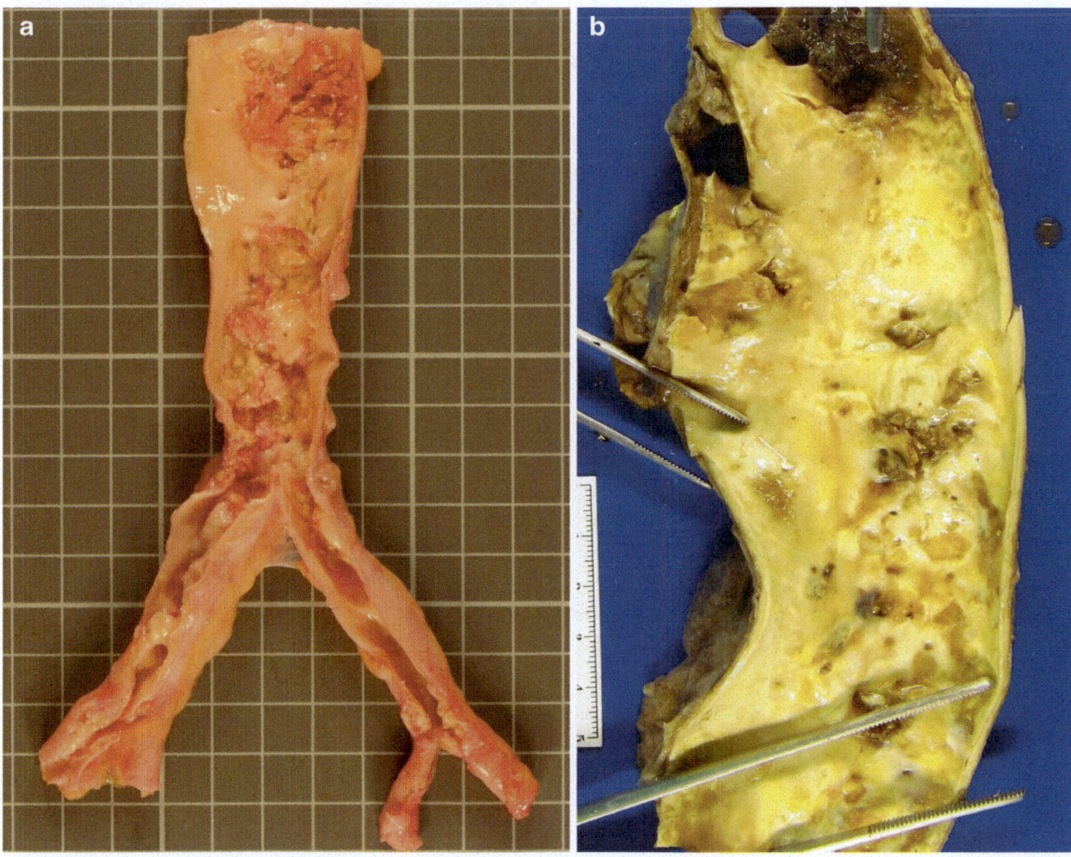

Fig. 3.4 A 61-year-old male with a previous history of hypercholesterolaemia and intermittent claudication in the lower extremities who died suddenly in the street from an acute coronary syndrome. (**a**) Complicated and calcified aortic atheromatosis involving the abdominal aorta and both iliac arteries. (**b**) Thoracic aorta with calcified and ulcerated plaques from a 94-year-old woman with generalised atherosclerosis

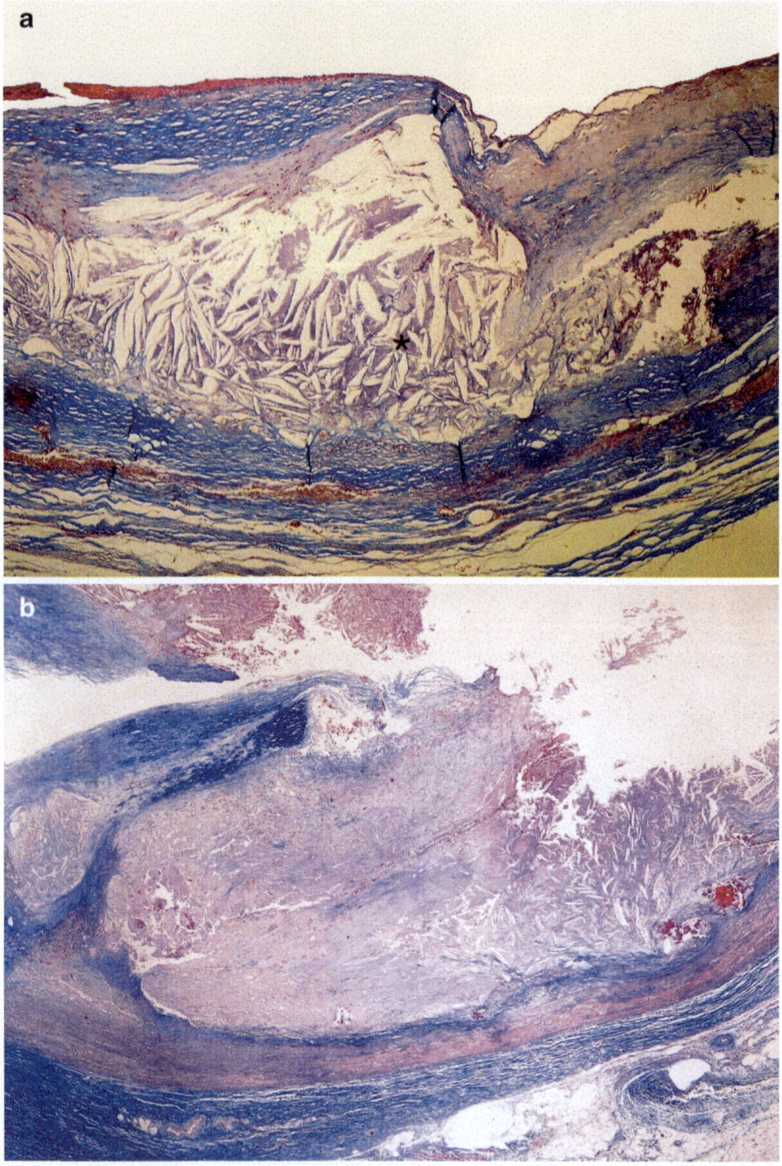

Fig. 3.5 *Complicated plaques.* (**a**) Histological image of an atheroma plaque with calcification and cholesterol crystals (*). (**b**) Microphotography of an ulcerated plaque (Masson's trichrome)

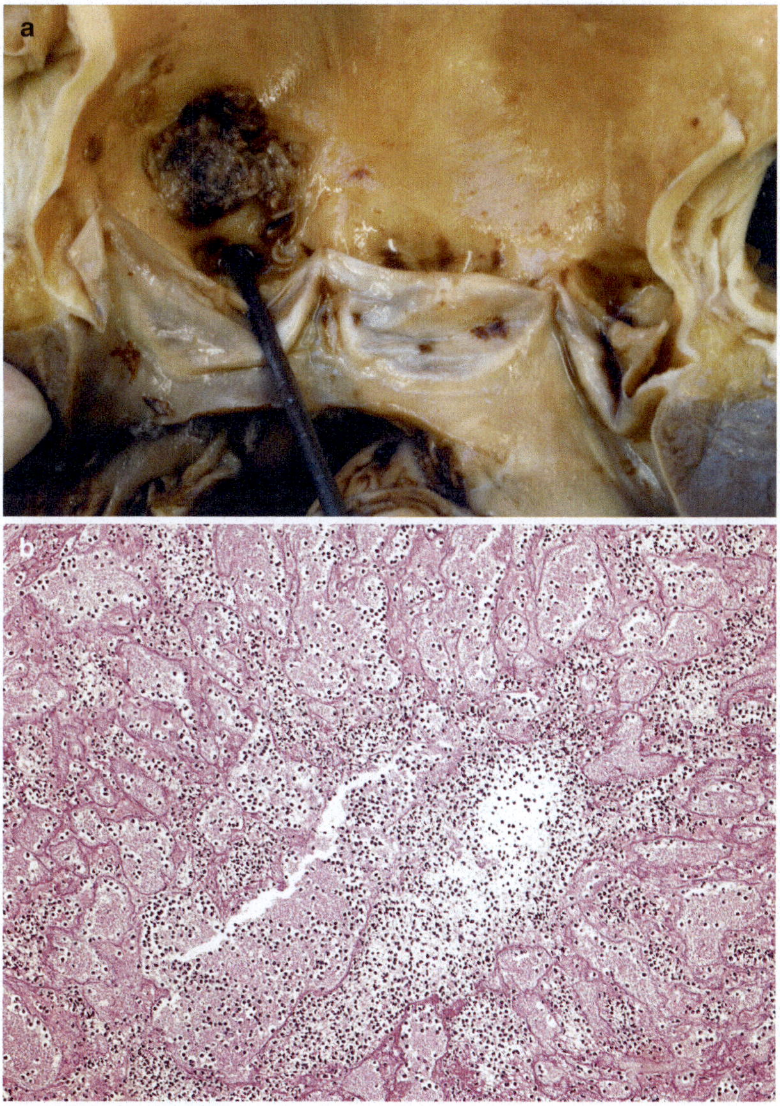

Fig. 3.6 A 56-year-old woman with pressure ulcers and gangrene and multiple ulcers caused by pressure and septic liver. (**a**) Septic mural thrombus over an atheroma plaque in the ascending aorta. (**b**) Microscopic image of the thrombus with numerous neutrophils

3.3 Aortic Aneurysms

An aneurysm is defined as an abnormal dilatation in a blood vessel or the cardiac wall. When an aneurysm is surrounded by arterial wall components or by an attenuated cardiac wall, the term true aneurysm is used. Conversely, those gaps in the vascular wall leading to an extravascular haematoma that freely communicates with the intravascular space ("pulsatile haematoma") are called false aneurysms or pseudoaneurysms.

Aortic aneurysms can be classified according to their aetiology, morphology and location (Table 3.3).

Infectious aneurysms can originate from supurative infections that invade, by proximity, the aortic wall; by septic thrombi embolization (mycosis and bacterial infections); or, by direct infection of the aortic wall caused by circulating microorganisms (e.g., syphilis). A non-infectious vasculitis, by bacterial or fungal infections or by formation of immunocomplexes or cross reactivity has also been described.

Traumatic aneurysms can be produced by direct action in cases of penetrating traumatisms affecting the arterial wall or by the action of bone fragments in some closed traumas. They can also be caused by indirect acceleration-deceleration mechanisms in cases of non-penetrating traumatisms. Aortic lesions caused by closed trauma have been observed in the thoracic aorta, mainly at the level of the aortic isthmus (between the origin of the left subclavian artery and the *ligamentum arteriosum*), since it is the anchor point between the fixed aortic arch and the relatively more mobile thoracic aorta, and therefore sensitive to traction forces exerted by its mobile parts during trauma. The aorta can also become damaged by direct compression against the spinal cord or even by an excessive increase in the intraluminal aortic pressure. These traumatic lesions tend to provoke partial or complete ruptures of the aortic wall (Fig. 3.7) and false aneurysms that can remain contained for several years, leading to ruptures or dissections.

Table 3.3 Classification of aortic aneurysms

By aetiology
Congenital
Inflammatory or infectious
Degenerative (the most frequent)
Associated with hereditary syndromes: Marfan, Beals-Hecht, Ehlers-Danlos or Loeys-Dietz syndromes
Traumatic
Iatrogenic
By morphology:
Saccular: only one part of the arterial wall is compromised
Fusiform: the entire circumference of the aortic wall is dilated
By location:
Abdominal
Thoracic
Annuloaortic ectasia

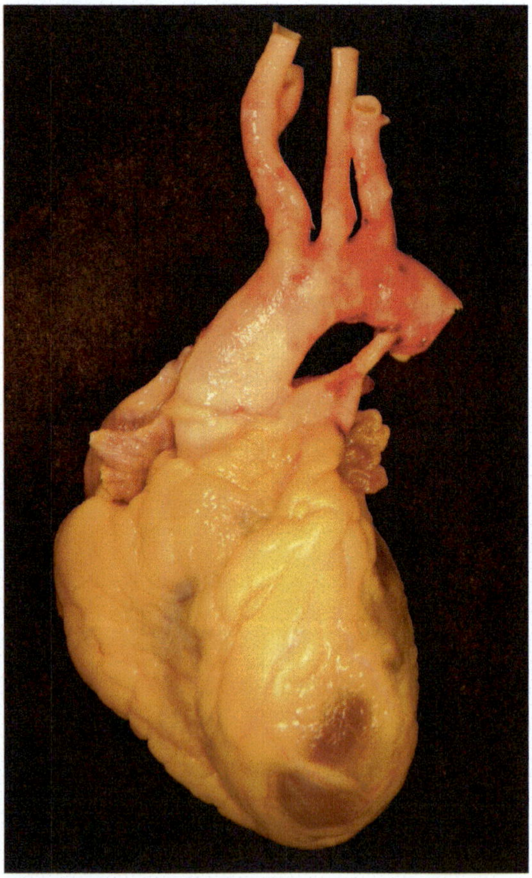

Fig. 3.7 Complete traumatic rupture of the thoracic aorta just below the origin of the subclavian artery, in a 39-year-old woman, caused by a motorbike traffic accident

3.3.1 Abdominal Aortic Aneurysms (AAA)

AAA constitutes the majority of aortic aneurysms and occurs in 2–5 % of the population over the age of 60, and in 10 % of the population over the age of 80. The majority of patients (75 %) are totally asymptomatic or present nonspecific symptoms such as epigastric or lumbar pain, gastrointestinal symptoms or a pulsatile abdominal mass, although the first clinical manifestation can be haemorrhagic shock and sudden death.

The majority of these aneurysms have an atherosclerotic origin and the main risk factors are: smoking, old age, hypertension, and male gender (4–8:1). There are also familial cases; however, AAA due to hereditary diseases such as Marfan's syndrome are very infrequent.

The major complication of AAA is rupture and free bleeding into the peritoneal cavity (aneurysms with anterior rupture) or into the retroperitoneal circundant tissues (Fig. 3.8). The clinical presentation with hypovolemic shock is usually faster in aneurysms with anterior intraperitoneal rupture. The risk of rupture is proportional to the size of the AAA. Thus, in AAA with a diameter of less than 4 cm, the risk of rupture is very low; between 4 and 4.9 cm, the risk is around 1 % per year; between 5 and 5.9 cm around 11 % annual; and with AAA larger than 6 cm, the risk is about 25 % per year. The progression of the diameter of the aneurysm depends principally on the blood pressure; therefore uncontrolled hypertension is associated with a faster progression in the size of the aneurysm.

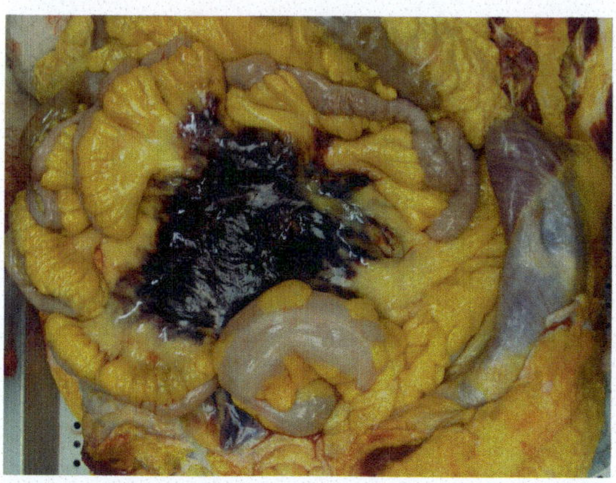

Fig. 3.8 Retroperitoneal haematoma secondary to the rupture of an abdominal aortic aneurysm. An 80-year-old woman who was dead on arrival to the hospital

Other less frequent complications include thromboembolic events due to embolia in ulcerated atheromas or mural thrombi, compression of adjacent structures (ureters, vertebrae), rupture with fistulization into the neighbouring organs, such as the inferior vena cava or the digestive tract (mainly the duodenum), or compression in arterial branches with ischaemic tissular lesion, particularly in the iliac (claudication), renal, mesenteric or vertebral (bone marrow ischaemia) arteries.

The atherosclerotic form gives rise to saccular or fusiform aneurysms, occasionally reaching up to 15 cm of diameter and with variable length. The arterial wall is thinned, with a destructured medial layer, making it more fragile and susceptible to spontaneous ruptures (Figs. 3.9 and 3.10). The aneurysm and adjacent aorta tend to have atheromatous ulcers covered by granular mural thrombi. The lumen in an aortic aneurysm is frequently occluded by a white-yellowish fibrin laminar thrombus. The destructured medial layer, together with the mural thrombi, can obstruct the lumen of the arterial branches, which might explain the acute and chronic distal ischaemic phenomena. In practice, the majority of ruptured atherosclerotic AAA are larger than 8 cm in diameter, are located at an infrarenal level, and extend toward the iliac bifurcation (Fig. 3.11). Occasionally, they can be associated with aneurysms in the primitive iliac arteries (Fig. 3.12).

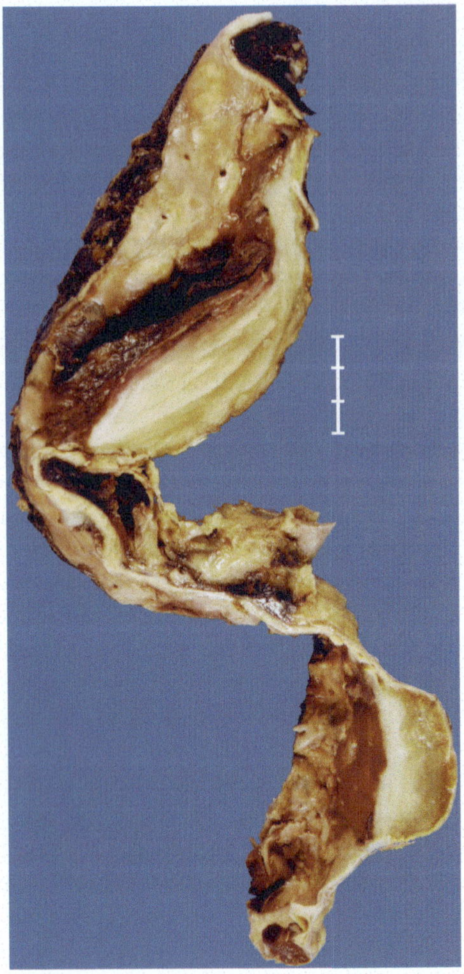

Fig. 3.9 An 81-year-old male who presented with two thrombosed atherosclerotic aneurysms, in thoracic and abdominal aorta respectively

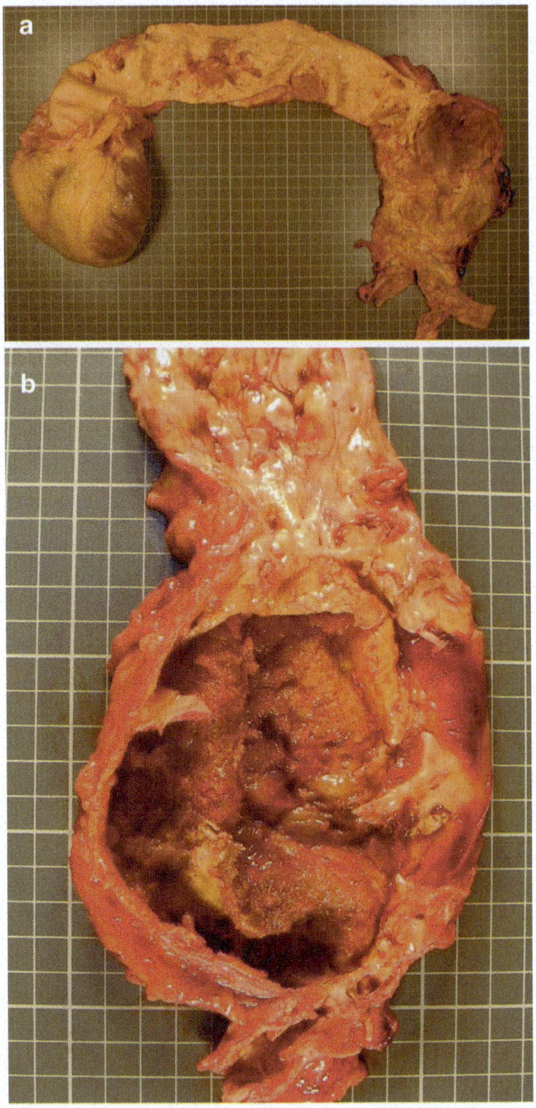

Fig. 3.10 (**a**) A tear in an atherosclerotic abdominal aortic aneurysm (AAA) in a 78-year-old male with a previous history of hypertension who died suddenly. (**b**) Detail of the atherosclerotic and fusiform abdominal aneurysm with an infrarenal location

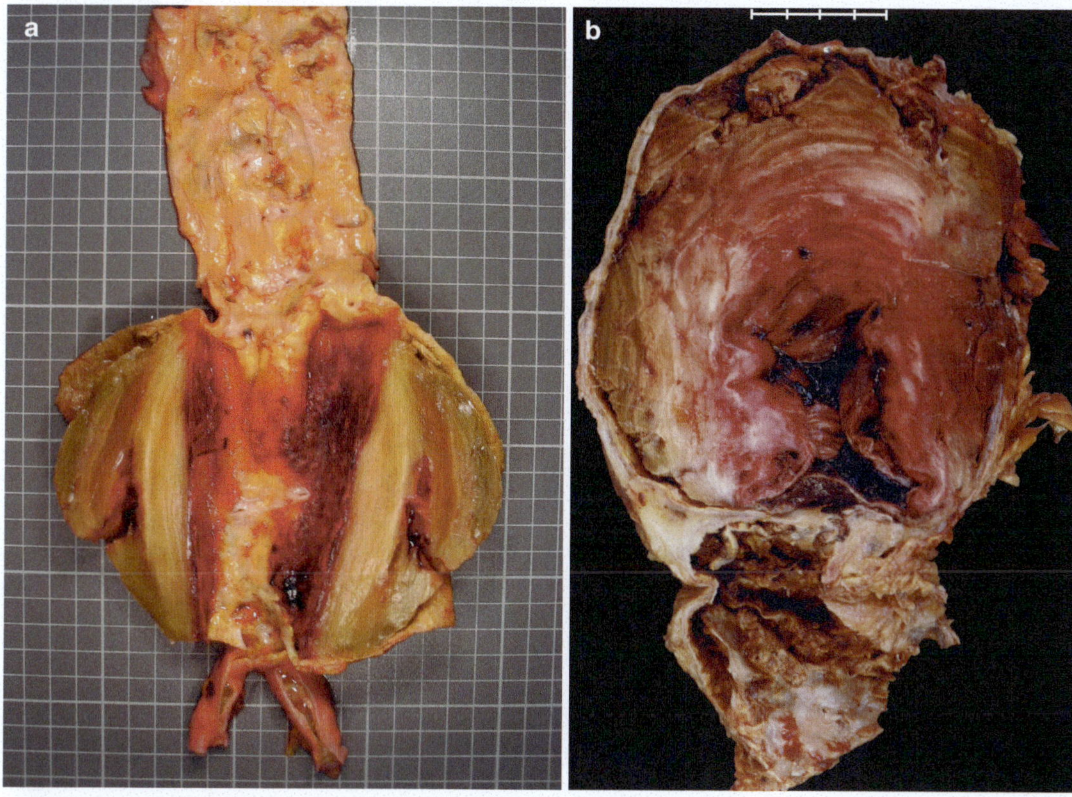

Fig. 3.11 (**a**) Longitudinal section of an abdominal aortic aneurysm showing laminar thrombosis. (**b**) Large thrombosed aneurysm in the abdominal aorta located below the origin of the renal arteries. Woman of 84 years with ischaemic cardiomyopathy

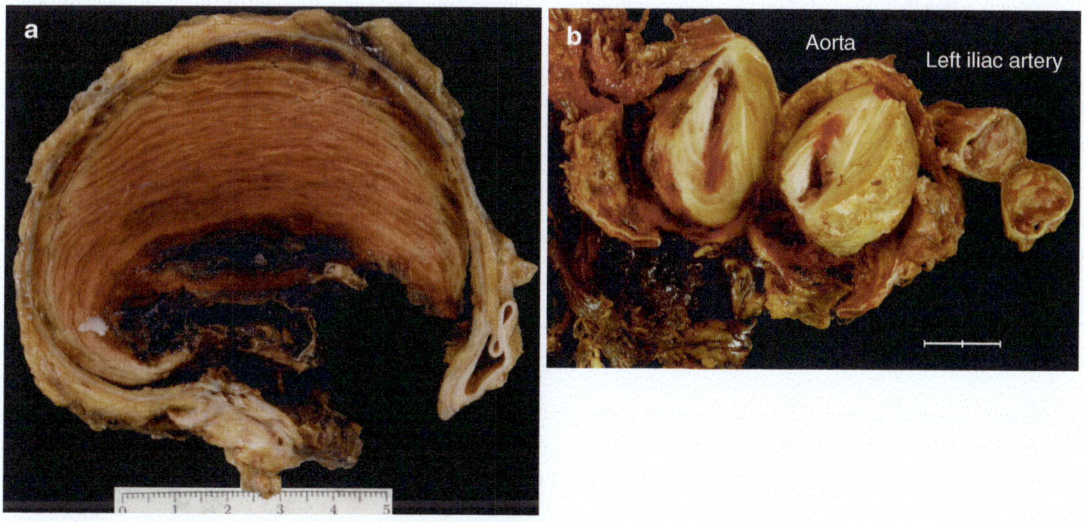

Fig. 3.12 (**a**) Subocclusive laminar thrombosis in a transversal section of a distal abdominal aortic aneurysm in a 76-year-old male. (**b**) Thrombosis in a ruptured abdominal aneurysm with circundant haemorrhage. The left iliac artery also presents a thrombosed aneurysm

3.3.2 Thoracic Aortic Aneurysms (TAA)

Thoracic aortic aneurysms comprise seven main categories (Table 3.4):

Aneurysms which are located in the proximal portion of the aorta can affect the aortic root (formed by the fibrous annulus, the Valsalva sinuses, the leaflets and the sinotubular junction) and/or the ascending aorta (from the sinotubular junction to the right brachiocephalic trunk).

3.3.2.1 Aneurysm of Sinus of Valsalva

Aneurysms in the sinuses of Valsalva are the least frequent and occur due to a weakness in the aortic wall that forms part of the sinus. This causes dilatation and the formation of a blind pocket in one of the aortic sinuses (usually the right sinus and less frequently the posterior one). The majority of sinus of Valsalva aneurysms are congenital lesions tightly associated with ventricular septal defects, although they can also be associated with aortic coarctation and with the presence of a bicuspid aortic valve. The age of onset ranges from 10 to 70 years with an average of 35 years.

Its clinical presentation is related to rupture that occurs between the ages of 20–40 years, although some cases of rupture have been described in childhood and also in the elderly. The physiological consequences of this rupture depend on the speed at which it is produced, the size of the ruptured orifice and the cavity it ruptures into. Aneurysms occurring in the right sinus generally rupture into the RV, those of the posterior sinus rupture into the right atrium (Fig. 3.13), and left sinus aneurysms rupture into the pulmonary artery, left ventricle, myocardium or epicardium. Some aneurysms of the posterior Valsalva sinus can rupture into the pericardial cavity causing sudden death by cardiac tamponade.

3.3.2.2 Annuloaortic Ectasia and Ascending Aorta Aneurysms

Annuloaortic ectasia is a symmetrical dilatation of the ascending aorta and the aortic root that can lead to a traction and consequent lack of contact of the valve leaflets, giving rise to aortic valve insufficiency (Fig. 3.14).

Aneurysms of the ascending aorta can present as an isolated pathology, be associated with a bicuspid aortic valve or be part of several syndromes such as Marfan syndrome, type IV Ehlers-Danlos or Loeys-Dietz syndromes. Histologically, a cystic medial degeneration (CMD) can be observed, which leads to weakening of the aortic wall and confers high risk of aortic dissection (Fig. 3.15). Exceptionally, aneurysms secondary to this type of wall lesion can affect the descending aorta (Fig. 3.16).

In the above-mentioned syndromes, CMD is caused by mutations in different genes; hence, a histopathological study is not sufficient for a definitive diagnosis. Thus, during the autopsy it is essential to perform a meticulous external and internal examination of the skin, skeleton, skull,

Table 3.4 Thoracic aortic aneurysms

Type of aneurysm	Location
Atherosclerotic/degenerative aneurysms	Descending aorta and aortic arch
Dissecting aneurysms	Ascending or descending aorta (*described later*)
Aneurysms associated with cystic medial degeneration	In aortic root (annuloaortic ectasia) and ascending aorta
Aneurysms caused by congenital weakness of an aortic sinus	Sinus of Valsalva
Pseudoaneurysms (generally secondary to trauma)	Close to the *ligamentum arteriosus*
Inflammatory aneurysms (secondary to Takayasu aortitis in the young and to giant cell aortitis in the elderly)	Ascending aorta (*described later*)
Infectious aneurysms (mycotic)	Descending aorta

face and all visceral organs in order to exclude a hereditary form. Table 3.5 describes the most frequent phenotypic features and the genes affected in these types of connectivopathies. Given the importance of the genetic study, it is necessary to collect blood and fresh tissue samples during the autopsy, and store them frozen for future analysis.

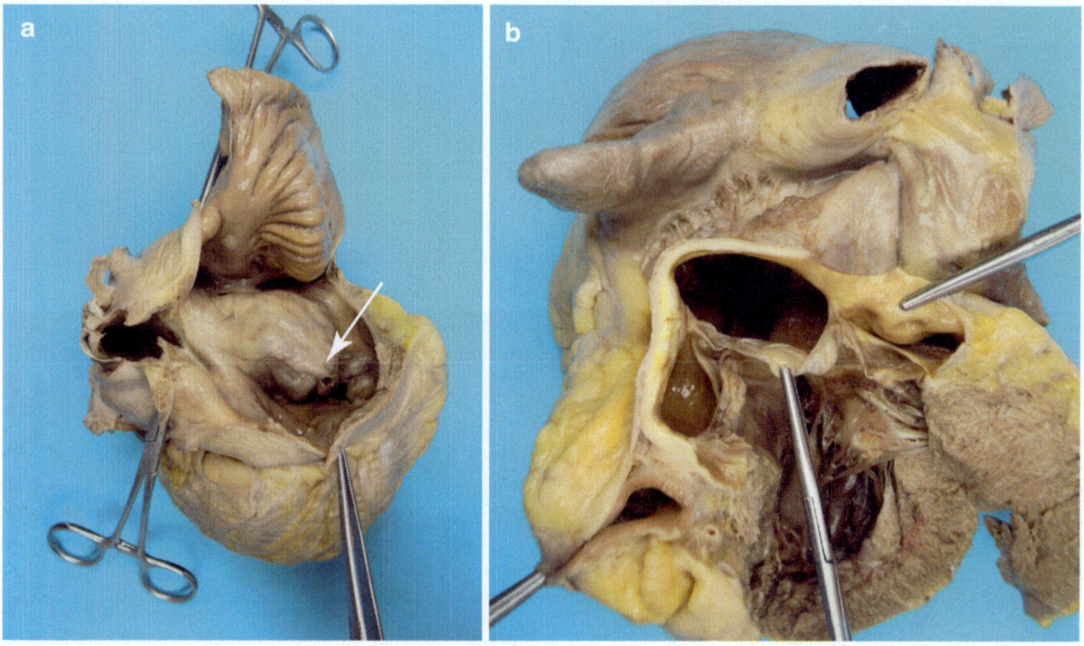

Fig. 3.13 (**a**) Giant ruptured aneurysm in the posterior Valsalva sinus bulging into the right atrium (the *arrow* points to the rupture). (**b**) Aneurysm viewed from the aorta. A 47-year-old woman who died suddenly during sexual intercourse

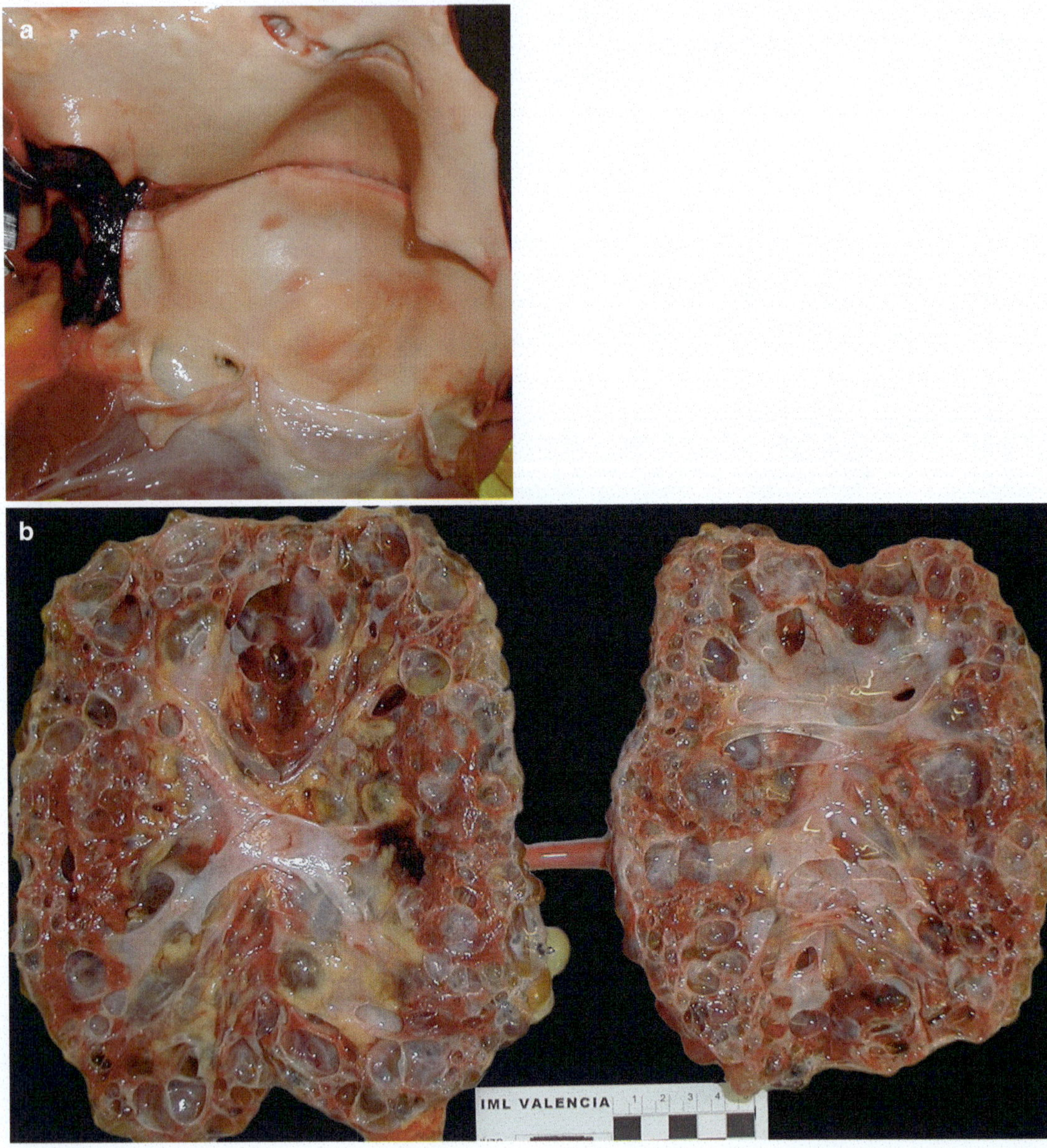

Fig. 3.14 A 39-year-old woman with a personal and family history of renal polycystosis without hypertension who died suddenly. (**a**) A large dilatation of the aortic root is observed (annuloaortic ectasia) associated with bicuspid aortic valve and transversal intimal dissection, origin of a parietal dissection (*arrows*). (**b**) Kidneys with a notable increase in size and weight (R: 1,140 g; L: 865 g) in which a replacement of the renal parenchyma with multiple cysts of various sizes can be observed (Molina et al. (2007). Reproduced with permission from the Editor of *EJ Autopsy*)

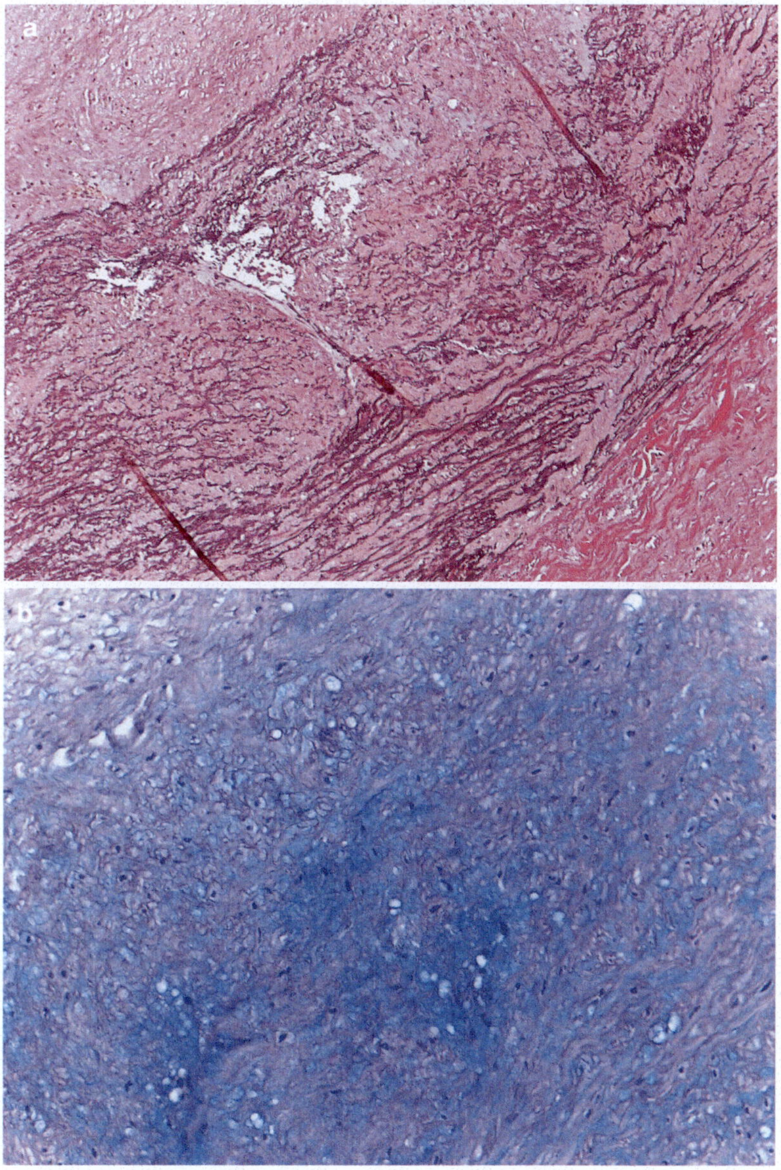

Fig. 3.15 *Cystic medial degeneration.* (**a**) Interruption of elastic fibres with formation of empty spaces (Weigert's stain). (**b**) Muccopolysaccharide deposit in the areas of elastic fibre loss (Alcian blue stain) (Courtesy of Maria Soledad Sánchez de León. INTCF. Madrid)

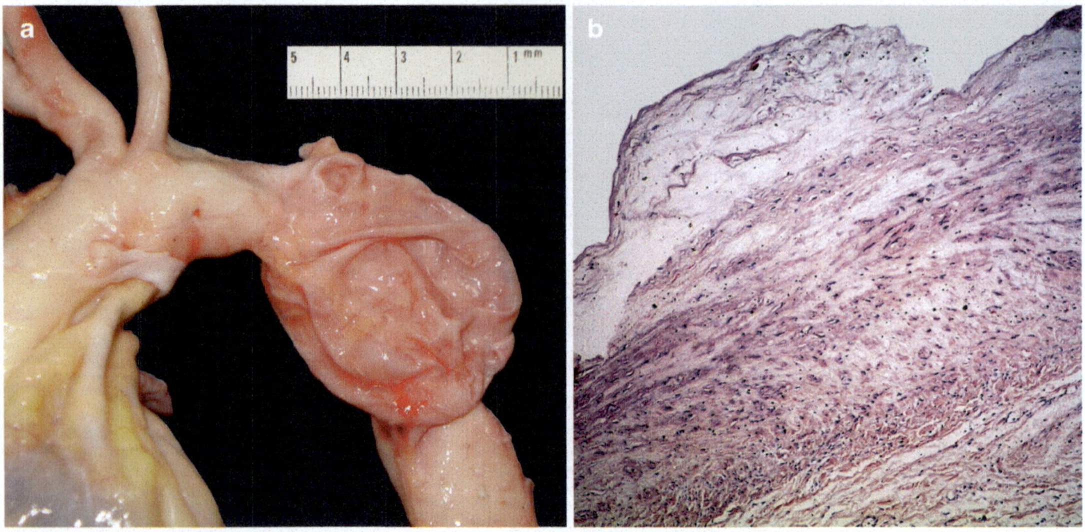

Fig. 3.16 (**a**) Saccular aneurysm in the descending thoracic aorta in a 15-year-old girl who died suddenly. The genetic study excluded the presence of genetic anomalies associated with Marfan syndrome. (**b**) Microscopically, the wall of the aneurysm showed loss of elastic and muscular fibres and replacement with a non-fibrilar basophilic matrix (CMD) (Masson's trichrome) (Pachar et al. (2010). Published with permission of the Chief Editor of Cuad Med Forense)

Table 3.5 Syndromes associated with aneurysm and aortic dissection

	Marfan syndrome	Loeys-Dietz syndrome	Type IV Ehlers-Danlos syndrome
Affected genes/heritage	FBN1/Aut. dominant	TGFBR1 (type 1) TGFBR2 (type 2) Aut. dominant	COL3A1
Cardiovascular signs	Dilatation of the aortic root Dilatation of the pulmonary artery	Aortic dissection (70 % of the patients) Dilatation of the pulmonary artery	Affects aorta and distal branches: renal, mesenteric, iliac and femoral arteries
Histopathological findings	Cystic medial degeneration with glycosaminoglycan deposit	Diffuse medial degeneration (separation of elastic fibres)	Minimal medial degeneration with interruption of elastic fibres and fibrous tissue organised between them
Skin signs	Hernias	Hernias	Velvety skin Translucid skin Ecchymosis Abnormal scarring Acrogeria
Ocular signs	Ectopia of the lens	Blue sclera	Blue sclera
Skeletal signs	Arachnodactyly *Pectus excavatum* Scoliosis Flat feet Dolichostenomelia	*Pectus excavatum* Joint hyperlaxity with dislocations	Ligament hyperlaxity Joint hypermobility
Craniofacial signs	Malar hypoplasia	Type 1: Craniosynostosis Cleft palate Bifid uvula Malar hypoplasia	
Vascular lesions excluding aorta		Tortuous vessels	Ruptures and dissections Postsurgical haemorrhages
Visceral signs	Emphysema Pneumothorax		Rupture and perforation of hollow viscerae

From Suárez-Mier and Morandeira (2012)

FBN1 fibrillin 1, *TGFBR* transforming growth factor β receptor, *COL3A1* collagen type III, α-1

3.4 Acute Aortic Syndrome

Acute aortic syndrome (AAS) encompasses three clinical entities that involve the aorta and have an acute clinical manifestation: aortic dissection (80 % of the cases), intramural haematoma (15 %) and penetrating atherosclerotic ulcer (5 %). Although these three clinical entities have distinct physiopathological mechanisms, their coexistence in the same patient is not unusual. Clinically, these conditions share common presenting features, marked by "aortic pain" that has been described as a very intense, acute, tearing, pulsatile and migratory chest pain. A retrosternal pain radiating to the neck or jaw suggests an involvement of the ascending thoracic aorta, whereas an abdominal or back pain suggests the involvement of the descending thoracic aorta or abdominal aorta, respectively.

Topographically, there are two different types, with defined clinical and surgical management:

- Proximal AAS, when the affected parts are the ascending aorta and aortic arch.
- Distal AAS, when the descending aorta is affected.

Mortality from acute ascending aorta dissection is extremely high and can range from 1 to 2 % per hour from the onset of symptoms. Treated medically, the mortality rate is 20 % in the first 24 h, 30 % at 48 h, 40 % at 7 days and 50 % over 1 month. Even with surgical treatment, mortality rates are very high (around 10 % at 24 h, 13 % at 7 days and almost 20 % at 30 days). In non-complicated descending aorta dissection, the mortality rate during the first 30 days is around 10 %, but if complications exist, such as limb or visceral ischaemia, renal insufficiency or contained rupture, mortality reaches 25 % in a month.

3.4.1 Aortic Dissection

Aortic dissection is a catastrophic disorder characterized by the entrance of blood in the media layer and the formation of a false channel full of blood within the aortic wall. If this channel ruptures through the outer aortic wall massive haemorrhage occurs, causing the death of the patient. Aortic dissection is not always associated with previous arterial dilatation (aneurysm); hence, many authors do not recommend using the classical term, dissecting aneurysm.

Morphologically, an aortic dissection involves the formation of a double aortic lumen (double-barrelled aorta) (Figs. 3.17 and 3.18), established by the original arterial lumen and the newly formed channel within the medial layer. This false channel communicates with the arterial lumen through an entry tear through which blood enters the channel, and a distal exit tear that communicates again with the aortic lumen.

The origin of aortic dissection is an intimomedial tear through which the blood enters the medial layer. Blood at high pressure dissects or splits the media causing a distal extension of the dissection (Fig. 3.19). The entry tear can occur in:

- Ascending aorta, around 1–5 cm above the Valsalva sinus (65 % of the cases).
- Descending aorta, below the origin of the left subclavian artery (20 % of the cases).
- Transversal aortic arch (10 % of the cases).
- Distal thoracoabdominal aorta (5 % of the cases).

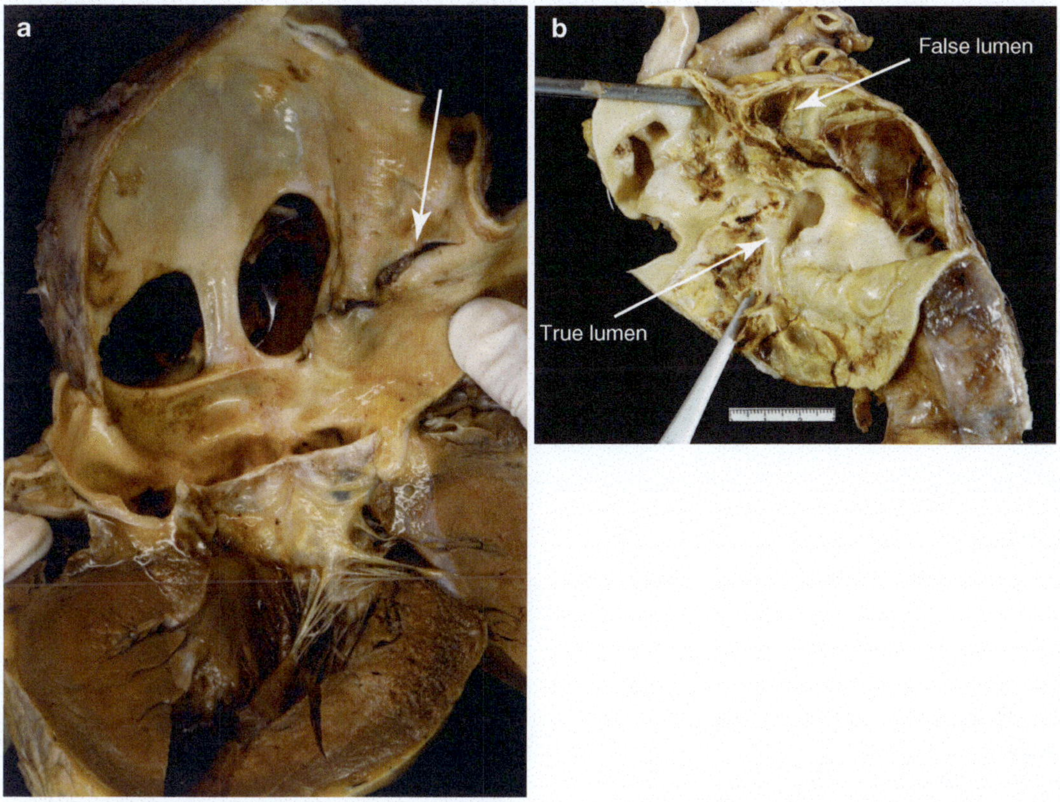

Fig. 3.17 A 90-year-old woman with a history of ischaemic cardiomyopathy who had a pacemaker. She died from cranioencephalic trauma as a result of a fall. A calcified aneurysmal dilatation of the ascending aorta can be observed with dissection and entry and exit tears at 2 cm from the aortic valve. A transversal intimal tear can be appreciated (*arrow*). (**b**) A 67-year-old male with a history of hypertensive cardiomyopathy with unstable angina and acute lung oedema, aortic valve replacement and coronary bypass. Image of the double-barrelled aorta: at the level of the aortic isthmus, two lumen can be appreciated, the true lumen on the left and the false lumen occupied with thrombi on the right (*arrows*). The end of the dissection was identified in the abdominal aorta above the cephalic trunk

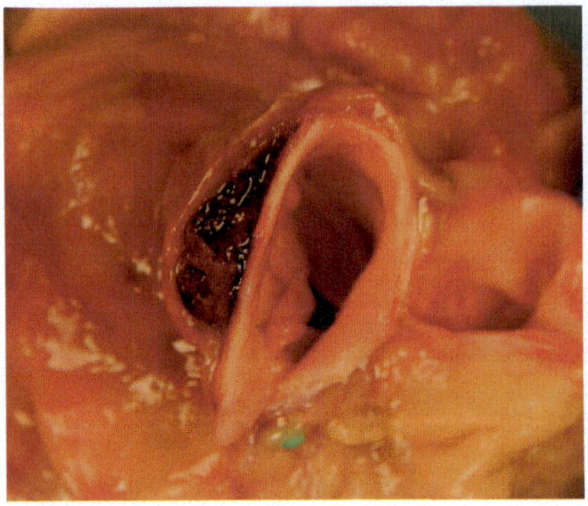

Fig. 3.18 Transversal section of an aortic dissection. The false lumen on the left, filled with blood, can be appreciated

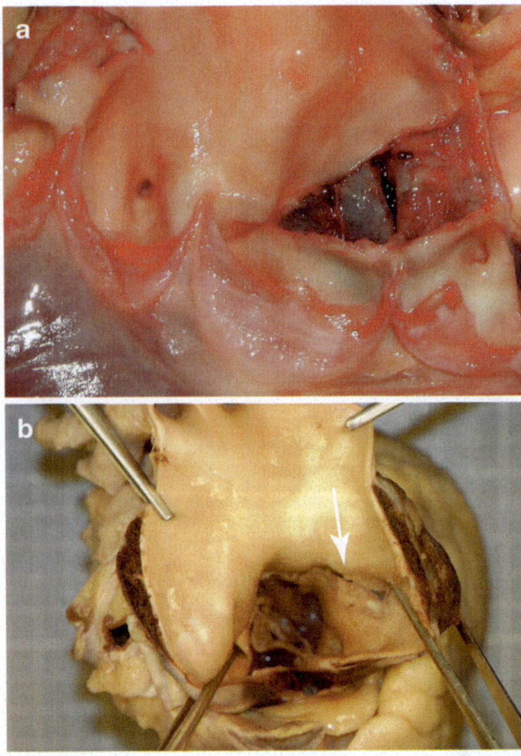

Fig. 3.19 Dissection in ascending aorta. (**a**) Intimomedial entry tear in sinotubular junction. (**b**) Intimal tear associated with bicuspid aortic valve (*arrow*)

Traditionally, aortic dissection has been classified according to the location of the original intimal tear and the extent of the dissection (DeBakey classification), or the involvement of the ascending aorta (Standford classification) (Table 3.6).

Table 3.6 Classification of aortic dissection

DeBakey classification (1965)	Standford classification (1970)
Type I. Intimal tear in ascending aorta that propagates to the aortic arch and descending aorta. It is the most frequent (54 %)	Type A: All dissections that involve the ascending aorta, independent of the location where they originate (62 %)
Type II. Intimal tear originates and is confined to the ascending aorta, without affecting the aortic arch (21 %)	Type B: The rest (38 %)
Type III. Intimal tear in distal aortic arch or proximal site of the descending aorta that propagates distally, although can extend retrogradely to the aortic arch and descending aorta (25 %)	

Type A dissections are associated with multiple risk factors including: hypertension, bicuspid aortic valve (Fig. 3.19b), Marfan syndrome, Loeys-Dietz syndrome, annuloaortic ectasia, aortic coarctaction (possibly by coexistence of hypertension and bicuspid aortic valve), trauma or can be iatrogenic following an aortic valve replacement and/or coronary revascularization surgery. Conversely, type B dissections are associated mainly with hypertension. However, in one quarter of patients there are no previous identified specific risk factors (Figs. 3.20 and 3.21).

Attending to the age, 90 % of cases appear in men aged 40–60 with a history of hypertension. In people over 40 years of age, hypertension is a risk factor in 71 % of the cases. In people younger than 40 there are other less common risk factors such as isometric sports (weightlifting, boxing, etc.), eclampsia, labour or pericardial abnormalities. An aortic medial dissection can also be associated with iatrogenic effects, due to complications in valve and aortic cardiovascular surgical treatments or after percutaneous coronary interventions.

Histology shows the dissection in the outer portion of the media. It is frequent to find cystic degeneration of the medial layer characterized by fragmentation, separation and loss of elastic fibres, which are replaced by mucopolysaccharide-filled spaces (Figs. 3.21 and 3.22). Nevertheless, in many cases there are no significant findings in the medial layer (Fig. 3.23). In cases of rupture of the ascending aorta, granulation tissue in the adventitia and periaortic epicardium can usually be observed, indicative of a certain degree of evolution, and which generally relates to symptomatology prior to death (Figs. 3.24 and 3.25).

In some cases death can occur due to a compromise of both coronary *ostia,* or by extension of the parietal dissection into the wall of the coronary arteries (Figs. 3.26 and 3.27).

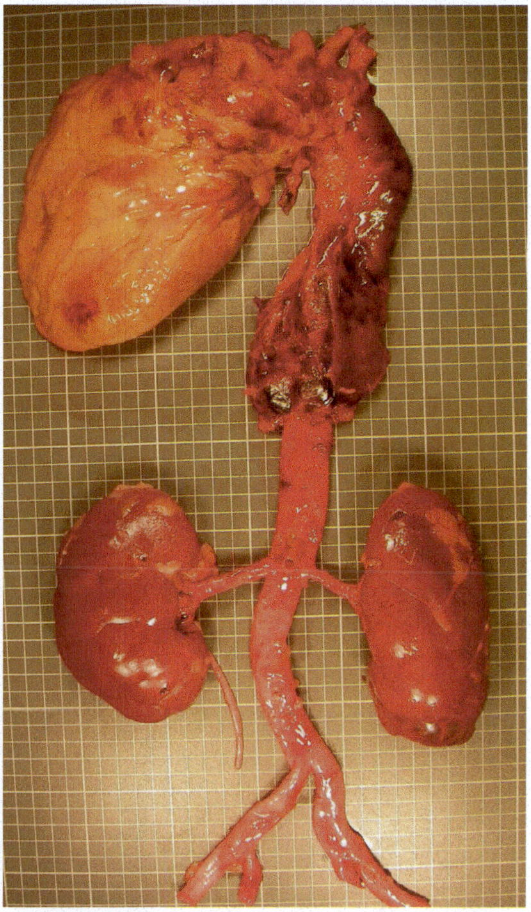

Fig. 3.20 Type I (DeBakey) aortic dissection extending from the ascending to the descending aorta. 48-year-old male who died suddenly at home several days after going to the emergency department with back pain that was considered mechanical. Heart weight of 750 g with evident concentric left ventricle hypertrophy

It is also possible that the intimomedial flap that separates both false and true aortic lumen invaginates initiating an aortic intussusception that clinically courses with an acute aortic obstruction at the level of the intussusception (Fig. 3.28).

Fig. 3.21 A 54-year-old sportsman who died suddenly in his house. Three days before, he attended the hospital emergency department with back pain. ECG and analytical determinations were normal and he was discharged. (**a**) Type II (DeBakey) aortic dissection confined to the ascending aorta. (**b**) Intimal tear (*arrow*) 1.5 cm above the sinotubular union of a bicuspid aortic valve. (**c**) Microscopic image of the aortic wall adjacent to the dissection, showing significant loss of elastic fibres (cystic medial degeneration) (*) (Weigert's stain)

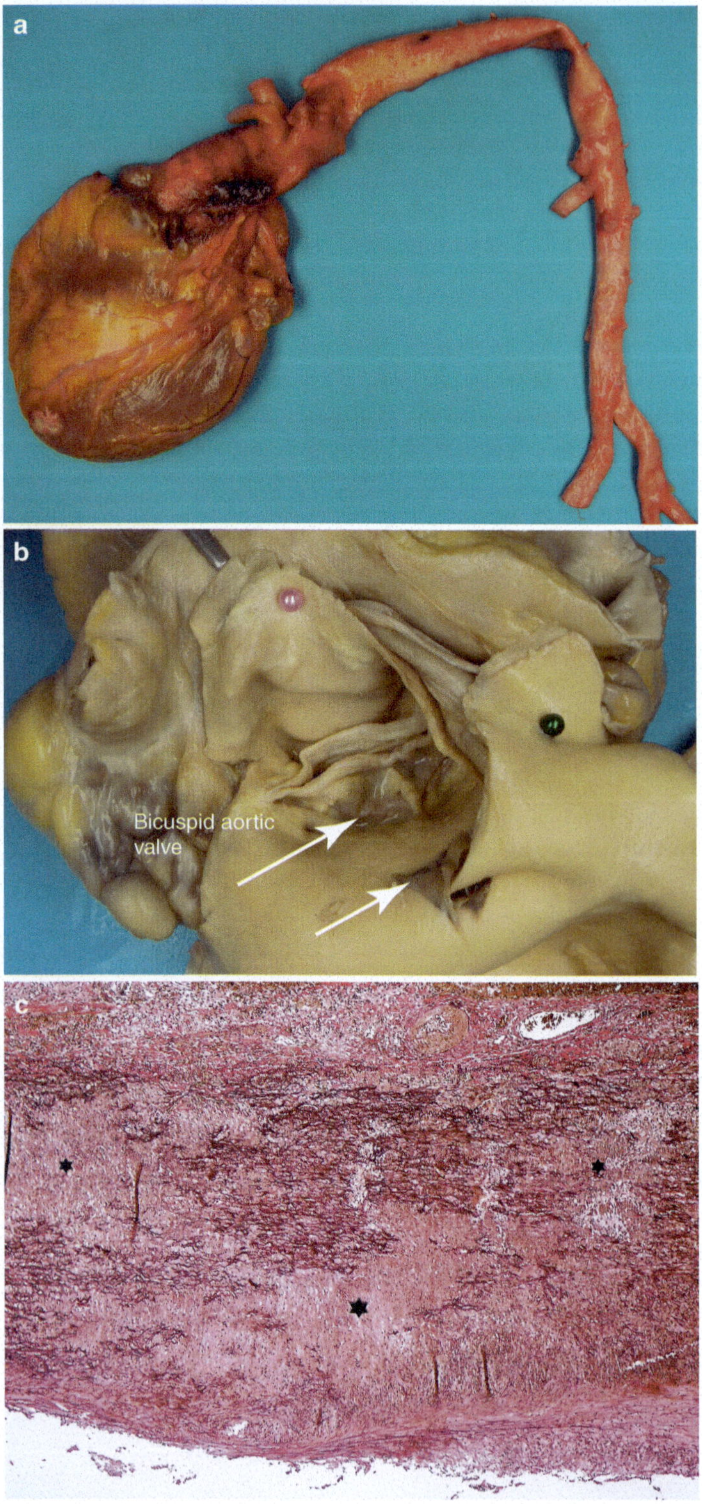

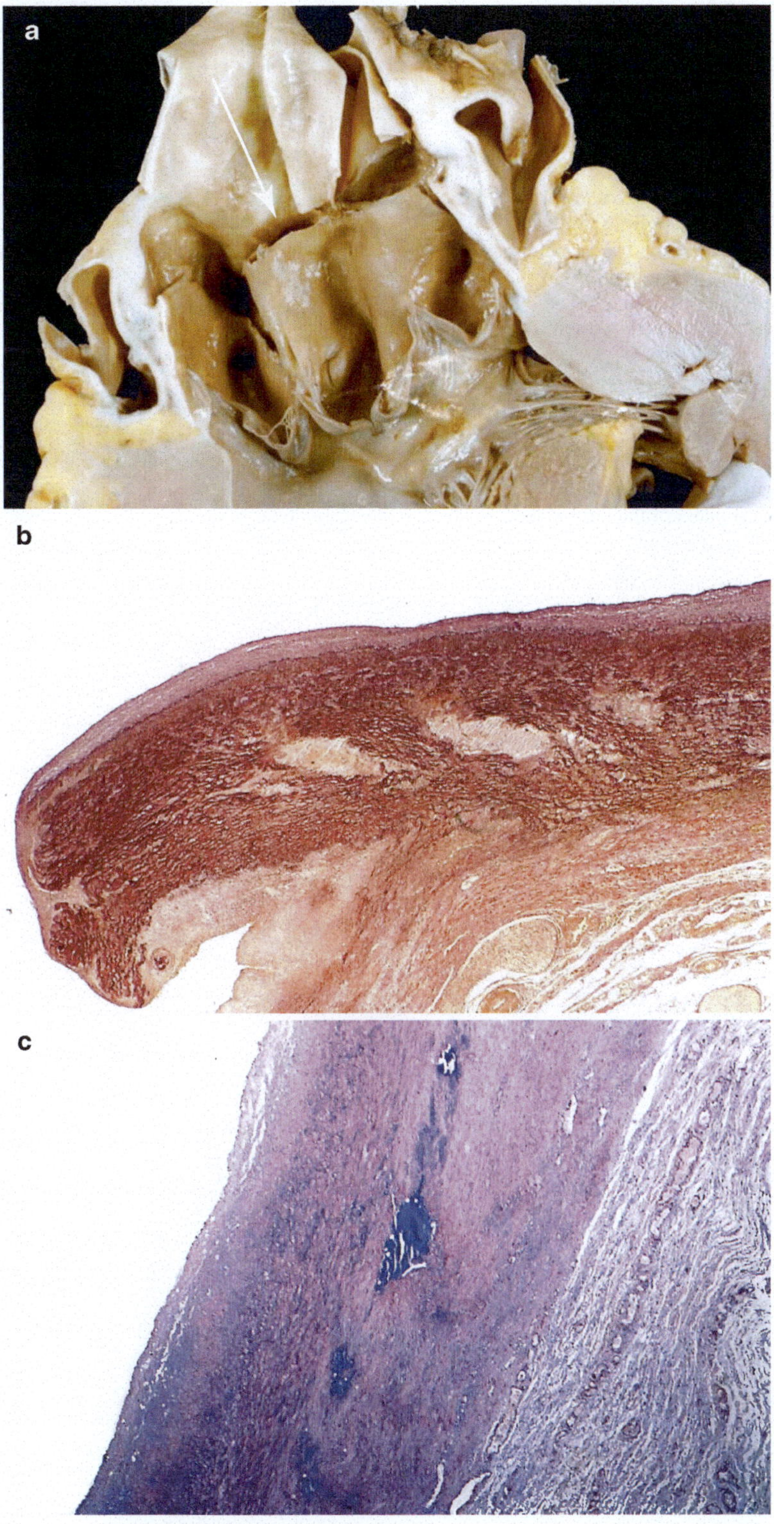

Fig. 3.22 Type II (DeBakey) dissection. A 20-year-old male, professional basketball player who died during a match. His height was 208 cm and he presented arachnodactyly and alterations in the rib cage (marfanoid signs). Autopsy revealed a haemopericardium of 750 cc secondary to aortic dissection and rupture. (**a**) Intimal transversal rupture practically circumferential (*arrow*). (**b**) Border of the intimal rupture with cystic medial degeneration (Weigert's stain). (**c**) Mucopolysaccharide deposit (Alcian blue stain)

Fig. 3.23 A 69-year-old male was examined in the emergency department with chest pain the day before his death. ECG, cardiac enzyme determination and thoracic X-ray were performed. Diagnosis was non-specific chest pain. The man died suddenly in his house the following day from cardiac tamponade. (**a**) The ascending aorta showed transmural rupture associated with a parietal dissection and haemorrhagic infiltration in the borders. There was a circumferential transversal intimal tear, 2.5 cm above the aortic valve that spared only 2 cm of the perimeter. (**b**) Dissection in the external part of the media without cystic degeneration. (**c**) Borders of the intimal tear (Weigert's stain)

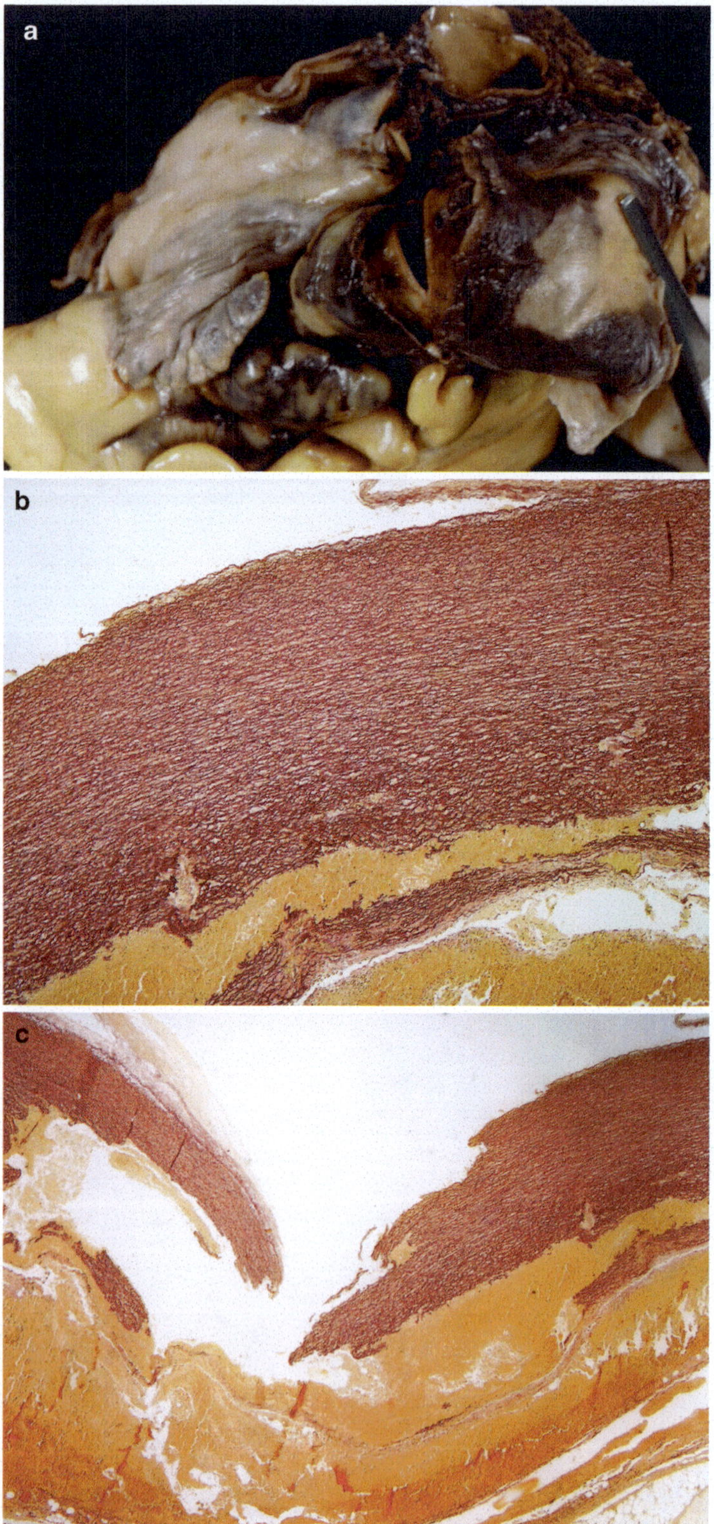

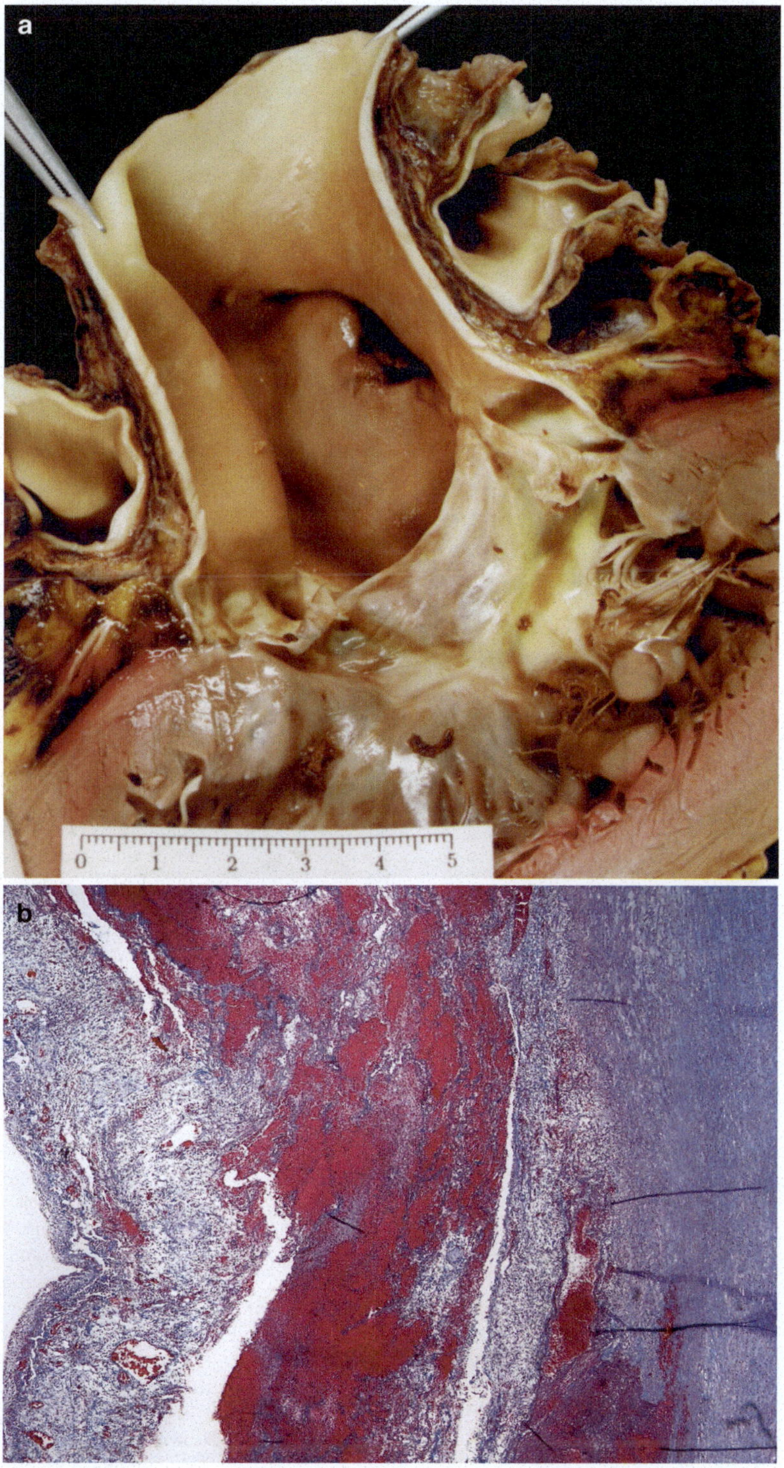

Fig. 3.24 A 19-year-old male suffered from pain and dizziness while skiing and was taken to the emergency department. After an ECG he was diagnosed with pericarditis. Several days later he died from cardiac tamponade. (**a**) Circular tear in ascending aorta associated with bicuspid aortic valve. (**b**) Microscopic image showing a dissection of the media on the right and a haemorrhage and granulation tissue on the left, indicative of a rupture with several days evolution

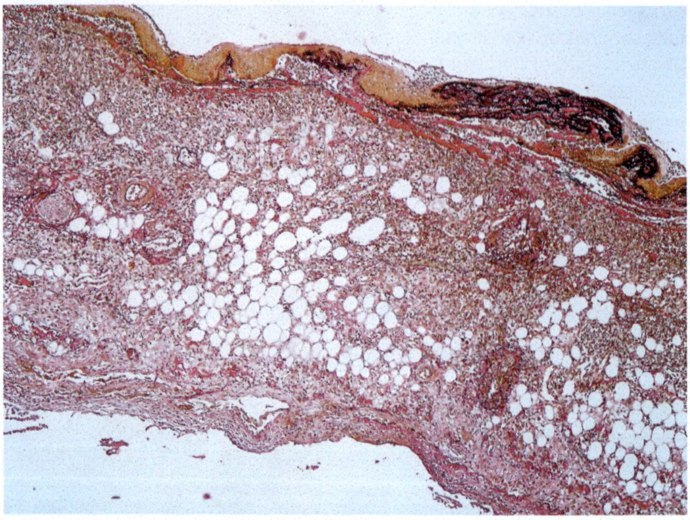

Fig. 3.25 Microscopic image of the external wall of an aortic dissection. Top, remains of the outer medial layer with elastic fibres and fibrin deposition. Bottom, periaortic epicardium with granulation tissue (Weigert's stain) (Courtesy of Mª Soledad Sánchez de León. INTCF. Madrid)

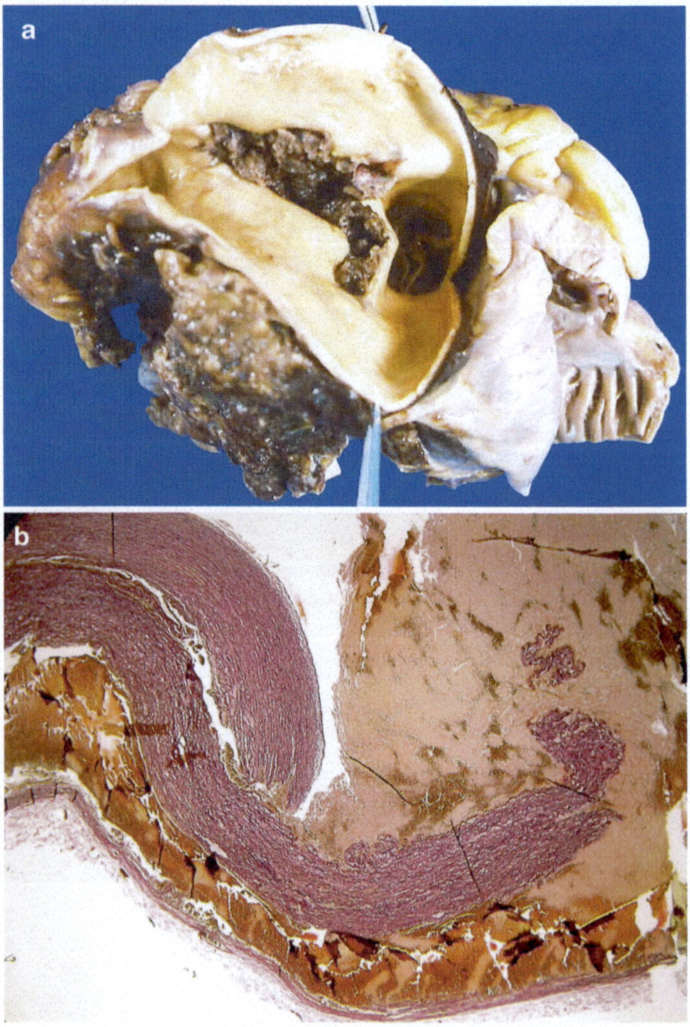

Fig. 3.26 A 59-year-old woman who started feeling unwell in the afternoon, with sweats and epigastric pain. She attended the GP surgery, was prescribed proton pump inhibitors and died soon after. Autopsy revealed haemopericardium and extensive haemorrhage in the adventitia of the aorta and pulmonary artery. Cardiac weight: 617 g. (**a**) Longitudinal intimal tear with anfractuous borders in the ascending aorta. (**b**) Histological image of the intimal rupture border with medial dissection and haemorrhage in the adventitia. (**c**) Stenosis >80 % of the lumen (no thrombosis) in the left anterior descending coronary artery. (**d**) Microphotography of the anterior wall of the LV with acute myocardial infarction of several hours' evolution. The infarct was attributed to the combination of compromised coronary ostia due to the dissection, and a severe stenosis of the left anterior descending coronary artery

Fig. 3.26 (continued)

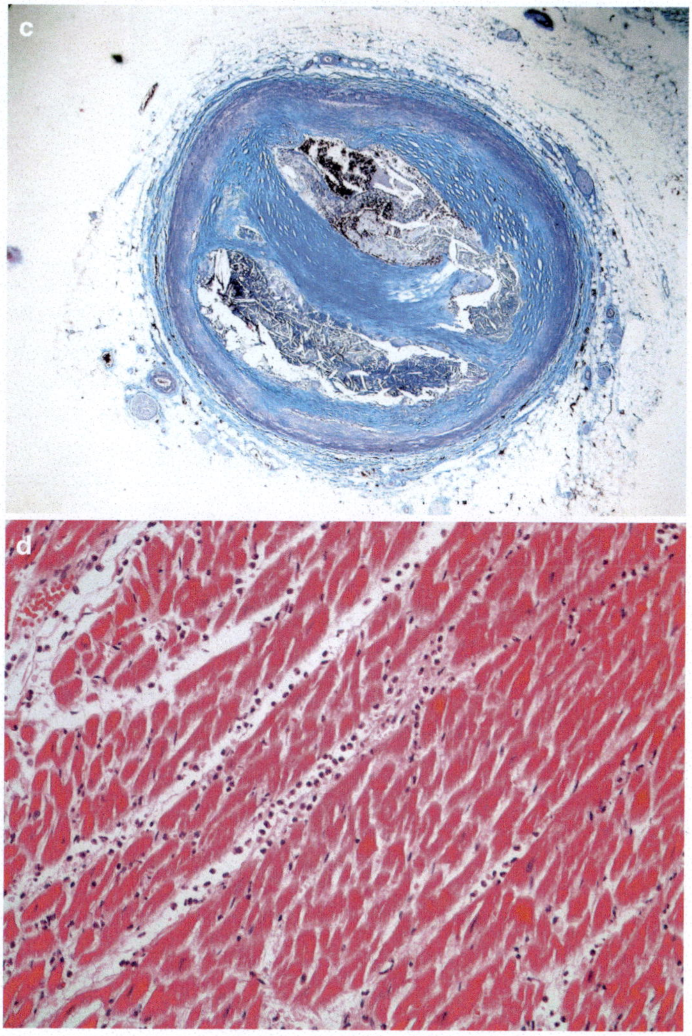

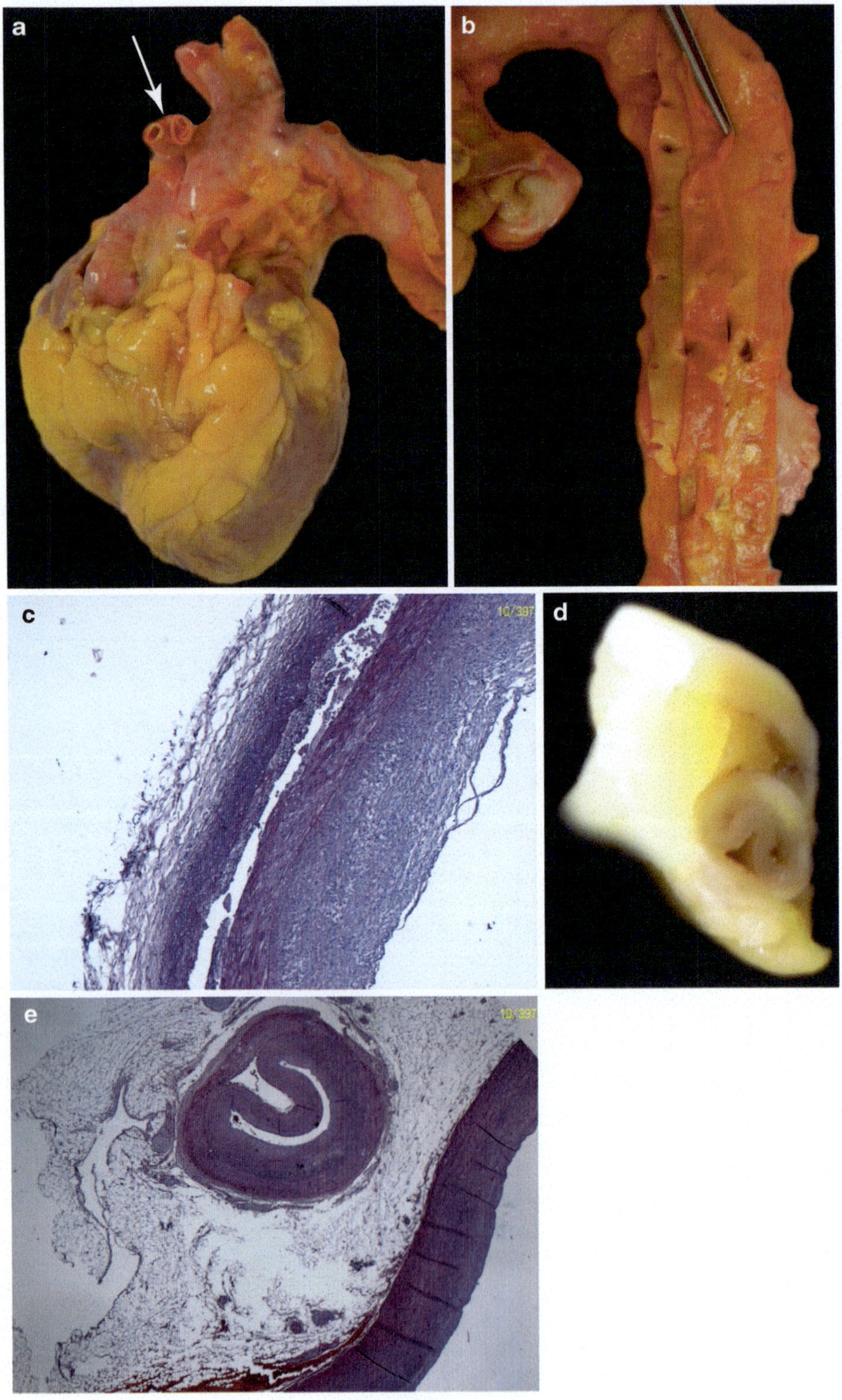
a
b
c
d
e

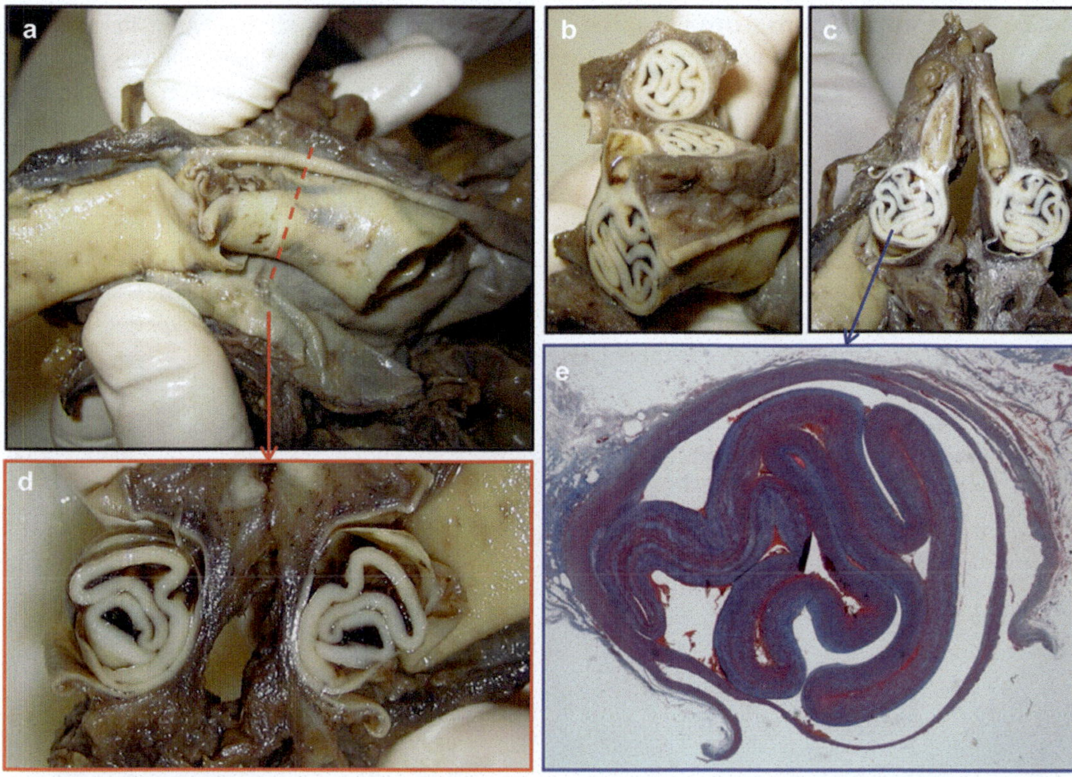

Fig. 3.28 A 15-year-old boy, diagnosed with Noonan syndrome, died suddenly in the street. The aortic valve was bicuspid and there was a severe dilatation of the aortic root. A complete circular dissection of the intima at the level of the annulus of the aortic valve was observed that was the origin of a complete detachment and intussusception of the inner wall of the aorta. The lesion was 5 cm long, ending at the aortic arch (**a**–**e**) (Cardona et al. (2012). Reproduced with permission from *Cuad Med Forense*)

Fig. 3.27 A 46-year-old male with no previous significant medical history, collapsed whilst taking a shower. (**a**) Type A dissection (Standford), without haematoma in the media, which causes a dissection in the right brachiocephalic trunk (*arrow*). (**b**) Dissection without haemorrhage in the descending aorta. (**c**) Microscopic detail of the aortic dissection where platelets and fibrin can be visualised. (**d**) Gross image of the main left coronary artery that was also affected by the aortic dissection. (**e**) Microscopic image of the medial dissection of the same coronary artery

3.4.2 Intramural Haematoma or Dissection with No Intimal Tear

This clinical entity is produced by the formation of a haematoma within the medial layer with no apparent intimal disruption (Fig. 3.29). It is attributed by the majority of authors to ruptures within the *vasa vasorum* of the medial layer, the blood vessels that penetrate the outer half of the aortic media from the adventitia in a superficial and deep plexus, and arborize within the media to supply the aortic wall. Other authors consider that penetrating aortic ulcers are a prerequisite for intramural bleeding. The haematoma may be reabsorbed or may progress to inward disruption of the intima, leading to communicating aortic dissection. The majority of haematomas occur in the descending aorta (70 %), and patients are older than those affected by classical aortic dissection. It is associated with hypertension and atherosclerotic risk factors.

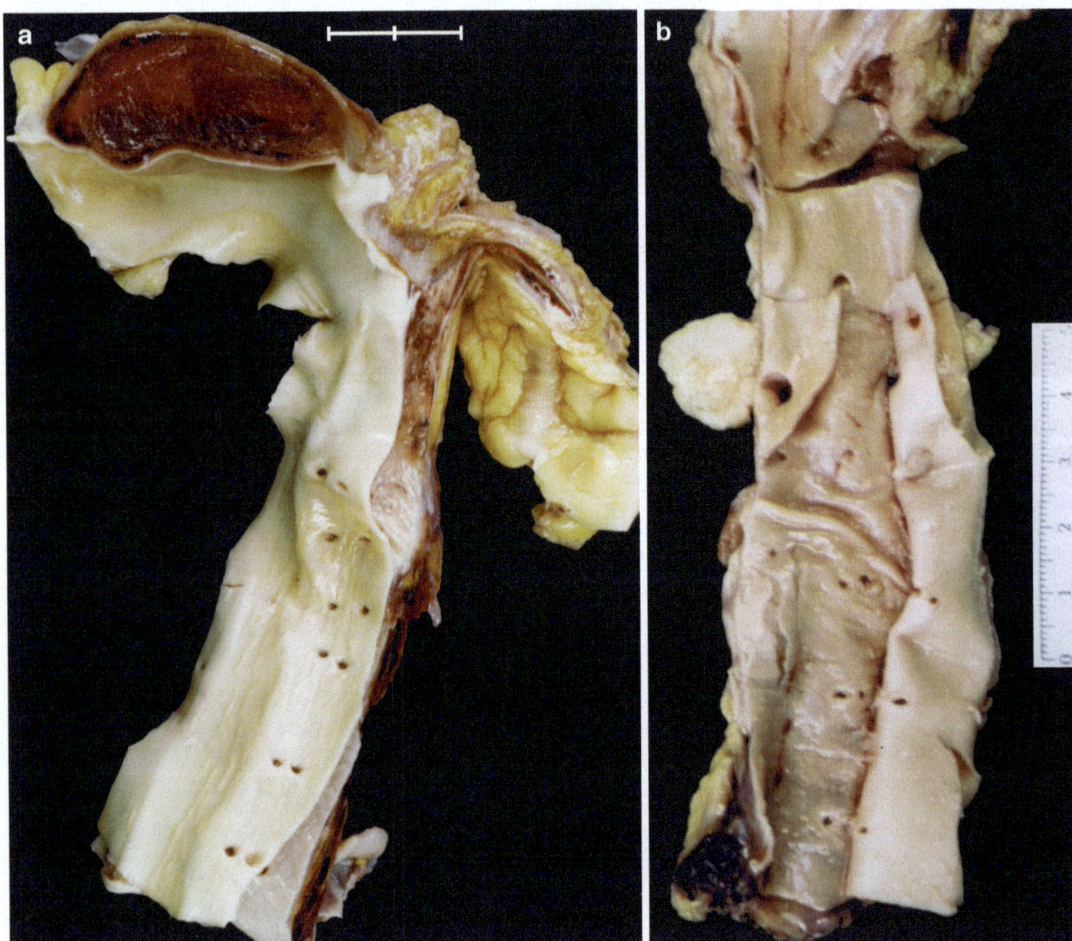

Fig. 3.29 A 23-year-old male operated for appendicitis 15 days before his death. (**a**, **b**) There was an intramural haematoma in the aortic arch of 25 mm thickness, and a type B dissection (Standford) in the descending aorta

3.4.3 Penetrating Aortic Ulcer

A penetrating aortic ulcer refers to an atherosclerotic plaque that, after ulceration and initial erosion, ruptures the internal elastic layer and penetrates into the aortic media. They can occur at any location in the aorta but are more frequent in the descending aorta. The most frequently affected area is the descending thoracic aorta (75 %) followed by the aortic arch and the ascending aorta. It occurs in elderly patients (70–80 years) and is associated with hypertension.

The most common clinical evolution of penetrating aortic ulcers is towards the formation of aneurysms or pseudoaneurysms. It is infrequent that the ulcer provokes a rupture in the aorta, but if this occurs, the haemorrhage could be contained by the haematoma produced following the rupture in the adventitia. This happens especially when the ulcers present at the abdominal aorta, where the retroperitoneal adjacent tissues help in the haemorrhage containment. Penetrating aortic ulcer can also lead to aortic dissection although the associated fibrosis and calcifications tend to make this progression difficult.

3.5 Coarctaction of the Aorta

Aortic coarctaction is a relatively frequent congenital pathology representing 5–6 % of the cardiovascular congenital abnormalities in children, and 15 % of those diagnosed in adults. It consists of a narrowing of the aorta that usually locates just distal to the origin of the left subclavian artery (Fig. 3.30), in the opposite wall to the *ductus arteriosus* (juxtaductal stenosis). Presentations with an atypical location in the descending and abdominal aorta have also been described in up to 2 % of the cases. These congenital atypical coarctactions of the aorta can be associated with diseases such as neurofibromatosis, tuberous sclerosis, Williams's syndrome, rubella or mucopolysaccharidosis. Some atypical forms are associated with acquired diseases such as Takayasu disease, tuberculosis, radiation and eosinophilia. In 58 % of the cases, other congenital abnormalities are also present, among them, bicuspid aortic valve (21 %) and patent *ductus arteriosus* (13 %) are the most frequent.

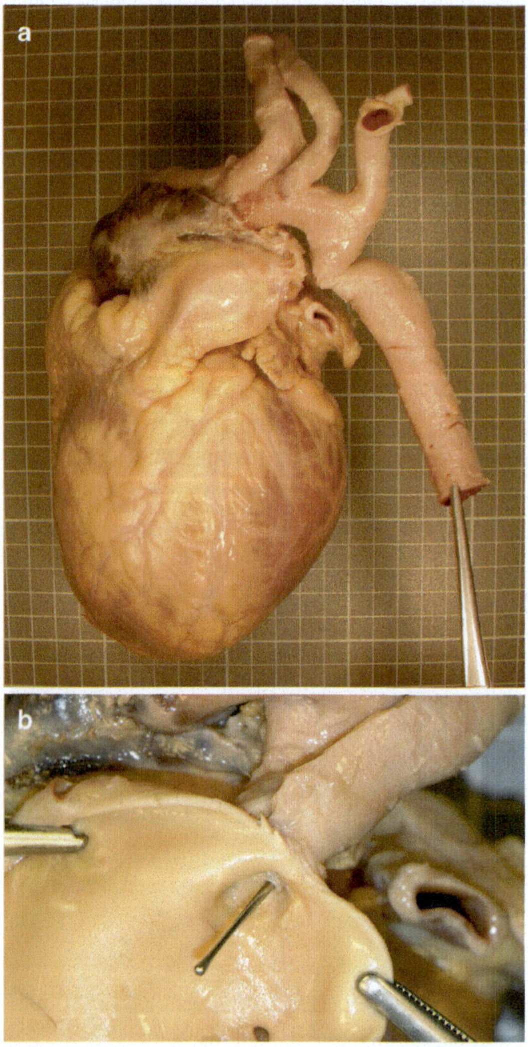

Fig. 3.30 Coarctaction of the aorta associated with bicuspid aortic valve in a 22-year-old male, with a history of untreated hypertension who died suddenly whilst driving. Autopsy also revealed ascending aortic dissection. (**a**) Juxtaductal stenosis can be appreciated (distal to the origin of the left subclavian artery) with pre and post-stenotic dilation. (**b**) Detail of the aortic lumen at the level of the stenotic area that just allows the passage of a probe (Rico et al. (2008). Reproduced with permission from *Cuad Med Forense*)

3.6 Aortitis

Aortitis is an inflammation of the aortic wall, with or without interruption of elastic fibres, parietal necrosis or fibrosis. By definition, aortitis implies the absence of underlying diseases that could produce secondary chronic inflammation, mainly atherosclerosis, although both entities can coexist. The classification of aortitis broadly includes infectious and non-infectious forms (Table 3.7). The three more important forms are Takayasu arteritis, syphilis and giant cell arteritis. Morphologically, they are similar so diagnosis requires a global assessment of the clinical characteristics and the pathological findings.

3.6.1 Takayasu Disease

Takayasu disease is a granulomatous inflammation of the aorta and its branches. The disease is characterized by an acute inflammatory reaction followed by repairing fibrotic processes of the large arteries that can lead to aneurysms or stenosis in the affected arteries. A pathological study

Table 3.7 Classification of aortitis

Non infectious
Takayasu arteritis (younger patients)
Giant cell arteritis (older patients)
Aortitis of collagen vascular disease (rheumatoid arthritis, ankylosing spondylitis, Reiter's syndrome, Behçet's disease, systemic lupus erythematosus etc.)
Inflammatory aneurysms of the abdominal aorta
Sarcoidosis
***Infectious*:**
Syphilitic aortitis
Tuberculosis
Pyogenic aortitis (bacterial or fungal)

From Virmani and Burke (2001)

reveals the existence of lesions located mainly in the aortic arch, although in one third of the cases other segments of the aorta are also affected. Macroscopically, there is an irregular thickening of the aortic wall by fibrosis and narrowing of the lumen. Occasionally some patchy areas of aneurysmal dilatation can be observed ("tree bark" appearance). Stenosis may be present in the lumen of the supra-aortic trunk, and also in the renal and splanchnic arteries. It can lead to dilatation in the ascending aorta, causing aortic valve insufficiency and stenosis of the coronary artery *ostia* (Fig. 3.31). The microscopic characteristics are summarised in Table 3.8.

Differential diagnosis includes several pathologies: inflammatory aortitis (syphilis, tuberculosis, lupus, rheumatoid arthritis, spondyloarthropathies, Behçet's disease, Kawasaki disease or giant cell arteritis), developmental abnormalities (coarctaction of the aorta, Marfan syndrome or Ehlers-Danlos syndrome) and other aortic pathologies such as ergotism or neurofibromatosis. Among these pathologies the most complex differential diagnosis is perhaps with tuberculous aortitis and, particularly, with giant cell arteritis with aortic involvement (10–15 % of the cases). Nevertheless, the tuberculous form tends to result in erosive lesions with less fibrosis and stenosis, both characteristic of Takayasu disease, and locates in the descending and abdominal aorta. The syphilitic form usually affects an older age group, with calcification, preserving the descending thoracic aorta, and stenosis is usually not a feature. Giant cell arteritis is associated with other features caused by involvement of the temporal artery that may serve to help in this difficult differential diagnosis.

3.6.2 Giant Cell Arteritis

Giant cell arteritis is a systemic panarteritis affecting elderly patients that presents with a focal granulomatous inflammation of large and medium-sized arteries, principally the temporal artery. The aorta is compromised in 15 % of the cases. The morphological findings are summarised in Table 3.9.

Table 3.8 Pathological features of Takayasu arteritis

Acute phase
Oedema, patchy necrosis, chronic inflammation
Scattered giant cells in the outer two thirds of the media, adventitia, adventitial fat and *vasa vasorum*
The *vasa vasorum* may show intimal proliferation with obliterations but no fibrinoid necrosis.
Chronic phase
Marked intimal and adventitial thickening of the vessels
Intimal thickening produces multisegmented circumferential stenosis, alternating normal and affected areas
Intimal hypocellularity with smooth muscle cells and scattered fibroblasts
Disorganised or absent elastic fibres replaced by collagen and granulation tissue
Areas of necrosis with giant cell infiltrates can persist

Table 3.9 Pathology of giant cell arteritis

Macroscopic
Aortic root dilatation with frequent medial dissection and rupture
The intima is wrinkled, with a "tree-bark" appearance
Microscopic: (Fig. 3.32)
Mononuclear inflammatory infiltrate: lymphocytes, plasma cells and histiocytes
Disruption of the internal elastic lamina with fragmentation and a giant cell-reaction of the Langhan's type, and necrosis
Mild-moderate intimal fibrosis with minimal inflammation in healed lesions

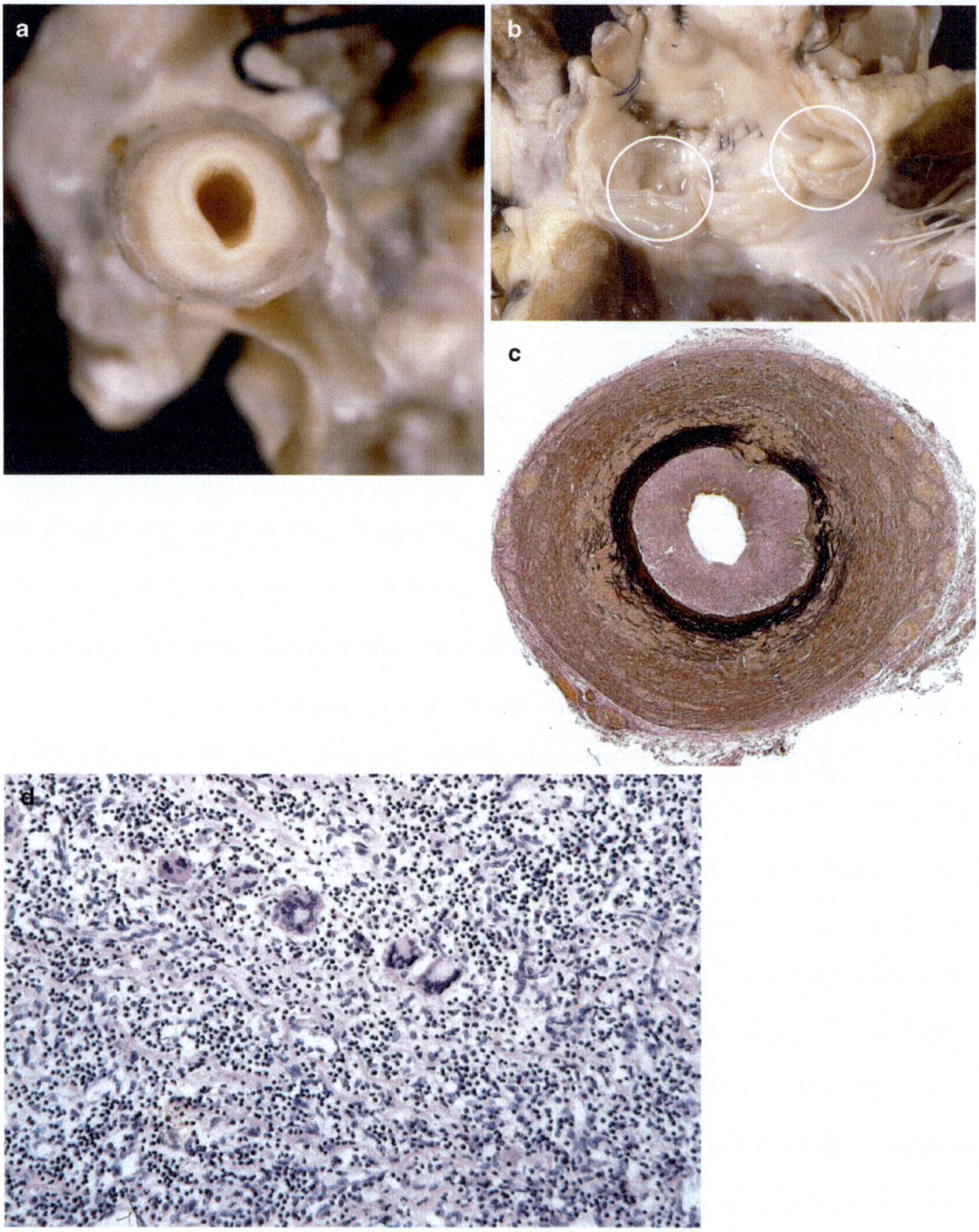

Fig. 3.31 A 14-year-old girl who died suddenly from myocardial ischaemia. (**a**) Concentric stenosis of the left internal carotid artery. (**b**) Aortic root with an increased wall thickness and stenosis of the coronary *ostia* (*circled*). (**c**) Histology of the carotid artery, showing periadventitial fibrosis, destruction of the elastic fibres in the tunica media and fibrous proliferation in the intima. (**d**) High magnification image shows a lymphocytic infiltrate associated with multinucleated giant cells (Courtesy of Dr. C. Basso and Dr. G. Thiene. Cardiovascular Pathology Unit. Department of Cardiac, Thoracic and Vascular Sciences. University of Padua Medical School, Italy. Basso et al. (1994))

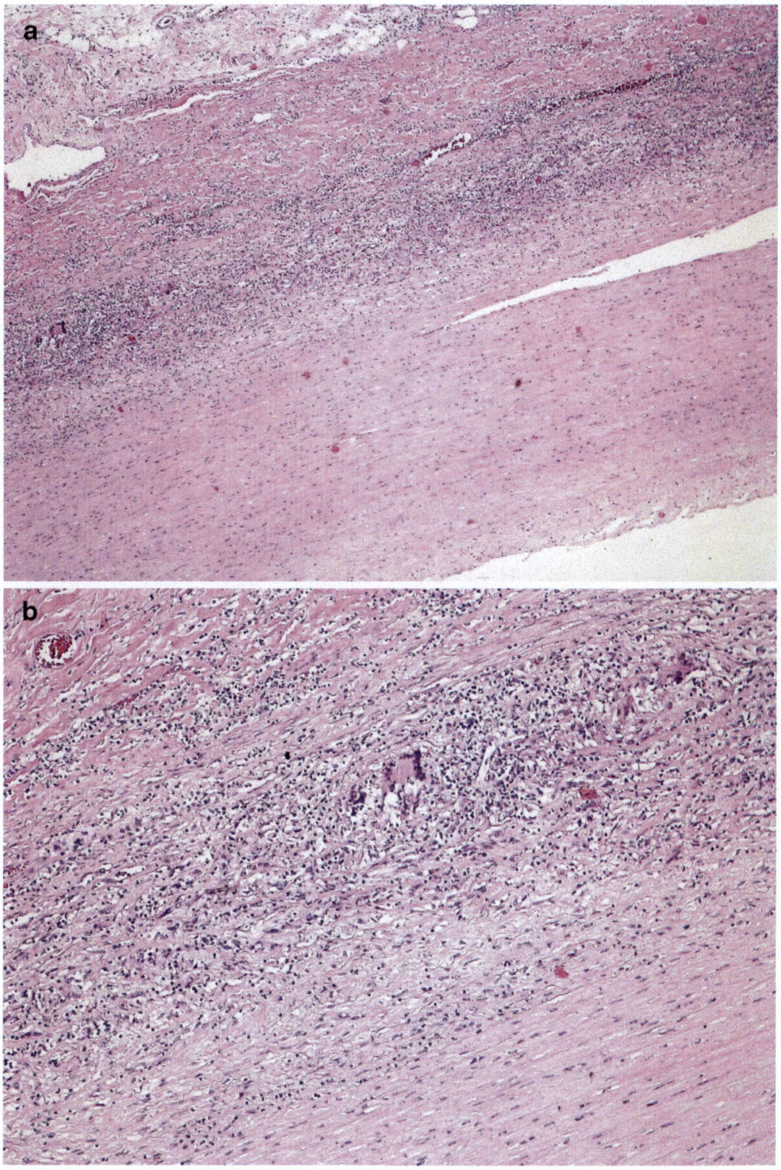

Fig. 3.32 An 81-year-old woman attended the emergency department with chest and abdominal pain. At admission, she suffered a sudden episode of loss of consciousness and arterial hypotension (90/60 mmHg). A brain CT scan and thoracic X-ray were performed. While she was in the radiology department she suffered a cardiac arrest. Autopsy revealed a transversal tear in the aorta 2 cm above the annulus and parietal dissection. (**a**) Panoramic view of the aorta showing intense mononuclear inflammatory infiltrate and dissection of the media layer. (**b**) High-power view of the infiltrate with multinucleated giant cells. The coronary arteries were also affected (see Fig. 6.14). A diagnosis of possible giant cell arteritis was made

The differences between giant cell arteritis and Takayasu's disease are: affects elderly patients, mild intimal thickening, lack of adventitial fibrosis, lack of inflammation with endarteritis obliterans, and affects more frequently the branches of the external carotid artery.

Table 3.10 summarizes the characteristics of the clinical entities associated with aortitis.

Table 3.10 Differential diagnosis of aortitis

Diagnosis	Histology	Preferred location
Syphilitic aortitis	Focal necrosis on aortic media. Perivascular lymphocytic infiltrations surrounding the *vasa vasorum*	Ascending aorta
Takayasu arteritis	Granulation tissue with multinucleated giant cells. Intimal fibrosis of the aorta. Degeneration and fragmentation of elastic fibres	Aortic arch and its branches
Giant cell arteritis	Giant cells (CD68 + ve) in granulation tissue, intimal fibrosis, alternating lesions, degeneration and fragmentation of elastic fibres	Aorta, initially large and medium-sized arteries. Coronary artery involvement is rare
Panarteritis nodosa	Fibrinoid necrosis followed by granulation tissue, and segmented involvement	Medium-sized arteries and arterioles
Coronary arteritis	Isolated foci of coronaritis without giant cells, with large lymphocytic infiltrates and germinal centres	Segmental involvement of the coronary arteries

From Dettmeyer et al. (2011)

Bibliography

Alomari IB, Hamirani YS, Madera G, Tabe C, Akhtar N, Raizada V. Aortic intramural hematoma and its complications. Circulation. 2014;129:711–6.

Basso C, et al. Sudden cardiac arrest in a teenager as first manifestation of Takayasu's disease. Int J Cardiol. 1994;43:87–9.

Burke A, Tavora F. Noninflammatory thoracic aortic aneurysms. In: Practical cardiovascular pathology. Philadelphia: Wolters Kluwer/Lippincott Williams & Wilkins; 2011. p. 73–85.

Camm AJ, Lüscher TF, Serruys PW. Tratado de medicina cardiovascular de la European Society of Cardiology. Barcelona: Ars Medica; 2008.

Cardona B, Giner S, Muñoz-Quirós JM, Molina P, Ortiz JM, Bañón R. Muerte súbita por disección aórtica circunferencial con intususpección intimo-intimal en un adulto joven con Síndrome de Noonan. Cuad Med Forense. 2012;18:89–94.

Dettmeyer RB. Vascular, cardiac valve, and metabolic diseases: chapter 14. In: Dettmeyer RB, editor. Forensic histopathology. Berlin-Heidelberg: Springer; 2011. p. 283–98.

Elefteriades JA, Pomianowski P. Practical genetics of thoracic aortic aneurysm. Prog Cardiovasc Dis. 2013;56:57–67.

El-Hammamsy I, Yacoub MH. Cellular and molecular mechanisms of thoracic and aortic aneurysms. Nat Rev Cardiol. 2009;6:771–86.

Feldman DN, Roman MJ. Aneurysms of the sinuses of Valsalva. Review. Cardiology. 2006;106:73–81.

Johnston SL, Lock RJ, Gompels MM. Takayasu arteritis: a review. J Clin Pathol. 2002;55:481–6.

Molina P, Ferrer P, Peyró R. Muerte súbita por disección aórtica en portador de enfermedad poliquística renal autosómica dominante. REA: E J Autopsy. 2007; 5(1):10–4.

Nathan DP, Ch X, Gorman JH, et al. Pathogenesis of acute aortic dissection: a finite element stress analysis. Ann Thorac Surg. 2011;91:458–63.

Pachar JV, Hidalgo J, Aguirre V, Alvarado O, Santos M. Aneurisma de arco aórtico secundario a degeneración de la media. A propósito de un caso de muerte súbita juvenil. Cuad Med Forense. 2010;16:161–5.

Patel HJ, Deeb GM. Ascending and arch aorta pathology, natural history, and treatment. Circulation. 2008; 118:188–95.

Rico A, Lucena J, Blanco M, Marín R, Barrero E, Santos M. Muerte súbita por disección de aorta torácica secundaria a coartación aórtica. Cuad Med Forense. 2008;14:55–60.

Ring WS. Congenital heart surgery nomenclature and database project: aortic aneurysm, sinus of valsalva aneurysm, and aortic dissection. Ann Thorac Surg. 2000;69:S147–63.

Ripperger T, Dieter Troger H, Schmidtke J. The genetic message of a sudden unexpected death due to thoracic aortic dissection. Forensic Sci Int. 2009;187:1–5.

RuDusky B. Aortic aneurysms and aortic dissection. In: Forensic cardiovascular medicine. Boca Raton: CRC Press; 2009. p. 35–44.

Schoen FJ. Vasos sanguíneos. In: Kumar V, Abbas AK, Fausto N, editors. Robbins and Cotran. Patologia estructural y funcional. 7th ed. Madrid: Elsevier; 2006. p. 517–60.

Suárez-Mier MP, Morandeira E. Sudden death in a young caused by rupture of an aneurysm above the right sinus of Valsalva [Muerte súbita en un joven por rotura de aneurisma sobre el seno de Valsalva derecho]. Cuad Med Forense. 2012;18(3–4):157–62.

Virmani R. Diseases of the aorta and large vessels. In: Cardiovascular pathology, Major problems in pathology, vol. 40. 2nd ed. Philadelphia: W.B. Saunders Company; 2001. p. 469–523.

Virmani R, Burke A. Nonatherosclerotic diseases of the aorta and miscellaneous diseases of the main pulmonary arteries and large veins. In: Silver MD, Gotlieb AI, Schoen FJ, editors. Cardiovascular pathology. 3rd ed. London: Churchill Livingstone; 2001. p. 107–37.

Pathology of the Pericardium

4

Pilar Molina, Beatriz Aguilera, Marta Cobo, and Pablo García-Pavía

Contents

P. Molina (✉)
Histopathology Section, Forensic Pathology Department, Institute of Legal Medicine, Valencia, Spain

University of Valencia, Valencia, Spain
e-mail: pilar.molina@uv.es

B. Aguilera
Histopathology Service, National Institute of Toxicology and Forensic Sciences, Las Rozas (Madrid), Spain

M. Cobo • P. García-Pavía
Heart Failure and Inherited Cardiac Diseases Unit, Cardiology Department, Hospital Universitario Puerta de Hierro, Majadahonda (Madrid), Spain
e-mail: pablogpavia@yahoo.es

4.1 Introduction

The pericardium is a fibroelastic sac surrounding the heart that influences its functionality, maintaining the heart in a mediastinal fixed position and protecting it against infections and tumoral invasions. The pericardium prevents excessive dilation of the heart, distributes hydrostatic forces and facilitates ventricular coupling during diastole.

Anatomically, it consists of a fibrous pericardium and a serous pericardium. The fibrous pericardium is a pyramidal-shaped sac that extends to 3–4 cm above the great vessels, fusing to their adventitial layer. The serous pericardium can be divided in two parts, visceral and parietal, separated by the pericardial cavity, which under normal conditions contains between 15 and 50 ml of a clear fluid. The parietal layer comprises a monolayer of mesothelial cells that covers the entire outer surface of the heart and the inner side of the fibrous pericardium. The visceral layer or epicardium is in intimate contact with the myocardium except in some areas (fibroadipose tissue in the atrioventricular and interventricular grooves), or when adipose tissue deposits increase with overweight or age (Fig. 4.1).

Epicardial adipose tissue is more than a fat depot since it is metabolically active and, given its intimate contact with the myocardium (there is no fascia between them), it has been recently associated with the normal physiology and structure of the coronary circulation. Epicardial fat

J.S. Lucena et al. (eds.), *Clinico-Pathological Atlas of Cardiovascular Diseases*,
DOI 10.1007/978-3-319-11146-9_4, © Springer International Publishing Switzerland 2015

can locally interact with and modulate the coronary arteries through an equilibrium between the paracrine or vasocrine secretion of pericardial inflammatory adipokines on the one hand (eg, tumor necrosis factor alpha, monocyte chemoattractant chemokines, interleukin-6, etc.) and, on the other hand with anti-inflammatory and anti-atherogenic adipokines such as adrenomoduline and adiponectin. Epicardial adipose tissue can play an important role in the inflammatory process at the atheroma plaque. When epicardial fat increases, it extends towards the anterior surface of the heart, and can eventually cover the entire heart with a layer of up to 2 cm in thickness (Fig. 4.2). The increase in epicardial fat in hemodynamic or metabolic pathologies may produce adverse lipotoxic, prothrombotic or proinflammatory effects.

4.2 Congenital Pathology

Congenital pathology of the pericardium is infrequent and includes complete or partial pericardial absence (agenesis) and congenital cysts.

4.2.1 Pericardial Congenital Defects

Pericardial congenital defects are observed at a frequency of 1/10.000 in autopsies. Complete or bilateral pericardial agenesis is the least frequent form and, in the absence of other associated congenital lesions, is generally asymptomatic. The main clinical manifestation is a mild nonspecific chest pain, likely caused by torsion of the great vessels due to excessive mobility of the heart. Among the partial defects, absence of the left pericardium is the most frequent (Fig. 4.3). Complications can occur when a portion of the myocardium herniates or strangulates at the level of the defect, which may compress the left coronary artery. Clinical symptoms include dyspnea, chest pain, syncope and/or sudden death. Moreover, these defects allow a greater mobility of the heart, which may increase the risk of suffering an aortic dissection after trauma.

4.2.2 Congenital Pericardial Cysts

Cysts can be uni- or multilocular with a diameter ranging from 1 to 5 cm. Patients are asymptomatic and the cysts are accidentally discovered during thoracic radiologic examinations (a radiodense homogeneous structure), or during autopsy. A differential diagnosis with acquired inflammatory pseudocysts and hydatid cysts should be performed.

Hydatid cysts in the pericardial cavity are generally asymptomatic but they can cause pseudo-angina symptoms. Also, if the cysts rupture and spill their contents into the pericardial cavity, they can produce pericardial effusion and cardiac tamponade.

Remarks

Pericardial pathology very rarely causes sudden cardiac death (SCD) apart from cardiac tamponade, although some cases of sudden death have been described associated with partial pericardial agenesis.

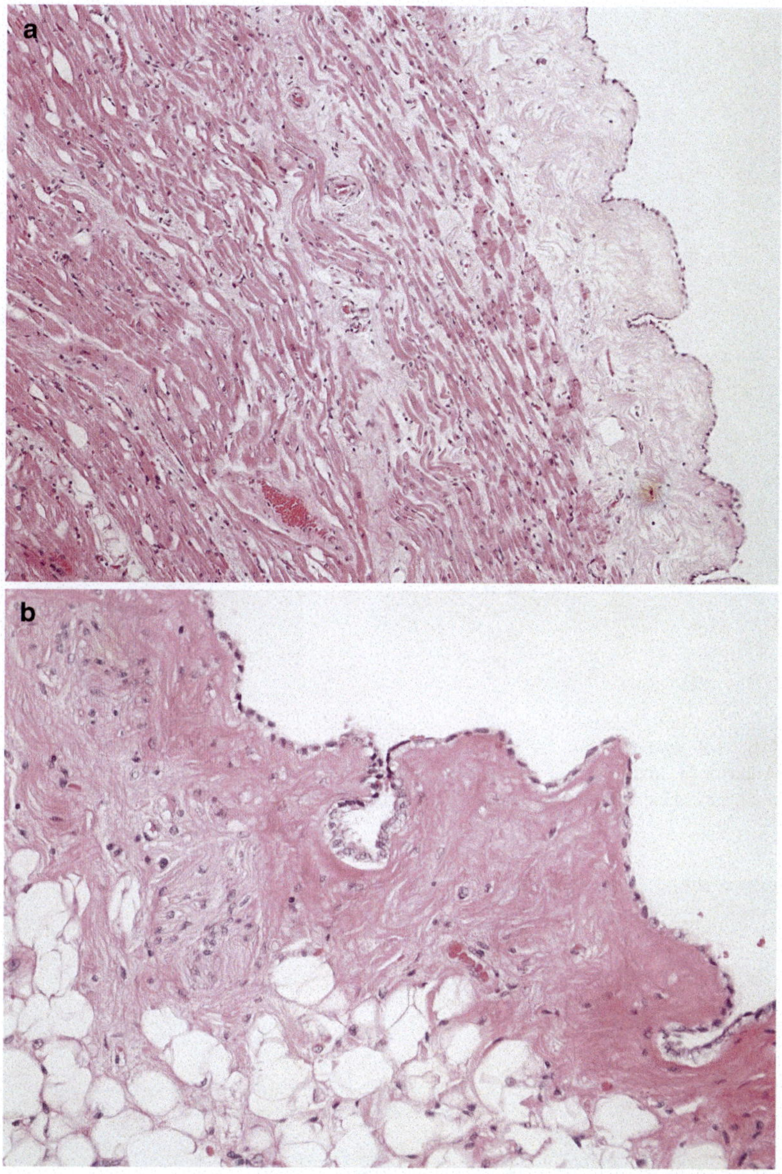

Fig. 4.1 *Histology of a normal pericardium.* Two photomicrographs of the visceral pericardium. (**a**) Next to the outermost layer of the myocardium. (**b**) A higher magnification, showing a mesothelial layer that coats loose connective and adipose tissues within which blood vessels and nerves are embedded

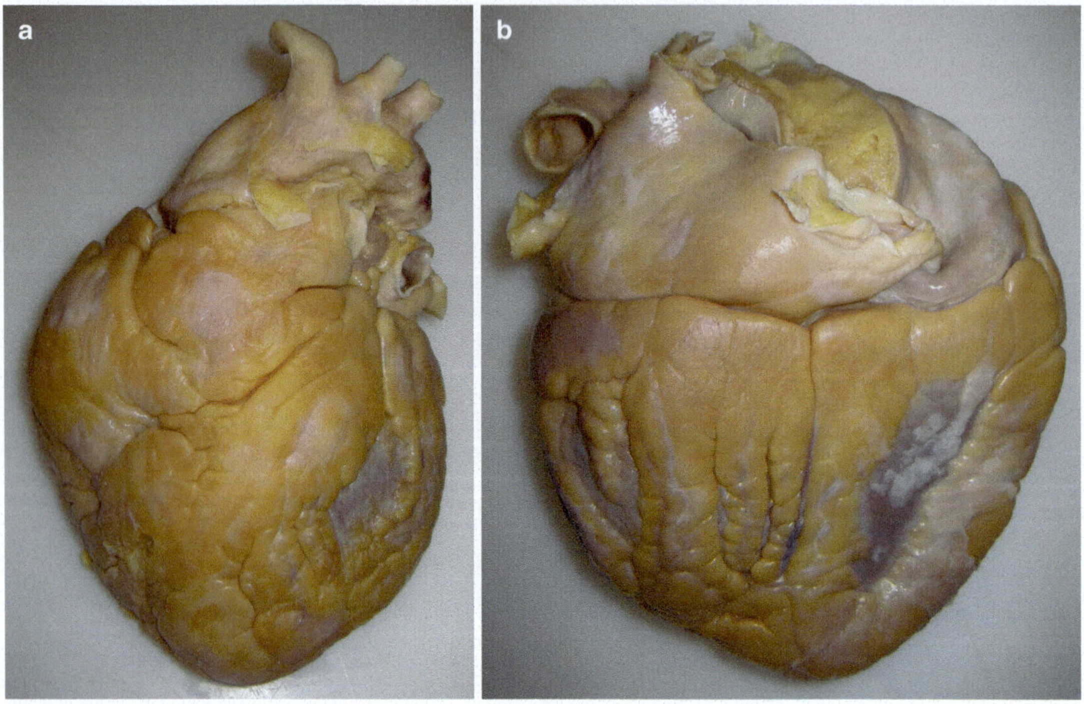

Fig. 4.2 *Increase of the epicardial adipose tissue*. Anterior (**a**) and posterior (**b**) views of a 516 g heart with a substantial increase of the epicardial fat, from a 52-year-old male who died suddenly at home. Autopsy revealed severe coronary atherosclerosis

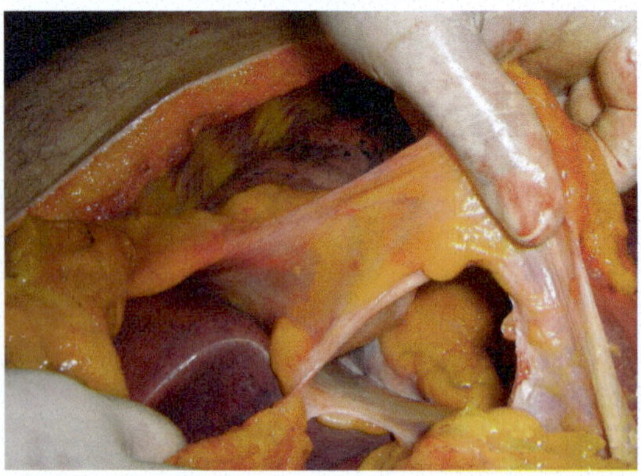

Fig. 4.3 *Partial pericardial agenesis*. The absence of the inferior pericardium is usually associated with a lack of the left hemidiaphragm, which implies a direct communication between the heart and the abdominal viscera (Courtesy of Dr. Iván Pávez. Medico-Legal Service, Santiago de Chile)

4.3 Pericardial Effusion. Cardiac Tamponade. Hemopericardium

Under particular circumstances, the fibrous or parietal pericardium distends due to liquid accumulation or pericardial effusion of diverse nature (exudate, transudate, blood). The clinical manifestations will depend on speed at which the fluid accumulates. Thus, effusions of no more than 200–300 cc, if the accumulation is fast (acute effusion), can lead to cardiac tamponade. In contrast, if the effusion is gradual (subacute/chronic effusion), up to 2 l of fluid can accumulate without significantly changing the intrapericardial pressure.

Cardiac tamponade is defined as an accumulation of pericardial fluid that produces an increase in the intrapericardial pressure, which in turn impedes ventricular filling during diastole and reduces the systolic volume. The most frequent pericardial causes of cardiac tamponade are metastatic carcinoma and idiopathic pericarditis. When a distended cardiac sac is found at autopsy the volume of fluid it contains should be determined, and its characteristics evaluated.

Hemopericardium refers to blood in the pericardial cavity, produced typically by a rupture in the myocardium, aortic wall or coronary artery, or due to alterations in the blood coagulation system. From a practical point of view, to correctly assess the dimensions of the blood clot it is convenient to measure the volume of blood and/or to weigh the clot contained within the pericardial cavity.

Serous effusions are most common in patients with cardiac insufficiency and hypoproteinemia, whereas bloody effusion most commonly results from ischaemic or traumatic cardiomyopathy, neoplasms or uraemic or infectious pericarditis. Chylous effusions are extremely rare and result from injury or obstruction of the thoracic duct (Fig. 4.4). Cardiac tamponade secondary to hemopericardium is typically caused by myocardial rupture due to ischaemia or trauma, and by aortic dissection (Fig. 4.5).

The pathological features of pericardial effusion are summarised in Table 4.1.

Table 4.1 Pathology of pericardial effusion

In chronic pericardial effusions there is a pericardial fibrous thickening with light or moderate inflammatory infiltration, mainly lymphocytic.
Subacute pathology responsible for the bleeding, including intramural haematoma on the aortic wall, can be observed in hemopericardium secondary to aortic dissection.
In myocardial rupture after acute myocardial infarction, the transmural extent of necrosis in the ischaemic myocardium must always be confirmed.

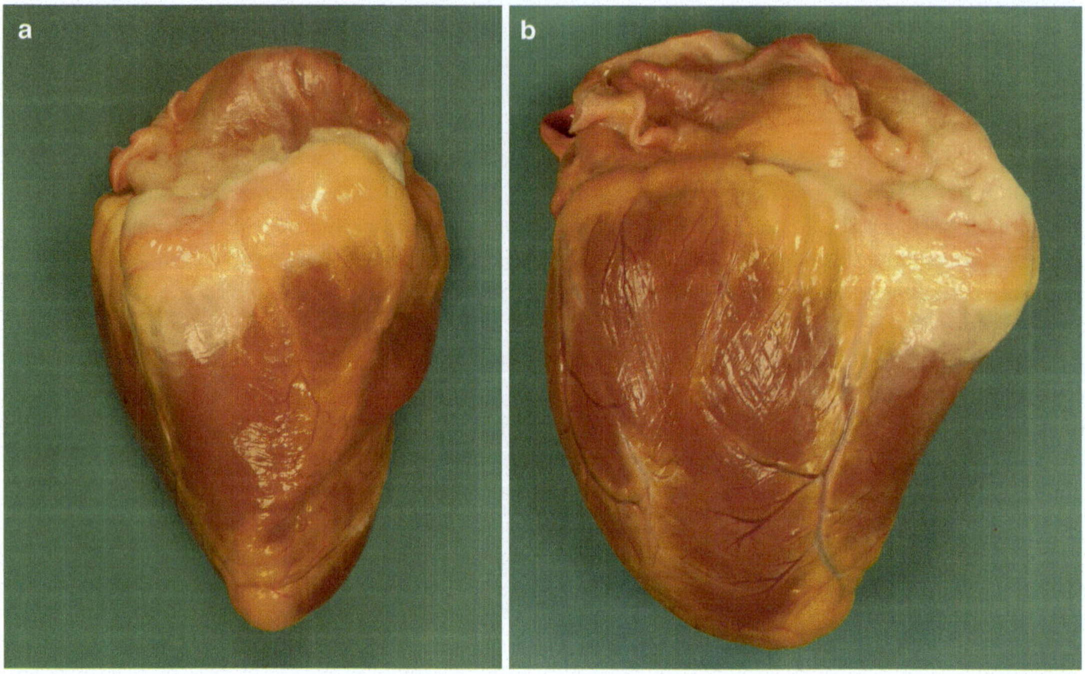

Fig. 4.4 *Chylopericardium*. Epicardial imbibition from the lympha as a result of chylopericardium due to an iatrogenic tear of the lymphatic thoracic duct. A 21-year-old male who was involved in a car accident and died from head injuries. (**a**) Right lateral view of the heart. (**b**) Posterior view of the heart

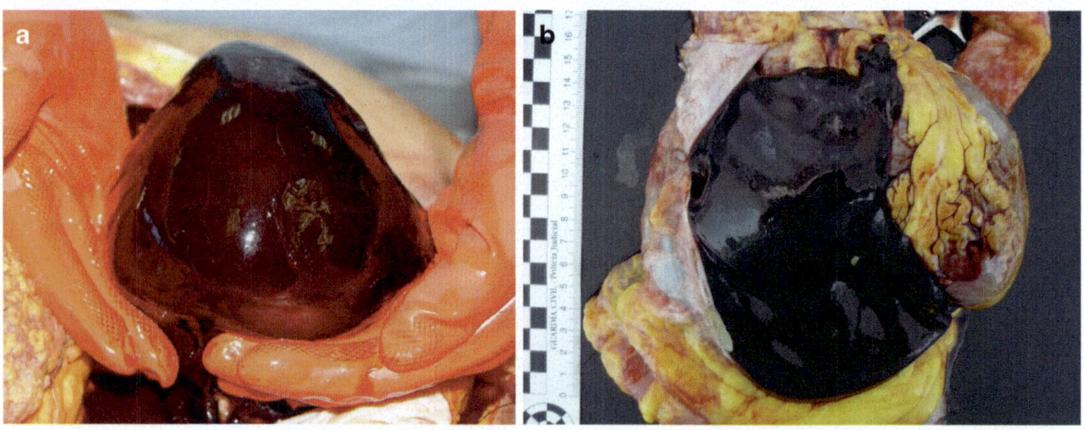

Fig. 4.5 *Cardiac tamponade due to hemopericardium*. (**a**) As the heart is removed, the blood clot adopts the shape of the pericardial sac (Courtesy of Dr. Juan Pedro Hernández Rincón. Institute of Legal Medicine, Murcia). (**b**) Sudden death by aortic dissection. An incision in the pericardial sac shows the pericardial cavity occupied by a large blood clot that displaces the heart

4.4 Pericarditis. Constrictive Pericarditis

The response of the pericardium to a lesion is an acute inflammatory reaction with fibrinous exudate (pericarditis), which in some instances can evolve into fibrous adhesions (Figs. 4.6 and 4.7) and septation of the effusion, constrictive fibrosis, dystrophic calcification and reactive mesothelial proliferation (in the latter case, a differential diagnosis with mesothelioma or metastatic adenocarcinoma can be difficult).

Pericarditis can be idiopathic or caused by infectious agents, by radiation, by trauma, iatrogenic (Fig. 4.8) or can be associated with an acute myocardial infarction, or secondary to systemic disease (rheumatoid arthritis, scleroderma, lupus, vasculitis, chronic renal insufficiency), endocrine disease (hypothyroidism, diabetes mellitus) or neoplasm. Macroscopic findings will depend upon the severity of the inflammatory process, the time from the onset and the causal agent.

Haemorrhagic pericarditis is generally produced by neoplastic infiltration (Figs. 4.14 and 4.15), infections (mainly tuberculosis), uraemia or trauma. Patients on anticoagulation treatments can develop haemorrhagic pericarditis with cardiac tamponade (Fig. 4.16).

Constrictive pericarditis is a chronic form of pericarditis whereby a fibrous or fibrocalcified thickening of one or both pericardial layers interferes with the proper filling of the heart chambers during diastole.

The clinical and pathological characteristics of the different types of pericarditis are summarised in Table 4.2.

Remarks

In pericarditis, a microscopic study should always include an analysis of the sinus node, since the pericardial inflammatory infiltrate can extend to the conduction system structures and provoke arrhythmias.

Table 4.2 Clinical and pathological characteristics of pericarditis

Viral pericarditis
Generally associated with myocarditis.
Findings are more microscopic than macroscopic (usually serohematic effusion).
Bacterial pericarditis (Figs. 4.9 and 4.10)
More frequent in children than in adults.
Thickening of both pericardial layers with fibrinopurulent exudate.
Fibrin and severe infiltration of polymorphonuclear leukocytes (neutrophils).
Tuberculous pericarditis (Figs. 4.11 and 4.12)
The most frequent form of pericarditis in underdeveloped countries. Also present in immunocompromised patients.
Acute phase similar to fibrinous or fibrinohaemorrhagic pericarditis. Granulomatous inflammation with/without granulomas in the subacute phase. Fibrous thickening, calcifications and constrictive pericarditis in the chronic phase.
Isolation of Koch's bacillus in the fluid or tissue is required.
Fungal or parasitic pericarditis
They are extremely rare. Macroscopic findings vary depending on the causal agent.
Pericarditis associated with AMI
It is usually associated with different degrees of fibrinous pericarditis. Adhesions between both pericardial layers in areas of transmural myocardial infarcts (microscopically: granulation tissue).
Uraemic pericarditis (Fig. 4.13)
Fibrinous pericarditis occurring in situations of renal insufficiency. The major complication, cardiac tamponade, appears in 17 % of the cases. Progression to constrictive pericarditis is infrequent.

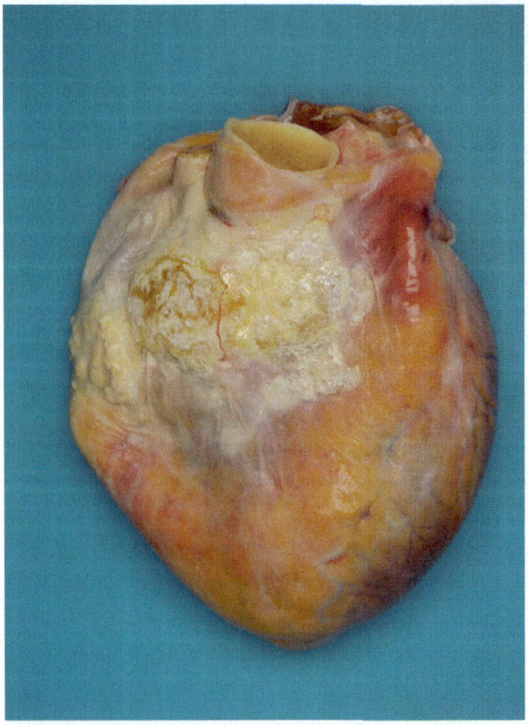

Fig. 4.6 *Pericardial fibrous plaque*. A 61-year-old woman who died suddenly at home from cardiac insufficiency. Thick pericardial whitish plaque, fibro-calcified in the atrium and right ventricle as a consequence of a previous pericarditis

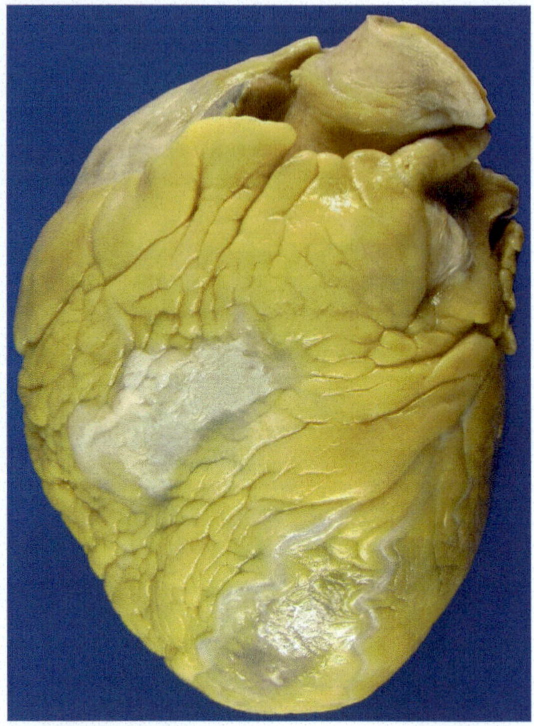

Fig. 4.7 *Pericardial fibrous plaque*. The most frequent pericardial lesion found at autopsy. Microscopically it is characterized by a thickened collagenous zone with lymphocytic infiltration in the basal area

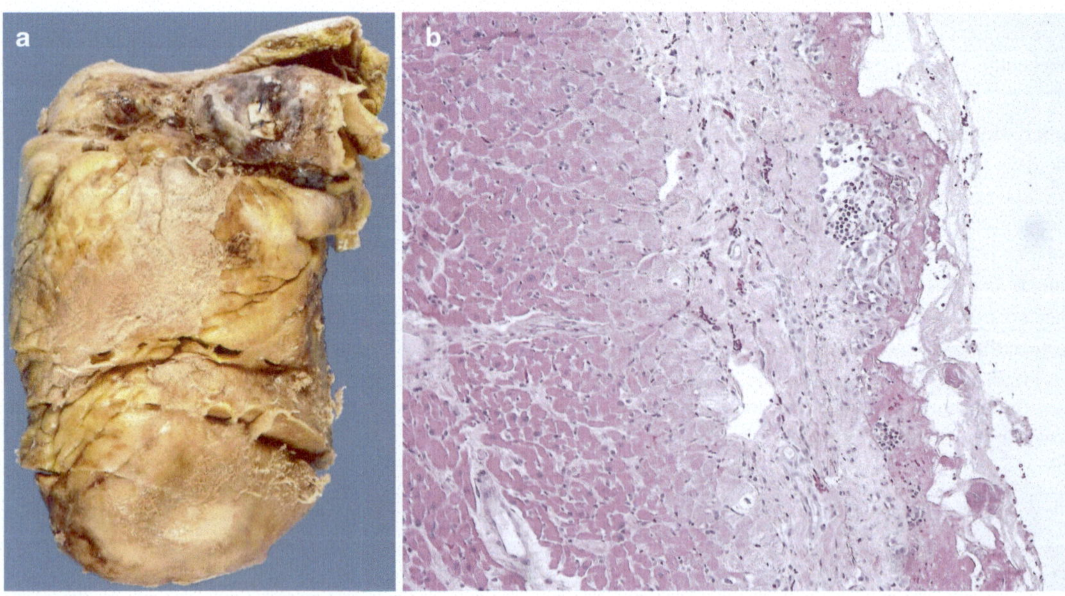

Fig. 4.8 *Post-surgical fibrinous pericarditis*. A 65-year-old woman who underwent surgery to replace the aortic valve. She died 10 days after intervention. (**a**) The visceral pericardium is covered by a mesh of fibrinous exudate. (**b**) Histologically, a superficial fibrin exudate over a layer of reactive mesothelial cells, with inflammatory polymorphonuclear leukocyte infiltration and pericardial fibrosis, is observed

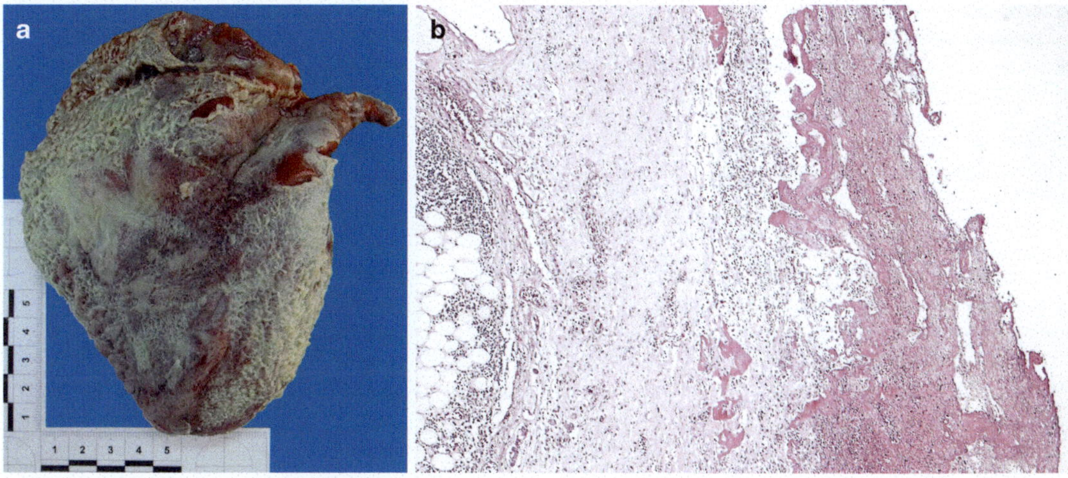

Fig. 4.9 *Purulent fibrinous pericarditis.* A 35-year-old male with advanced AIDS. Systemic infection with pericarditis and meningitis. (**a**) The cardiac surface is covered by a thick layer of a fibrinopurulent exudate appearing as a whitish granular-like material, classically termed "bread and butter" appearance, which extends between both pericardial layers. (**b**) Histologically, it corresponds to extensive fibrin and polymorphonuclear leukocyte exudates in superficial areas with oedema and active granulation tissue in depth

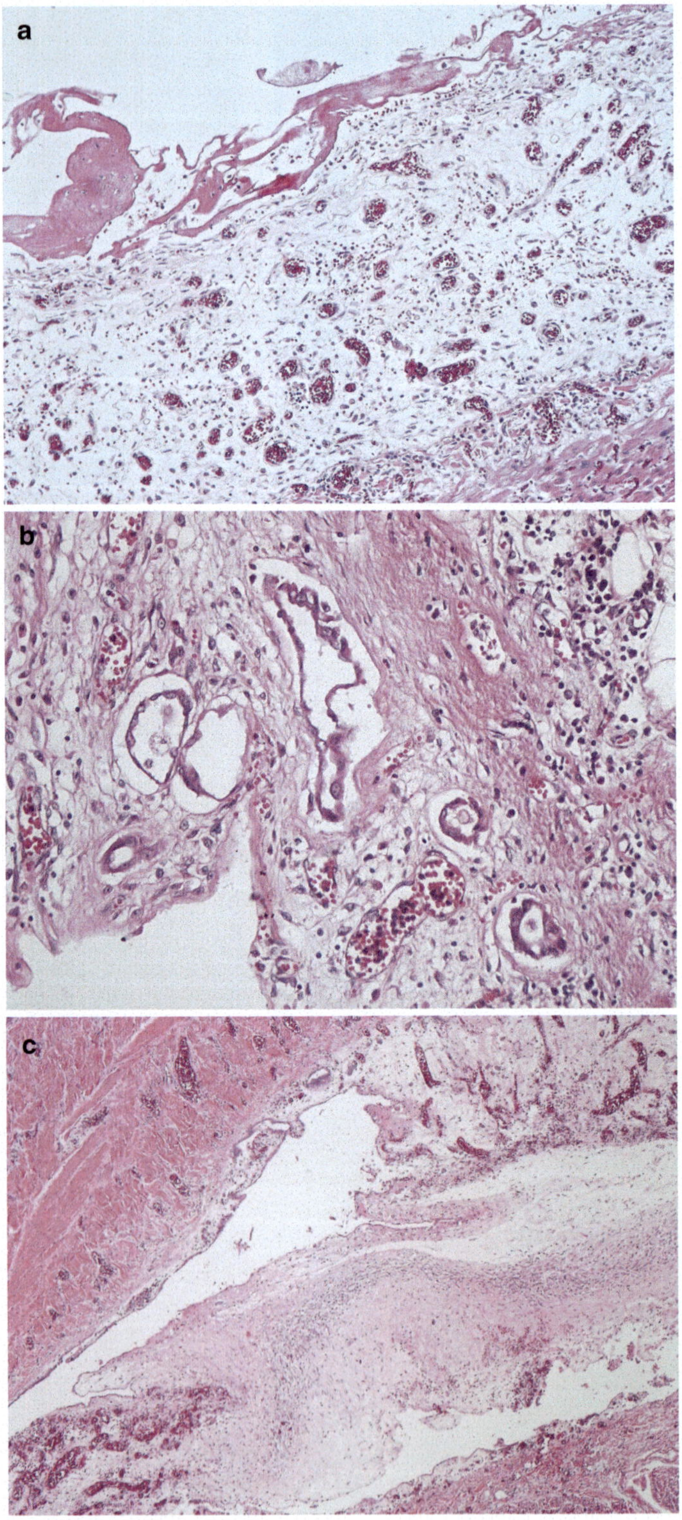

Fig. 4.10 *Evolution of fibrinous pericarditis*. Fibrinous perdicarditis can resolve or, as in many cases, evolve to form adhesions between the pericardial layers. (**a**) Granulation tissue with few inflammatory cells is formed under the fibrinous exudate. (**b**) The mesothelium regenerates and glandular-shaped inclusions can be observed. A differential diagnosis must be made with metastatic adenocarcinoma. (**c**) If the inflammatory process persists, fibrous adhesions may form between the two pericardial layers. In the image, the parietal fibrous pericardium is at the top and the visceral pericardium at the bottom

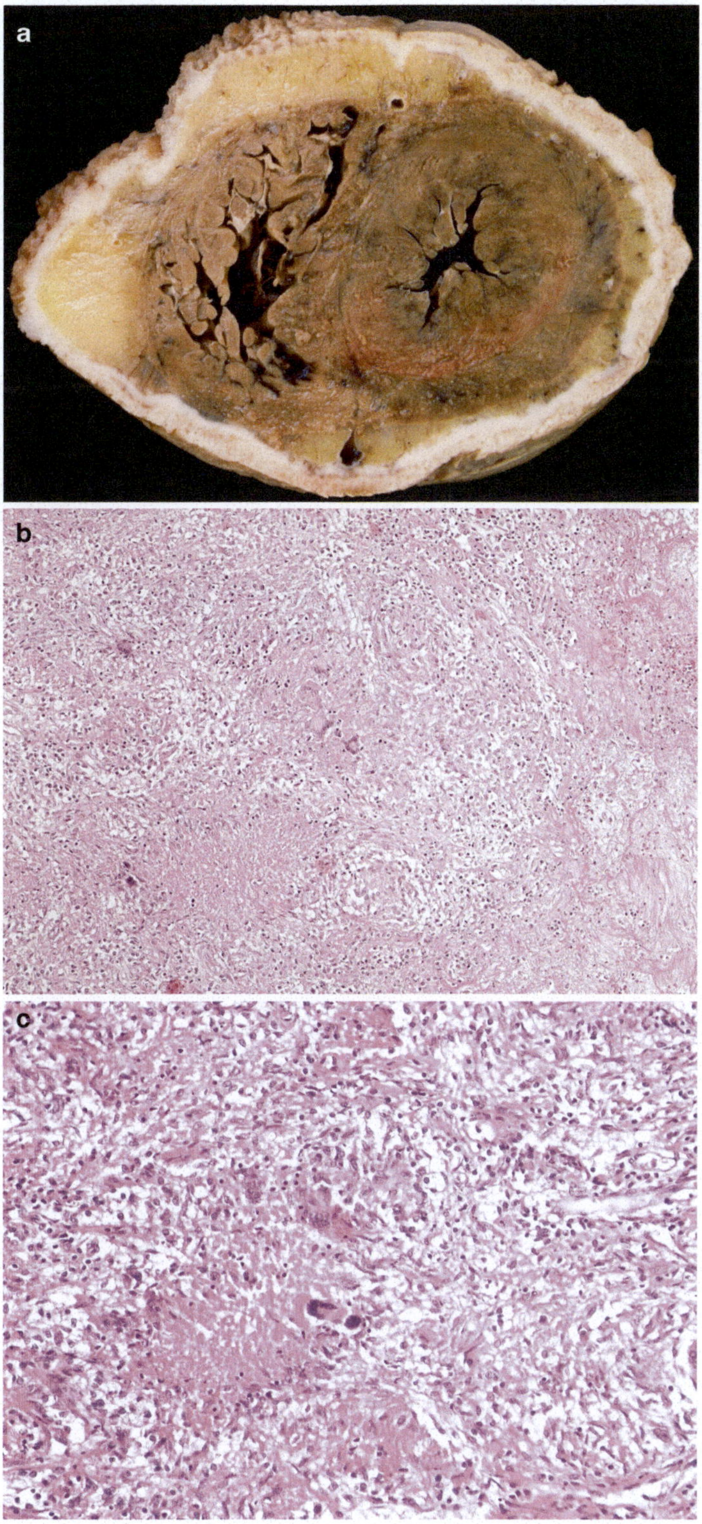

Fig. 4.11 *Constrictive pericarditis due to tuberculosis.* A 75-year-old male. (**a**) Transversal section corresponding to the apical-third portion of both ventricles. Very thickened pericardium with an irregular surface, forming a shield of 0.5 cm thickness surrounding the heart. (**b**) Microscopic study reveals an extensive inflammatory infiltrate in the visceral pericardium with necrotizing granuloma formation (with scarce acid-alcohol-resistant bacilli) surrounded by a leukocytic infiltrate with abundant fibrinous exudate on the surface. (**c**) Detail of a granuloma with central necrosis and giant and epithelioid cells in the periphery

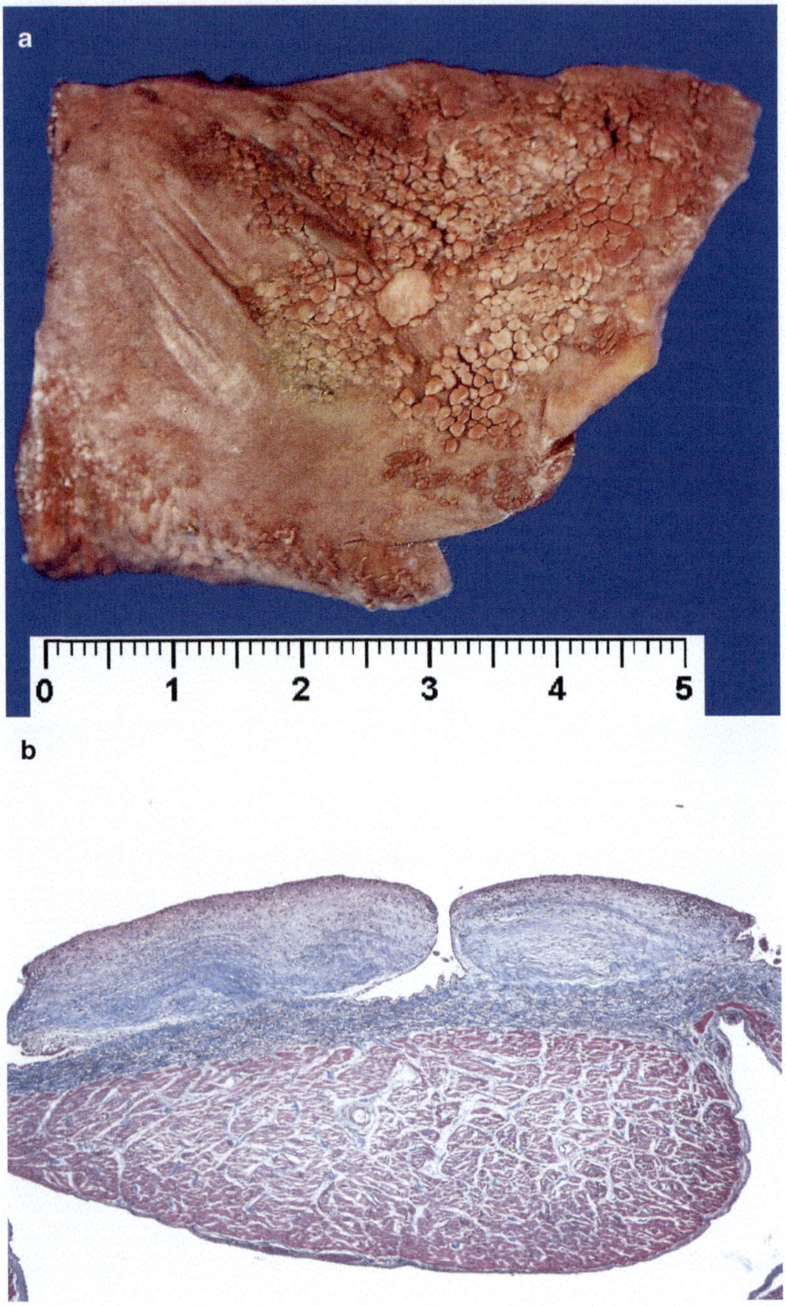

Fig. 4.12 *Villous appearance of the pericardium of the right atrium.* (**a**) The lateral wall of the right atrium occasionally presents a villous or slightly nodular appearance, referred to as "milk-spots". (**b**) Histologically, these nodules correspond to dense subepicardial collagen deposits with scarce lymphocytic inflammation

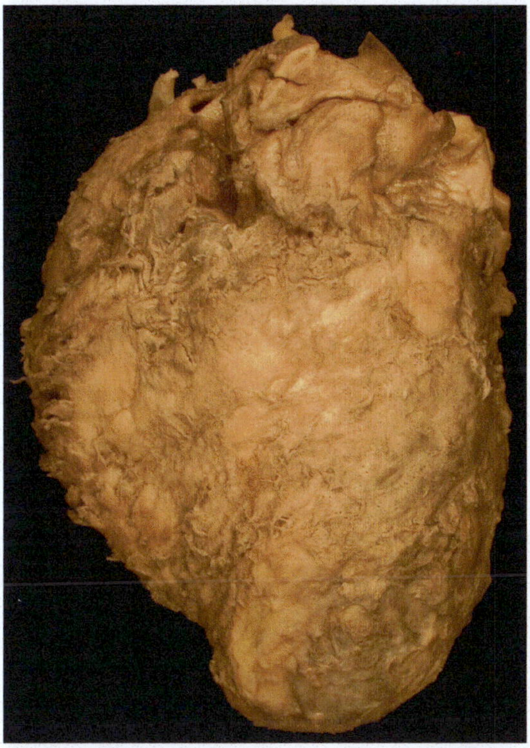

Fig. 4.13 *Uraemic pericarditis*. Fibrinous pericarditis with fibrinous exudate with the typical “bread and butter” appearance. A 42-year-old woman, with a history of chronic renal insufficiency and progressive dyspnea, who died at home

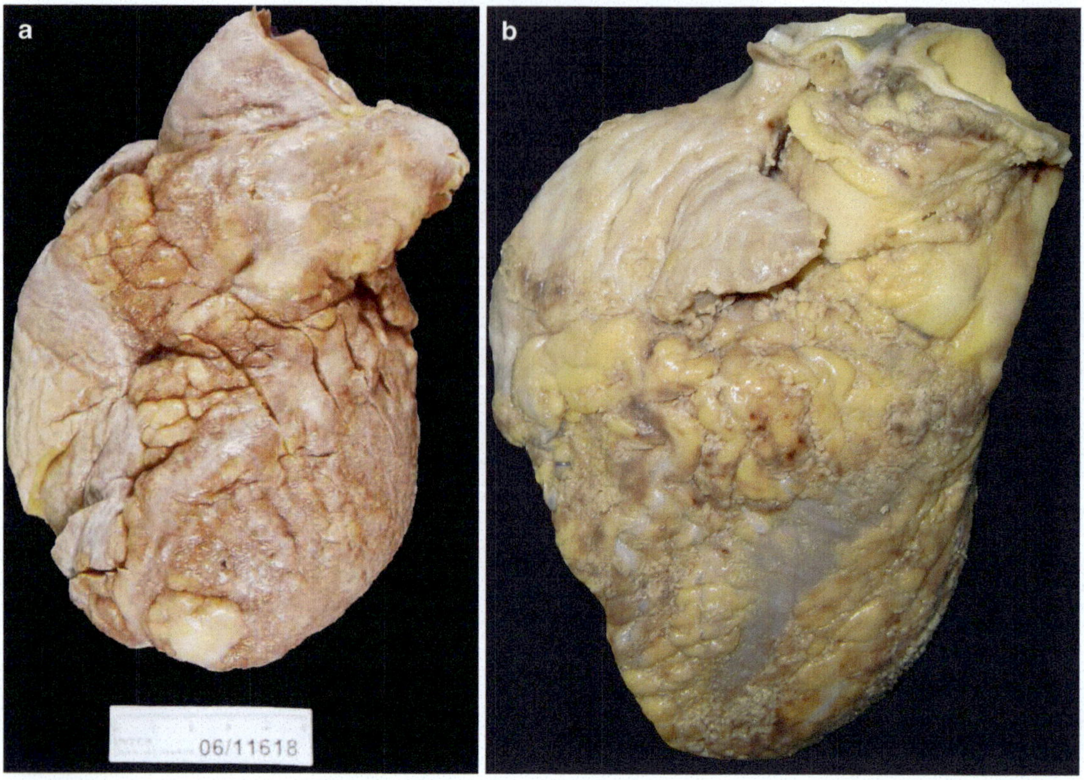

Fig. 4.14 *Carcinomatous pericarditis*. Carcinomatous pericarditis can offer several different macroscopic images. Histology is therefore essential to confirm a metastasis diagnosis. (**a**) Pericarditis with tumoural nodules in the pericardium. (**b**) Note the shaggy appearance of the pericardium, with haemorrhagic foci

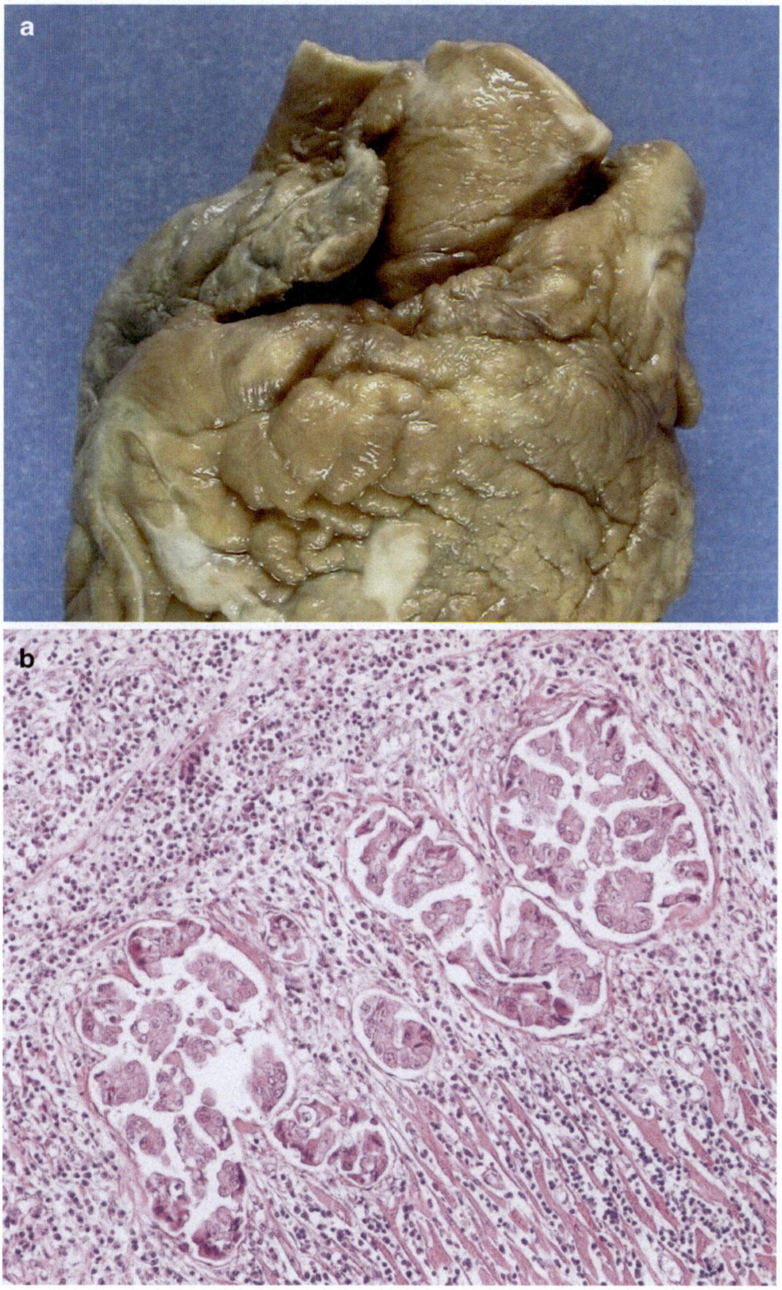

Fig. 4.15 *Carcinomatous pericarditis*. A 62-year-old male who died in the street. (**a**) Nodular epicardial surface. (**b**) Microscopically, the epicardial lymphatic vessels are occupied by adenocarcinoma emboli, surrounded by a leukocytic infiltrate compatible with carcinomatous lymphangitis

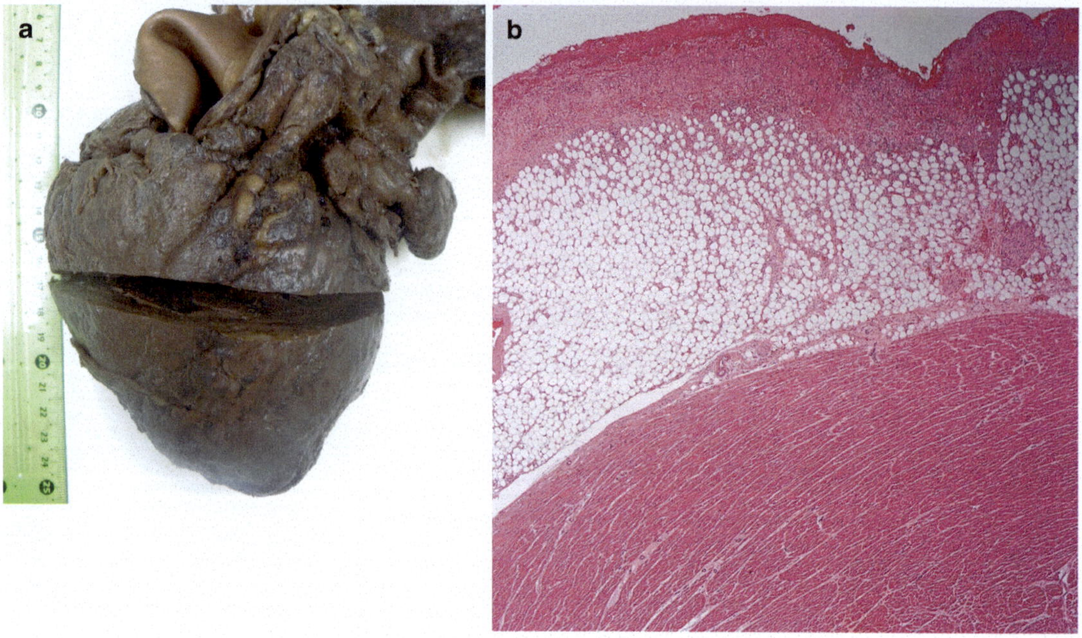

Fig. 4.16 *Fibrinohaemorrhagic pericarditis*. A 46-year-old male who died in the hospital from cardiac tamponade. (**a**) Heart of 570 g with a fibrinohaemorrhagic surface. (**b**) Microscopically, a thickening of the pericardium with haemorrhage and fibrosis can be appreciated (Courtesy of Dr. Ricardo González-Cámpora, Department of Pathology. HU Virgen Macarena, Sevilla)

Bibliography

Burke A, Tavora F. Pericardial disease. In: Practical cardiovascular pathology. Philadelphia: Wolters Kluwer/ Lipincott Williams & Wilkins; 2011. p. 270–80.

Butany J, Woo A. The pericardium and its diseases. In: Silver MD, Gotlieb AI, Schoen FJ, editors. Cardiovascular pathology. 3rd ed. New York: Churchill Livingstone; 2001. p. 375–401.

Dudzinski DM, Mak GS, Hung JW. Pericardial diseases. Curr Probl Cardiol. 2012;37:75–118.

Iacobellis G, Bianco AC. Epicardial adipose tissue: emerging physiological, pathophysiological and clinical features. Trends Endocrinol Metab. 2011;22:450–7.

Kosar F, Aksoy Y, Sahin I, Redil N. Pericardial hydatid cyst mimicking acute coronary syndrome. Tex Heart Inst J. 2005;32:570–2.

Maisch B, Seferovic P, Ristic A, Erben R, Rienmüller R, Adler Y, Tomkowski W, Thiene G, Yacoub M. Guidelines on the diagnosis and management of pericardial diseases executive summary the task force on the diagnosis and management of pericardial diseases of the European Society of Cardiology. Eur Heart J. 2004;25:587–610.

Mestres A, Toshani A, Hemdan A, Alewa A, Bernal J. Hydatid pericardial tamponade: a grape soup. Lancet. 2011;377:1862.

Peebles CR, Shambrook JS, Harden SP. Pericardial disease – anatomy and function. Br J Radiol. 2011;84(Spec No 3):S324–37.

Uzün I, Büyük Y, Pakiş I, Doğru A, Calk AU. Sudden death due to congenital pericardial defect: an autopsy case. Am J Forensic Med Pathol. 2008;29:242–4.

Ventura F, Neri M, De Stefano F, Fineschi V. Muerte súbita cardiaca en un caso de agenesia pericárdica completa no diagnosticada. Rev Esp Cardiol. 2010;63:1103–5.

Virmani R, Burke A, Farb A, Atkinson JB. Pericardial disease. In: Cardiovascular pathology, Major problems in pathology series, vol. 40. 2nd ed. Philadelphia: Saunders Company; 2001. p. 386–423.

Coronary Atherosclerotic Disease

5

Joaquín S. Lucena, Pedro M. Garamendi,
and Luís Nombela-Franco

Contents

J.S. Lucena (✉)
Department of Forensic Pathology,
Institute of Legal Medicine, Seville, Spain

Forensic Medicine, University of Cadiz, Cadiz, Spain
e-mail: joaquin.lucena@uca.es

P.M. Garamendi
Department of Forensic Pathology,
Institute of Legal Medicine, Huelva, Spain

L. Nombela-Franco
Cardiovascular Institute, Hospital Clínico San Carlos,
Universidad Complutense de Madrid, Madrid, Spain

5.1 Introduction

Atherosclerosis is a chronic disease characterized by abnormal stiffness or loss of elasticity in the elastic and muscular medium- and large-sized arteries. Although arteriosclerosis is regarded etiologically as a multifactorial disease, the most frequent cause is the deposition or infiltration of lipid-rich material (atheroma plaques) or atherosclerosis in the arterial wall, induced by local and systemic factors. Hypercholesterolemia and especially high plasma concentrations of low density lipoproteins (LDL) are a significant risk factor for the premature onset of atherosclerosis and ischaemic heart disease.

Interactions between the arterial intima and blood components play a key role in the formation of intimal lesions that ultimately will become a fibro-lipid atheroma plaque. As the plaque develops there is a progressive reduction of the arterial lumen, with invasion of the medial layer. Depending on the location of the lesion, different organs can be affected (heart, brain, kidney, etc.). More specifically, if the affected vessel is one of the coronary arteries, a reduction in the blood flow to the myocardium will occur, leading to acute or chronic ischaemia, depending on the speed of the process.

Coronary atherosclerotic disease (CAD) is one of the most frequent lesions that a forensic pathologist encounters in his daily practice, either as a cause of death or as an incidental finding. In 99 % of the cases, CAD is the origin of ischaemic

J.S. Lucena et al. (eds.), *Clinico-Pathological Atlas of Cardiovascular Diseases*,
DOI 10.1007/978-3-319-11146-9_5, © Springer International Publishing Switzerland 2015

heart disease. Complications of an atheroma plaque, such as mural or occlusive thrombosis due to rupture or erosion, represent the major substrate for acute coronary syndromes (ACS) and sudden death (SD). One half of the patients with ischaemic heart disease die suddenly and SD is the first manifestation of CAD in 25 % of the cases.

In the well-known study of Davies (1992), using *postmortem* arteriography and histology in a series of 168 cases of SD caused by CAD, occlusive thrombosis was observed in 30 % of cases and mural thrombosis in 43 %. A solitary plaque fissure accounted for 8 % of the cases, and acute lesions could not be found in 19 %. According to these results, at least 73 % of SD caused by ACS is associated with an unstable plaque and thrombosis, which represents the anatomopathological substrate that predisposes to the onset of an episode of acute myocardial ischaemia.

5.2 Characteristics and Types of Atheroma Plaques. Plaque Evolution

The formation of an atheroma plaque is the fundamental lesion in CAD. Classically, as Virchow stated in 1858, the term atheroma refers to a dermoid cyst, in other words to an encapsulated fatty mass. The atheroma plaque increases in size with time, causing a progressive stenosis of the vessel lumen.

Several morphological classifications exist to describe the evolution of an atheroma plaque. Among them, we consider that proposed by Virmani and collaborators in 2000 to be the simplest and most complete description. This is a simplification of the classification scheme from the American Heart Association (1994, 1995) and ranges from initial or pre-atherosclerotic lesions (intimal hyperplasia and intimal xanthoma), to atherosclerotic lesions (pathological intimal hyperplasia, fibrous cap atheroma, calcified nodule and fibrocalcific plaque) (Table 5.1).

At the level of the coronary arteries, these lesions can compromise one, two or the three epicardial coronary arteries. Autopsy findings are very diverse since the involvement of just one vessel ranges from 15 to 26 %, of two vessels between 27 and 39 % and of three vessels from 33 to 62 %.

In deceased young people, only one vessel tends to be affected and the lesion is located mainly in the proximal third of the left anterior descending (LAD) coronary artery, just distal to the left coronary trunk bifurcation. There is a haemodynamic explanation for this favoured location since at this point the speed of the blood flow is critically reduced and the vessel wall experiences more traction. In contrast, CAD in adults and the elderly usually affects several of the coronary arteries, the LAD coronary artery, the right coronary artery (RCA) and the left circumflex artery (LCx).

Plaques seen in young people who die from sudden cardiac death (SCD) due to CAD consist of a combination of cellular and dense fibrous tissues in the intima (fibrocellular plaque) (Figs. 5.8, 5.9, 5.10, 5.11, 5.12, 5.13, and 5.14) (Table 5.2). In contrast, plaques examined in adults and the elderly are characterized by a large lipid core (fibrolipid plaque) that can calcify and lead to a fibrocalcific plaque (Fig. 5.7).

The degree of calcification, determined by coronary CT angiography, has been used as a marker for atherosclerosis. These studies show a strong correlation between the levels of calcium in the coronary plaque and the risk of suffering cardiovascular events. Nevertheless, the presence of calcification is a poor predictor of the degree of coronary stenosis in a specific coronary segment.

Table 5.1 Morphologic evolution of a coronary atherosclerotic lesion

	Description	Thrombosis
Non-atherosclerotic Intimal Lesions		
Intimal hyperplasia	Accumulation of smooth muscle cells in the intima, in the absence of lipids or macrophage foam cells (Fig. 5.1).	Absent
Intimal xanthoma (fatty streak)	Intimal accumulation of foam cells without a necrotic core or fibrous cap. These lesions usually regress (Fig. 5.2).	Absent
Progressive atherosclerotic lesions		
Pathologic intimal hyperplasia *(Juvenile atheromatosis)*	Smooth muscle cells in a matrix rich in proteoglycans with areas of extracellular lipid accumulation without necrosis (Fig. 5.3).	Absent
Erosion	Luminal thrombosis in direct contact with the intima over an area with no endothelium. Plaque with the same characteristics as above.	Thrombus mostly mural and non-occlusive
Fibrous cap atheroma *(Stable plaque)*	Well-defined necrotic core with overlying fibrous cap (Figs. 5.4 and 5.5).	Absent
Erosion	Luminal thrombosis in direct contact with the intima over an area with no endothelium, but without communication with the necrotic core. Plaque with the same characteristics as above.	Thrombus mostly mural and non-occlusive
Thin fibrous cap atheroma *(Vulnerable plaque)*	Thin fibrous cap infiltrated by macrophages and lymphocytes, with few smooth muscle cells and an underlying necrotic core (Fig. 5.6).	Absent. Intraplaque haemorrhage may be present
Plaque rupture	Fibroatheroma with fibrous cap rupture. Occlusive thrombosis in communication with the necrotic core	Thrombus, generally occlusive
Calcified nodule	Calcified nodule which protrudes through the fibrous cap to the lumen. There is neither endothelium nor inflammatory infiltrates in that area.	Thrombus usually non-occlusive
Fibrocalcific plaque	Collagen-rich plaque with significant luminal stenosis. Lesions usually contain large areas of calcification with few inflammatory cells. A necrotic core may be present (Fig. 5.7).	Absent

Modified from the classification from AHA and Virmani et al. (2000)

Table 5.2 Clinico-pathological correlation of coronary atherosclerosis

Stable angina
Stenosis greater than 75 % of the lumen.
Unstable angina
Plaque rupture with mural thrombus.
Acute infarction
Plaque rupture and occlusive thrombus.
Sudden death
One or more of the following lesions can be found:
Stenosis greater than 75 % in one or more of the main coronary artery trunks.
Rupture or erosion of the plaque with mural or occlusive thrombosis.
Acute myocardial infarction.
Scarring from a healed myocardial infarction.

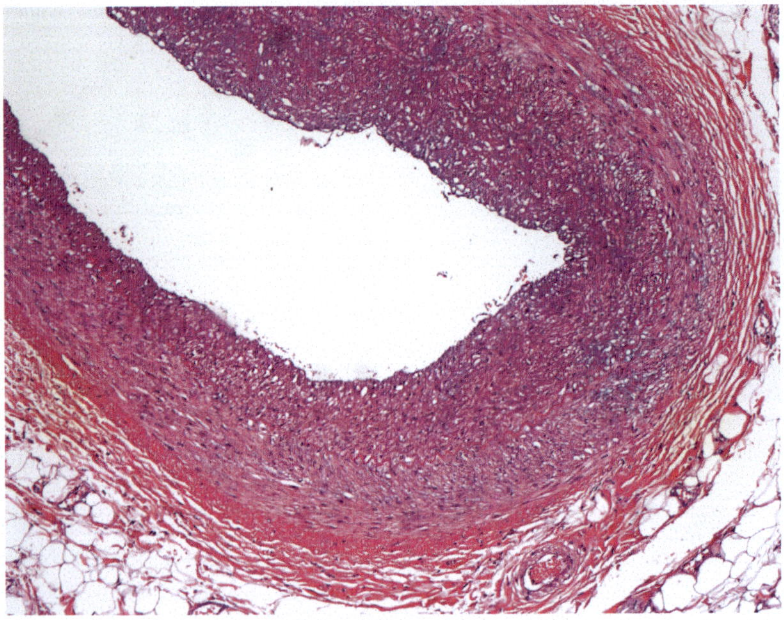

Fig. 5.1 *Intimal hyperplasia*

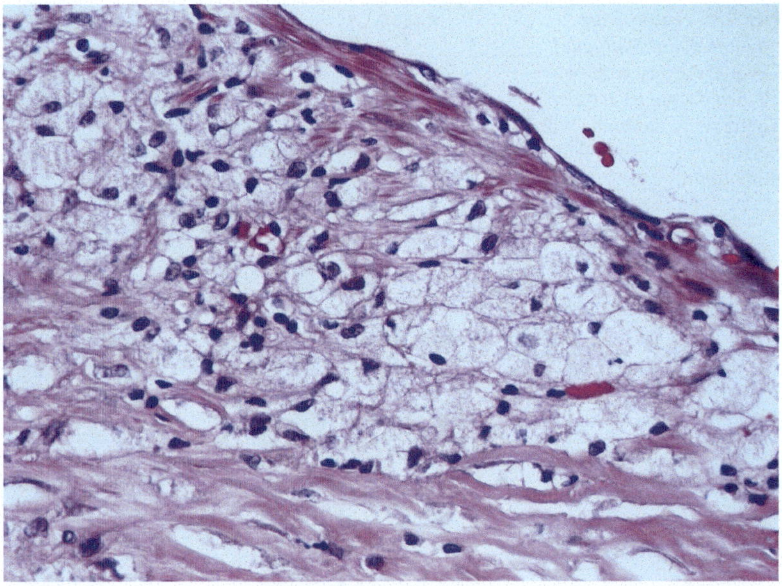

Fig. 5.2 *Intimal xanthoma*

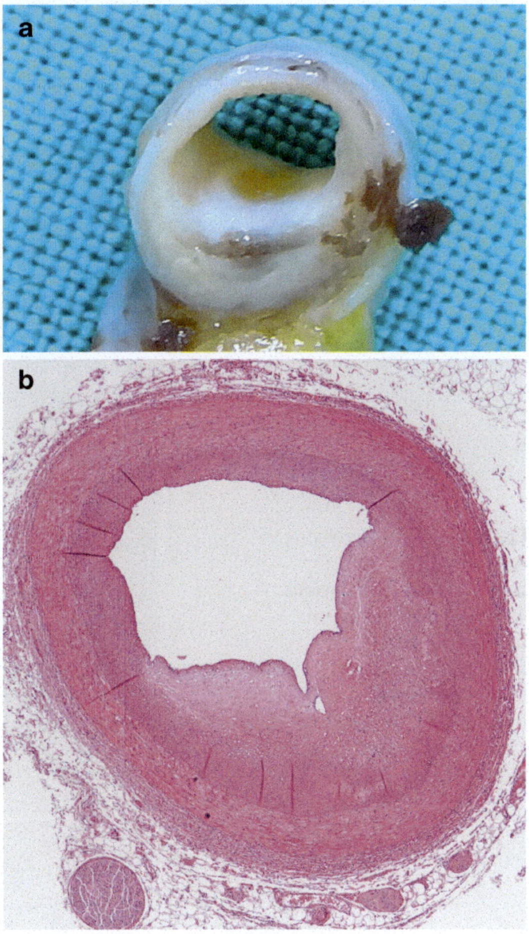

Fig. 5.3 *Pathological intimal hyperplasia.* (**a**) Macroscopic image. (**b**) Microscopic image

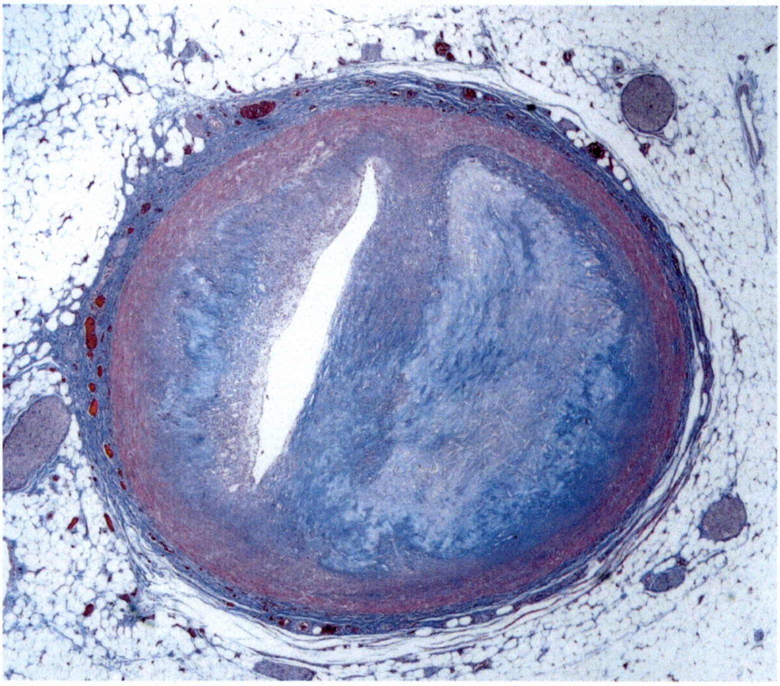

Fig. 5.4 *Fibrous cap atheroma (stable plaque).* Coronary artery with two plaques facing each other. To the right, atheroma with necrotic core surrounded by a thick fibrous cap

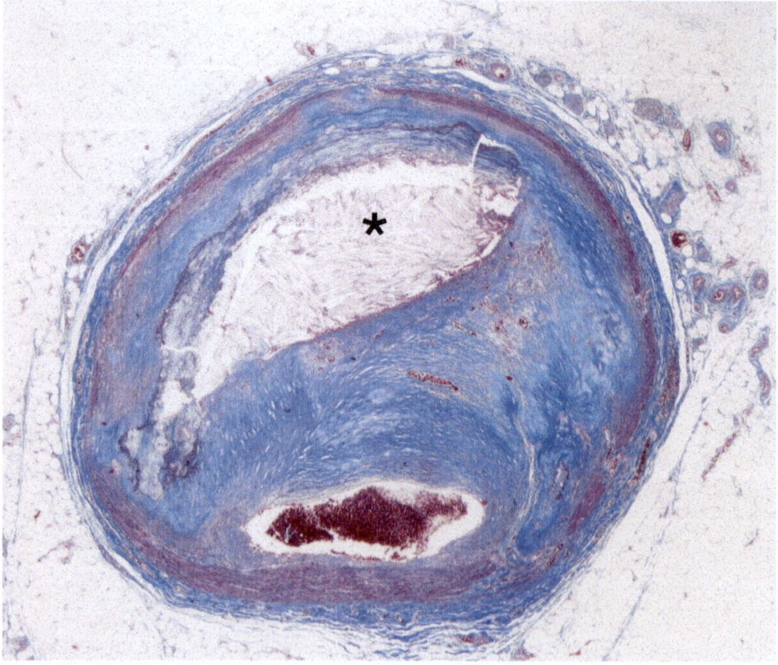

Fig. 5.5 *Fibroatheroma (stable plaque).* More developed atheroma plaque, with a large necrotic core focally calcified (*) containing numerous cholesterol crystals and overlaid by a thick fibrous cap. Lumen reduction is >75 %

Fig. 5.6 *Atheroma with thin fibrous cap (vulnerable plaque).* Necrotic core protruding towards the lumen, overlaid by a very thin yet still complete fibrous cap

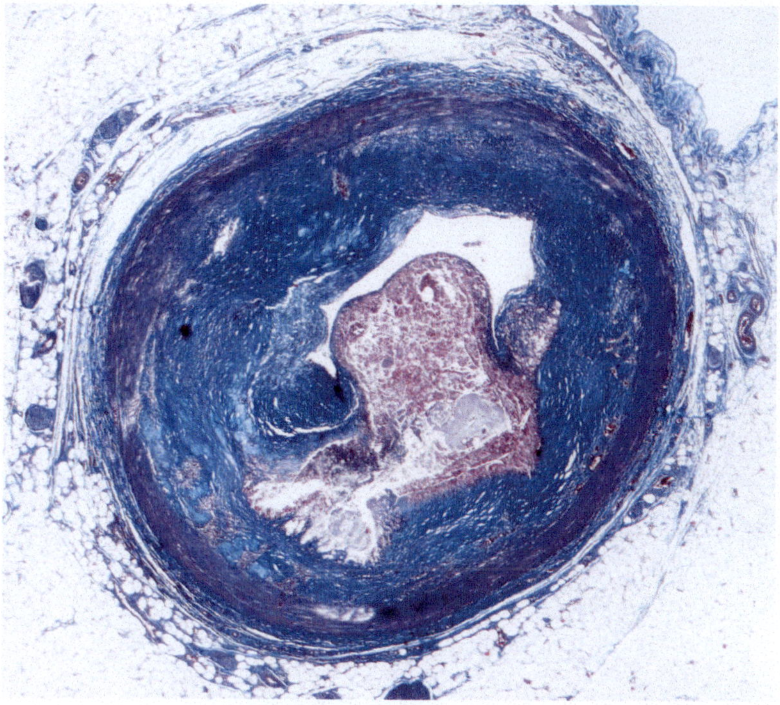

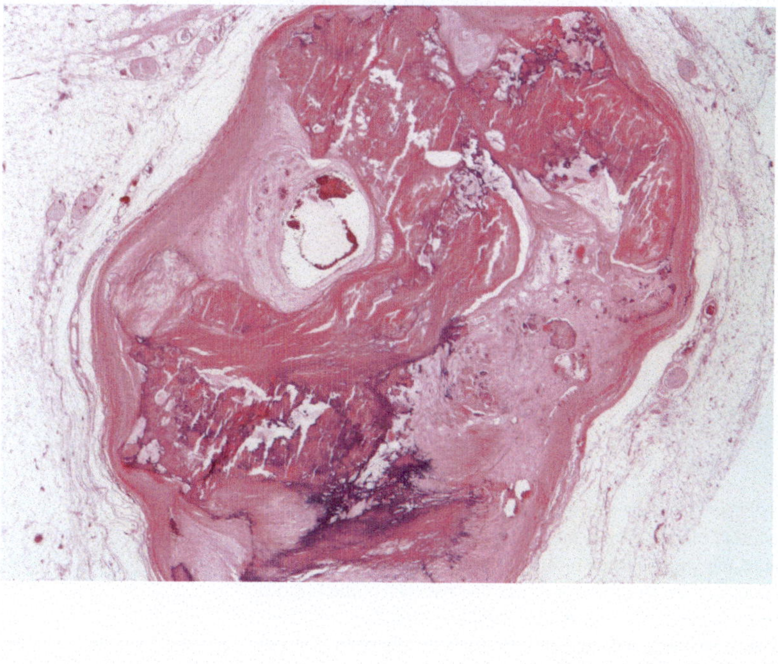

Fig. 5.7 *Fibrocalcific plaque.* Almost circumferential calcific plaque that causes a lumen reduction >85 %

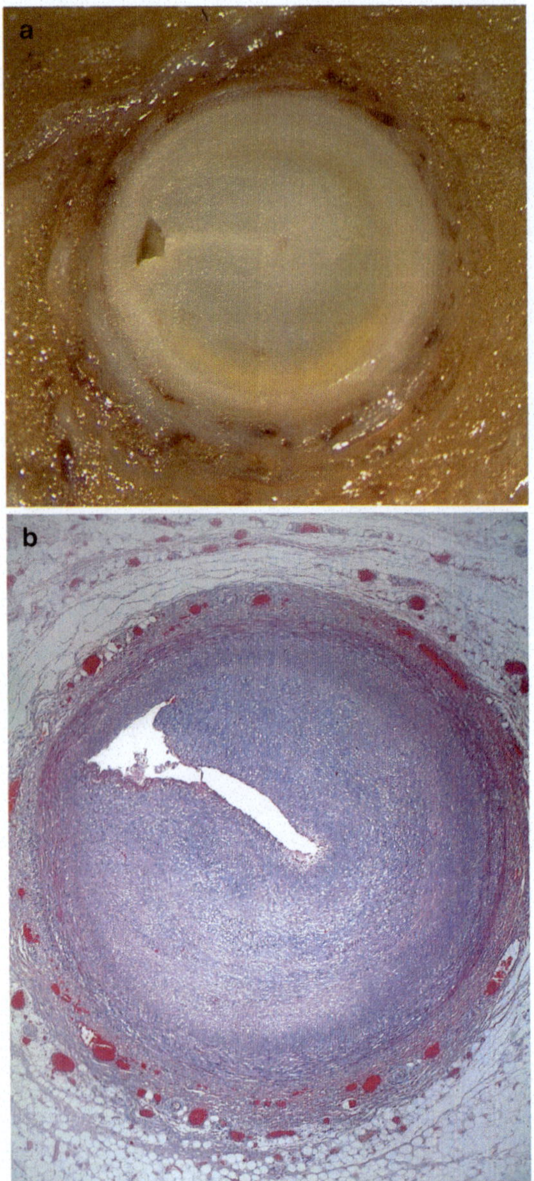

Fig. 5.8 *Fibrocellular plaque.* Fibrotic plaque in a 31-year-old male who died suddenly whilst at rest. (**a**) A whitish plaque with a fibrotic appearance can be appreciated, causing a lumen reduction greater than 90 %. A very small yellowish fatty streak can be observed at the bottom of the image. (**b**) Significant intimal hyperplasia consisting of smooth muscle cells with little fibrosis. A small mural thrombus due to plaque erosion can be appreciated in the vessel lumen (Masson's trichrome)

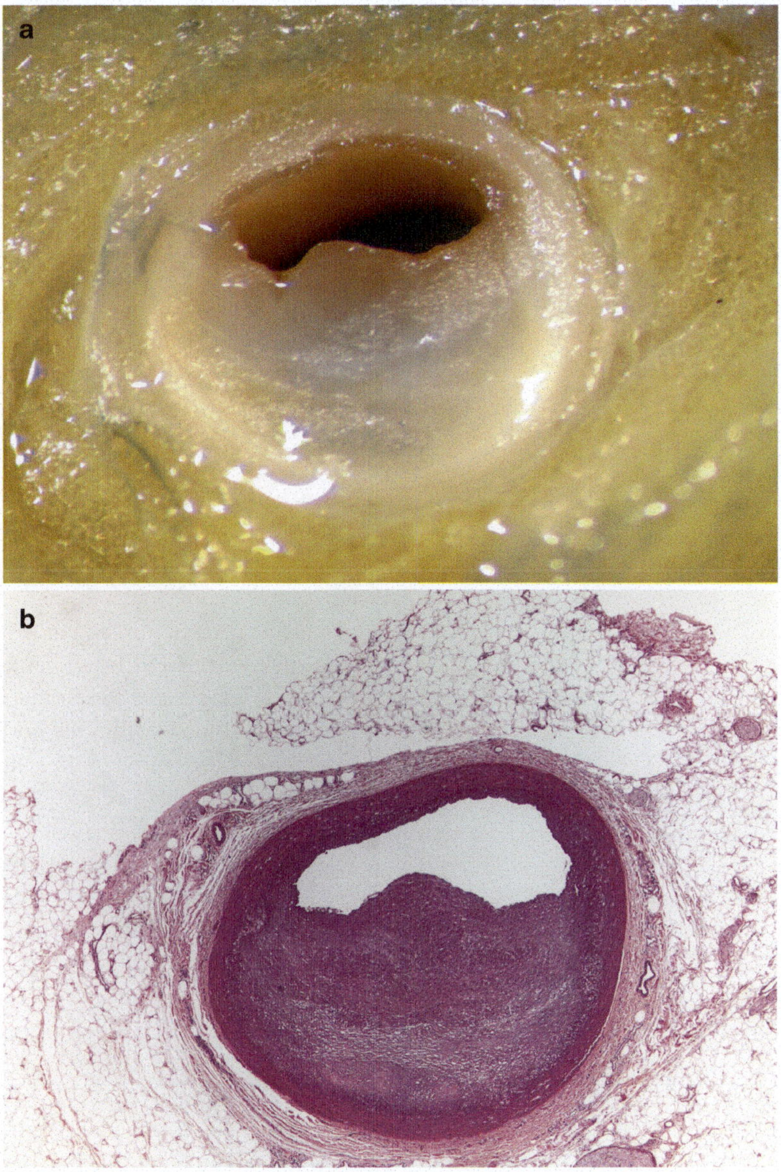

Fig. 5.9 *Pathological intimal hyperplasia.* A 33-year-old woman with previous personal and familial history of hypercholesterolaemia, who died whilst sleeping. (**a**) A whitish plaque can be appreciated in the proximal third of the LAD, which causes luminal stenosis greater than 75 %. (**b**) Histologically, pathological intimal hyperplasia with a macrophage-formed small lipid core can be observed

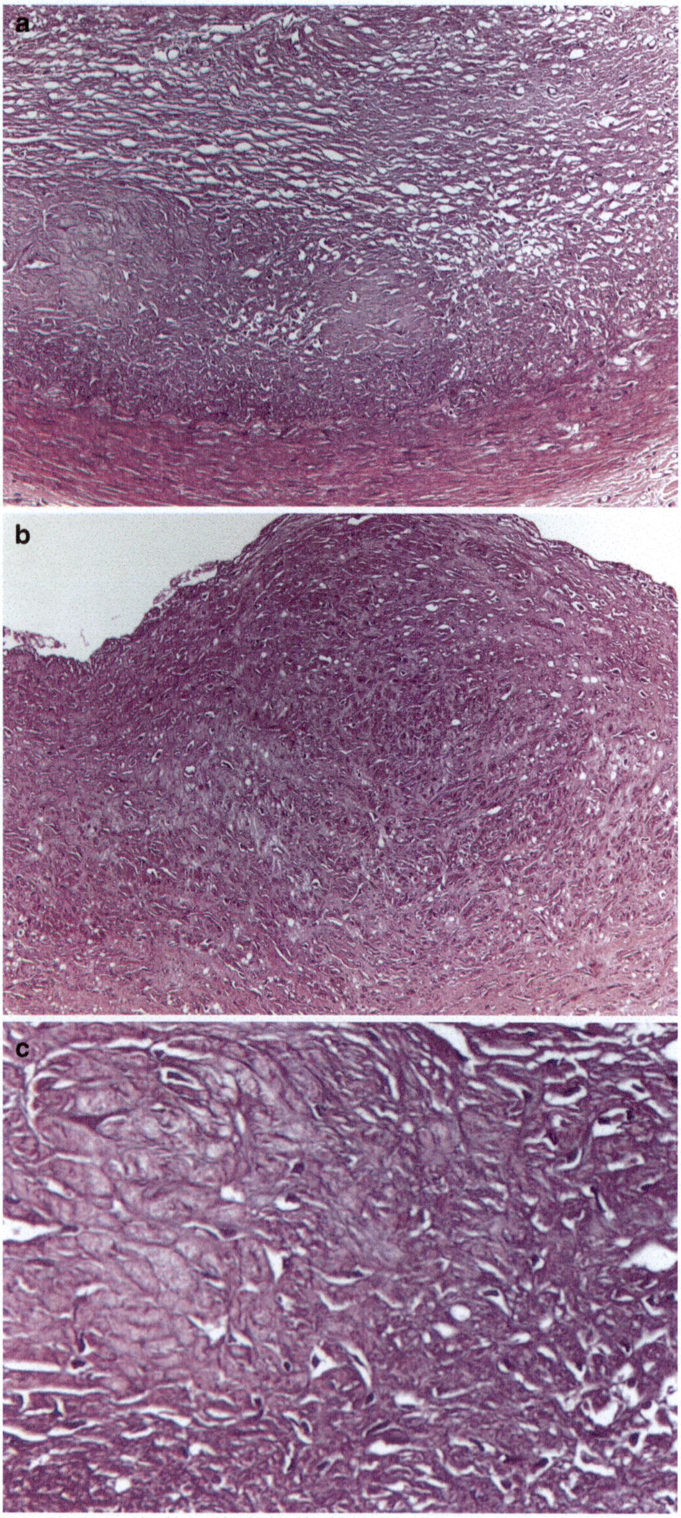

Fig. 5.10 (**a**) Detail of the medial layer with a small lipid core. (**b**) Detail of intimal hyperplasia. (**c**) Higher magnification of the lipid cells and some apoptotic smooth muscle cells

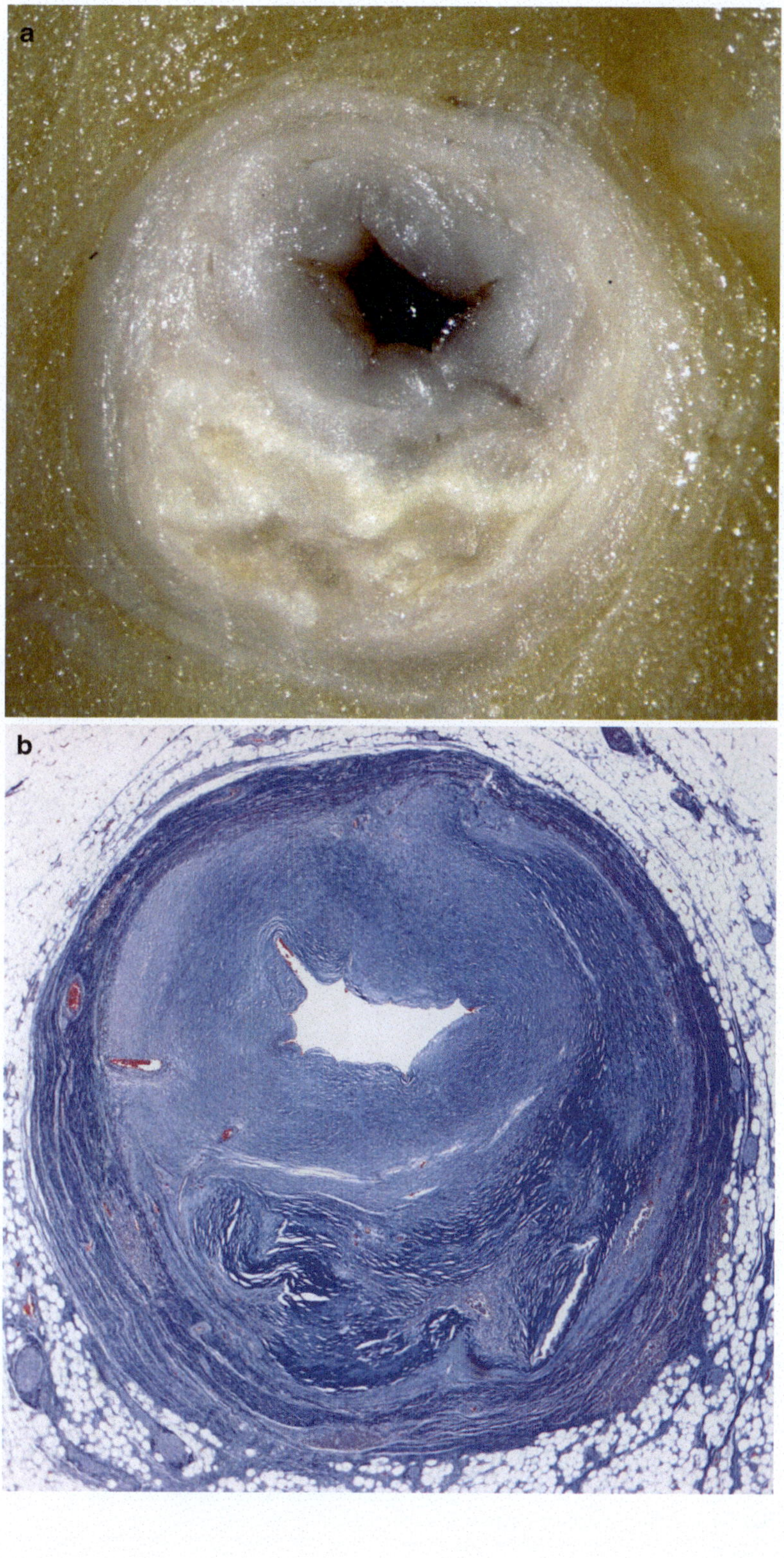

Fig. 5.11 A 32-year-old male with a history of cocaine abuse who died suddenly at home. A toxicological analysis excluded the presence of cocaine. (**a**) Plaque in the RCA with a significant lipid core and severe luminal stenosis. (**b**) Microscopic image (Masson's trichrome)

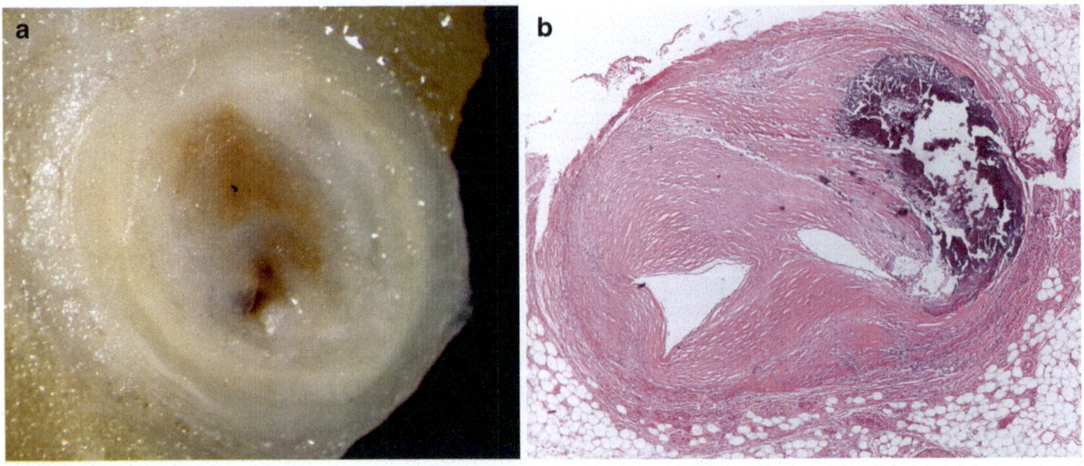

Fig. 5.12 A 35-year-old male with a history of morbid obesity, dyslipidemia and heavy smoking who died suddenly whilst having sexual intercourse. (**a**) Luminal stenosis of around 95 % in the LAD. (**b**) Microscopic image, where a calcific plaque with haemorrhage, thick fibrous cap and severe stenosis can be observed

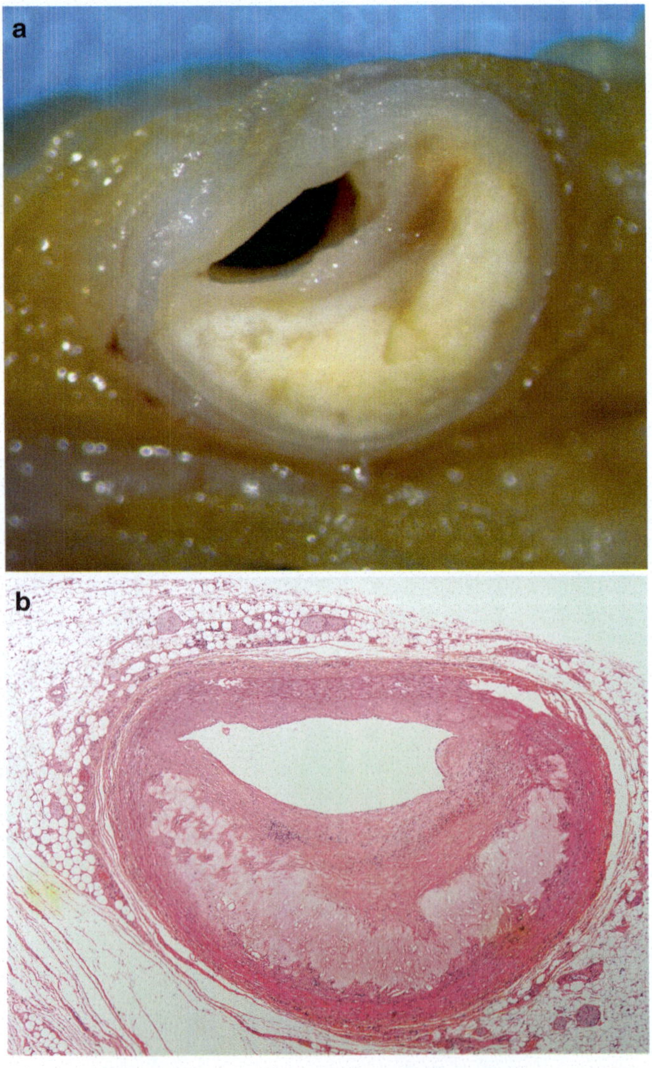

Fig. 5.13 A 46-year-old woman who died suddenly whilst sleeping. (**a**) Plaque with lipid appearance that causes luminal stenosis greater than 75 % in the LAD. (**b**) Fibrolipid plaque with large necrotic core and thick fibrous cap (stable plaque)

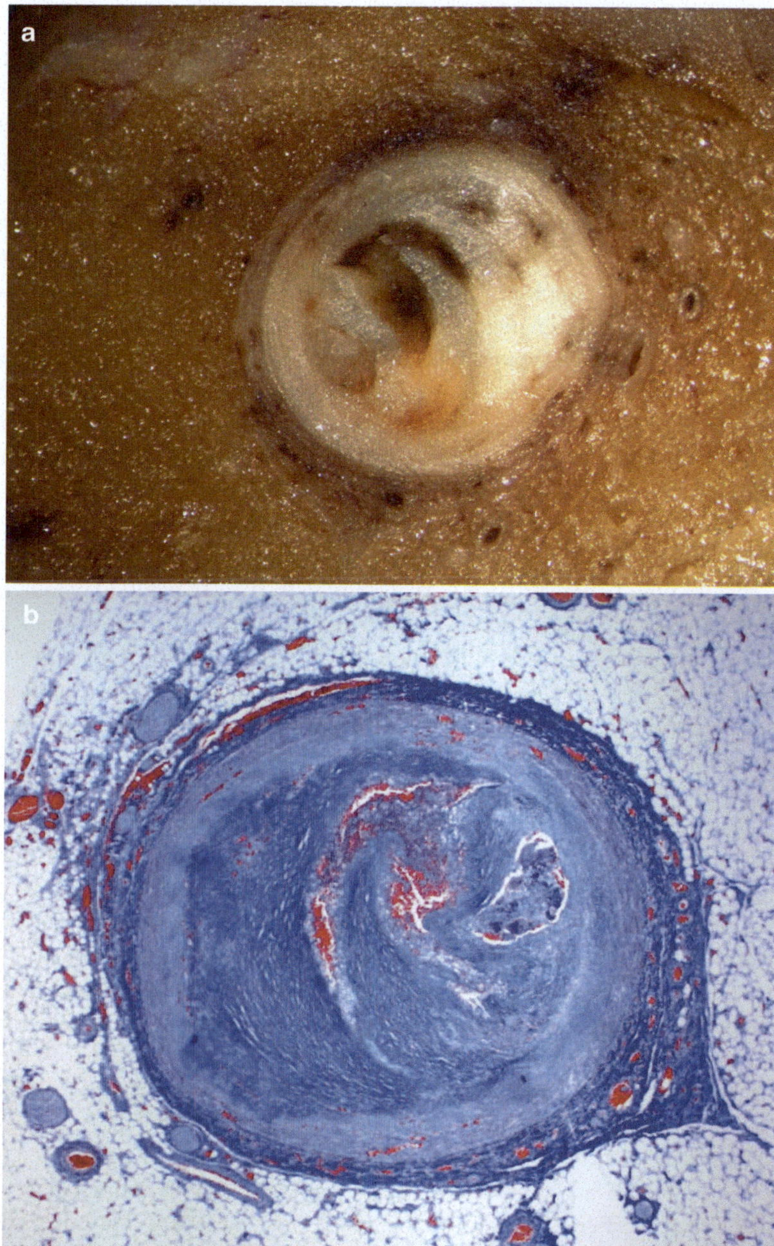

Fig. 5.14 A 51-year-old male with a history of schizophrenia who died suddenly whilst sleeping. (**a**) A fibrous plaque can be observed in the proximal tract of the LCx artery which causes luminal stenosis of around 90 %. The RCA also presented with a similar type of plaque (two-vessel disease). (**b**) Histological image of an occlusive fibrous plaque with haemorrhagic foci (Masson's trichrome)

5.3 Atheroma Plaque Complications

As previously mentioned, an atheroma plaque, as an encapsulated fatty mass, is analogous to an abscess contained within a fibrous capsule, and as such, can suffer erosion or rupture. Erosion or rupture of the fibrous cap exposes the atheromatous material to the flowing blood components, initiating the process of platelet aggregation and coagulation that will ultimately lead to thrombus formation.

To fully understand the complications of the atheroma plaque it is important to mention the concept of vulnerable or unstable plaque, a term that has been used to define those atherosclerotic plaques that are especially prone to rupture. The vulnerable plaque has a thin fibrous cap overlying a large lipid core (Fig. 5.6). Accordingly, the mechanisms of plaque rupture or erosion are related to thinning and inflammation of the fibrous cap. Both erosion and rupture of a plaque are the basis for the development of a platelet thrombus that can be either mural or occlude the entire vessel lumen (occlusive).

The percentage of coronary thrombosis in different autopsy series studying ischaemic sudden death ranges from 20 to 70 %. These wide variations can be due to the population studied (timeframe between the onset of symptoms and death, presence of acute myocardial infarction, type of prodromic symptoms) and the type of pathological examination performed (macroscopic or histologic). Based on the experience from the National Institute of Toxicology and Forensic Sciences in Madrid, thrombosis was observed in 37 % of the cases of ischaemic heart disease; however, this value increased to 56 % in people under 40 years of age.

Coronary thrombosis can be secondary to three distinct processes:

(a) *Plaque erosion*: Characterized by acute thrombosis in direct contact with the intima. It is more frequent in people aged 30–50 years, with no gender preference, and is associated with tobacco, especially in premenopausal women, although there is also a heritable component. Generally the process of plaque erosion occurs in stenosis of less than 70 % of the lumen (Figs. 5.15, 5.16, 5.17, 5.18, 5.19, and 5.20). It is observed in 70–75 % of SCD due to coronary thrombosis.
(b) *Plaque rupture*: Characterized by an interruption in the fibrous cap with acute thrombosis in continuity with the underlying lipid core. According to some authors, plaque rupture is present in 25–30 % of SD that occurs from coronary thrombosis, and is more frequent in males than in females between the ages of 35–80 years. It tends to be associated with dyslipidemia, diabetes and tobacco (Figs. 5.21 and 5.22).
(c) *Calcified nodule*: More infrequent than the two previous processes. It refers to a heavily-calcified plaque in which there is a rupture of the fibrous cap and the thrombus is associated with a protruding and calcified nodule. It appears in the elderly, and it is more common in men than in women. This lesion, more frequent in the carotid arteries, tends to be related with the onset of intraplaque haemorrhage.

It is important to be aware that in up to 50 % of the cases of SCD, coronary thrombosis settles in plaques with less than 75 % luminal stenosis. On the other hand, it has also been demonstrated that in 10 % of SCD from CAD, there is neither necrotic core nor fibroatheroma formation, and this situation is much more frequent in young people.

Other lesions may occur in the atheroma plaques, not related to thrombosis such as intraplaque haemorrhage (Figs. 5.23 and 5.24), healed erosion/rupture and recanalization of an old occlusive thrombus (Figs. 5.25 and 5.26).

An uncommon complication of coronary atherosclerosis is the formation of an aneurysm. The term coronary aneurysm refers to the dilation of a coronary segment of more than 1.5 times the diameter of the normal adjacent vessel. The prevalence of coronary aneurysms, based on angiography studies, ranges from 0.3 to 5.3 % and the causes are multifactorial:

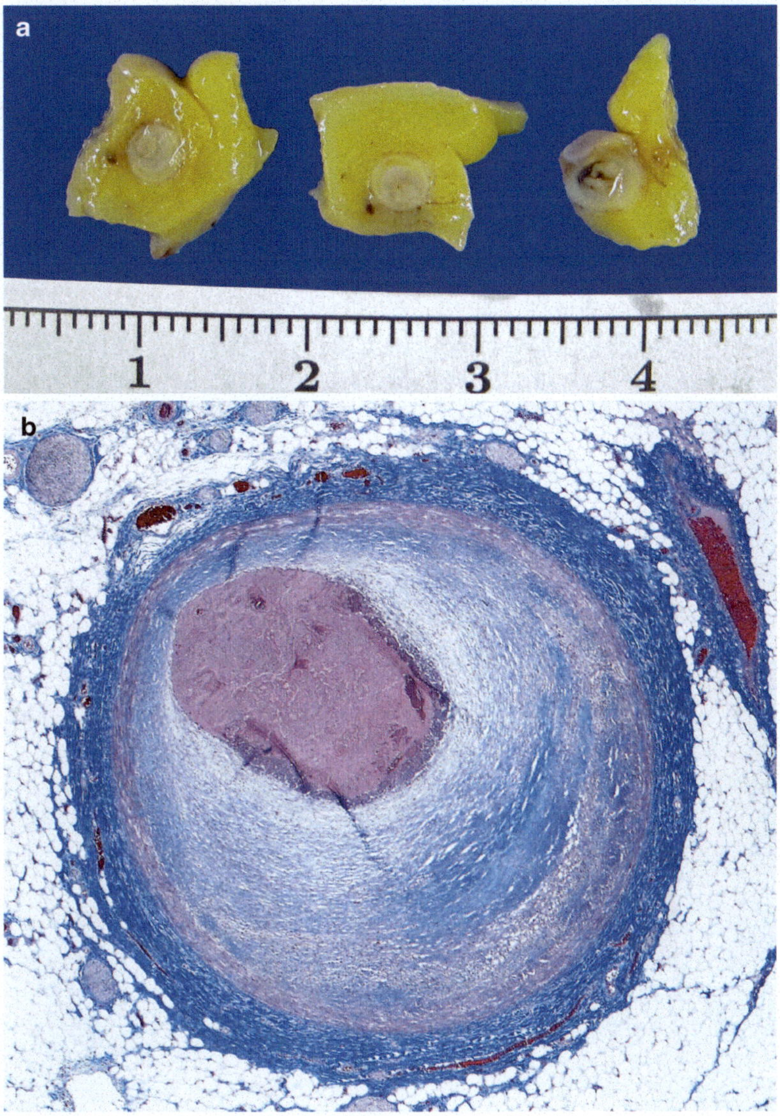

Fig. 5.15 *Plaque erosion.* A 34-year-old woman who died after presenting with chest pain radiating to the shoulders. (**a**, **b**) There was an occlusive platelet thrombus over a fibrous cap with intimal thickening and without necrotic core in the LAD coronary artery. The anterior wall of the LV showed an early acute myocardial infarction (see Fig. 5.28)

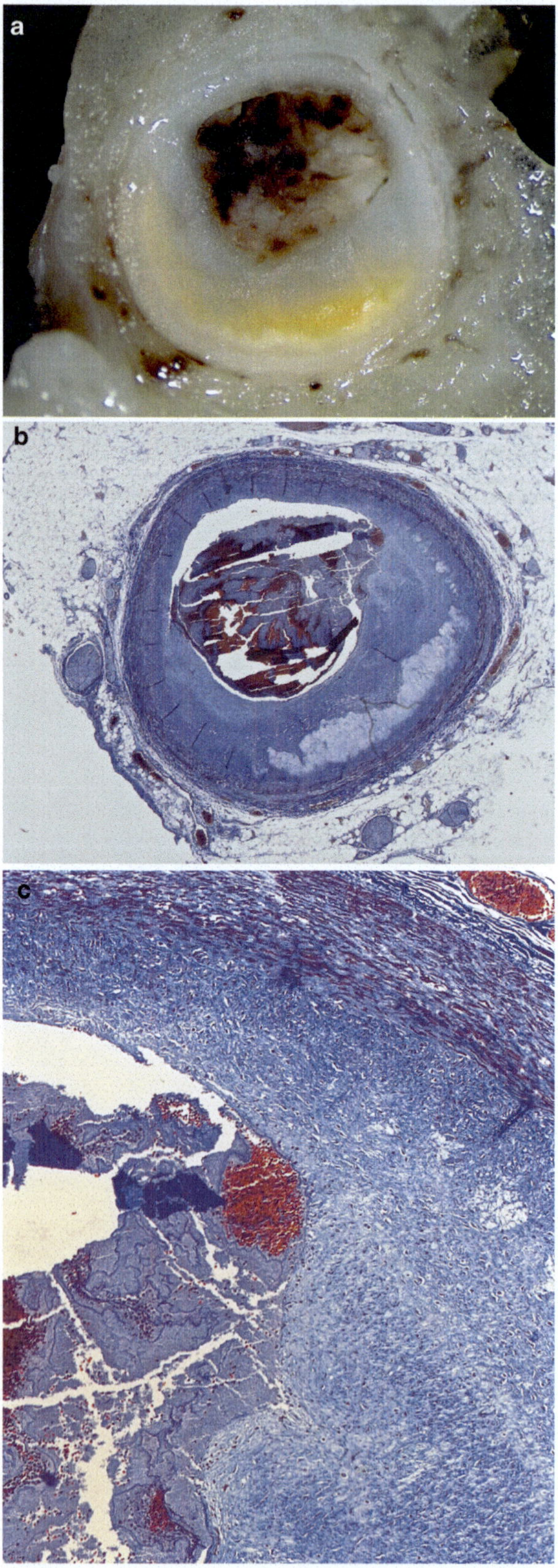

Fig. 5.16 A 34-year-old male with no previous relevant medical history who died suddenly at home. (**a**) Image of a fresh fibrolipid atheroma plaque with occlusive thrombosis in the proximal third of the LAD coronary artery. (**b**) Erosion in the plaque edges can be appreciated, which determines an acute occlusive thrombosis. (**c**) Detail of the erosion in one of the plaque edges (Masson's trichrome)

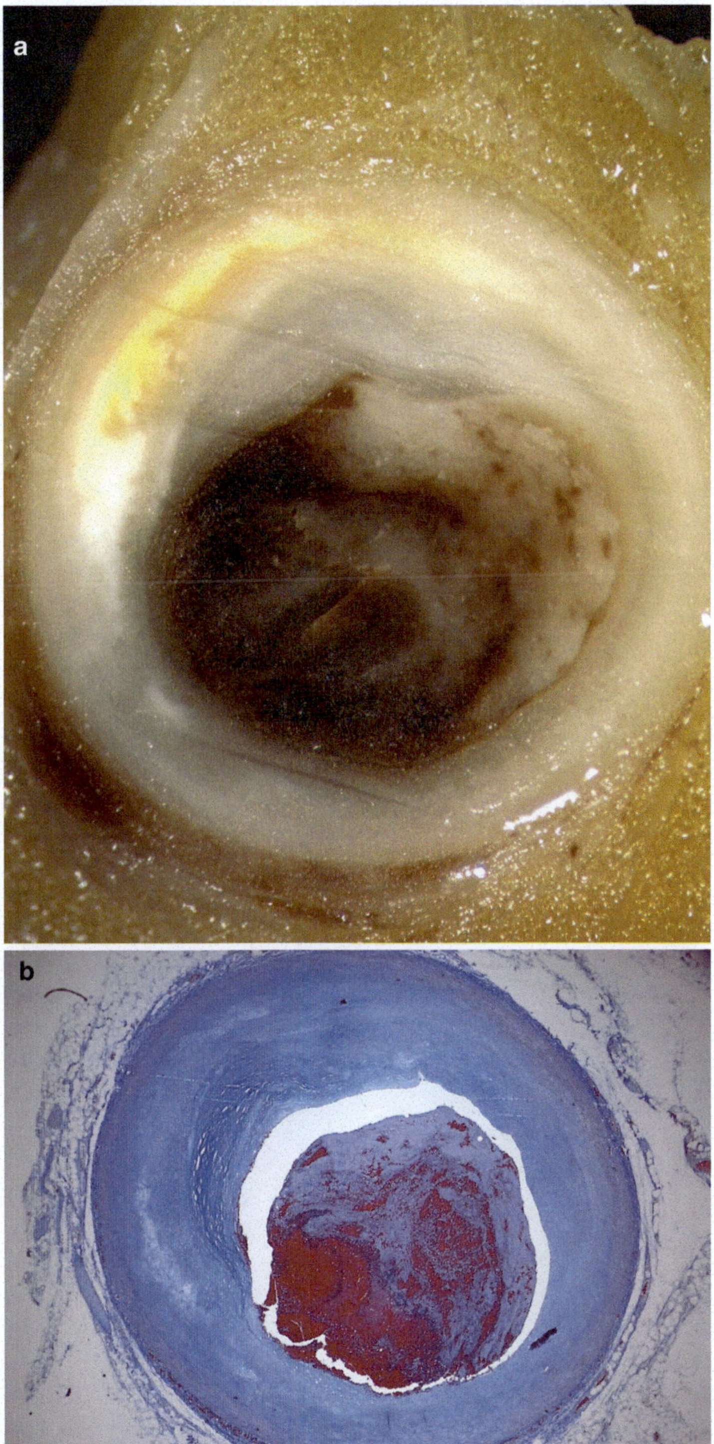

Fig. 5.17 A 44-year-old male, smoker, who died suddenly in the street. (**a**) An eccentric atheroma plaque with a small lipid core and acute occlusive thrombosis in the proximal third of the LAD coronary artery. (**b**) Microscopic image of the acute occlusive thrombosis due to plaque erosion (Masson's trichrome)

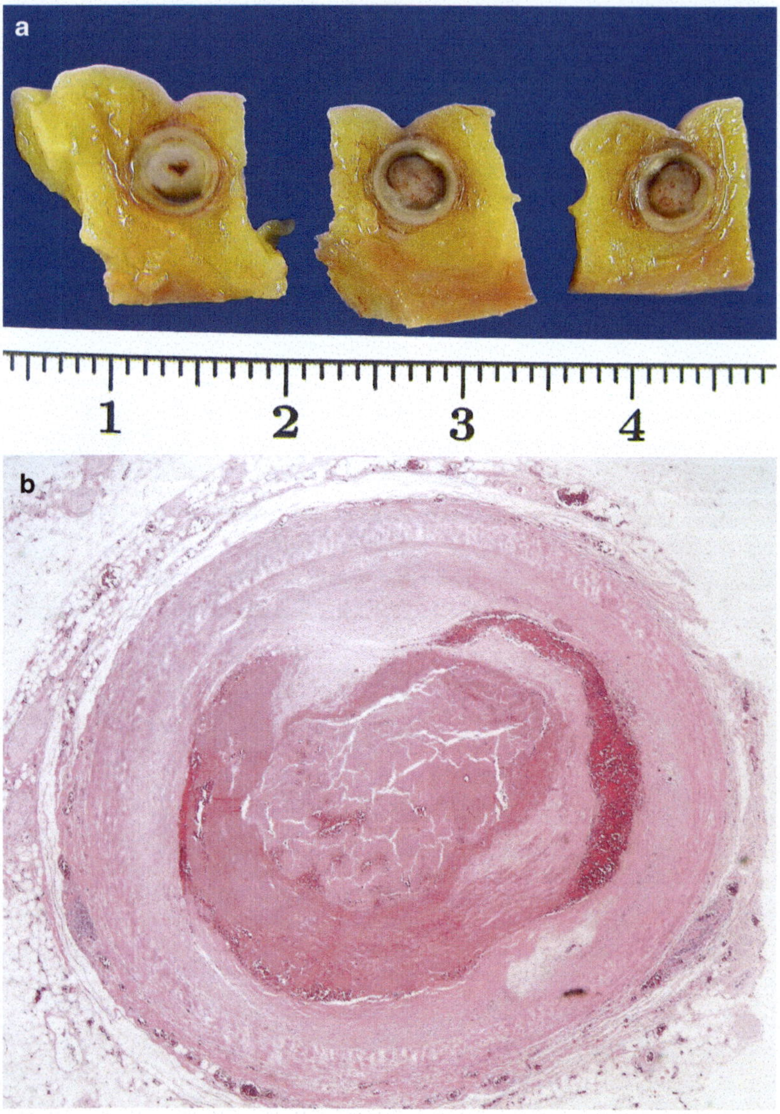

Fig. 5.18 A 43-year-old male, in prison, with a history of drug abuse, who died suddenly. (**a**) Serial sectioning of the LAD coronary artery shows eccentric haemorrhage and occlusive thrombosis along a portion of 2 cm. (**b**) Microscopic examination reveals a haemorrhage, a fibrin thrombus attached to the intima, and a superimposed occlusive platelet thrombosis

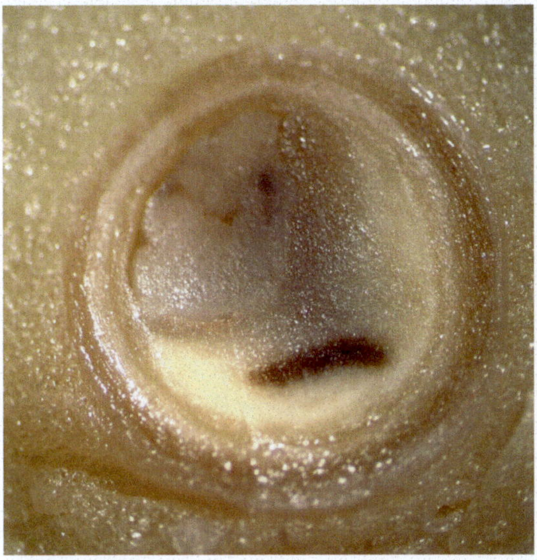

Fig. 5.19 A 54-year-old woman, without known medical history, who died suddenly in bed. A small intraplaque haemorrhage and acute occlusive thrombosis in the proximal third of the LAD coronary artery can be appreciated

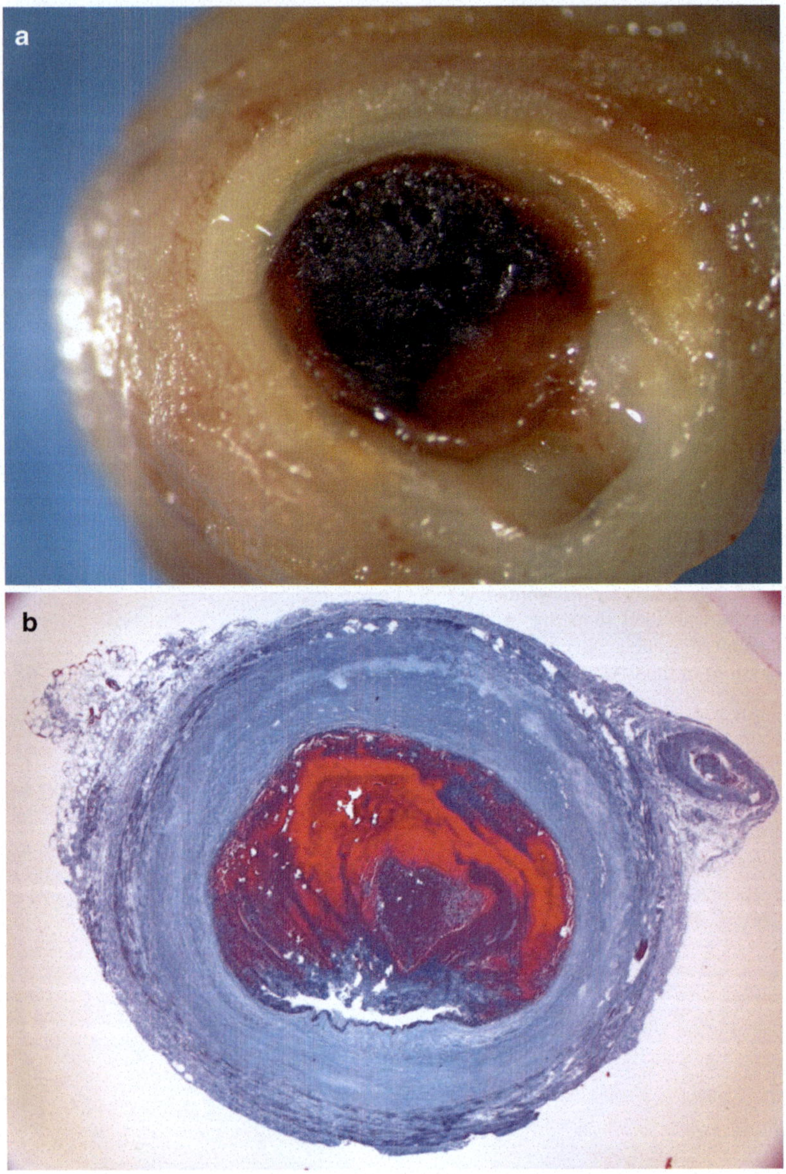

Fig. 5.20 A 63-year-old male, with a history of dyslipidemia, who died suddenly whilst driving. (**a**, **b**) Acute occlusive thrombosis due to plaque erosion can be appreciated in the proximal third of the RCA (Masson's trichrome)

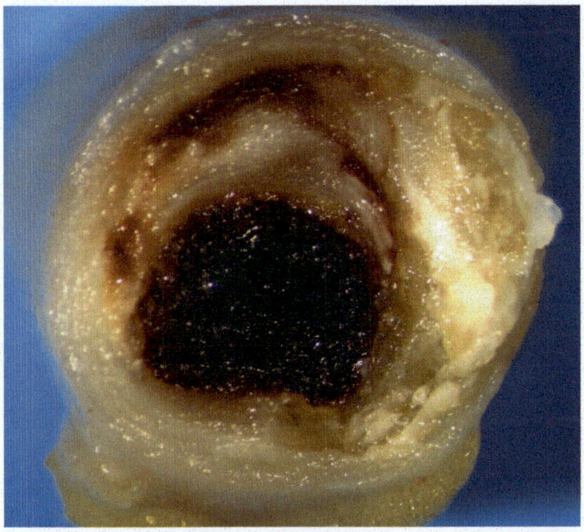

Fig. 5.21 *Plaque rupture*. An 86-year-old male, smoker, without other relevant personal history, who died suddenly whilst shaving in the bathroom. A calcific plaque in the RCA with necrotic core and acute thrombosis due to plaque rupture can be appreciated. There is continuity between the thrombus and the necrotic lipid core

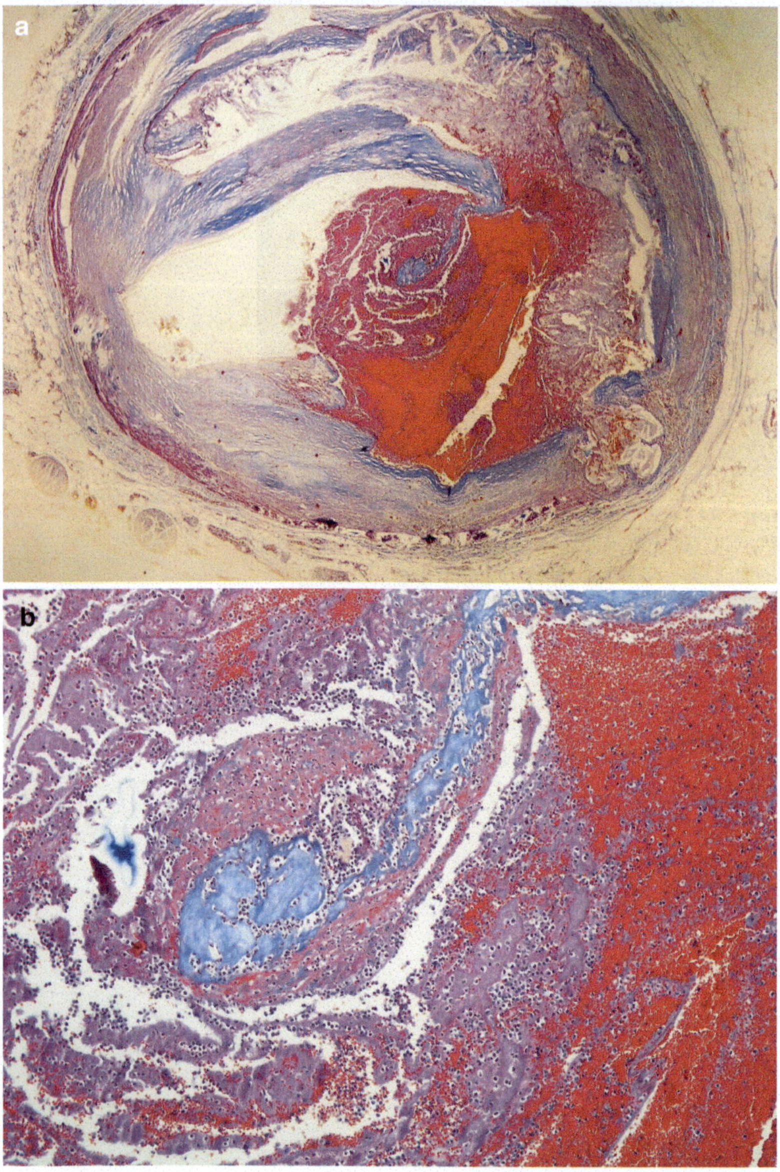

Fig. 5.22 *Plaque rupture.* A 77-year-old male, with hypertension, who was found dead in the bathroom. (**a**) Low magnification image of the RCA showing a large plaque rich in cholesterol crystals with intraplaque haemorrhage and rupture of the thin fibrous cap. (**b**) Higher magnification image where a platelet thrombus around the area of fibrous cap rupture can be appreciated (Masson's trichrome). The stenosis of the coronary artery was >80 % and there was AMI in the posterior wall of the LV and a healed infarction in the anterior wall

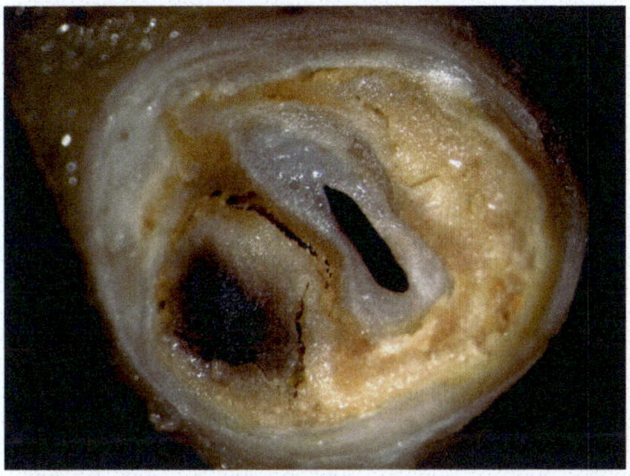

Fig. 5.23 *Fibroatheroma with intraplaque haemorrhage.* A 64-year-old male, with a history of hypertension, who died suddenly after playing a paddle game. A heavily calcific plaque and intraplaque haemorrhage can be appreciated in the RCA. Luminal stenosis greater than 75 %

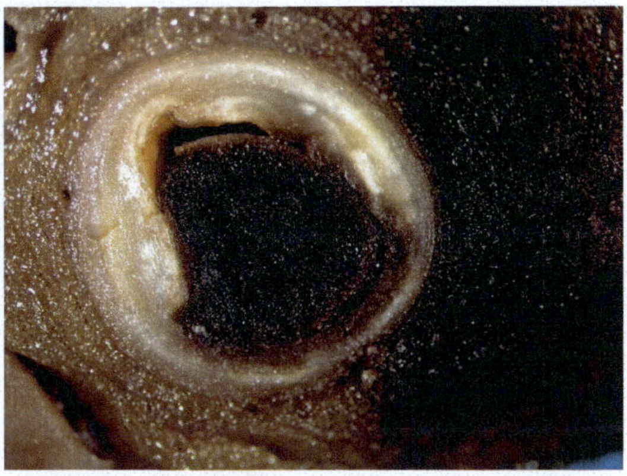

Fig. 5.24 *Intraplaque haemorrhage.* A 45-year-old male, heavy smoker (40 cigarettes/day), who died suddenly at home. He had suffered from chest pain radiating to the left shoulder for several days and, because of this, he was attended at the hospital emergency department. Diagnosis was "atypical thoracic pain". A massive intraplaque haemorrhage, almost occlusive, can be appreciated in the proximal third of the posterior descending coronary artery, together with a haemorrhagic infiltrate in the epicardial white adipose tissue

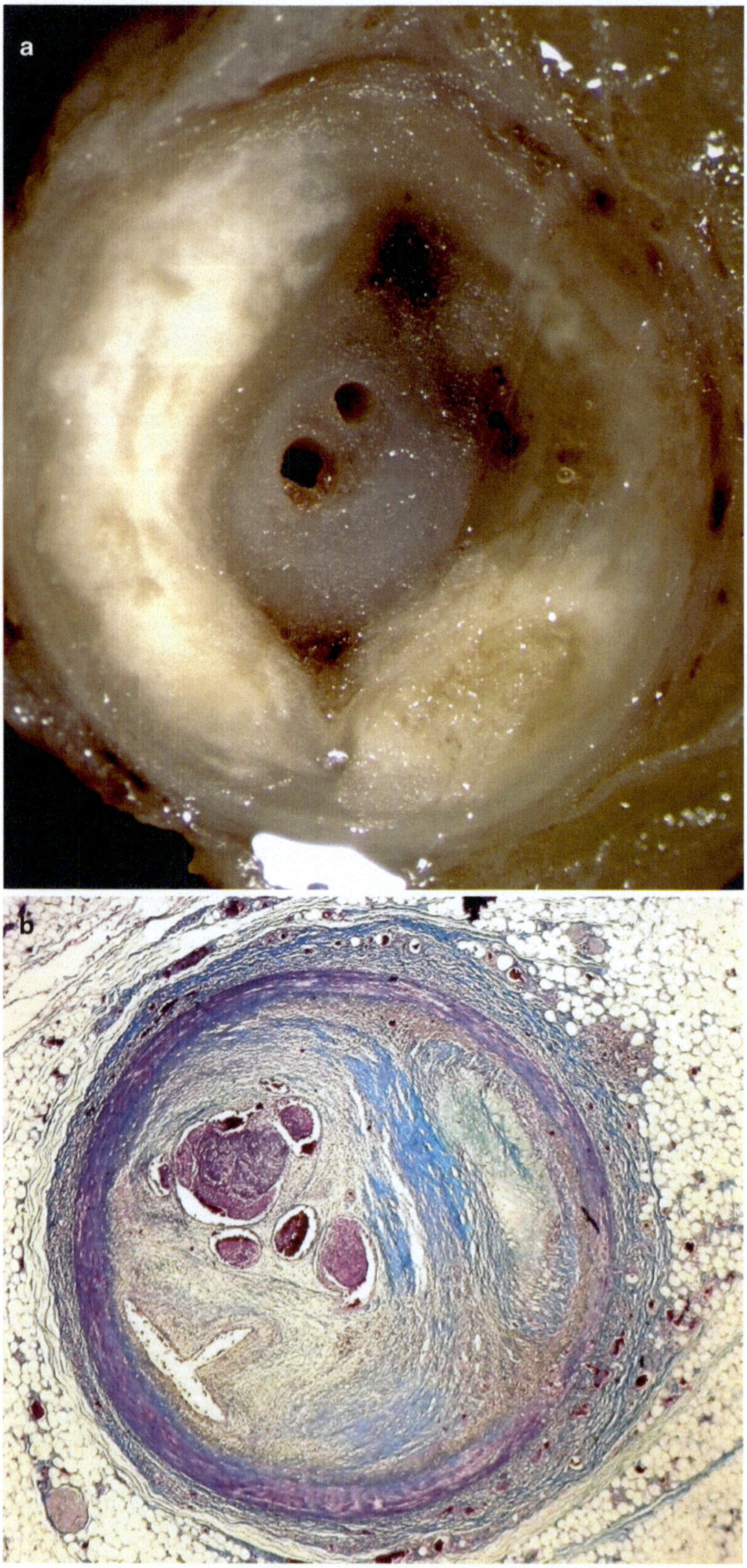

Fig. 5.25 *Recanalized thrombosis.* (**a**) A 45-year-old woman, without known previous medical history, who died at home. Fibrolipid plaque with necrotic core in the LAD coronary artery, intraplaque haemorrhage and recanalized thrombus. (**b**) Histological image from another case of recanalized thrombosis (Masson's trichrome)

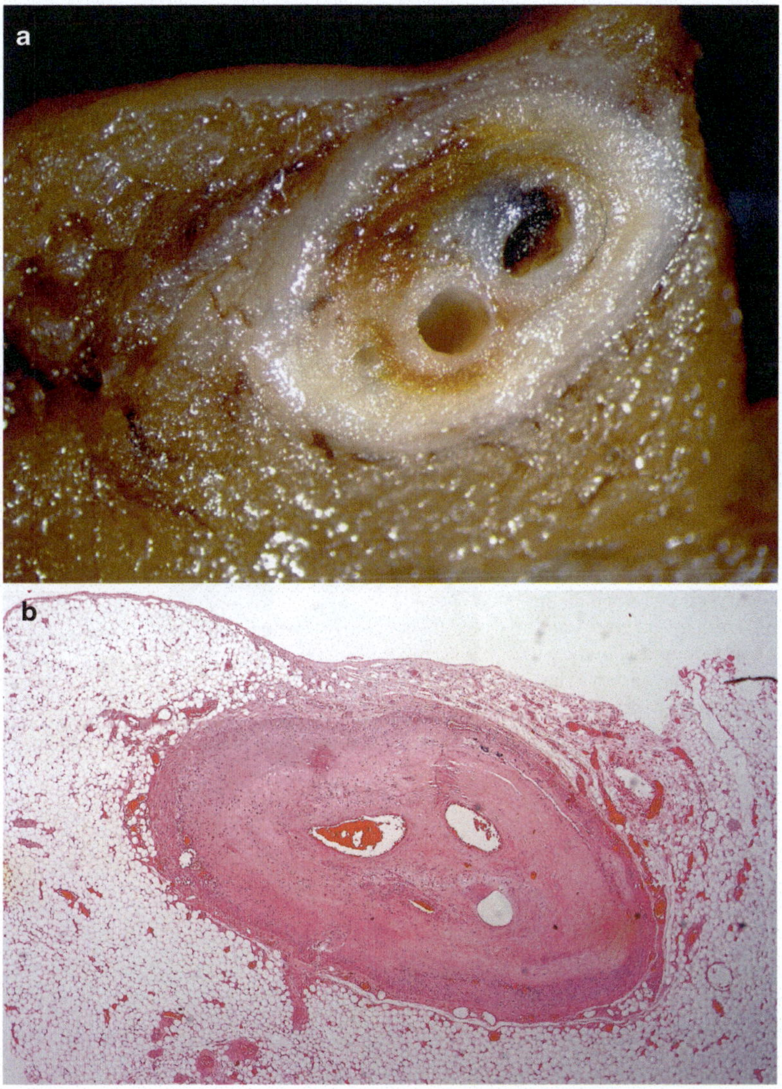

Fig. 5.26 *Recanalized thrombosis*. A 59-year-old male, without known previous medical history, who was found dead in bed. (**a**, **b**) He presented with a recanalized thrombosis in the posterior descending coronary artery and a healed postero-septal myocardial infarction

- 50 % are caused by atherosclerosis, mainly in the elderly
- 20–30 % are congenital
- 10–20 % are due to inflammatory or connective tissue disorders (Kawasaki disease, Takayasu arteritis, lupus, rheumatoid arthritis, etc.).

From a histological point of view, atherosclerotic aneurysms present the same alterations in the intimal and medial layers as those observed in patients with atherosclerotic lesions, suggesting a similar physiopathological mechanism. The most frequently affected coronary artery is the RCA and it can cause sudden death from myocardial ischaemia as a consequence of a rupture or thrombosis (Fig. 5.27).

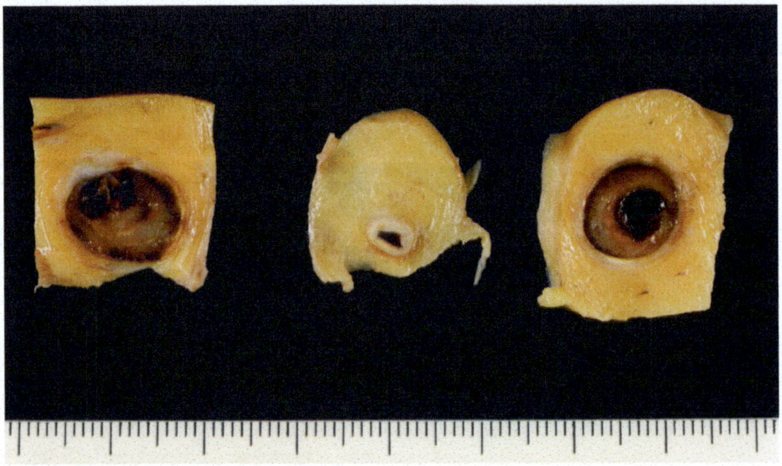

Fig. 5.27 *Atherosclerotic aneurysm.* A 41-year-old male with no known pathological history who died suddenly at home. The RCA presented with two thrombosed atherosclerotic aneurysms of 18 and 20 mm of length, respectively, and mural thrombosis in the LAD coronary artery. The posterior wall of both ventricles showed signs of an early acute myocardial infarction

5.4 Myocardial Ischaemia

Ischaemic heart disease (IHD) is a broad term encompassing a group of clinico-pathological entities caused by an imbalance between myocardial oxygen supply and demand. The term includes acute coronary syndromes (unstable angina, non ST-segment elevation myocardial infarction (NSTEMI), ST-segment elevation myocardial infarction (STEMI) and chronic stable angina.

From a forensic point of view, there are instances when myocardial ischaemia shows no clinical manifestations, and is found unexpectedly during autopsy in cases of sudden death, or even when people die from non-cardiac causes.

In this section we will focus on the anatomopathological aspects of acute myocardial infarction (AMI) and the healed infarct.

5.4.1 Acute Myocardial Infarction

An acute myocardial infarction is the result of a sustained myocardial ischaemia which leads to coagulation necrosis of cardiomyocytes followed by a polymorphonuclear leukocyte response. In 90 % of the cases of death from AMI, autopsy reveals an acute thrombosis over an atherosclerotic plaque. Coronary occlusion determines the necrosis of myocardial tissue, which progresses in parallel with the duration of ischaemia and extends from the subendocardium to the subepicardium, ultimately affecting the entire thickness of the ventricular wall (transmural infarct).

The distribution of the infarcted area, depending of the coronary artery involved, is summarised in Table 5.3. In 10 % of cases, MI follows a different distribution due probably to an occlusion in a coronary artery which supplies blood through collateral branches of the territory of another artery chronically occluded (for instance, an anterior infarct after thrombosis in the RCA that collaterally irrigated the anterior territory due to a chronic occlusion of the LAD coronary artery).

Table 5.3 Infarct location according to the coronary artery involved

Anterior
Left coronary artery trunk or left anterior descending coronary artery
Antero-septal
Left anterior descending coronary artery or left diagonal coronary artery
Antero-lateral
Left anterior descending coronary artery or left circumflex coronary artery
Lateral
Left oblique marginal or left circumflex coronary arteries
Posterior
Right coronary artery or left circumflex coronary artery (when left dominance occurs)

The main clinical presentation of AMI is a constant chest pain, described as burning, squeezing or as a weight in the central chest or left hemithorax. The pain may radiate to the neck, jaw or left upper arm and is associated with vegetative symptoms (nausea, vomiting, and profuse sweating). Onset of AMI is usually sudden but there could be triggering factors (physical or emotional) that have been related to the rupture of an atheroma plaque. Electrocardiogram (ECG) is the best method to confirm the diagnosis and to identify the coronary artery involved. From a clinical point of view, the differentiation from other types of thoracic pain (mechanical or pleuritic) is of critical importance, as well as the differential diagnosis with other potentially fatal pathologies such as pulmonary thromboembolism and acute aortic syndrome.

From an anatomical point of view, five criteria are used to describe and classify MI with ventricular involvement:

1. Distribution along the ventricular thickness:
   (a) Transmural infarcts are those that involve more than 50 % of the ventricular thickness, from the subendocardium to the epicardium. They are caused by occlusive thrombosis of the coronary artery that irrigates the infarcted area.
   (b) Subendocardial infarctions are usually caused by severe multi-vessel coronary atherosclerosis with mural thrombosis that leads to different degrees of occlusion. They can also be secondary to situations of low cardiac output as in shock states.
2. Location in the transverse plane:
   (a) Anterior, involving the anterior part of the free wall of the LV and the anterior two thirds of the septum (50 %).
   (b) Lateral, when the infarcted area lies between the anterolateral and posteromedial papillary muscles (15 %).
   (c) Posterior, involving the posterior wall of the LV and the septum (25 %). Occasionally, this type of infarct can affect the posterior wall of the RV.
   (d) Circumferential, involving the entire ventricular circumference (10 %).
3. Location in the longitudinal plane:
   (a) Base
   (b) Mesocardium
   (c) Apex
4. Location in the left ventricle, right ventricle or in both
5. Temporal evolution of the infarct:
   (a) Early: less than 6 h.
   (b) Acute: less than 7 days.
   (c) Subacute or in the process of healing: older than 7 days and less than 30 days.
   (d) Old or healed: Starts from 1 month until completion between 2 and 3 months.

The temporal evolution of an infarct is critical from a medico-legal point of view, especially in instances where professional malpractice claims are raised concerning diagnosis and treatment which result in the death of the patient (Table 5.4).

Table 5.4 Temporal evolution of non-reperfused myocardial infarction

	Macroscopic features	Microscopic features
<4 h	No alterations. Only visible with enzymatic techniques (Fig. 5.28a)	Usually no alterations. Waviness of myocardial fibres can be observed at the border
4–12 h	No alterations or dark mottling	Initiation of coagulation necrosis. Interstitial edema and haemorrhage (Figs. 5.28b and 5.29)
12–24 h	Red-grayish mottling	Ongoing coagulation necrosis (hypereosinophilic changes of cardiomyocytes with pyknosis of nuclei) that progresses with the loss of nuclei. Neutrophilic infiltration begins (Fig. 5.30)
1–3 days	Mottling with yellow-tan infarct centre	Coagulation necrosis with loss of nuclei and striations. Interstitial infiltrate of neutrophils (Figs. 5.31 and 5.32)
3–7 days	Yellow-tan softened centre very well delineated by a hyperemic border	Phagocytosis of dead cells (cardiomyocytes and neutrophils) by macrophages begins at infarct borders (Fig. 5.33)
7–10 days	Maximum evolution of the yellow-tan centre with depressed red-tan margins	Well-developed phagocytosis of dead cells. Beginning of granulation tissue formation at infarct margins (Fig. 5.34)
10–14 days	Red-gray depressed infarct borders	Necrotic cardiomyocytes have disappeared. Well-established granulation tissue with new blood vessels in the periphery
2–8 weeks	Gray scar that slowly progresses to white from the border towards the core of the infarct	Increased collagen deposition with decreased cellularity. Acellular areas with a geographic appearance (Fig. 5.35)
>2 months	Scarring complete (Figs. 5.36a and 5.37)	Dense collagenous scar formed (Fig. 5.36)

Modified from Robbins (2004)

It is estimated that half of the people who die from AMI do so within the first hour and hence there will be no gross alterations in the myocardium. It is possible, however, to highlight the area of necrosis by exposing fresh myocardium slices to enzymatic stains such as triphenyltetrazolium chloride (TTC) or nitro blue tetrazolium (NBT). The enzymatic stain imparts a brick-red/brown colour to intact, noninfarcted myocardium where the dehydrogenase activity is preserved. In contrast, because this enzyme is depleted in the area of ischaemic necrosis, the infarcted area is revealed as an unstained pale zone (Fig. 5.28a).

Haemorrhagic MI is more frequent in hearts that have been reperfused with thombolytic therapy, although in 5 % of cases it can occur spontaneously. Reperfusion can alter the macroscopic and microscopic appearance of the necrotic area. Macroscopically, a blood infiltrate in the myocardium can be observed (Figs. 5.38 and 5.39a). Histologically, there is an area of coagulation necrosis and contraction band necrosis of the cardiomyocytes, together with extensive interstitial haemorrhage (Fig. 5.39b).

Infarction of the right ventricle tends to be accompanied by an infarct in the posterior wall of the LV. This occurs in up to 25 % of the cases (Fig. 5.40). Infarction of only the right ventricle occurs in less than 5 % of the MI encountered at autopsy.

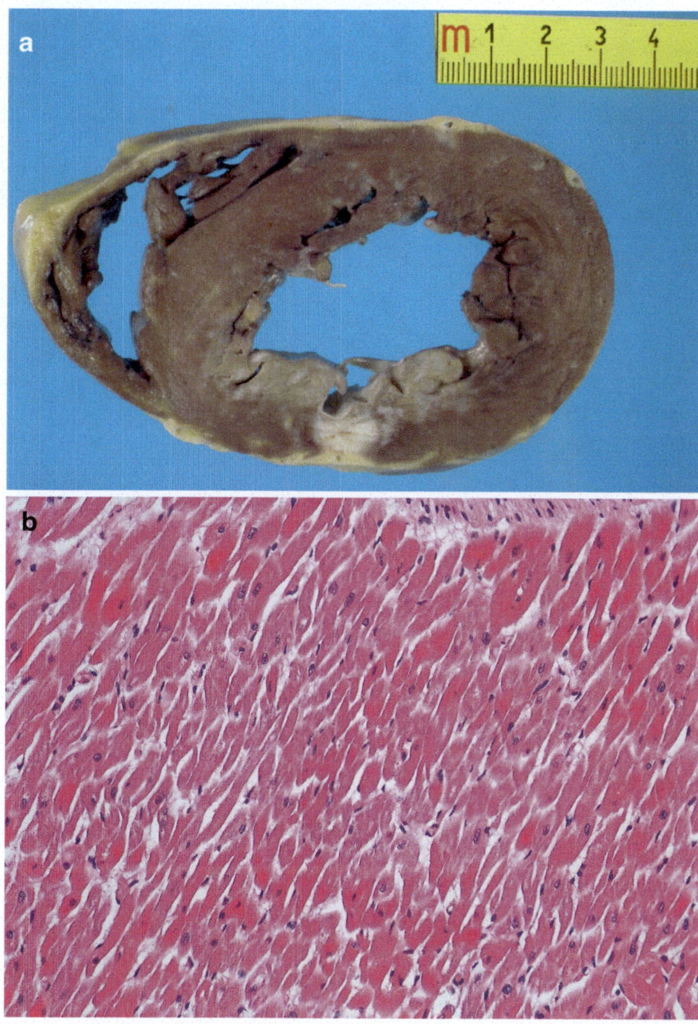

Fig. 5.28 *Early infarction.* (**a**) A very recent subendocardial myocardial infarction in the posterior wall of the LV, as visualised by the enzymatic NBT stain. A pale area contrasting with the normal brown myocardium can be appreciated. A whitish scar from a healed transmural infarct in the centre of the posterior wall of the LV can also be observed (Courtesy of Dr. Gaetano Thiene and Dr. Cristina Basso. *Cardiovascular Pathology Unit*. Department of Cardiac, Thoracic and Vascular Sciences. University of Padua Medical School, Italy). (**b**) Same patient as in Fig. 5.15 where myocytic hypereosinophilia and pyknosis can be seen in the anterior wall of the LV

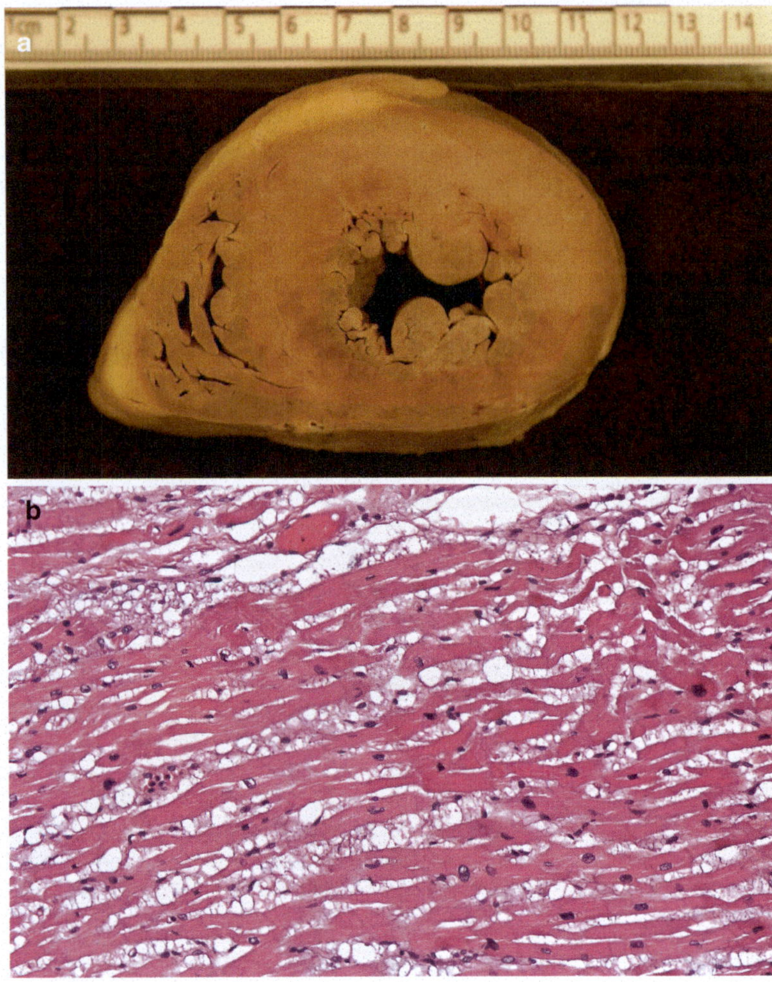

Fig. 5.29 *Early infarction.* A 42-year-old male who died in the street. (**a**) Dark colour in the posterior wall of the LV and in the septum. There was an acute occlusive thrombosis in the RCA. (**b**) Waviness of cardiomyocytes with hypereosinophilia and interstitial edema

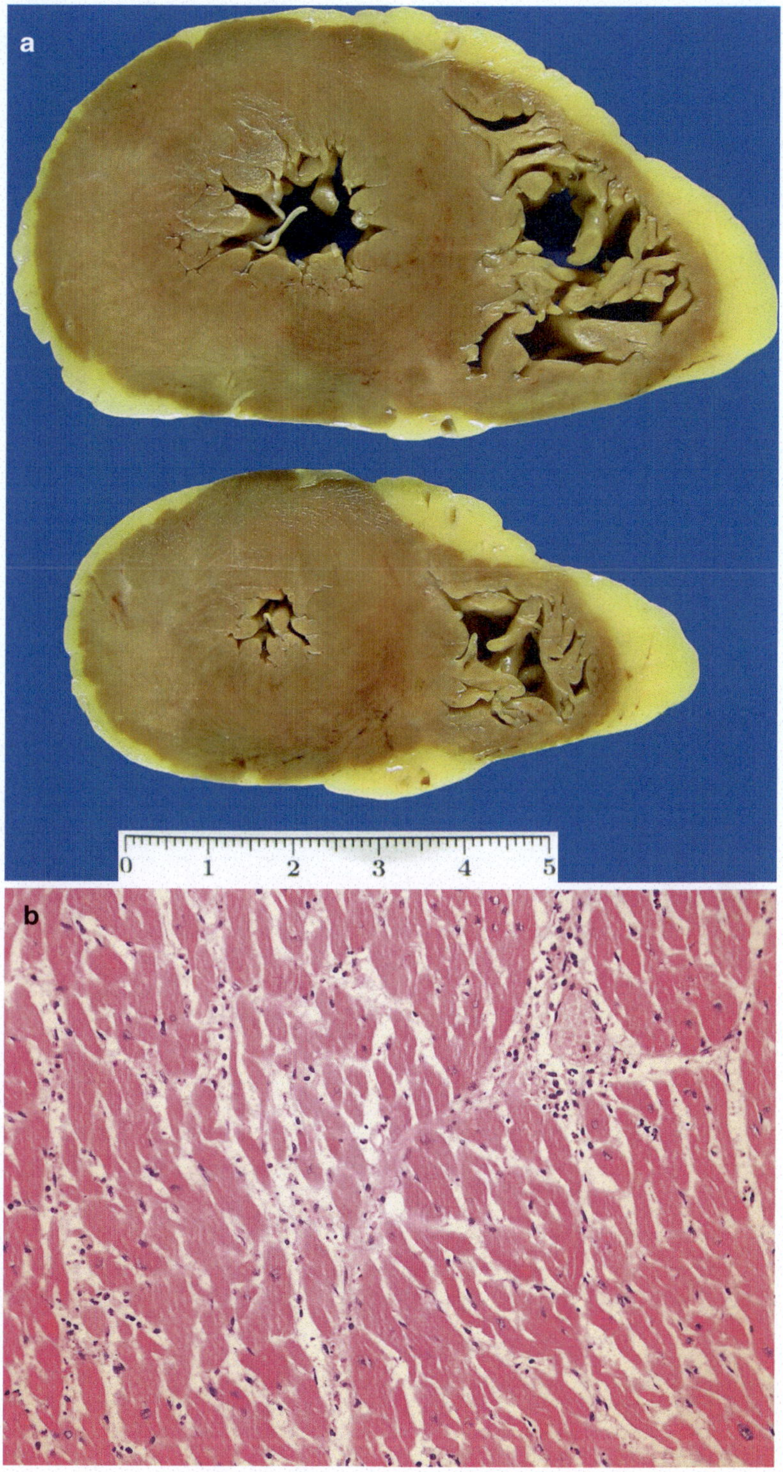

Fig. 5.30 *AMI of 12–24 h.* (**a**) In the apical third a red-tan colour can be seen macroscopically in the septum and circumferential in the apex. (**b**) Incipient polymorphonuclear inflammatory infiltrate

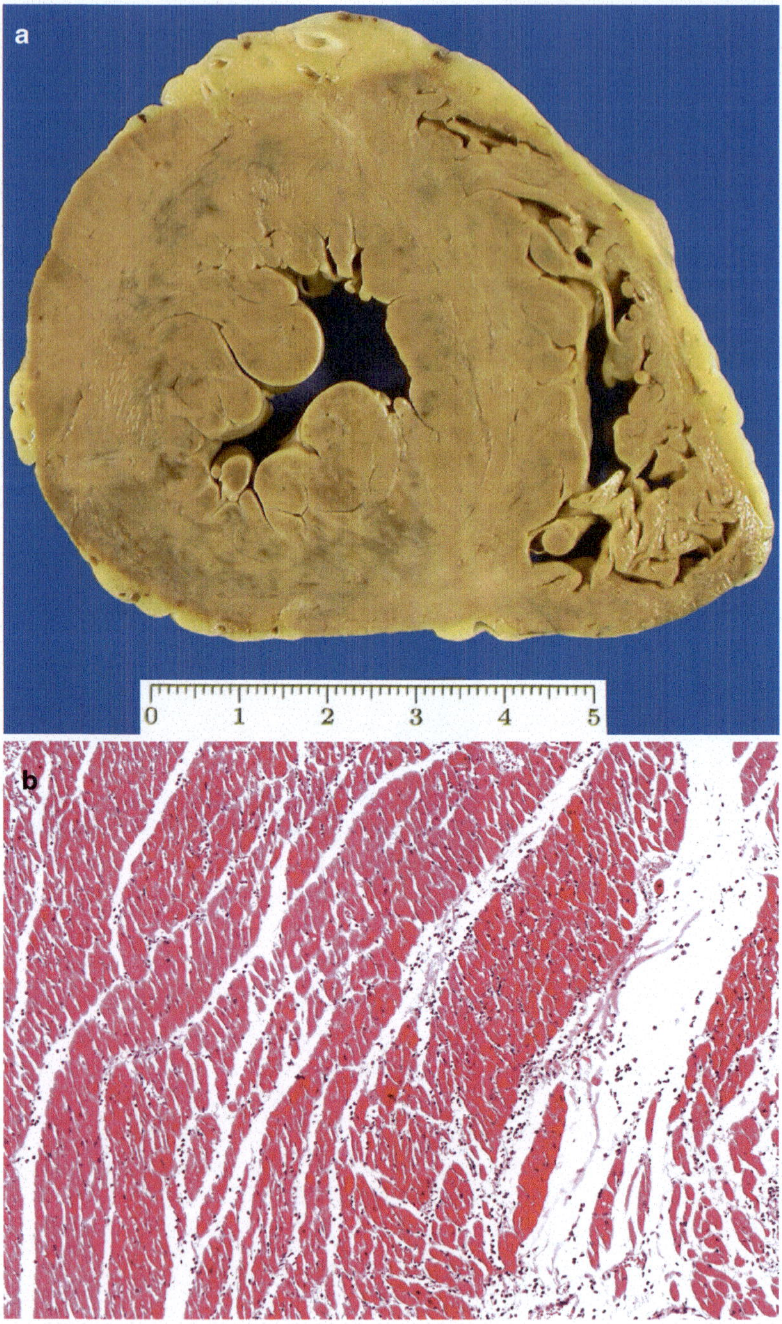

Fig. 5.31 *AMI of more than 24 h.* A 42-year-old male who died suddenly. The right coronary artery presented with occlusive thrombosis. (**a**) Grayish mottled appearance in the posterior wall of the LV. (**b**) Transmural coagulation necrosis can be observed, with contraction bands and interstitial inflammatory infiltrate of polymorphonuclear leukocytes

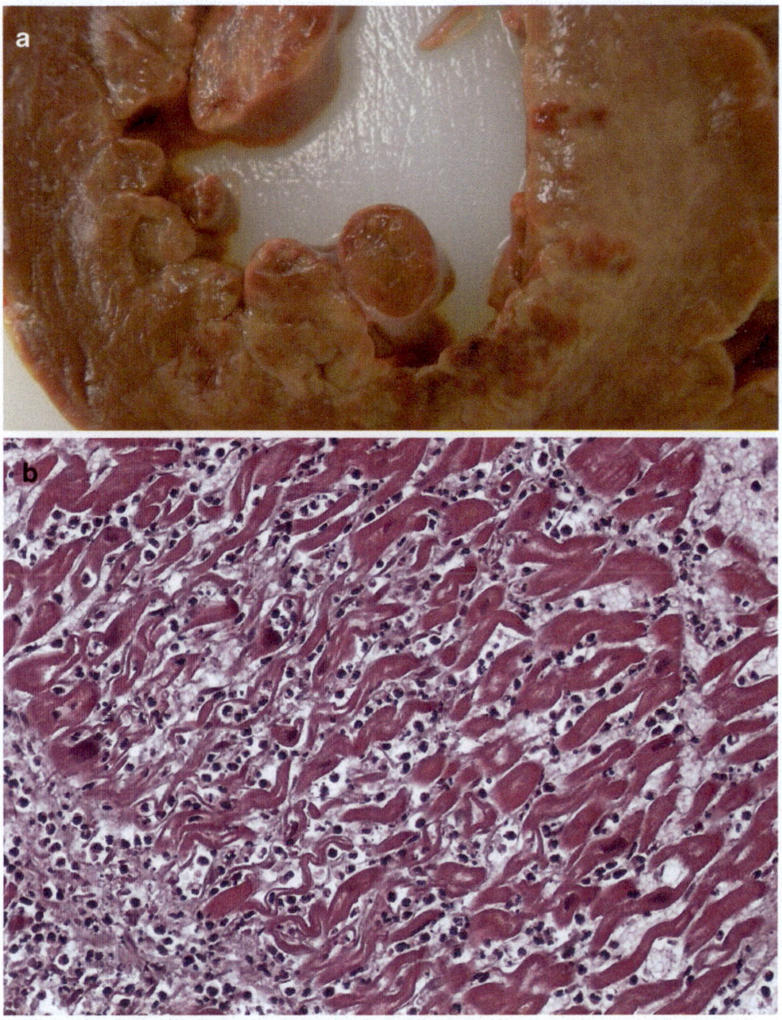

Fig. 5.32 *AMI of 1–3 days.* A 43-year-old male, with a history of insulin-dependent diabetes and a non-specified psychiatric disorder, who dies suddenly at home. (**a**) Gross examination reveals hemorrhagic areas in the subendocardium of the posterior wall of the LV. (**b**) Coagulation necrosis of the cardiomyocytes with loss of nuclei, striations and interstitial infiltrate of neutrophils

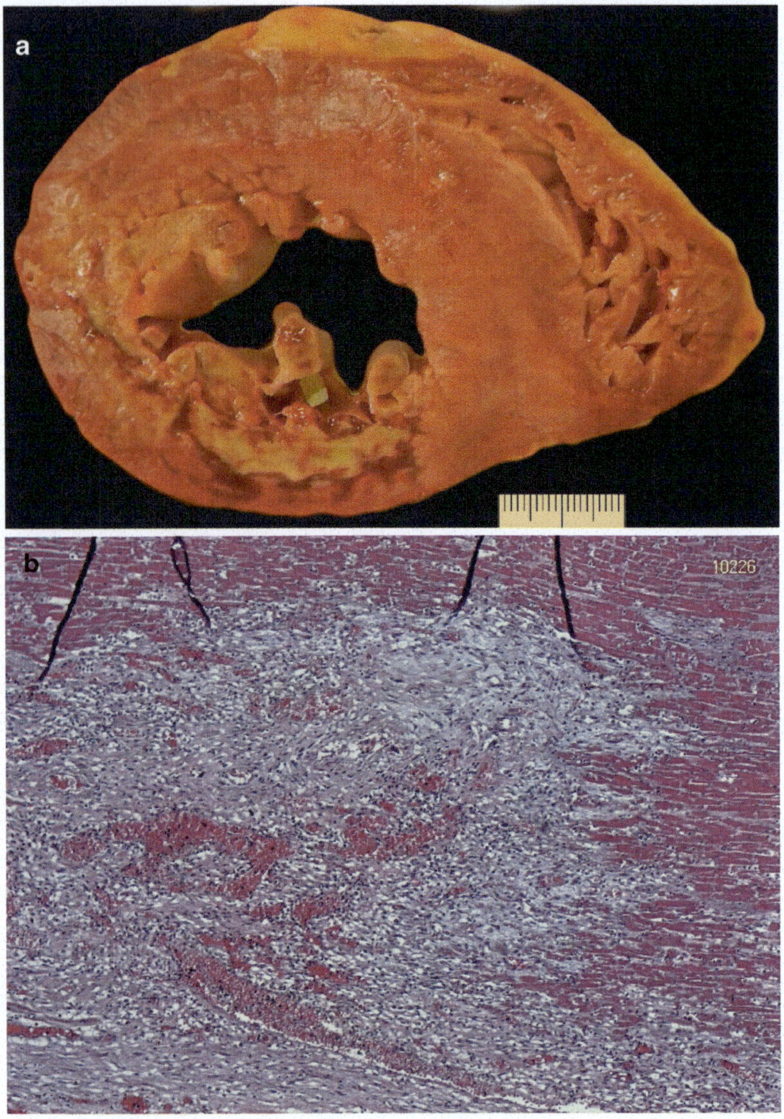

Fig. 5.33 *AMI of 5–7 days*. A 52-year-old male, without relevant medical history, who died suddenly at home. (**a**) Gross examination shows a necrotic yellow-tan centre surrounded by hyperemic borders in the posterior wall of the LV. (**b**) Phagocytosis of dead cardiomyocytes and neutrophils by macrophages begins

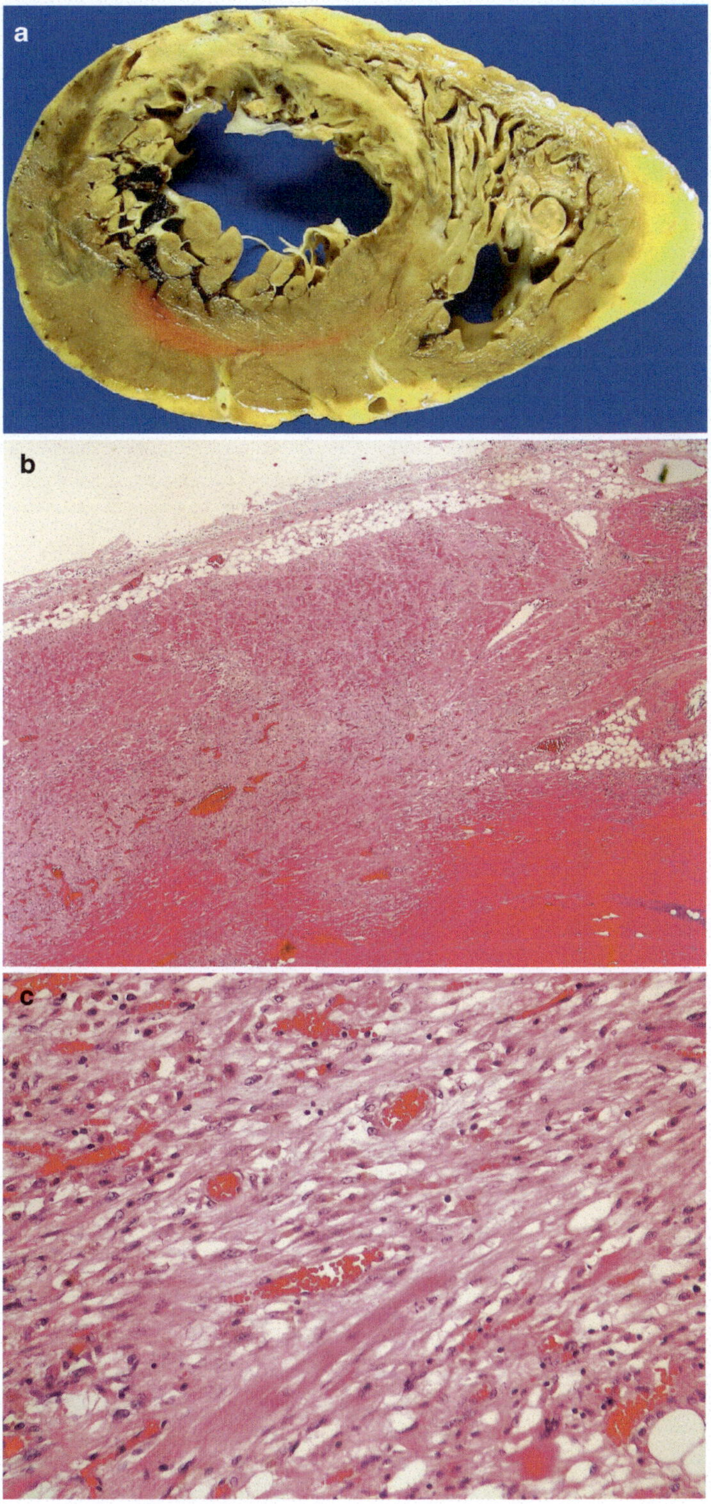

Fig. 5.34 *AMI of 7–10 days.* A 61-year-old woman who died suddenly at the hospital. (**a**) Gross examination shows a thinning in the anterior wall of the LV and septum, and a yellowish myocardium with dark borders. (**b**) Microscopically, granulation tissue surrounding mummified necrotic myocytes can be observed. Fibrinous pericarditis in the pericardium. (**c**) Detail of the borders where fibroblasts and newly-formed vessels can be seen

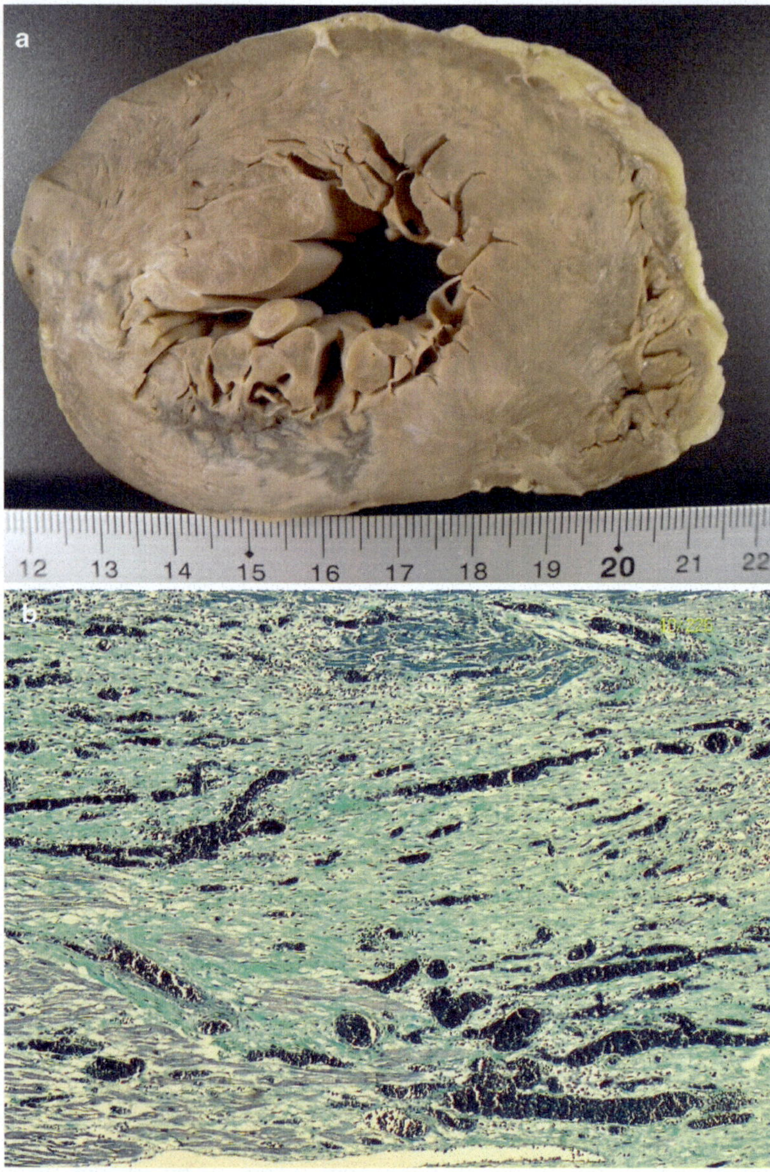

Fig. 5.35 *AMI of 4–6 weeks*. A 50 year-old male who complained of precordial pain a week before he was found dead at home. (**a**) Gross examination shows a gray scar in the posterior wall of the LV that is surrounded by whitish areas in the lateral wall. (**b**) Myocardium replaced by collagen tissue with fibroblasts and very dilated vessels (Masson's trichrome)

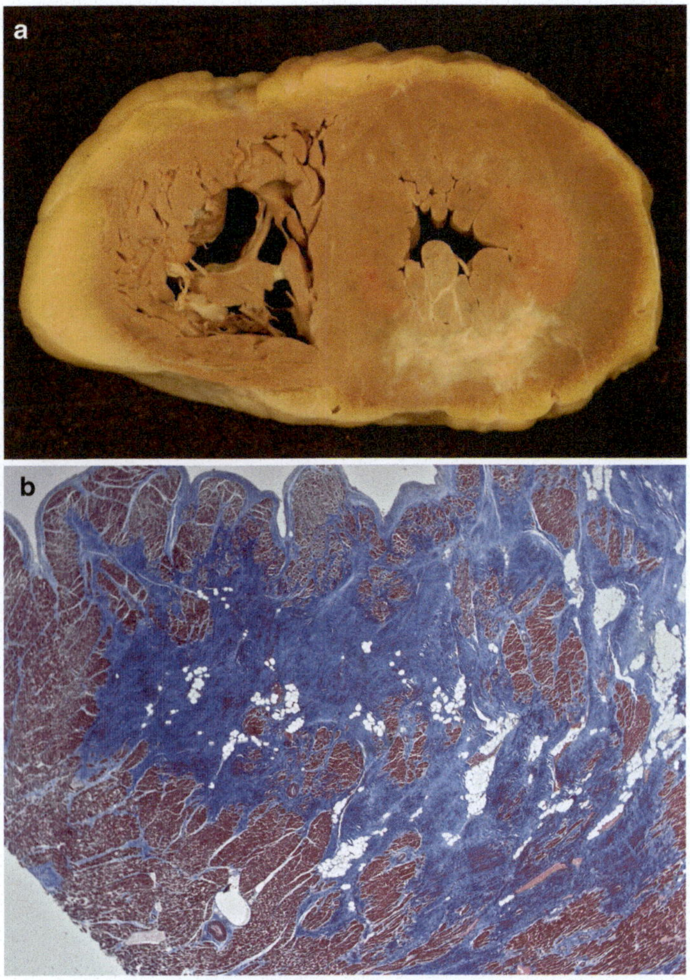

Fig. 5.36 *Healed MI.* A 89-year-old woman with a history of atrial fibrillation, who died from a haemorrhagic stroke. (**a**) Autopsy reveals a healed infarction in the posterior wall of the LV that had been clinically dismissed. (**b**) Large scar consisting of dense collagen tissue (with patches of adipose tissue) that occupies almost the entire thickness of the ventricular wall (transmural) and has replaced the necrotic myocardium (Masson's trichrome)

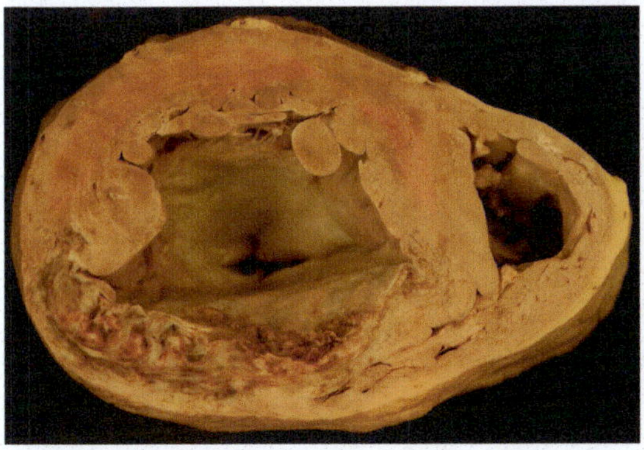

Fig. 5.37 *Healed MI.* A 38-year-old male, in prison, who died suddenly. Scarred infarct with thinning of the posterior wall of the LV. Associated mural thrombosis can also be appreciated. There was an organised and recanalized thrombus in the RCA

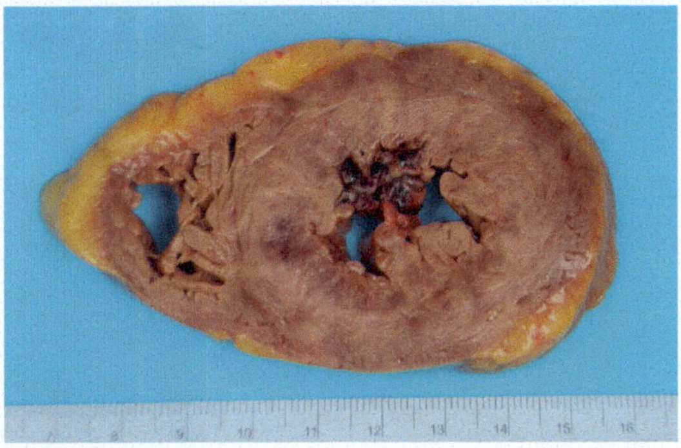

Fig. 5.38 *Reperfusion haemorrhage following AMI.* A 45-year-old male with a history of chest pain and vegetative syndrome lasting for 2 days, who suffered sudden cardiac arrest at home. He was taken to hospital and underwent coronary recanalization and unsuccessful advanced cardiopulmonary resuscitation manoeuvres. Fresh heart where the circumferential haemorrhagic infiltrate can be appreciated in the subendocardium of the LV as well as a mural thrombus in the anterior wall

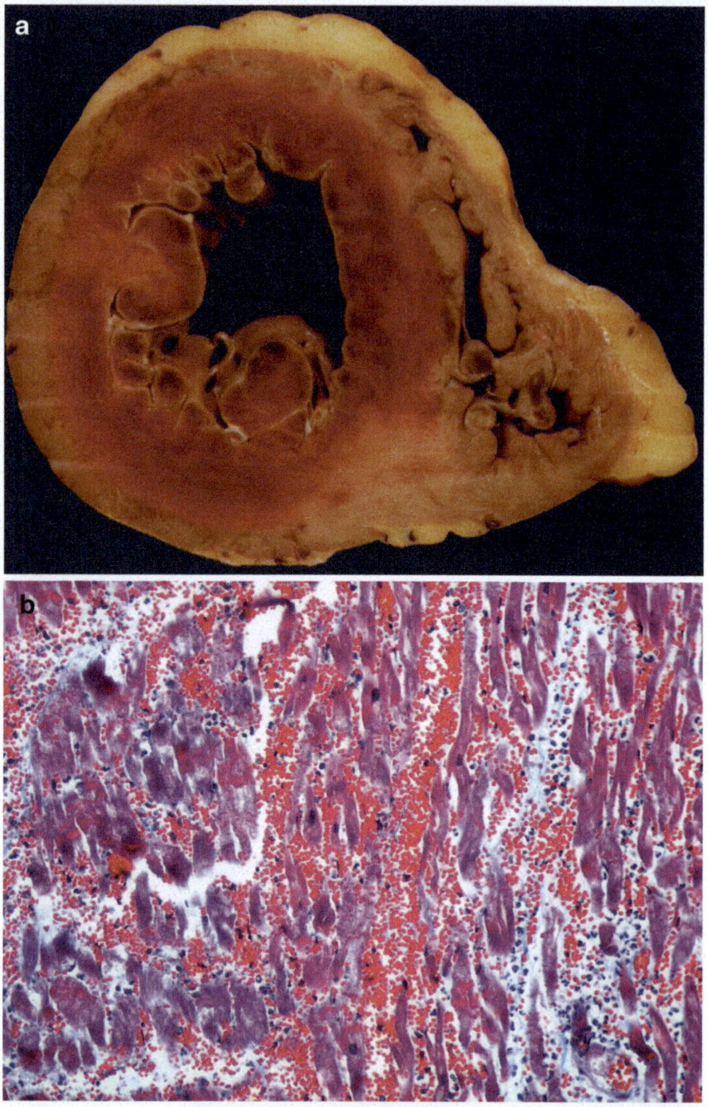

Fig. 5.39 *Reperfusion haemorrhage following AMI.* A 43-year-old male who died 2 days after suffering electrocution. (**a**) Haemorrhagic circumferential AMI due to ischaemia-reperfusion. (**b**) Coagulation necrosis with contraction bands and extensive interstitial haemorrhagic infiltrate

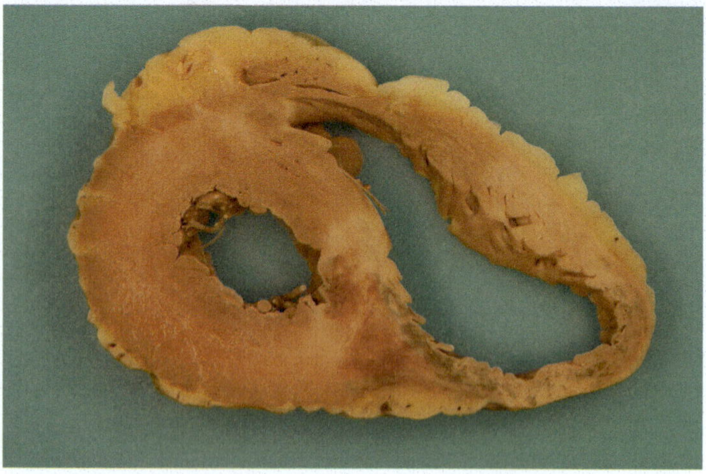

Fig. 5.40 *Infarct in the RV.* A 44-year-old male who died suddenly in the street. A postero-septal AMI in the LV extending to the posterior wall of the RV can be appreciated

5.4.2 Healed Myocardial Infarction

Myocardial ischaemia, when prolonged with time, evolves to a white and firm scarred lesion located in the area of distribution of the MI. It can be single or multiple, depending on the number of infarcts the patient has suffered during his/her lifetime. In the autopsy setting, there are many cases where an infarction scar is seen but there is no clinical history of a previous MI; this is especially frequent in infarcts affecting the posterior wall of the LV (Figs. 5.36 and 5.37).

5.4.3 Complications of Acute Myocardial Infarction

5.4.3.1 Pericarditis

Pericarditis, usually focal, is a rare complication of AMI, only observed in 5 % of the infarcts, especially in transmural MIs (Fig. 5.41).

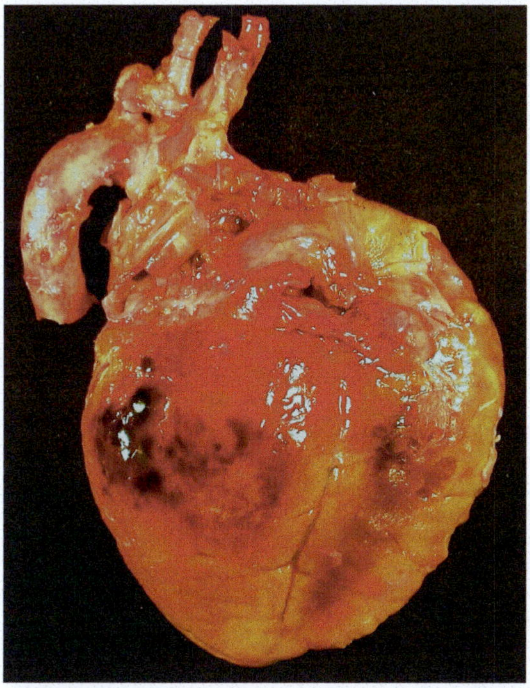

Fig. 5.41 *Fibrinohaemorrhagic pericarditis.* A 35-year-old male, with a history of Tetralogy of Fallot surgically corrected in his childhood, dies suddenly. He presented with a transmural AMI in the posterior wall of the LV

5.4.3.2 Myocardial Rupture

Rupture of the left ventricular free wall is an uncommon (between 1 and 5 % of MIs, depending on the series) but serious and frequently fatal complication of MI that results in cardiac tamponade and cardiogenic shock. Risk factors for free wall rupture include age older than 70 years, female gender, extensive infarction with little collateral circulation and no prior history of angina, and fibrinolytic reperfusion therapy instead of primary angioplasty. Rupture occurs most commonly within 5–7 days after infarction, the period in the healing process when reabsorption of necrotic tissue is maximal. We can encounter three types of myocardial rupture:

- Type I: Tear of the ventricular wall without thinning (Fig. 5.42).
- Type II: Tear of the necrotic myocardium (Fig. 5.43).
- Type III: Severe thinning of the necrotic wall with aneurysm formation (Fig. 5.47).

The other two mechanical complications associated with MI are papillary muscle rupture leading to acute mitral valve insufficiency and interventricular septal rupture leading to free communication between both ventricles. Both complications require emergency surgical repair (Fig. 5.44).

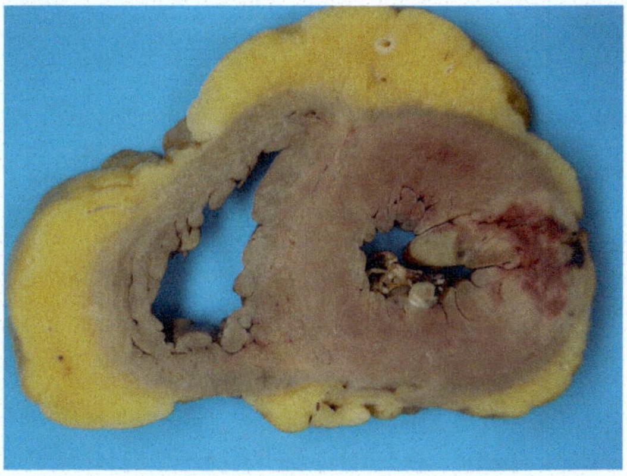

Fig. 5.42 Transmural AMI with rupture of the free wall. A 47-year-old male who died suddenly whilst in the bathroom. Lateral transmural AMI with wall rupture that led to cardiac tamponade (Type I). He presented with an acute occlusive thrombosis in the left circumflex coronary artery

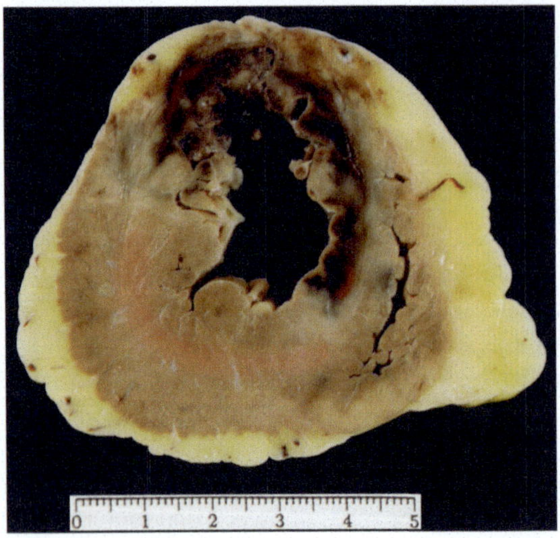

Fig. 5.43 Transmural AMI with rupture of the anterior wall. A 72-year-old woman who suffered AMI and underwent stent surgery in the LAD coronary artery. She died suddenly at home 3 h after being discharged from hospital. Anterior transmural AMI with rupture of the necrotic wall leading to cardiac tamponade (Type II)

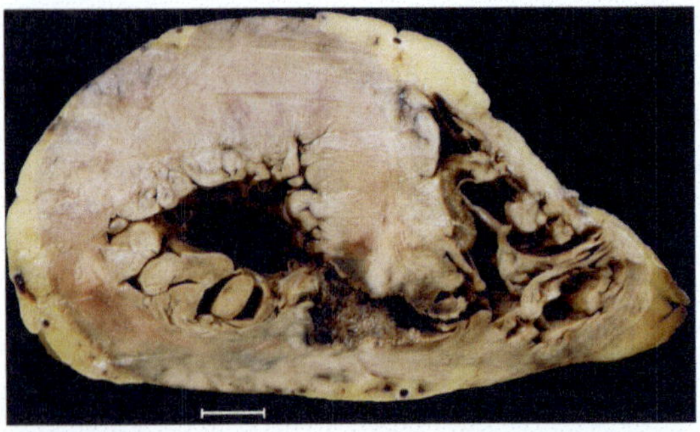

Fig. 5.44 Transmural AMI with rupture of the septum. A 61-year-old male died suddenly in the evening after undergoing surgical intervention in the morning due to a cutaneous lesion in his hand. AMI with perforation in the posterior segment of the interventricular septum, leading to extensive communication between the two ventricular chambers

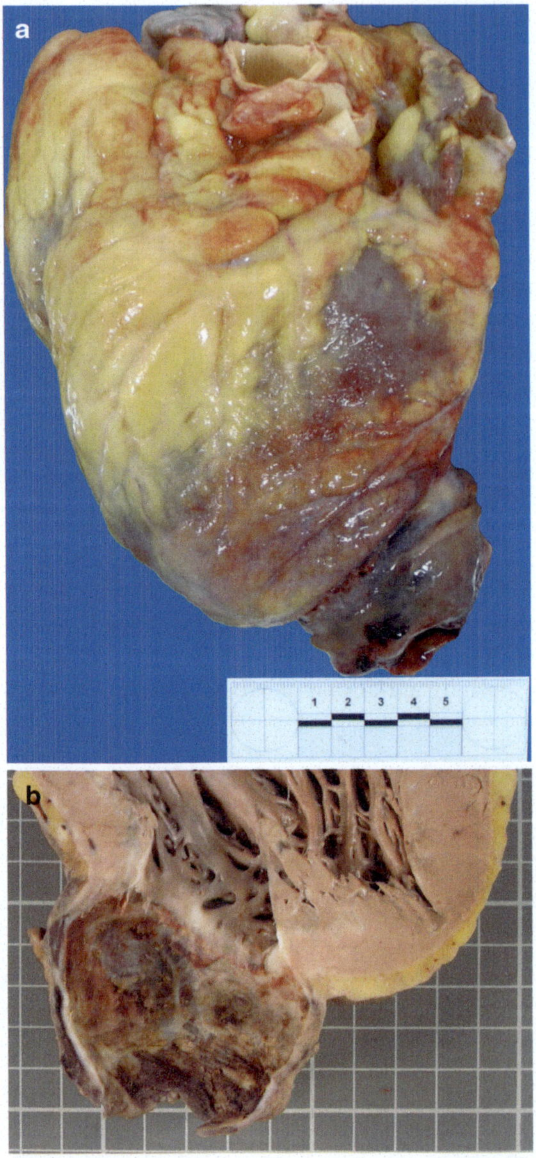

Fig. 5.45 *Aneurysm with wall rupture. A 55-year-old male who died suddenly in a sauna.* (**a**) Rupture of an aneurysm in the apex as a complication of a MI. The marked thinning of the aneurysm's wall determines its ulterior rupture and cardiac tamponade (Type III). (**b**) Detail of an aneurysm consisting mainly of fibrous tissue. The aneurysmal cavity was filled with thrombi that were removed to better appreciate the wall

5.4.3.3 True and False Aneurysms (with or Without Thrombosis)

True aneurysms can develop as a complication in 5–10 % of MIs, especially in large transmural infarcts located anteriorly or in the apex. From a pathological point of view, the wall of a true aneurysm consists of fibrous tissue with cardiomyocyte foci and occasionally white adipose tissue and calcium. The aneurysm's cavity may contain thrombi, predisposing to embolic complications. The most serious complication of an aneurysm is rupture leading to sudden death by cardiac tamponade (Fig. 5.45).

False aneurysms or pseudoaneurysms appear when ventricular free wall rupture is contained by the adjacent pericardium, leading to a subacute clinical presentation. Without surgical intervention pseudoaneurysms might eventually rupture causing sudden death by cardiac tamponade (Fig. 5.46).

5.4.3.4 Mural Thrombosis in the Left Ventricle

Mural thrombosis is a complication in up to 20 % of myocardial infarcts, especially transmural infarcts involving the anterior wall and apex (Figs. 5.47 and 5.48). The major risk in mural thrombosis is detachment of the blood clots leading to systemic embolism.

5.4.3.5 Left Ventricular Chamber Dilatation

Scar fibrosis may determine a notable thinning in the wall, with ventricular chamber dilatation that clinically manifests as congestive cardiac insufficiency by ventricular dysfunction, also known as ischaemic cardiomyopathy or ischaemic dilated cardiomyopathy (Fig. 5.49).

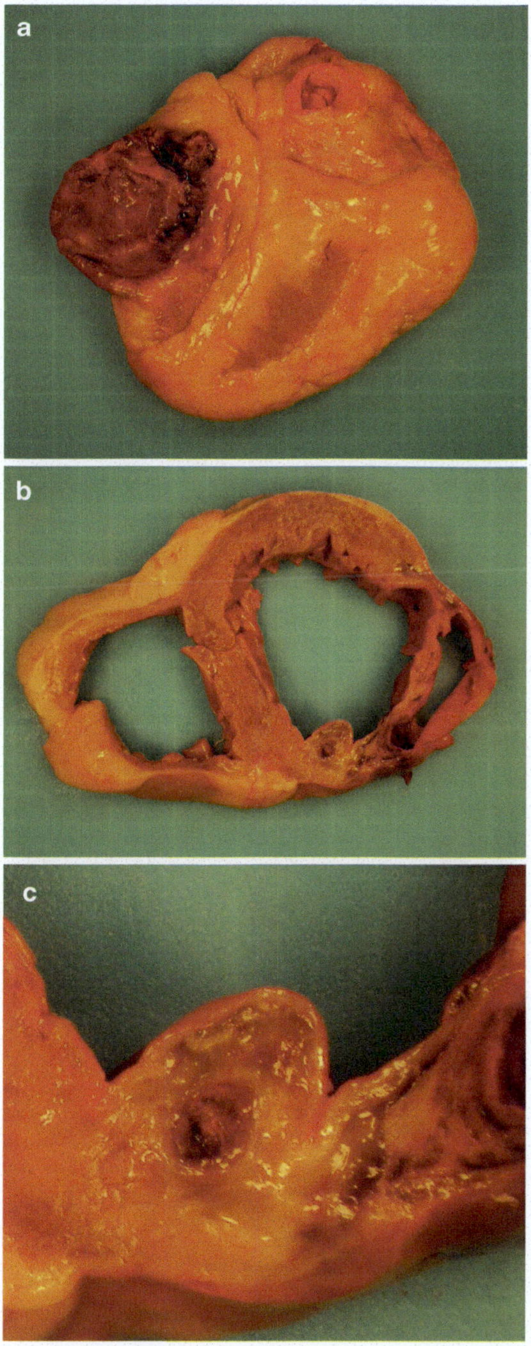

Fig. 5.46 *Pseudoaneurysm with wall rupture.* A 62-year-old male who was found dead in bed. (**a**) Rupture of a pseudoaneurysm as a complication of an old MI in the lateral and posterior walls of the LV. (**b**) Detail of the pseudoaneurysm in a transverse section. (**c**) Detail of the RCA where acute plaque rupture with occlusive thrombosis can be appreciated

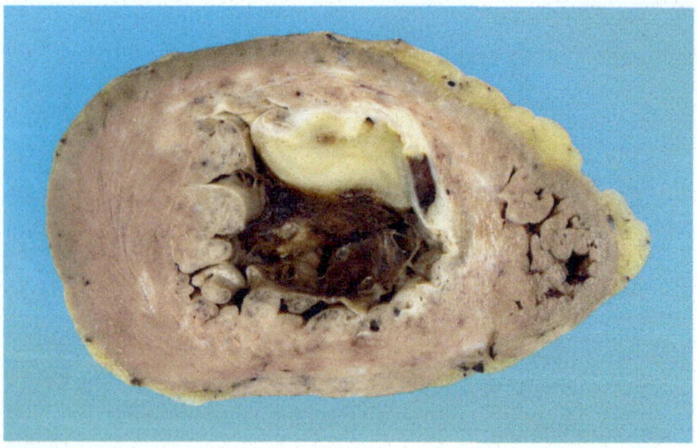

Fig. 5.47 A 63-year-old male died suddenly in the street. Circumferential MI with wall thinning due to scar fibrosis and associated mural thrombosis. Severe coronary atherosclerosis affecting the three main arterial trunks

Fig. 5.48 A 52-year-old male, with no known pathologies, who lost control of his car in a city avenue and collided with a tree. Antero-septal MI with wall thinning and associated mural thrombosis

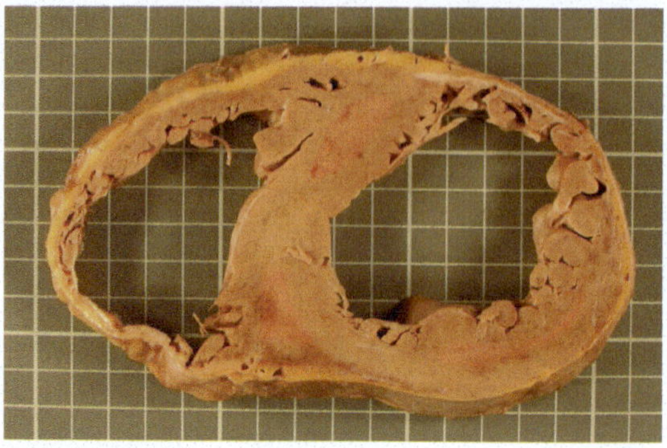

Fig. 5.49 Ischaemic dilated cardiomyopathy. Biventricular dilation secondary to a myocardial infarction, leading to the almost complete disappearance of the myocardium in the anterior wall of the LV and the posterior wall of the RV

5.5 Pathology of Coronary Revascularization with Stent Implantation

Currently, the use of stents has become the standard technique for coronary revascularization. There are many types of balloon expandable stents for intracoronary use, with different metallic scaffolds (steel, chrome-cobalt, platinum, etc.) or even with biomaterials that can be reabsorbed and eliminated with time (bioresorbable scaffolds). Furthermore, stents can be coated with anti-proliferative drugs (placlitaxel, sirolimus, everolimus, zotarolimus, etc.), which allow slow drug elution into the coronary wall to avoid neo-intimal proliferation and restenosis. For autopsy of hearts carrying stents, refer to Chap. 1 (cardiovascular autopsy technique) and remember to carry out radiology of the entire heart before sectioning the coronary artery in order to visualise correctly the location of the stent.

In this chapter we will briefly explain the pathology associated with stents (restenosis, thrombosis and neoatherosclerosis) and the iatrogenic effects related to their implantation.

5.5.1 Restenosis

The angiographic definition of restenosis is a reduction of 50 % in the vessel lumen that occurs typically in the first months post-implantation. The frequency of restenosis varies depending on the clinical and angiographic characteristics of the patient, ranging from 10 to 30 % with uncoated stents and from 3 to 6 % with coated stents. From a pathological point of view, restenosis is related to the healing process of the vessel's intimal layer which, when exacerbated, promotes intimal thickening (neointimal proliferation) and ultimately luminal stenosis. Autopsy studies in humans highlighted an association between the lesion in the medial layer, inflammation, angiogenesis and the penetration of the lipid core with the neointimal proliferation and restenosis that occur after stent implantation (Fig. 5.50).

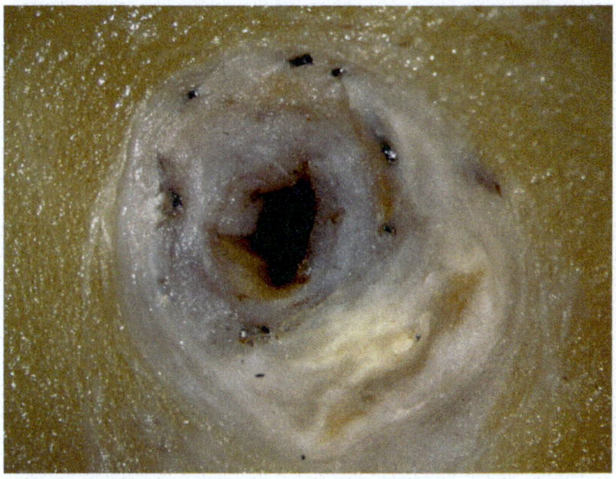

Fig. 5.50 A 32-year-old male with a history of obesity and two MIs that resulted in the implantation of a stent in the RCA. He died suddenly at home. Stent restenosis with intimal proliferation and a plaque with a significant lipid core can be appreciated

5.5.2 Thrombosis

Stent thrombosis can be classified as acute (first 24 h post-implantation), subacute (1–30 days), late (1–12 months) or very late (>12 months). The approximate incidence of thrombosis in the first month or in the first year is around 1 %, decreasing after at a rate of 0.5 % annual (Fig. 5.51). Multiple thrombosis risk factors have been identified associated with the patient's intrinsic factors (diabetes, genetic factors, acute myocardial infarction, renal insufficiency, ventricular dysfunction) or with the procedure (infra-expansion of the stent, dissection at the stent borders, length, decrease in the distal flow, coated stent). Nevertheless, the most important risk factor is the premature interruption of the antiagregant therapy.

5.5.3 Iatrogenic Complications of Stent Implantation

The procedure of percutaneous coronary angioplasty (PCA) is not exempt from complications related to any of the components used (catheter, guide, balloon or stent) during implantation. The major complications are dissection, perforation and coronary artery occlusion (Fig. 5.52), which can lead to iatrogenic myocardial infarction or cardiac tamponade. In an angiographic study carried out by Biondi-Zoccai et al. (2006), the authors showed that coronary dissection due to PCA with stent implantation occurred in 2.8 % of the patients and the coronary artery most frequently involved was the LDA (70 %) followed by the RCA in 15 % of the cases.

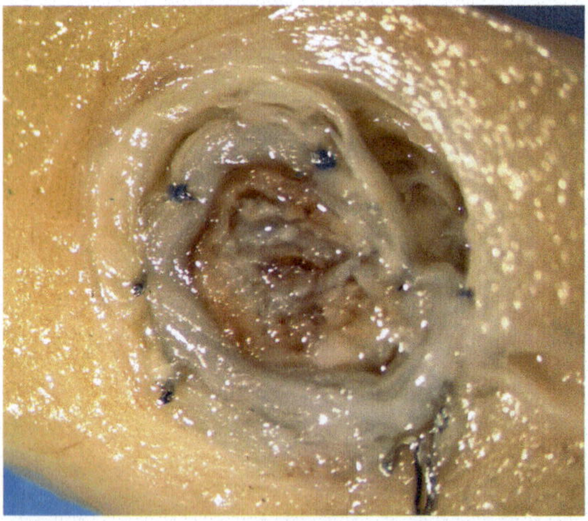

Fig. 5.51 A 29-year-old woman with a previous MI at the age of 21 that led to the implantation of a stent in the LDA. She suffered precordial pain and died suddenly at home. Complete luminal occlusion due to occlusive stent thrombosis can be appreciated

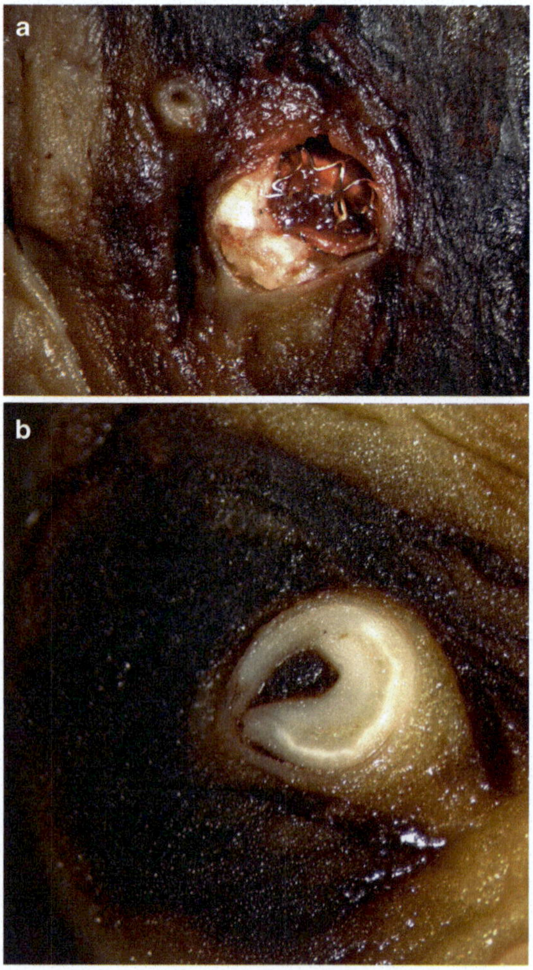

Fig. 5.52 (**a**) A 63-year-old woman with severe atherosclerosis in the RCA and the LDA coronary artery who suffered wall perforation during stent implantation in the RCA, which caused AMI in the posterior wall of the LV. (**b**) Blood occluded the vessel's lumen and invaded the epicardial adipose tissue due to the rupture of the vessel's wall

Bibliography

Badimón L, Vilahur G, Padró T. Lipoproteinas, plaquetas y aterotrombosis. Rev Esp Cardiol. 2009;62:1161–78.
Baroldi G. Myocardial cell death, including ischemic heart disease and its complications. In: Silver MD, Gottlieb AI, Schoen FJ, editors. Cardiovascular pathology. 3rd ed. London: Churchill Livingstone; 2001. p. 198–255.
Basso C, Thiene G. The pathophysiology of myocardial reperfusion: a pathologist's perspective. Heart. 2006;92:1559–62.
Basso C, Rizzo S, Thiene G. The metamorphosis of myocardial infarction following coronary recanalization. Cardiovasc Pathol. 2010;19:22–8.
Biondi-Zoccai G, Agostoni P, Sangiorgi GM, Airoldi F, et al. Incidence, predictors, and outcomes of coronary dissections left untreated after drug-eluting stent implantation. Eur Heart J. 2006;27:540–6.
Burke A, Tavora F. Practical cardiovascular pathology of coronary atherosclerosis. In: Burke A, editor. Tavora practical cardiovascular pathology. Philadelphia: Wolters Kluwer/Lippincott Williams & Wilkins; 2011. p. 73–85.
Cizek SM, Bedri S, Talusan P, Silva N, Lee H, Stone JR. Risk factors for atherosclerosis and the development of preatherosclerotic intimal hyperplasia. Cardiovasc Pathol. 2007;16:344–50.
Corrado D, Basso C, Poletti A, Angelini A, Valente ML, Thiene G. Sudden death in the young. Is acute coronary thrombosis the major precipitating factor? Circulation. 1994;90:2315–23.
Crea F, Camici PG, Bairey Merz CN. Coronary microvascular dysfunction: an update. Eur Heart J. 2014; 35:1101–11.
Davies MJ. Anatomic features in victims of sudden coronary death. Circulation. 1992;85:19–24.
Farb A, Weber DK, Kolodgie FD, Burke AP, Virmani R. Morphological predictors of restenosis after coronary stenting in humans. Circulation. 2002;105:2974–80.
Farb A, Burke AP, Kolodgie FD, Virmani R. Pathological mechanisms of fatal late coronary stent thrombosis in humans. Circulation. 2003;108:1701–6.
Iqbal J, Onuma Y, Ormiston J, Abizaid A, Waksman R, Serruys P. Bioresorbable scaffolds: rationale, current status, challenges, and future. Eur Heart J. 2014;35: 765–76.
Libby P. Mechanisms of acute coronary syndromes and their implications for therapy. N Engl J Med. 2013;368:2004–13.
Norris RM. The natural history of acute myocardial infarction. Heart. 2000;83:726–30.
Patel MR, Meine TJ, Lindblad L, Griffin J, Granger CB, Becker RC, Van de Werf F, White H, Califf RM, Harrington RA. Cardiac tamponade in the fibrinolytic era: analysis of >100000 patients with ST-segment elevation myocardial infarction. Am Heart J. 2006;151:316–22.
Rogers JH, Lasala JM. Coronary artery dissection and perforation complicating percutaneous coronary intervention. J Invasive Cardiol. 2004;16:493–9.
Schoen FJ. The heart. In: Kumar V, Abbas AK, Fausto N, editors. Robbins and Cotran: pathologic basis of disease. 7th ed. Philadelphia: Elsevier Saunders; 2005. p. 555–618.
Silvanto A, Noronha SV, Sheppard MN. Myocardial infarction with normal coronaries: an autopsy perspective. J Clin Pathol. 2012;65:512–6.
Stary HC, et al. A definition of initial, fatty streak, and intermediate lesions of atherosclerosis: a report from the Committee on Vascular Lesions of the Council on Arteriosclerosis, American Heart Association. Special Report. Arterioscler Thromb. 1994;14:840–56.
Stary HC, et al. A definition of advanced types of atherosclerotic Lesions and a Histological Classification of Atherosclerosis. A report from the Committee on Vascular Lesions of the Council on Arteriosclerosis, American Heart Association. Circulation. 1995;92: 1355–74.
Suarez-Mier MP, Merino JL. False lumen stent placement during iatrogenic coronary dissection. Cardiovasc Pathol. 2013;22:176–7.
Subirana MT, Juan-Babot JO, Puig T, Lucena J, Rico A, Salguero M, Borondo JC, Ordóñez J, Arimany J, Vázquez R, Badimon L, Thiene G, Luna AB. Specific characteristics of sudden death in a Mediterranean Spanish population. Am J Cardiol. 2011;107:622–7.
Robbins & Cotran Pathologic Basis of Disease, Seventh Edition. Saunders: Philadelphia; 2004.
Tavora F, Cresswell N, Li L, Ripple M, Fowler D, Burke A. Sudden coronary death caused by pathological intimal thickening without atheromatous plaque formation. Cardiovasc Pathol. 2011;20:51–7.
Virchow R. Cellular pathology. As based upon physiological and pathological histology. Lecture XVI-Atheromatous affection of arteries. 1858. Nutr Rev. 1989;47:23–5.
Virmani R, Kolodgie FD, Burke AP, Farb A, Schwartz SM. Lessons from sudden coronary death: a comprehensive morphological classification scheme for atherosclerotic lesions. Arterioscler Thromb Vasc Biol. 2000;20:1262–75.
Waller BF. The pathology of acute myocardial infarction: definition, location, pathogenesis, effects of reperfusion, complications, and sequelae. Cardiol Clin. 1988;6:1–28.
Yang EH, Kapoor N, Gheissari A, Burstein S. Coronary and intracerebral arterial aneurysms in a young adult with acute coronay syndrome. Tex Heart Inst J. 2012;39:380–3.

Non-Atherosclerotic Coronary Artery Disease

6

Pilar Molina, Benito Morentin, and Roberto Barriales-Villa

Contents

P. Molina (✉)
Histopathology Section, Forensic Pathology Department, Institute of Legal Medicine, Valencia, Spain

University of Valencia, Valencia, Spain
e-mail: pilar.molina@uv.es

B. Morentin
Forensic Pathology Department, Institute of Legal Medicine, Bizkaia, Spain

Forensic Medicine, Basque Country University, Bizkaia, Spain
e-mail: morentin.b@aju.ej-gv.es

R. Barriales-Villa
Inherited Cardiovascular Diseases Unit, Cardiology Department, Complejo Hospitalario Universitario A Coruña, Coruña, Spain

6.1 Introduction

The normal pattern of the coronary artery tree consists of two main coronary arteries (right and left), which originate in the right and left aortic sinuses, respectively. The main left coronary artery branches to give rise to the left anterior descending and circumflex coronary arteries. They arise perpendicularly to the aortic root without coursing between the pulmonary and aortic trunks. Right or left dominance (depending on where the posterior descending artery originates) is considered an individual variation with no pathological significance (see Chap. 1).

Although occurring at low frequency, several congenital or acquired disorders are associated with sudden cardiac death (SCD): in relation to the origin (number, position, shape and size of the *ostium*, angle of origin), the proximal (aortic intramural) or distal course (fistulas, vasculitis, dissections), and the intramural branches. In one half of the cases, the initial diagnosis is *post-mortem*, since they tend to debut with sudden death. Retrospectively, it's possible to discover some syncope or chest pain symptomatology prior to death.

Among the non-atherosclerotic coronary diseases, the most frequent and worth mentioning disorders are anomalous origin, spontaneous dissection, intramyocardial bridge and vasculitis. Other less frequent pathologies are: fistulas, embolisms, fibromuscular dysplasia, idiopathic infantile arterial calcification and tumoral pathology.

J.S. Lucena et al. (eds.), *Clinico-Pathological Atlas of Cardiovascular Diseases*,
DOI 10.1007/978-3-319-11146-9_6, © Springer International Publishing Switzerland 2015

Remarks
There are familial cases of anomalous origin of the coronary arteries, in which a family screening would be recommended
Anomalies of the coronary arteries can be the cause of death with different degrees of certainty:
Certain cause: anomalous origin from the pulmonary artery
Probable cause: anomalous left anterior descending or left coronary artery origin from the right aortic sinus with an interarterial course
Uncertain cause: anomalous right coronary artery origin from the left aortic sinus, anomalous left anterior descending coronary artery origin from the right aortic sinus without an interarterial course, high take-off origin, intramyocardial bridge

6.2 Congenital Anomalies

Congenital anomalies of the coronary arteries constitute the most frequent cause of sudden death (SD) from non-atheriosclerotic coronary diseases (second cause of death in athletes in the USA). In the majority of cases, SD occurs during or immediately after exercise. There are several classifications and their incidence varies depending on the source of the published study (whether based on necropsies, coronary angiographies, echocardiographies, multisection CT, cardiac MRI, etc.), although in general it can be assumed that the incidence ranges from 0.3 to 1.3 %.

The clinical spectrum and presentation is variable, from asymptomatic patients to patients presenting with angina, acute myocardial infarction and sudden death. The major concern for clinicians is to establish which type of coronary artery anomaly may lead to SD. In young people (<30 years), when the anomalous coronary artery is the dominant one or when the course is interarterial (between the aorta and pulmonary artery) or intramural, the risk of sudden death increases.

The most frequent causes are summarised in Table 6.1.

6.2.1 Anomalous Origin of the Coronary Arteries from the Opposite Aortic Sinus

This is the most common congenital anomaly. There are different variations according to the location of the *ostium*, to whether they originate from a common orifice or from two separate ones, and to the course of the anomalous coronary artery until it reaches its dependent territory. Thus, an interarterial course (between the aorta and pulmonary artery) is associated with a greater compromise in arterial blood flow and consequently with increased lethality. Some authors suggest that anomalous courses are associated with a higher presence of atherosclerotic disease.

6.2.2 Anomalous Origin of the Coronary Arteries from the Pulmonary Trunk

Although there are some instances where both coronary arteries originate from the pulmonary artery, the most frequent situation is anomalous left coronary artery from the pulmonary artery, and in 95 % of the cases from the left pulmonary sinus. Until recently, diagnosis of this anomaly in an infant was associated with high mortality, although currently the use of non-invasive imaging techniques has allowed the diagnosis of new cases in the adult population with an average age of 41 years. The right or left anterior descending coronary arteries can also arise from the pulmonary artery but the frequency is very low and these anomalies are generally asymptomatic, although in some cases they have been described associated with sudden death.

6.2.3 High Take-off Origin

This congenital anomaly is an infrequent cause of sudden death. The coronary *ostia* are usually located in the Valsalva sinuses although they can also be situated above them, especially the left *ostium*. A location of the *ostium* up to 2–3 mm above the sinotubular junction is considered normal. Although there is no conclusive measurement it is generally accepted that high take-off origin occurs when the distance to the Valsalva sinus is greater than 1 cm. In these cases, the initial course of the coronary artery is vertical and intramural with an acute take-off angle that hinders diastolic filling. The *ostium* is buttonhole-shaped.

6.2.4 Myocardial Bridges

Typically, the main coronary arteries are located in the epicardium, but in some occasions a segment can tunnel through the myocardium during part of its course. Myocardial bridges are most commonly located in the proximal and middle third segment of the left anterior descending coronary artery and, with less frequency, in the posterior descending or circumflex arteries. Although this finding does not imply significant haemodynamic alterations in the majority of instances, in a small percentage of cases, myocardial bridging causes a transient reduction in the arterial lumen during systolic contraction ("milking effect"). It is important to take into account that myocardial bridges are part of the phenotype of hypertrophic cardiomyopathy and they are frequently found in these patients (between 15 and 28 % in angiography series).

On rare occasions, myocardial bridging has been associated with sudden death. In the majority of these cases, death occurs after an episode of intense physical exercise or emotional stress.

Although there is no consensus about the diagnostic criteria, some guidelines and considerations have been published by several cardiovascular pathologists in order to establish myocardial bridging as the cause of sudden death:

- Virmani et al. (2001) recommend to register the length and depth of the intramyocardial segment and to look carefully for ischaemic involvement in the anterior wall and septum of the LV. When the tunnel depth is greater than 5 mm, it can be considered as a possible cause of sudden death. The presence of disarray and fatty degeneration in the myocardium above the tunnel supports the diagnosis, as well as its association with a small posterior descending coronary artery and/or tunneling of other epicardial arteries.
- Thiene et al. (2001) establish pathological relevance when the dimensions of the tunnelled segment range from 2 to 3 cm in length and 2–3 mm in depth, and the morphology of the surrounding myocardium is similar to a sphincter with concentric disposition around the artery. As the previous authors highlighted, myocardial disarray and fibrosis in the surrounding myocardium supports the diagnosis.

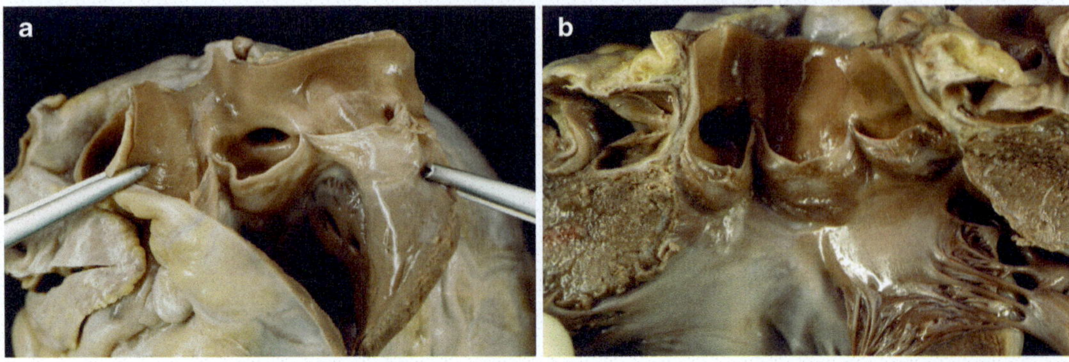

Fig. 6.1 *Anomalous origin of the left coronary artery from the pulmonary artery.* An 11-year-old boy who died suddenly after a race. (**a**) Origin of the left coronary artery from the pulmonary artery. (**b**) Very dilated origin of the right coronary artery in the normal location (right aortic sinus)

Table 6.1 Congenital anomalies most related to SD

Anomalous origin of the left coronary artery from pulmonary artery (Fig. 6.1)
Ostium in left pulmonary sinus (most frequent)
Dilated and tortuous right coronary artery and its branches
Signs of chronic ischaemia and left ventricular hypertrophy are frequently found
Anomalous origin of the left coronary artery from the right aortic sinus (Figs. 6.2 and 6.3)
Anomalous ostium close to the aortic valve commissure, and frequently slit-like. Acute angle between the aorta and the coronary artery
The course can be interarterial (towards the right in the epicardium or within the aorta adventitia), retroaortic (towards the left) or within the septum
Signs of chronic ischaemia and left ventricular hypertrophy are frequently found
Anomalous origin of the right coronary artery from the left aortic sinus (Figs. 6.4 and 6.5)
Anomalous ostium similar to the above
Signs of myocardial ischaemia are generally not found
Single coronary ostium and artery (Fig. 6.6)
Single left or right coronary artery originating the rest of the branches
High take-off origin (Figs. 6.5, 6.6 and 6.7)
Ostium located more than 1 cm above the sinotubular junction
Usually buttonhole-shaped with an initial intramural course
Myocardial bridging (Figs. 6.8, 6.9 and 6.10)
Depth of the tunnel more than 5 mm or depth of 2–3 mm for 2–3 cm in length
Morphology of the surrounding myocardium similar to a sphincter
Perivascular myocardial disarray and fat degeneration

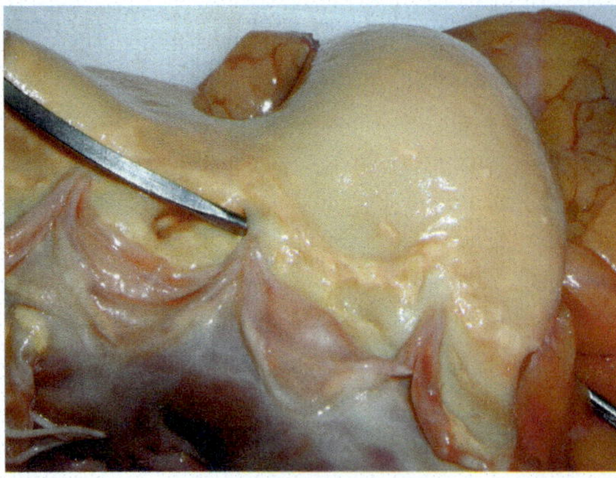

Fig. 6.2 *Anomalous origin of the left coronary artery from the right aortic sinus.* Sudden death after prolonged exertion in a 47-year-old male. The heart showed atherosclerotic plaques in the aorta, severe coronary atherosclerosis in the proximal course of the LAD coronary artery and subendocardial scarring fibrosis in the LV. In this case, the cause of death should be attributed to the coronary atherosclerosis rather than to the congenital anomaly. However, the anomalous origin of the coronary artery may speed up the atherosclerotic process given the turbulence in the blood flow caused by the presence of acute angles in its course

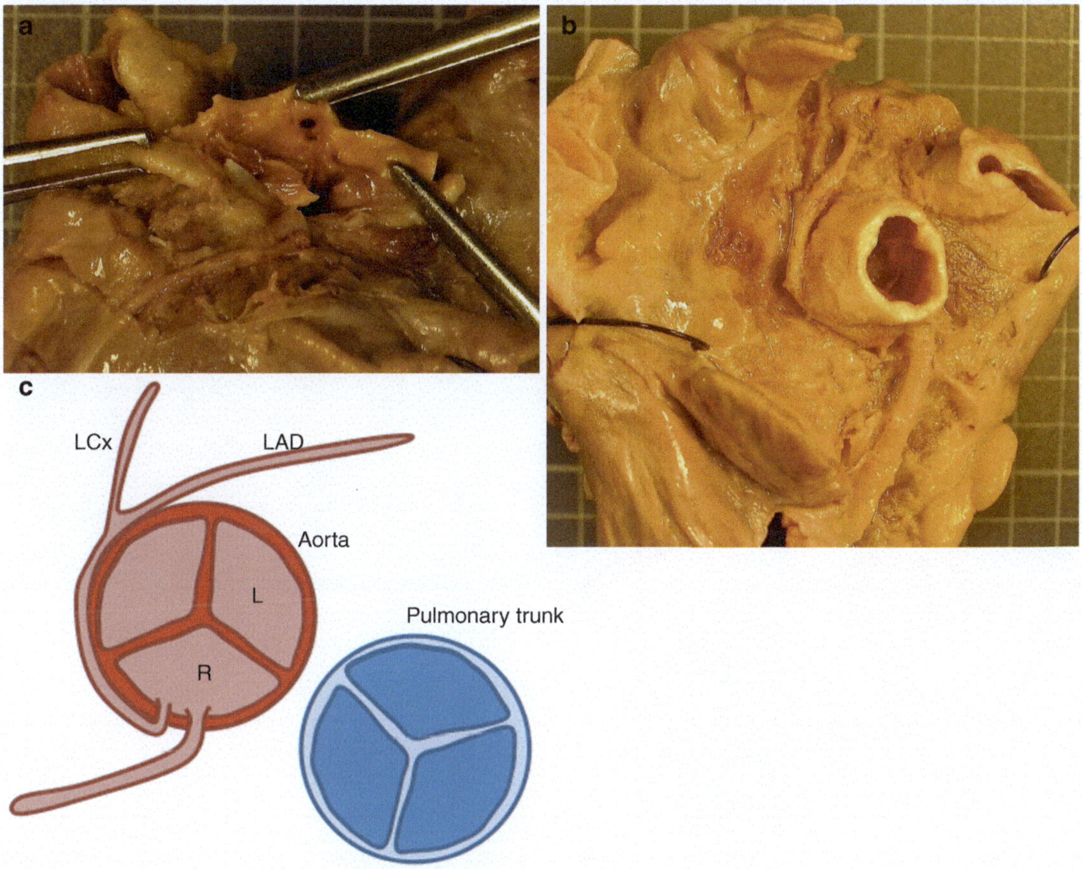

Fig. 6.3 *Anomalous origin of the left coronary artery from the right aortic sinus.* An 8-year-old boy who suffered a cardiac arrest episode when coming off a fairground attraction. (**a**) Both coronary arteries arise from the right coronary sinus, the left coronary *ostium* is located below the right one and its calibre is smaller. (**b**) The right coronary artery follows its usual path, but the left coronary artery turns to the right and backwards, originating an acute take-off angle, continuing its course within the aorta's adventitia and surrounds the posterior side of the aorta up to its bifurcation (Saldaña et al., (2009). Published with permission from Cuadernos Medicina Forense). (**c**) Diagram representing this anomaly

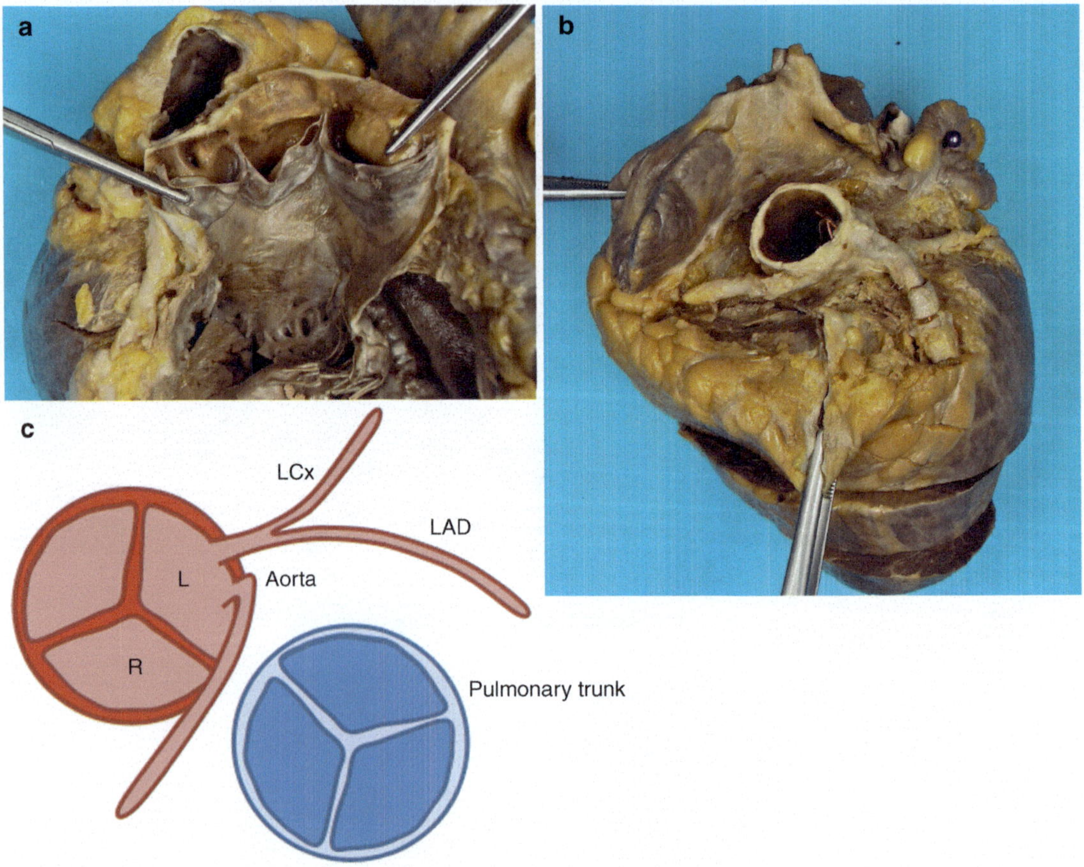

Fig. 6.4 *Anomalous origin of the right coronary artery from the left aortic sinus.* A 43-year-old male who died suddenly after complaining for several days of pain in the left arm. (**a**, **b**) Anomaly in the RCA that arises from the left sinus and has an intramural aortic course. (**c**) Diagram representing this anomaly

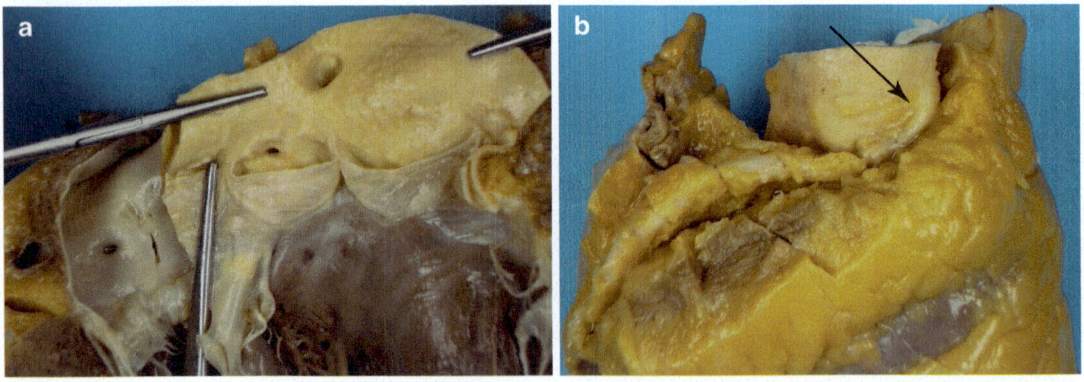

Fig. 6.5 *High take-off and displacement to the left of the right coronary ostium.* A 49-year-old male who died in hospital while undergoing biopsy of a tumoural tongue lesion. (**a**) Right *ostium* 1.5 cm above the sinotubular junction and displaced laterally to the left over the anterior commissure, at the junction between the right and the left Valsalva sinuses. (**b**) Detail of the high take-off of the right coronary artery with an acute angle take-off and an intramural aortic course of 2 cm (*arrow*) until it reaches its natural path underneath the right atrial appendage

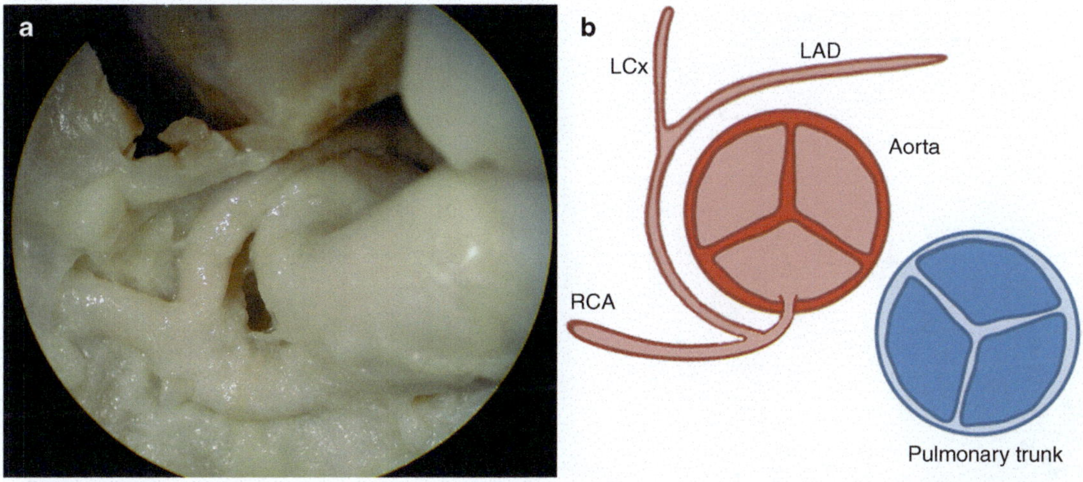

Fig. 6.6 *Common origin of both right and left coronary arteries in the right aortic sinus (single coronary artery).* A 4-month-old infant male with malformation syndrome of unknown aetiology with karyotype 47XYY. Sudden death at home. (**a**) Single coronary *ostium* in the right coronary sinus from where a common trunk arises that immediately bifurcates in both left and right coronary arteries. The RCA keeps its normal course whereas the LCA surrounds the aortic arch in its posterior side and courses to reach the anterior part of the heart, where it originates the circumflex and LAD arteries. (**b**) Diagram representing this anomaly

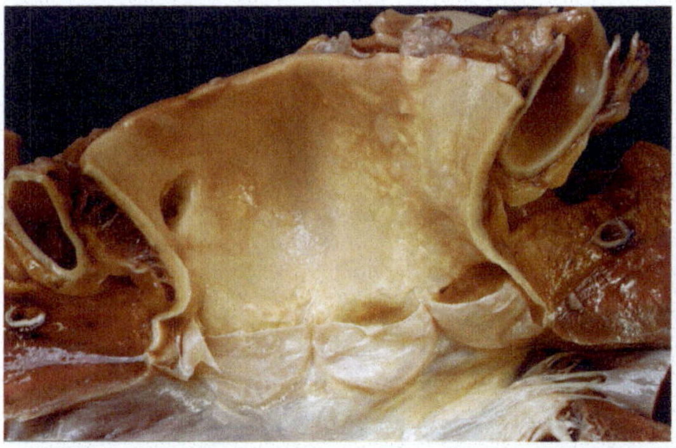

Fig. 6.7 *High take-off origin of the right coronary artery.* A 60-year-old woman who died suddenly at home. Autopsy: hydrocephalia with subependymal gliosis and dilated cardiomyopathy (heart weight: 815 g. LV: 5 cm in diameter)

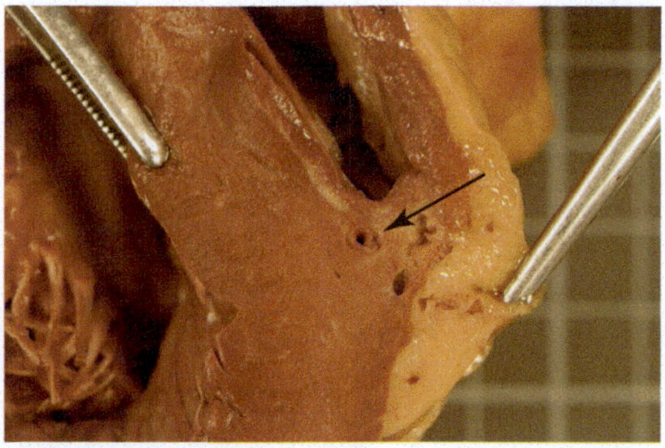

Fig. 6.8 *Intramyocardial course of the left anterior descending coronary artery.* A 47-year-old male who died suddenly at home after an episode of chest pain. Heart weight of 418 g with biventricular hypertrophy (RV wall thickness: 0.5 cm; LV wall thickness: 1.5 cm. The LAD coronary artery was dilated in its proximal portion and from the mesocardium, the course turns intramyocardial for 4 cm in which it tunnels through the interventricular septum at a depth of 7 mm from the epicardium (*arrow*), emerging again to it. Areas of fibrosis and perivascular adipose infiltration can be appreciated in the myocardium surrounding the bridge

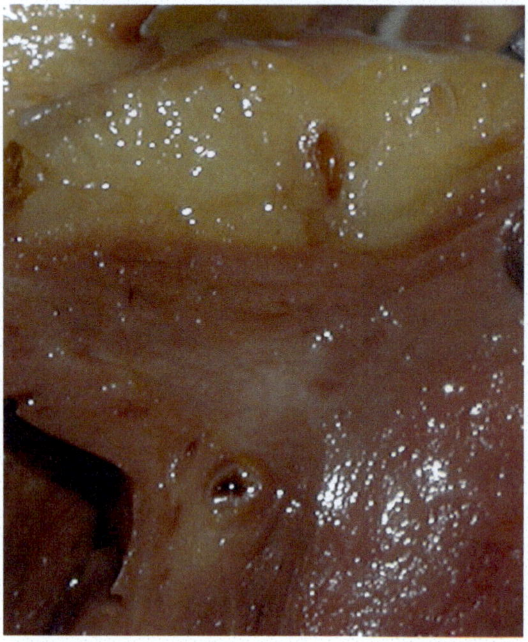

Fig. 6.9 *Intramyocardial course of the left anterior descending coronary artery.* Sudden death in a 27-year-old male. The middle third portion of the LAD coronary artery dips into the myocardium, at a depth of 0.7–0.8 cm and for a length of 1.3 cm. Areas of fibrosis and perivascular adipose infiltration in the myocardium surrounding the bridge

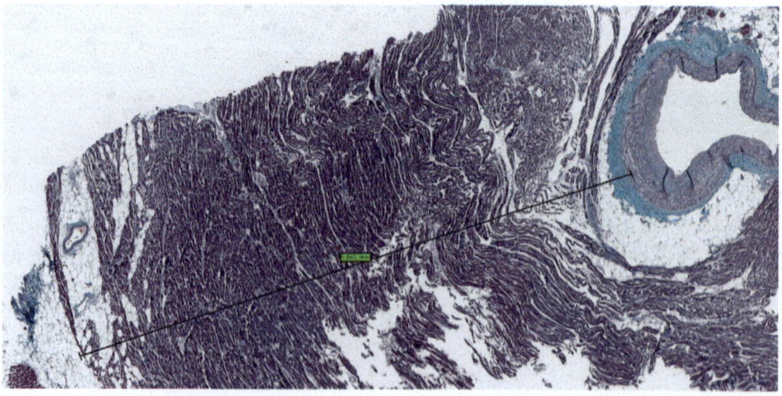

Fig. 6.10 *Histological section of a myocardial bridge.* Sudden death in a 33-year-old male. Intramural course of the middle third of the left anterior descending coronary artery (myocardial bridge) of 5.065 mm depth and 20 mm length. Band of subepicardial myocardial tissue enveloping the coronary artery, causing compression of the artery during systole (Masson's trichrome)

6.3 Coronary Artery Dissection

Coronary artery dissection is characterized by the formation of a haematoma in the arterial wall between the media and adventitia that compresses the vessel lumen, provoking acute ischaemia. Spontaneous forms appear mainly in young women and one third of these are related to pregnancy. Dissection can be spontaneous or iatrogenic. Depending on its extension, the clinical presentation can be angina, AMI, arrhythmias or sudden death. The coronary artery most frequently involved is the LAD in women and the RCA in men. Exceptionally, several arteries can be affected. A plethora of predisposing factors have been described: atherosclerosis, hormonal imbalance in the postpartum period or due to the use of contraceptive drugs (progesterone-dependent changes in the connective tissue), or some collagenopathies such as Marfan or Ehlers-Danlos syndromes. Hemodynamic triggering factors, such as intense physical exercise or cocaine action, all provoke an acute increase in the perfusion pressure of the coronary arteries. Some authors have described a possible association between coronary artery dissection and fibromuscular dysplasia in other non-coronary vascular beds.

Histology reveals a normal medial layer, however an inflammatory infiltrate mainly composed of eosinophils can be seen in the adventitia of the dissected area. The significance of this eosinophilic infiltrate may have a double interpretation. The presence of haemosiderin, macrophages and lymphocytes is indicative of a chronic state, which together with a possible lytic action by proteases released from the eosinophilic granules would imply a causal link. However, it is strange that the infiltrate is confined only to the adventitia, in contrast with true vasculitis. An alternative explanation would be that the infiltrate results from a gradual process of several hours' evolution, where the separation between the arterial medial and outer elastic layers would impede the transfer of these cells from the *vasa vasorum* of the adventitia. Later, a small tear in the arterial intima would allow blood to flow through, prompting an acute occlusion of the arterial lumen. The absence of this infiltrate in other areas would favour this latter theory. The pathological features are summarised in Table 6.2.

Table 6.2 Pathology of coronary artery dissection

Macroscopic features:
The most frequently involved coronary artery is the LAD (80 %) and the dissection locates to the outermost third of the arterial wall
Intramural haematoma with vessel lumen obliteration: the newly-formed blood clot fills the false lumen simulating in many occasions coronary thrombosis (Figs. 6.11 and 6.12)
The heart usually does not show any other pathology except for acute ischaemia in the more evolved cases
Microscopic features: (Fig. 6.13)
The haematoma is located between the muscular layer and the outermost elastic layer
Inflammatory infiltrate in the adventitia mainly composed of eosinophils
Degenerative or atherosclerotic pathology is usually not found
An intimal tear may be observed

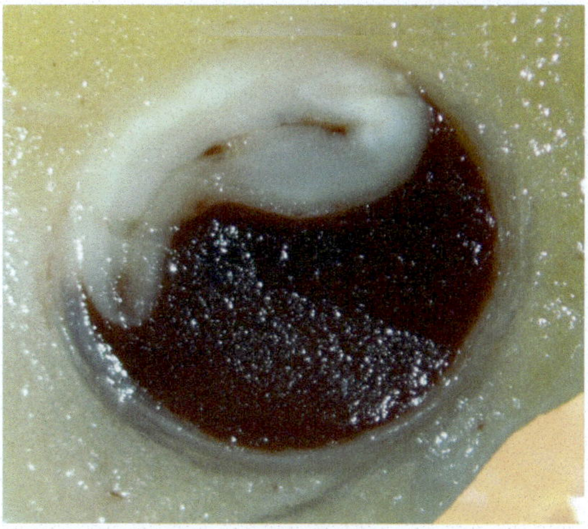

Fig. 6.11 *Dissection in the left anterior descending coronary artery.* Sudden death in a 46-year-old woman after an episode of acute dyspnea and syncope. Dissection in the LAD coronary artery (3 cm in length) and the first diagonal

Fig. 6.12 *Dissection in the left anterior descending coronary artery.* A 38-year-old woman who died suddenly while having a family meal 18 days after giving birth. Transverse section of the artery showing a recently-formed haematoma that has dissected the arterial medial layer, displacing and obliterating the vessel lumen

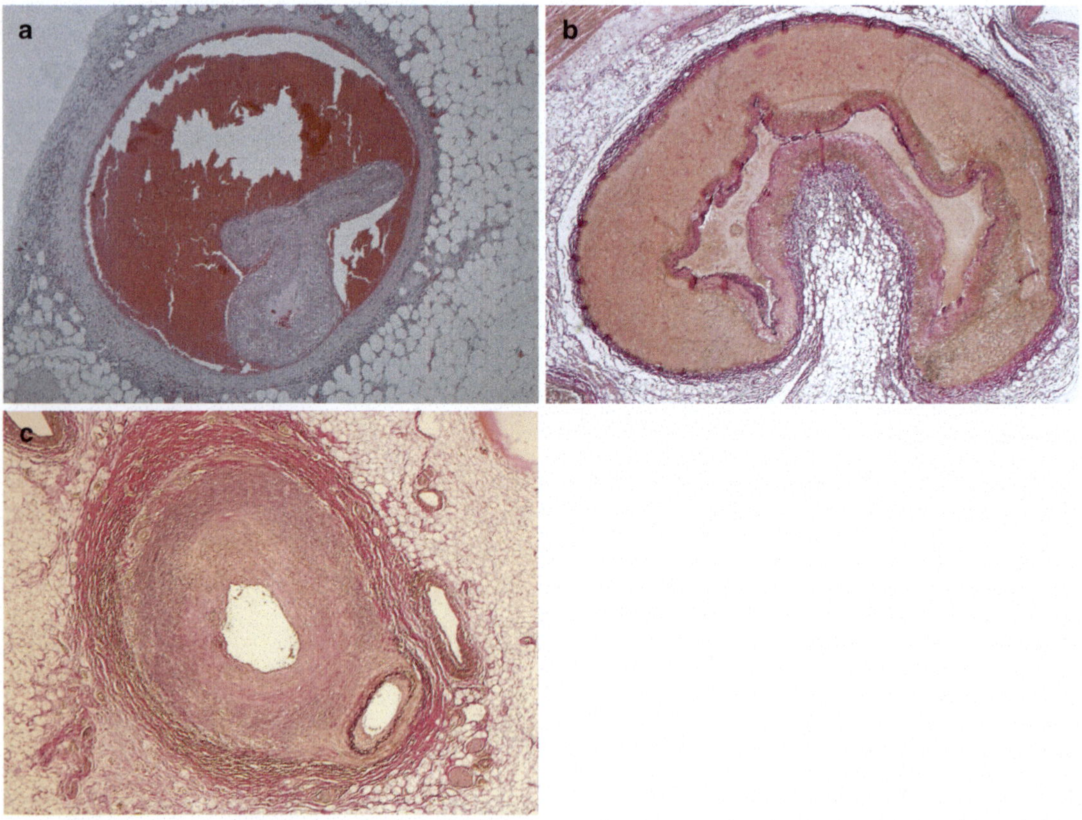

Fig. 6.13 *Histological sections of coronary artery dissection. (***a***)* Acute haemorrhage between the arterial media and adventitia that collapses the vessel lumen. Inflammatory infiltrate in the adventitia formed mainly by eosinophils. (**b**) In this section, the dissection spares a portion of the arterial wall (Weigert's stain). (**c**) Old recanalized dissection. The specific staining of the elastic fibres (Weigert's stain) allows the identification of the true lumen surrounded by the internal elastic layer (*), and its differentiation from the recanalized haematoma (+)

6.4 Vasculitis

Coronary vasculitis is a rare cause of sudden cardiac death. It is defined as the presence of inflammatory cells in the vessel walls, causing structural damage. The loss of the vessel's integrity can lead to bleeding or to a compromise of the vessel's lumen prompting ischaemia or necrosis. Coronary vasculitis can occur as an isolated entity or as a manifestation of systemic processes such as Kawasaki, Takayasu's, Churg-Strauss or Wegener's disease or systemic lupus erythematosus, among others. In adulthood, Takayasu arteritis is the type of vasculitis that most affects the coronary arteries, preferentially the coronary *ostia* in women. In children and the young, sequelae of a previous vasculitis in the context of Kawasaki disease, with thombosed aneurysms, is the most common aetiology and affects preferentially the male gender. The pathological characteristics are summarised in Table 6.3.

Table 6.3 Pathology of vasculitis

Macroscopic features:
Proximal involvement, close to the coronary *ostia*
Frequent concentric obstruction without atheroma plaques
The heart can present with acute ischaemia or myocarditis
It can co-exist with aortic dissection
Microscopic features: (Figs. 6.14, 6.15 and 6.16)
Intimal fibrous proliferation with a destructured medial layer due to an inflammatory process in which there may be necrosis, granuloma formation and different types of inflammatory cells (polymorphonuclear leukocytes, plasma cells and mulitnucleated cells)

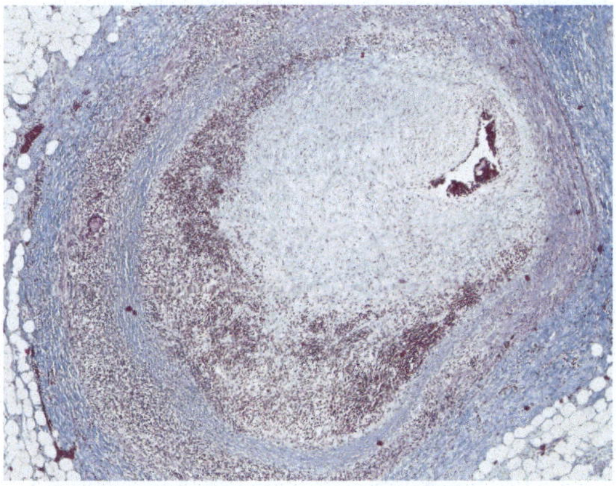

Fig. 6.14 *Aortic dissection and coronary artery involvement caused by giant cell arteritis.* An 81-year-old woman who attended the emergency department with chest and abdominal pain, hypotension, and died not long after. Intimal tear in the ascending aorta with parietal dissection (same case as Fig. 3.32). There was luminal stenosis greater than 80 % in the RCA and LAD coronary artery (not shown), arteritis with marked intimal fibrous proliferation and an intense inflammatory process of the media, composed of lymphocytes, plasma cells and multinucleated giant cells, and also fibrosis (Masson's trichrome)

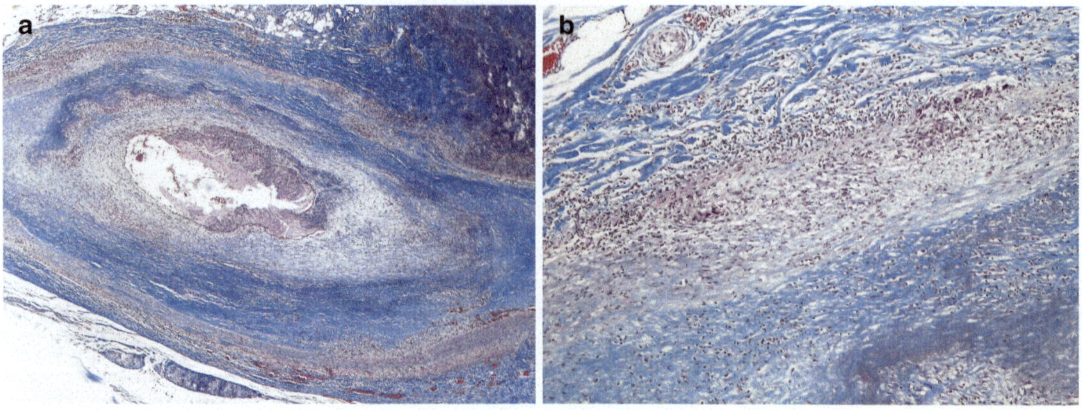

Fig. 6.15 *Takayasu arteritis.* A 24-year-old woman with no previous medical history who died suddenly during her daily routine. (**a**) Occlusion in the left main coronary artery near the *ostium* caused by severe wall fibrosis and mural thrombosis. (**b**) Inflammatory infiltrate in the adventitia with multinucleated giant cells surrounded by lymphocytes. There are also whitish plaques in the ascending aorta that, in the microscopic analysis, correspond to giant cell aortitis. (Masson's trichrome)

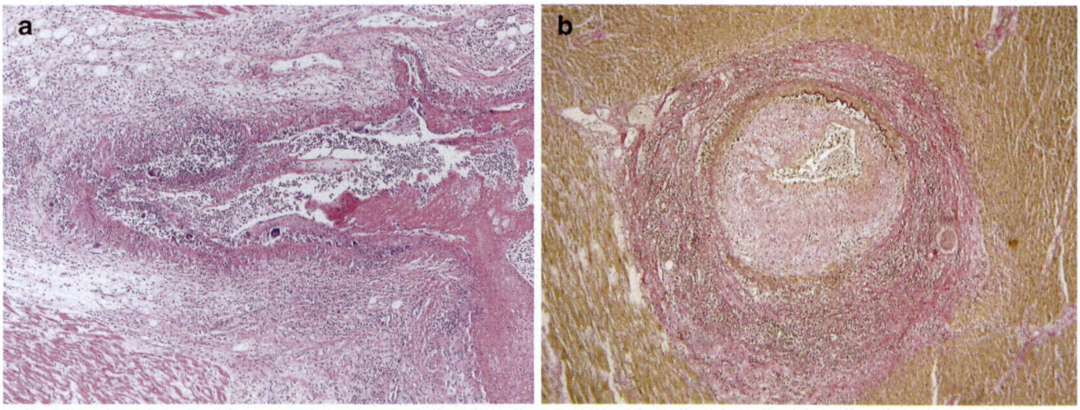

Fig. 6.16 *Wegener's granulomatosis.* Sudden death in a 45-year-old male admitted to hospital with fever, productive cough, occasionally with haemoptoic expectoration, light dyspnea and malaise. Chest X-ray: Bilateral pulmonary infiltrates. Results from c-ANCA test were received *post-mortem* and the values were high: c-ANCA = 1:320 (N > 1:40). This type of vascultis is characterised by the presence of lesions, which are found in different stages of evolution. (**a**) Intramural coronary artery with intimal thrombus formation, parietal necrosis and mixed inflammatory infiltrate of polymorphonuclear cells, lymphocytes, plasma cells and giant cells that extends to the adventitia. (**b**) Evolved lesion with marked intimal fibrosis. Weigert's stain, specific for elastic fibres, demonstrates the partial destruction of the medial layer

6.5 Intramyocardial Small Vessel Disease

The coronary artery tree consists of epicardial arteries (conducting vessels), intramyocardial arteries (resistance vessels), arterioles, capillaries, venules and veins. Intramyocardial small vessel disease is the term given to describe the pathology associated with the coronary microcirculation, which includes the coronary vessels with a diameter less than 300 μm. There is ever increasing evidence that coronary microcirculation plays a key role in the physiopathology of numerous cardiac diseases.

Pathology of intramyocardial small vessels has been described associated to a large number of cardiac diseases: hypertrophic or dilated cardiomyopathy, Friedreich's ataxia, scleroderma, long QT syndrome, Marfan syndrome, muscular dystrophy, aortic stenosis, mitral valve prolapse, hypertension, diabetes mellitus, hypercholesterolemia or syndrome X.

It can be the only pathological finding in cases of sudden death, and when this occurs it is difficult to establish whether intramyocardial small vessel disease is *per se* the cause of death or just an epiphenomenon. The physiopathology of this vascular dysfunction is multifactorial and complex. It is thought that the mechanisms by which it could lead to sudden death are due to ventricular arrhythmias secondary to the fibrosis produced in the territory irrigated by the small vessel. The pathological features are summarised in Table 6.4.

Camici and Crea (2007) proposed the following classification for intramyocardial small vessel disease:

1. **Coronary microvascular dysfunction in the absence of obstructive large vessel coronary artery disease or myocardial disease.** These cases correlate with the presence of cardiovascular risk factors (tobacco, hypertension, diabetes and insulin resistance). It is not exactly known whether this dysfunction affects the arterioles or the arteries between 100 and 500 μm that are located between the epicardial vessels and the arterioles.
2. **Coronary microvascular dysfunction in the presence of myocardial disease.** It occurs in primary (dilated or hypertrophic cardiomyopathy) or secondary (hypertensive and valvular) cardiomyopathies. It is caused by an adverse remodelling of the intramural arterioles.
3. **Coronary microvascular dysfunction in the presence of obstructive large vessel coronary artery disease.** It can occur in the context of stable coronary artery disease or in acute coronary syndrome. It is responsible for the "no re-flow phenomenon" associated with a worse prognosis during percutaneous reperfusion.
4. **Iatrogenic coronary microvascular dysfunction.** It occurs after reperfusion of the coronary lesion and may be due to vasoconstriction or distal embolism.

Table 6.4 Pathology of intramyocardial small vessel disease

Macroscopic features:
The heart may show signs of acute or chronic ischaemia or left ventricular hypertrophy, although it is more frequent that the heart appears normal on gross examination
Microscopic features **(Figs. 6.17, 6.18 and 6.19):**
Thickening of the arterial wall
Muscular hyperplasia or dysplasia of the medial layer
Elastosis in the adventitial layer
Perivascular and interstitial fibrosis surrounding the arterioles

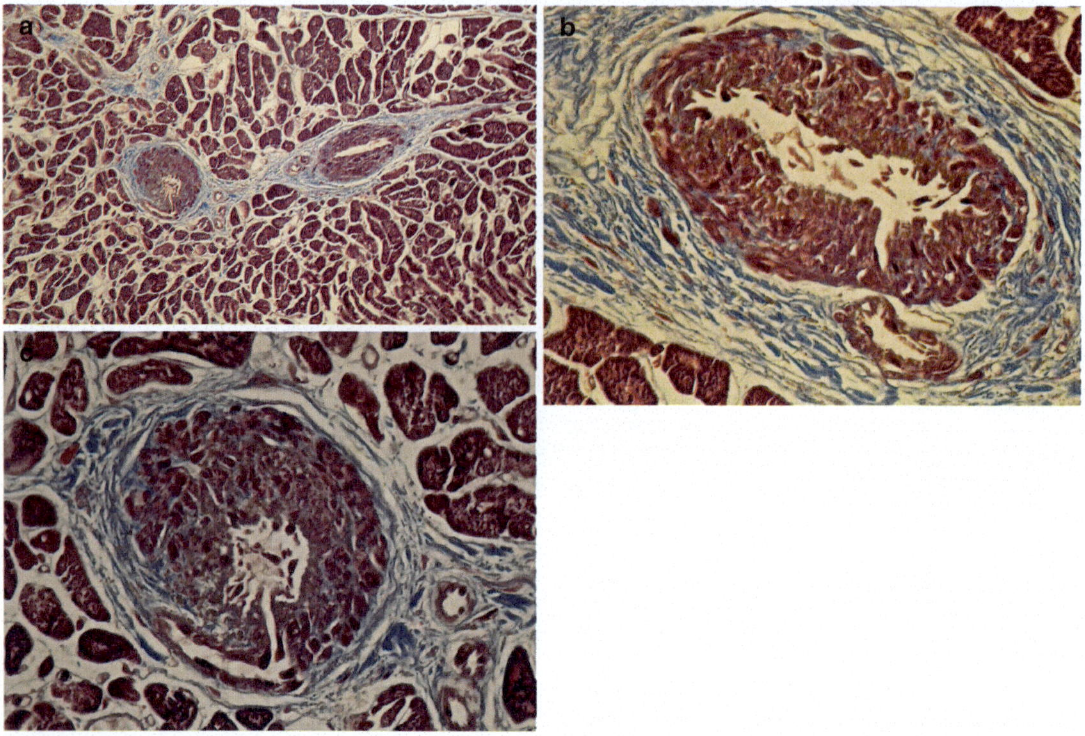

Fig. 6.17 *Intramyocardial small vessel disease.* A 23-year-old male who died suddenly while having a family meal. BMI 22.5, heart 305 g without gross anomalies. Toxicology was negative. (**a**) Parietal thickening of the arterial wall caused by muscular hyperplasia. (**b**, **c**) Higher magnification images of the lesion (Masson's trichrome)

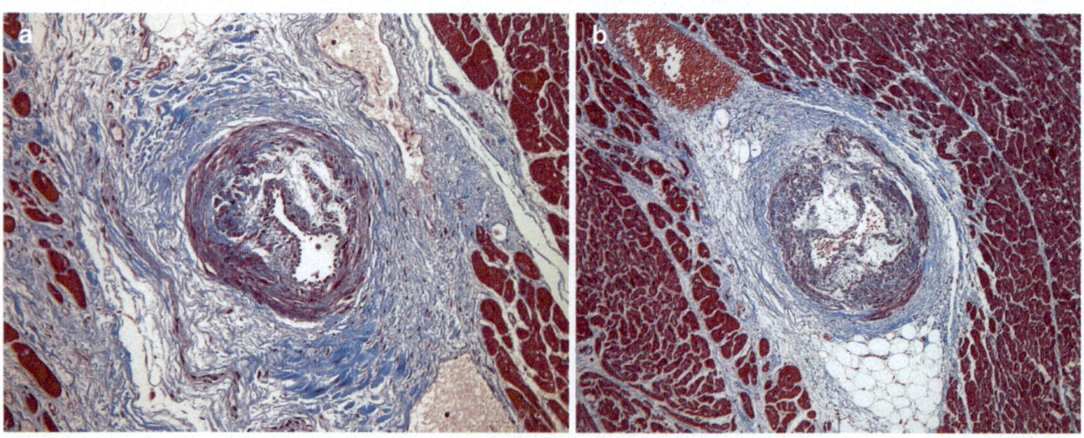

Fig. 6.18 *Intramyocardial small vessel disease.* Sudden death in a 54-year-old woman without previous medical history. *Post-mortem* cardiac examination revealed an increased heart weight with hypertrophy of the left ventricle, bilateral atrial dilation, subendocardial infiltration of adipose tissue in both ventricles, and mild atherosclerosis in the LAD coronary artery. (**a**, **b**) Intramyocardial small vessel disease in intramural arteries, with dysplastic thickening of the medial layer and significant perivascular fibrosis (Masson's trichrome)

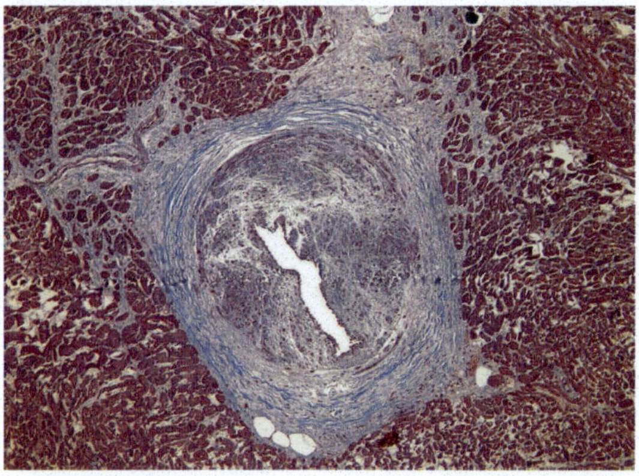

Fig. 6.19 *Intramyocardial small vessel disease.* A 44-year-old male with no previous medical history died suddenly whilst working in the field. *Post-mortem* study: heart weight 483 g (mean value related to body weight: 290 g (range 198–425 g). Left ventricular wall thickness: 1.5 cm, interventricular septum: 1.5 cm and right ventricle: 0.2 cm. Valves, myocardium and coronary arteries with no gross lesions. Microscopically: small vessel lesion (fibromuscular dysplasia) in the summit of the interventricular septum and in the free wall (Masson's trichrome)

Bibliography

Aguilera B, Suárez-Mier MP. Muerte súbita cardiaca. REA:E J Autopsy. 2003;1:21–34.

Angelini P. Coronary artery anomalies—current clinical issues. Definitions, classification, incidence, clinical relevance, and treatment guidelines. Tex Heart Inst J. 2002;29:271–8.

Barriales-Villa R, Morís C. Anomalías congénitas de las arterias coronarias con origen en el seno de Valsalva contralateral: ?qué actitud se debe seguir? Rev Esp Cardiol. 2006;59:360–70.

Basso C, Morgagni GL, Thiene G. Spontaneous coronary artery dissection: a neglected cause of acute myocardial ischemia and sudden death. Heart. 1996;75:451–4.

Basso C, Corrado D, Thiene G. Congenital coronary artery anomalies as an important cause of sudden death in the young. Cardiol Rev. 2001;9:312–7.

Basso C, Thiene G, Mackey-Bojack S, Frigo AC, Corrado D, Maron BJ. Myocardial bridging, a frequent component of the hypertrophic cardiomyopathy phenotype, lacks systematic association with sudden cardiac death. Eur Heart J. 2009;30:1627–34.

Brothers JA, Stephens P, Gaynor JW, et al. Anomalous aortic origin of a coronary artery with an interarterial course. J Am Coll Cardiol. 2008;51:2062–4.

Camici PG, Crea F. Coronary microvascular dysfunction. N Engl J Med. 2007;356:830–40.

Carratalá F, Ortiz JM, Presentación C, Molina P. Patología coronaria no arteriosclerótica como causa de muerte súbita en adultos: Casuística del Instituto de Medicina Legal de Valencia (1997–2005). Cuad Med Forense. 2005;42:307–16.

Cheitlin MD, MacGregor J. Congenital anomalies of coronary arteries: role in the pathogenesis of sudden cardiac death. Herz. 2009;34:268–79.

Daniotti A, Povolo G, Barchitta A, Abudureheman A, Cardaiolia P, Basso C. Spontaneous coronary artery dissection mimicking aortic dissection. Cardiovasc Pathol. 2004;13:330–3.

Herrmann J, Kaski JC, Lerman A. Coronary microvascular dysfunction in the clinical setting: from mystery to reality. Eur Heart J. 2012;33(22):2771–82b.

Hill SF, Sheppard MN. Non-atherosclerotic coronary artery disease associated with sudden cardiac death. Heart. 2010;96:1119–25.

Lanza GA, Crea F. Primary coronary microvascular dysfunction: clinical presentation, pathophysiology, and management. Circulation. 2010;121:2317.

Malagò R, Pezzato A, Barbiani C, Alfonsi U, Nicolì L, Caliari G, Pozzi Mucelli R. Coronary artery anatomy and variants. Pediatr Radiol. 2011;41:1505–15.

Morentin B, Peciña T, Aguilera B. Sudden death due to an anomalous origin of the left anterior descending coronary artery from the pulmonary trunk in a 9-year-old boy. Rev Esp Cardiol. 2000;53:294–6.

Saldaña L, Santos M, Rodríguez JC, Rico A, Pachar JV, Lucena J. Origen anómalo de la arteria coronaria izquierda en el seno aórtico derecho y muerte súbita: a propósito de dos casos. Cuad Med Forense. 2009;15:67–75.

Thiene G, Basso C, Corrado D. Cardiovascular causes of sudden death. In: Silver MD, Gotlieb AI, Schoen FJ, editors. Cardiovascular pathology. 3rd ed. New York: Churchill Livingstone; 2001. p. 326–74.

Tweet MS, Hayes SN, Pitta SR, Simari RD, Lerman A, Lennon RJ, Gersh BJ, Khambatta S, Best PJ, Rihal CS, Gulati R. Clinical features, management, and prognosis of spontaneous coronary artery dissection. Circulation. 2012;126:579–88.

Van der Wall A. Coronary artery pathology. Heart. 2007; 93:1484–9.

Veinot JP, Johnston B, Acharya V, Healey J. The spectrum of intramyocardial small vessel disease associated with sudden death. J Forensic Sci. 2002;47:384–8.

Virmani R, Burke A, Farb A, Atkinson JB. Cardiovascular pathology, Major problems in pathology series , vol. 40. 2nd ed. Philadelphia: Saunders company; 2001.

Yau JM, Singh R, Halpern EJ, Fischman D. Anomalous origin of the left coronary artery from the pulmonary artery in adults: a comprehensive review of 151 adult cases and a new diagnosis in a 53-year-old woman. Clin Cardiol. 2011;34:204–10.

Pathology of the Heart Valves

7

M. Paz Suárez-Mier, Benito Morentin, Marta Cobo, Evaristo Castedo, and Pablo García-Pavía

Contents

7.1 Introduction

Valvular pathology may have a congenital or acquired origin and can affect any one of the four heart valves, although the most frequently involved are the mitral and aortic valves. Valve disease results in stenosis (difficulty in opening), regurgitation (closure incompetence), or both. The most common abnormalities are aortic stenosis and mitral stenosis.

Clinical presentation depends on the severity of the valve damage, the speed of onset of the pathological changes and the compensatory mechanisms from the cardiovascular system. Usually the deleterious effect progresses slowly and is accompanied by adaptive changes of the heart.

Until a few decades ago, the most significant cause of valve disease was rheumatic fever, although at present there has been a considerable increase in the number of valvulopathies secondary to age-related degenerative alterations.

M.P. Suárez-Mier (✉)
Histopathology Service, National Institute of Toxicology and Forensic Sciences, Las Rozas (Madrid), Spain
e-mail: mariapaz.suarez@justicia.es

B. Morentin
Forensic Pathology Department, Institute of Legal Medicine, Bizkaia, Spain

Forensic Medicine, Basque Country University, Bizkaia, Spain

M. Cobo • P. García-Pavía
Heart Failure and Inherited Cardiac Diseases Unit, Cardiology Department, Hospital Universitario Puerta de Hierro, Majadahonda (Madrid), Spain

E. Castedo
Cardiac Surgery Department, Hospital Universitario Puerta de Hierro, Majadahonda (Madrid), Spain

J.S. Lucena et al. (eds.), *Clinico-Pathological Atlas of Cardiovascular Diseases*,
DOI 10.1007/978-3-319-11146-9_7, © Springer International Publishing Switzerland 2015

7.2 Aortic Valve Pathology

The aortic valve normally presents tricuspid or trileaflet morphology. It consists of three leaflets (cusps or valvular veils): right, left and non-coronary; each of them anchored to the corresponding Valsalva sinus following a parabolic line. Usually the two coronary arteries, right and left, arise from the right and left sinuses, respectively (Fig. 7.1) (see Chap. 1).

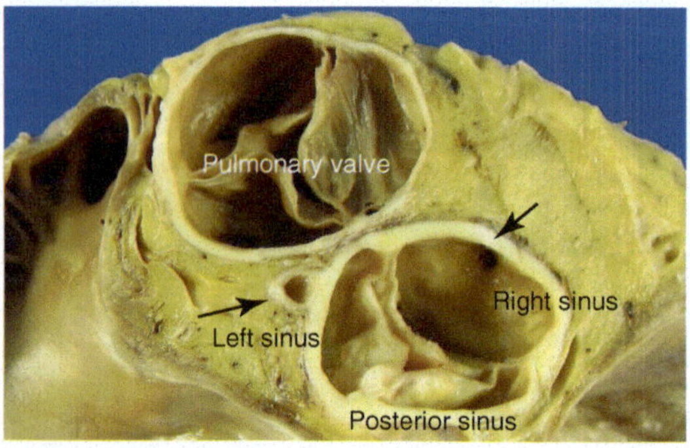

Fig. 7.1 Cross-sectional view at the level of the semilunar valves. In the aortic valve, the right coronary artery *ostium* can be observed in the right sinus and the beginning of the left coronary artery trunk next to the left sinus (*arrows*)

7.2.1 Aortic Stenosis

Aortic stenosis is the most common cardiac valvulopathy in Europe and North America. The most frequent aetiology of aortic stenosis is valve degeneration and calcification associated with aging, followed by congenital conditions, mainly bicuspid aortic valve, which is more common in younger patients. Rheumatic aetiology is rare. Depending on the location of the stenosis, aortic stenosis can be valvular (at the level of the leaflets), supravalvular (above the leaflets) and subvalvular (below the annulus) (Table 7.1).

Aortic stenosis produces pressure overload in the left ventricle characterized by left ventricular hypertrophy without dilation, although in very advanced stages ventricular dilation and dysfunction may occur. Patients may be asymptomatic for many years despite having a morphologically-severe stenosis. When cardiovascular symptoms appear (syncope, angina, dyspnea or heart failure), valve replacement is usually required, in the majority of cases by surgical intervention. The technical development of percutaneous aortic valve replacement offers a new hope for patients at high risk for surgical intervention.

Table 7.1 Morphological classification of aortic stenosis

Valvular	Supravalvular	Subvalvular
Congenital	Membranous	Membranous
Bicuspid	Hourglass deformity	Tunnel-like
Unicuspid	Hypoplastic	Obstructive hypertrophic cardiomyopathy
Acquired		
Senile calcific		
Postinflammatory		
Mixed		

7.2.1.1 Aortic Valve Stenosis

Congenital Conditions

Bicuspid Aortic Valve

Bicuspid aortic valve (BAV) is one of the most common congenital heart diseases, with an incidence of 1–2 % in the general population and a strong male predominance (4:1). Several familial cases have been described in an autosomal dominant pattern with variable penetrance, and it is frequent in people affected by Turner's syndrome. BAV occurs due to alterations in the process of valvulogenesis and leaflet formation. The two cusps are generally asymmetric and 75 % of these present a *raphe* or abortive commissure in the midline of the larger cusp. This *raphe* does not reach the aortic wall and contains elastic fibres, features that allow its differentiation from a post-inflammatory commissural fusion.

In up to 80 % of cases, fusion takes place between the right and the left coronary cusps in such a way that the leaflets are orientated in an anterior and posterior manner. The *raphe* is usually found in an anterior position with the coronary artery *ostia* located at both sides of the *raphe* (Subtype A) (Fig. 7.2). Coarctaction of the aorta is associated with this type of malformation.

Less frequently, the commissure between the right-coronary and non-coronary cusps is absent and the leaflets are orientated to the right and to the left (Subtype B) (Fig. 7.3). The *raphe* usually locates in the right leaflet (Fig. 7.4). This morphology is associated with a greater severity of valve dysfunction (stenosis and aortic valve regurgitation).

The most common complication of BAV is aortic valve stenosis caused by the progressive and degenerative process of dystrophic calcification (Figs. 7.4, 7.5 and 7.6). Stenosis of the bicuspid aortic valve occurs approximately two decades earlier than in the tricuspid valve. In cases of severe calcification, the differentiation between a bicuspid aortic valve and a process of calcification on a valve with post-inflammatory fibrosis becomes a very difficult task.

In up to 20 % of patients with congenital aortic stenosis, other concomitant cardiovascular abnormalities exist. For example, an association with aortic dissection is common (Fig. 7.7) and complications such as infective or rheumatic endocarditis may occur, causing mixed forms of aortic stenosis (Fig. 7.8).

Unicuspid Aortic Valve

There are two types of congenital unicuspid aortic valves: (1) the *acommissural* or dome-shape unicuspid valve, with three aborted commissures (or *raphes*), and (2) the *unicommissural* unicuspid aortic valve in which adjacent cusps of two commissures are congenitally fused. In the majority of cases, the only functional commissure is the left-non coronary, which originates an eccentric orifice in the shape of a teardrop that extends towards the left, posterior to the annulus, resembling an "exclamation mark".

Patients with unicuspid aortic valve, as occur in cases of BAV, are at higher risk of infective endocarditis and aortic dissection. This is especially true for unicommissural valves, and symptoms appear once calcification occurs (Figs. 7.9 and 7.10). It is commonly associated with leaflet dysplasia. Accomisural valves are often associated with aortic atresia and hypoplastic left heart syndrome.

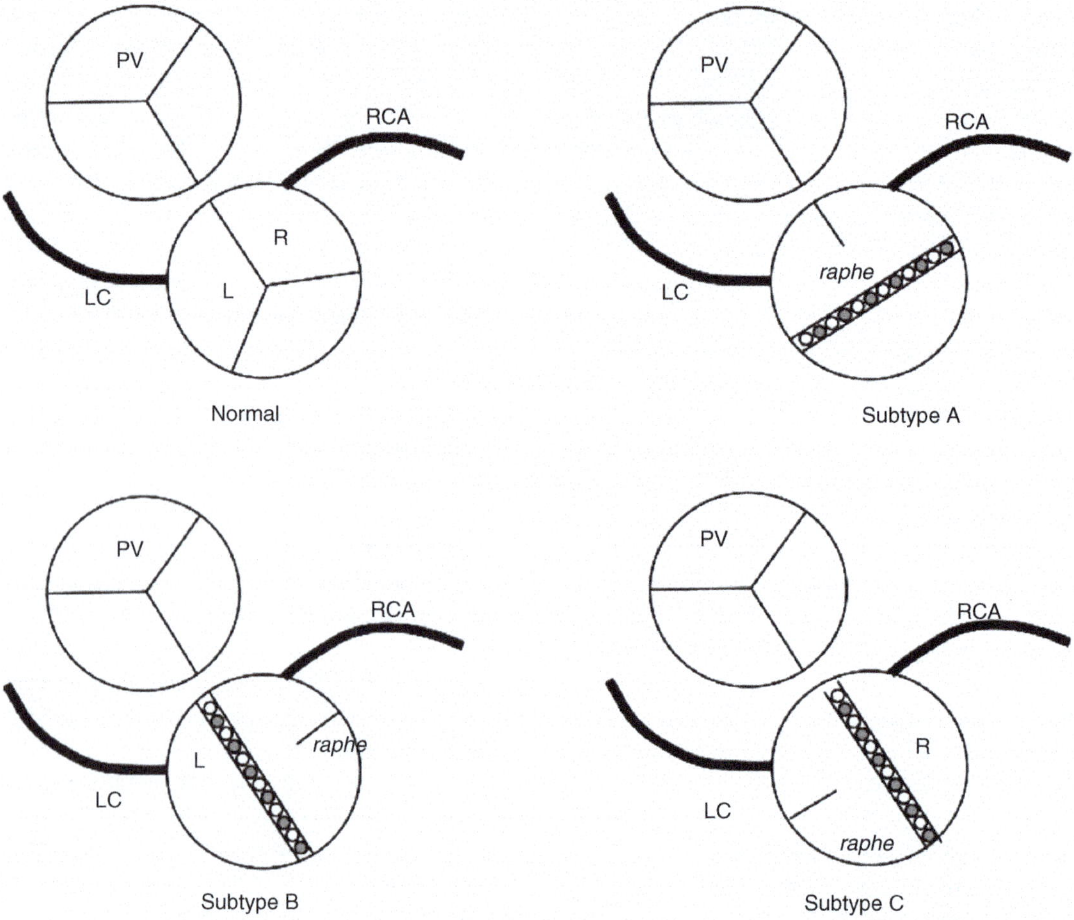

Fig. 7.2 Schematic illustration of the different modalities of bicuspid aortic valve (Reproduced with permission from Fernández et al. 2010)

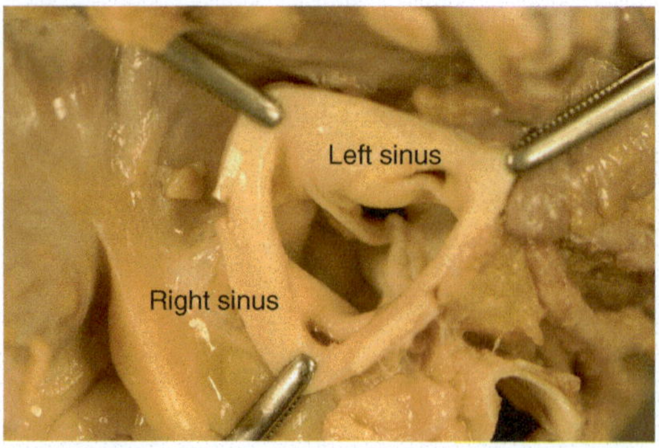

Fig. 7.3 *Bicuspid aortic valve*. The origin of the coronary arteries can be appreciated in each of the sinuses. BAV was found during autopsy of a 41-year-old male who died from ischaemic heart disease

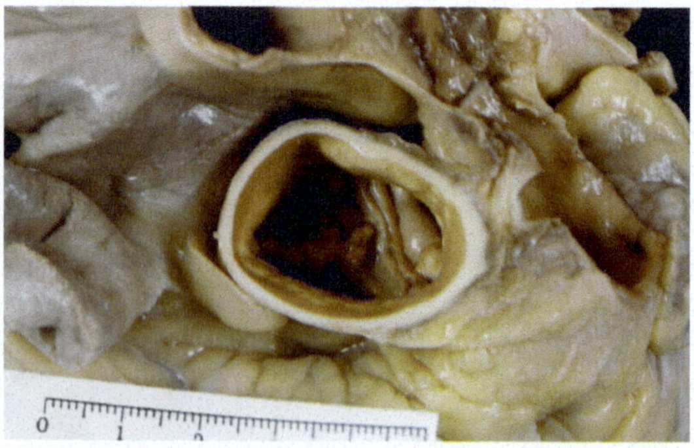

Fig. 7.4 *Calcification of a bicuspid aortic valve*. Sudden death in a 55-year-old male. Bicuspid valve with a calcified *raphe* located between the right and the posterior sinuses (Subtype B). Heart weight was 594 g and there was LV hypertrophy

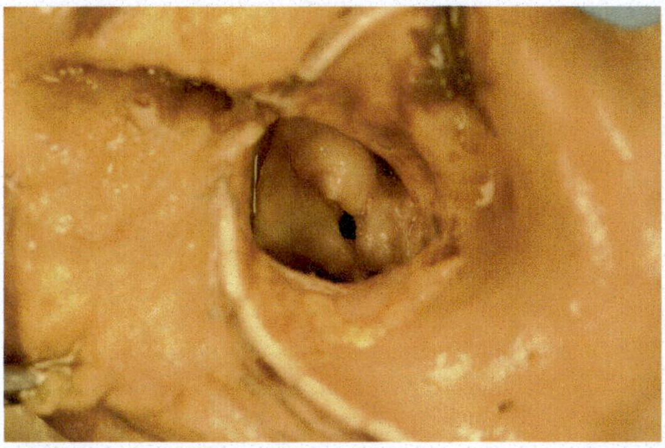

Fig. 7.5 *Bicuspid aortic valve with dysplastic and partially calcified leaflets*. A 31-year-old male who died suddenly while playing football. The heart weighed 640 g and showed severe concentric LV hypertrophy

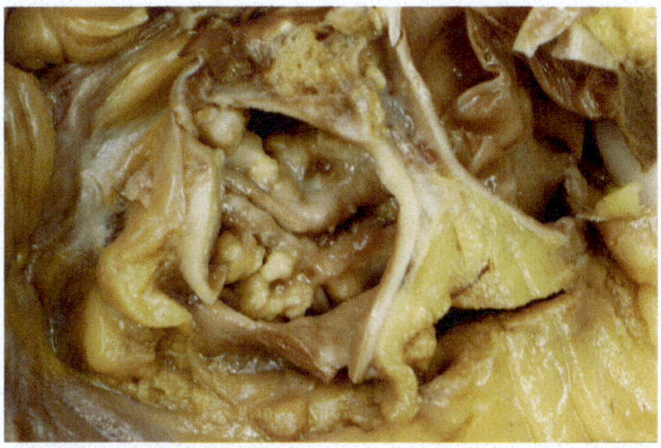

Fig. 7.6 *Calcification of a bicuspid aortic valve*. Sudden death in a 47-year-old male. A bicuspid aortic valve with heavily calcified leaflets, causing a significant obstruction of the LV outflow tract, which ultimately led to cardiac hypertrophy with a heart weight of 760 g

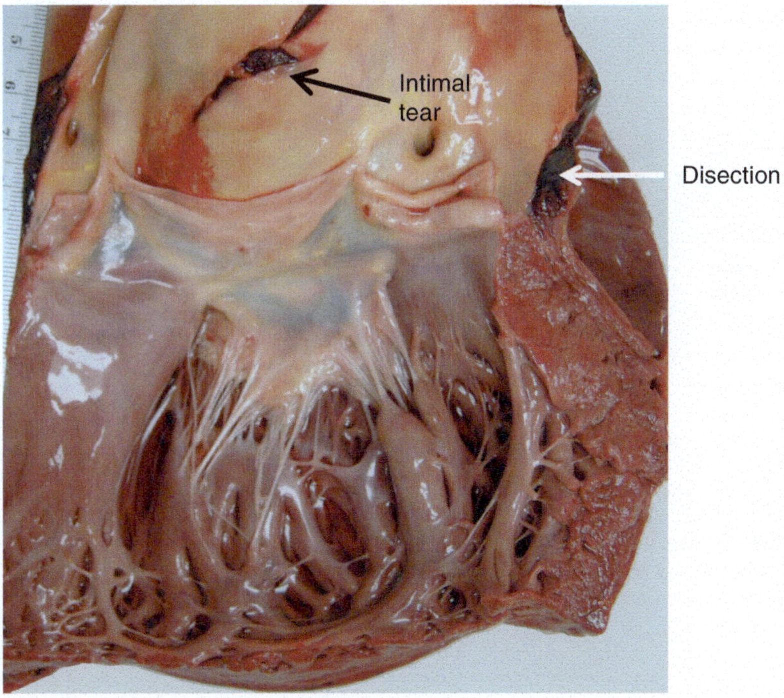

Fig. 7.7 *Aortic dissection associated with bicuspid aortic valve*. A 30-year-old male. In the hours prior to death he presented prodromal symptoms such as chest pain. The cause of death was cardiac tamponade due to haemopericardium secondary to a type A aortic dissection. Dissection of the aortic wall, with intimal tear and dilatation of the aortic root can be appreciated in the image

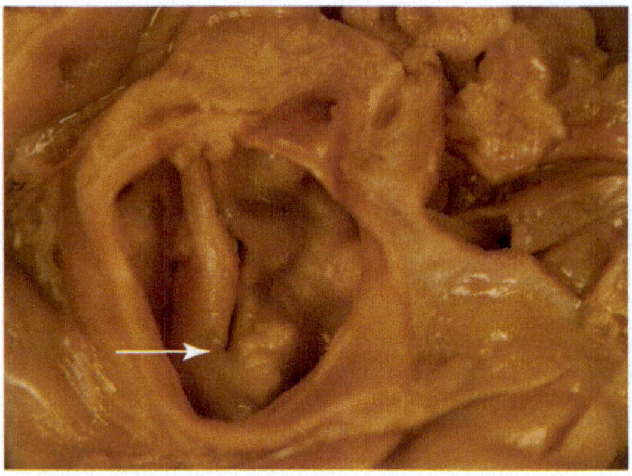

Fig. 7.8 *Mixed aortic stenosis*. A 61-year-old male who suddenly fainted while walking in the street. A bicuspid aortic valve in addition to commissural fusion (*arrow*), probably secondary to an inflammatory process. Heart weight: 345 g, with LV wall thickness: 18 mm

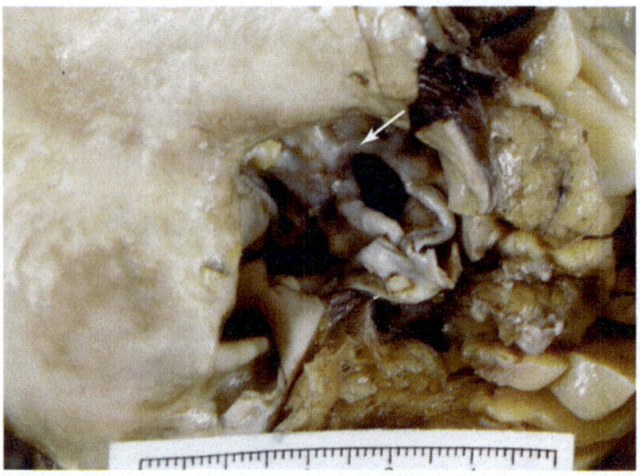

Fig. 7.9 *Unicuspid aortic valve*. The valve has a unique commissure (*arrow*) and extensive calcification. A 53-year-old male who died while playing raquetball. Severe LV hypertrophy (heart weight: 600 g). Severe coronary atherosclerosis affecting two vessels was also diagnosed (RCA and LAD coronary artery)

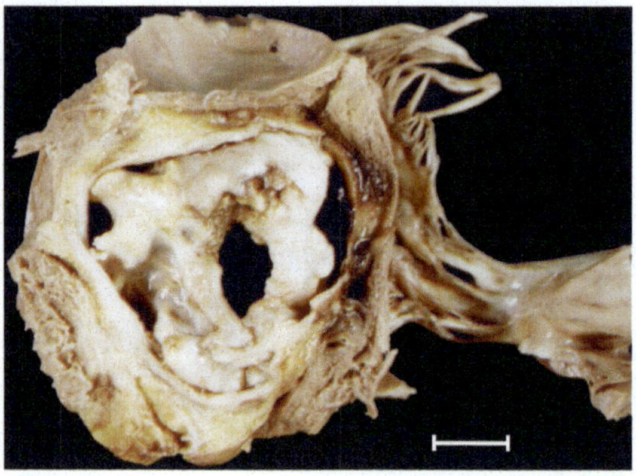

Fig. 7.10 *Aortic stenosis and regurgitation in a unicuspid aortic valve*. A 39-year old-male found dead at work. Severe cardiac hypertrophy was observed (heart weight of 924 g)

Acquired Pathology

Degenerative Calcific Aortic Stenosis

Degenerative calcific aortic stenosis is the most common type of valvular heart disease that requires valve replacement, and constitutes the valvulopathy that most frequently causes sudden death in older adults.

Morphologically, it is characterized by the deposition of nodular calcific masses on the leaflets, mainly at their base (Figs. 7.11 and 7.12A). The disease can be asymptomatic or result in left ventricular outflow obstruction, which causes a progressive increase in the LV pressure that in turn leads to concentric hypertrophy (Fig. 7.12b).

Fig. 7.11 *Degenerative calcific aortic stenosis.* A 54-year-old woman who was found dead at home. View of the aortic valve from the aortic side. Tricuspid valve with marked calcification (whitish nodules) at the base of the leaflets, with no commissural involvement. Necropsy also revealed moderate coronary atherosclerotic disease and severe cardiac hypertrophy

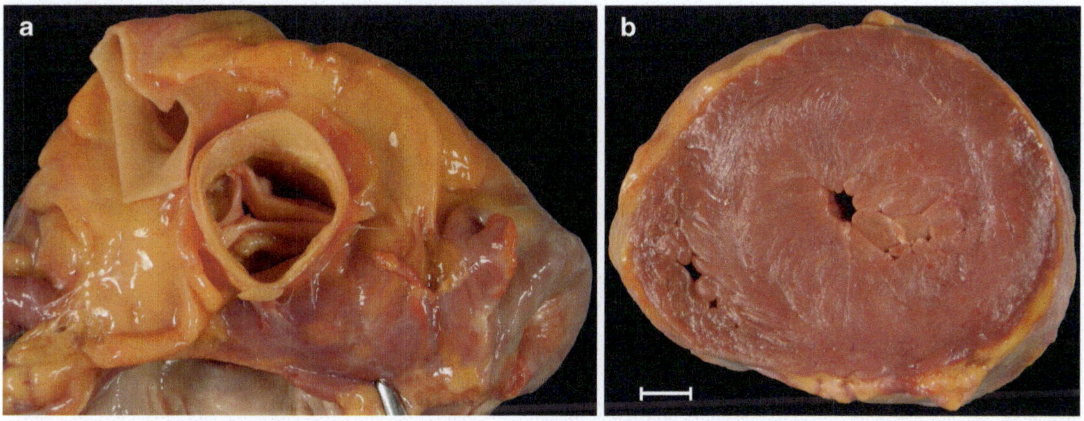

Fig. 7.12 *Degenerative calcific aortic stenosis.* A 74-year-old woman, with a history of chronic renal failure treated with haemodialysis, who died suddenly at home. (**a**) Tricuspid aortic valve with leaflet calcification. (**b**) Secondary concentric left ventricular hypertrophy. Heart weight: 495 g and wall thickness: 2 cm

Post-Inflammatory Aortic Stenosis

Inflammatory aortic valve disease has a mostly rheumatic origin and is associated with mitral valvulopathy (see below).

Fibrous thickening of the leaflets occur, which determines valve regurgitation in addition to stenosis, and commissural fusion (Fig. 7.13). With time, calcification may develop, initiating mixed types of aortic stenosis with features of degenerative and rheumatic stenosis (Fig. 7.14).

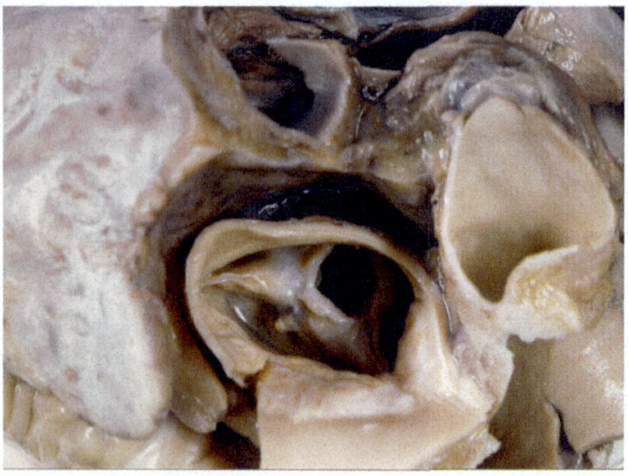

Fig. 7.13 *Post-inflammatory aortic stenosis*. Sudden death in a 57-year-old male with a previous history of osteomuscular problems, treated with NSAIDs. The aortic valve showed commissural fusion and calcification, which reduced the lumen to 7 mm in diameter (0.38 cm^2). Heart weight: 570 g

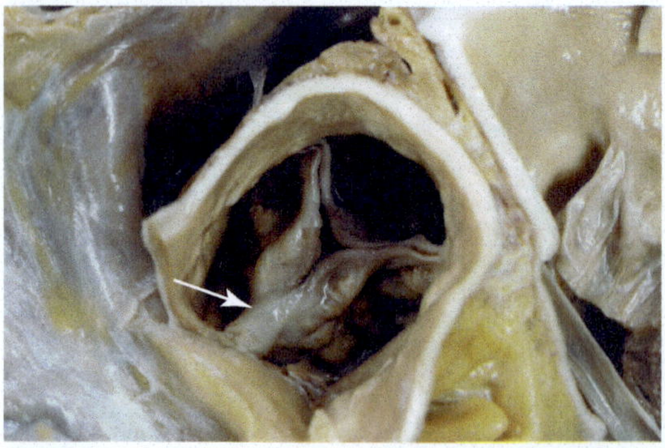

Fig. 7.14 *Mixed aortic stenosis*. A 68-year-old male with hypertension. The aortic valve showed nodular calcification in the leaflets and commissural fusion (*arrow*). Mitral valve was also involved

7.2.1.2 Supravalvular Aortic Stenosis

Congenital supravalvular aortic stenosis is the least common form of left ventricular outflow tract obstruction. It usually manifests during childhood, and in some instances is part of Williams syndrome (also associated with characteristic facial anomalies, mental delay and hypercalcaemia).

Morphologically, the narrowing can be located in the sinotubular junction, creating an hourglass-type supravalvular stenosis (Figs. 7.15 and 7.16); or there can also be a diffuse narrowing for a variable distance along the ascending aorta (Fig. 7.17), or even in the aortic arch vessels. The leaflets tend to be dysplastic and it is often associated with a bicuspid aortic valve and coronary artery lesions.

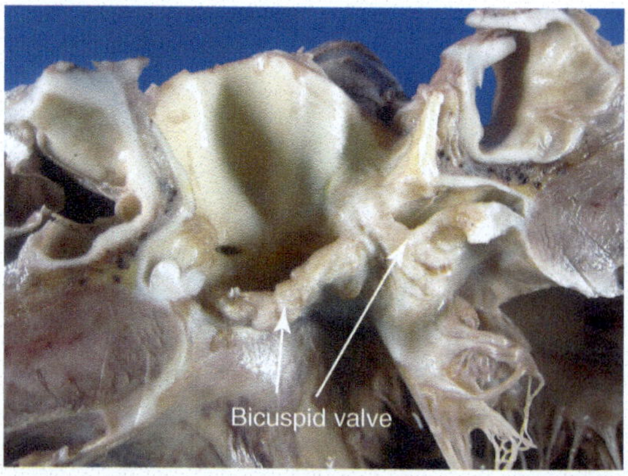

Fig. 7.15 *Supravalvular aortic stenosis associated with a bicuspid aortic valve*. A 13-year-old boy diagnosed with bicuspid aortic valve underwent percutaneous valvuloplasty at the age of 9 years. During his last year the disease evolved to moderate aortic regurgitation and dilation and dysfunction of the LV. He died suddenly in a park. Image shows a bicuspid valve with thickened and rough leaflets (compatible with valvuloplasty), sinotubular junction stenosis, and post-stenotic dilation creating an hourglass deformity. Heart weight was 476 g, with a dilated LV and scar tissue from a circumferential subendocardial infarction

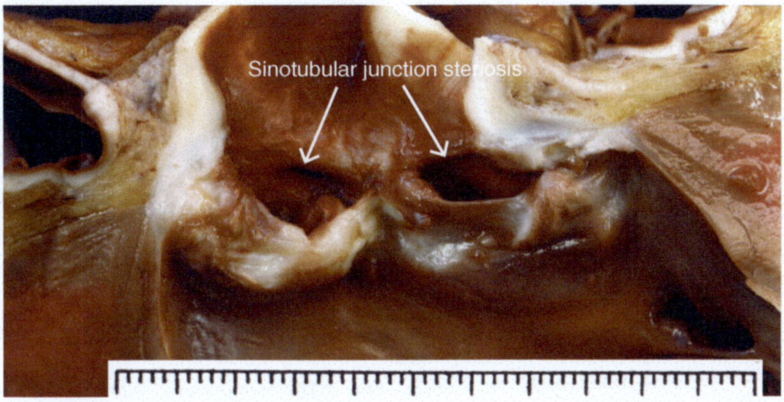

Fig. 7.16 *Supravalvular aortic stenosis associated with a bicuspid aortic valve*. A 12-year-old boy who died suddenly during physical education class. Morphologically, the aorta is similar to the above image (Image reproduced with permission from Suarez-Mier and Aguilera 2002)

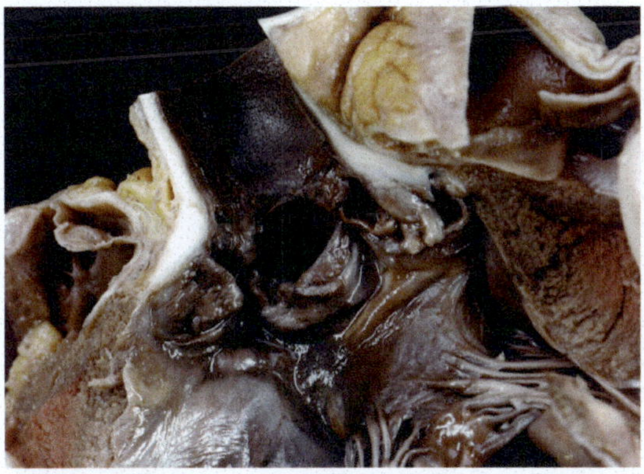

Fig. 7.17 *Supravalvular aortic stenosis of the tubular type*. Sudden death in a 27-year-old woman with Williams's syndrome. Note that the ascending aorta is tubular (2 cm in diameter) with a thickened wall. Dysplastic and dilated Valsalva sinuses (Image reproduced with permission from Suárez-Mier and Morentin 1999)

7.2.1.3 Subvalvular Aortic Stenosis

Subvalvular aortic stenosis (subaortic stenosis) affects 0.1/1,000 live births and is more common in the male gender. It is associated with other malformations such as BAV, ventricular septal defects, patent ductus arteriosus or coarctation of the aorta. Morphologically, there are three types (Table 7.1): thin membranous type (90 % of the cases), tunnel type and fibromuscular collar type that relates to obstructive hypertrophic cardiomyopathy with extensive hypertrophy in the anterior portion of the basal septum (see Chap. 8). It can also be caused by abnormalities in the mitral valve implantation.

It has been regularly observed that this type of obstruction develops and progresses over time, so for many experts this is an acquired rather than a congenital valvular disease.

The membranous type implies the formation of a fibrous membrane that extends from the interventricular septum onto the anterior leaflet of the mitral valve (Figs. 7.18 and 7.19). The clinical presentation ranges from a stable and asymptomatic disease that can last for many years to a rapidly progressive disease. It is frequently associated with aortic regurgitation and infective endocarditis and can relapse after surgical resection.

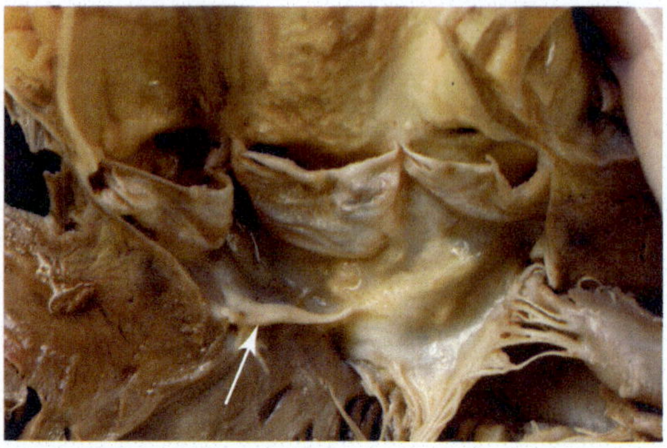

Fig. 7.18 *Membranous subaortic stenosis*. A 36-year-old male with a history of subaortic stenosis. He died of ischaemic cardiopathy with a three vessel disease and occlusive thrombosis in the RCA. A fibrous membrane can be appreciated below the aortic valve annulus that reaches the anterior leaflet of the mitral valve (*arrow*)

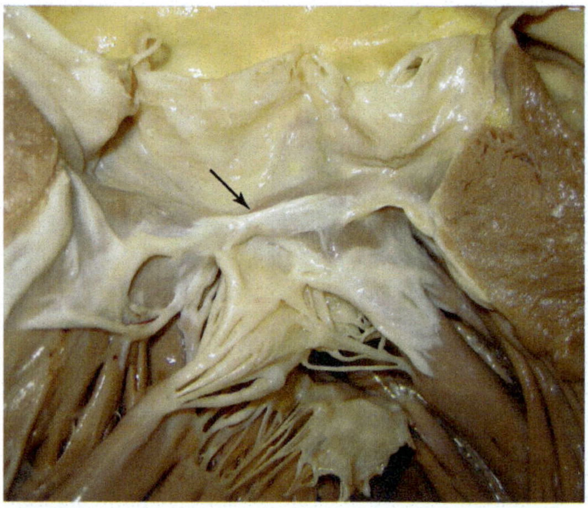

Fig. 7.19 *Membranous subaortic stenosis*. A 52-year-old male, diabetic, with no cardiac history, who died while sleeping. Heart weight: 724 g. View of the LV outflow tract where a subaortic fibrous membrane can be appreciated associated with an anomalous mitral valve, with shortening of the *chordae tendineae* of the anterior pillar (*arrow*)

7.2.2 Aortic Regurgitation

Aortic regurgitation is produced when the aortic valve does not close adequately. The acute form may be seen in cases of type A aortic dissection or valve disease due to endocarditis. The chronic form, which is more frequent, is due usually to valve pathology or aortic annulus distortion caused by dilation in the ascending aorta. Table 7.2 describes the classification of aortic regurgitation based on the mechanism.

The most frequent causes of aortic regurgitation in the Western world are secondary to aortic dilation and bicuspid valve involvement. The pathology of the aorta is described in Chap. 3.

Table 7.2 Types of aortic regurgitation

Type I: Dilation of the aortic root with normal cusps and normal motion (in patients over 60 years it is idiopathic or associated with hypertension; in younger patients it is related to Marfan syndrome)
Type II: Secondary to valve prolapse with mixoid or excessive tissue and commissural disruption
Type III: Secondary to restricted cusp motion with valve structural lesions (rheumatic valve disease, endocarditis, bicuspid valve, etc.)

Taken from El Khoury et al. (2005)

7.3 Mitral Valve Pathology

Normal function of the mitral valve requires the structural integrity of all its components. Table 7.3 summarises the macroscopic and microscopic features of the mitral valve (see Chap. 1).

Table 7.3 Anatomy of the mitral valve

Gross examination: (Figs. 7.20a and 7.20b)
Semicircular anterior leaflet: Rough zone that coapts with the posterior leaflet Smooth zone
Fenestrated posterior leaflet with indentations Rough zone that coapts with the anterior leaflet Membranous zone Basal zone
Anterolateral and posteromedial commissures
Tendinous cords (*Chordae tendinae*)
Microscopic examination (three layers): (Fig. 7.20d)
Atrialis: thin layer of aligned elastic and collagen fibres in continuity with the left atrial endocardium
Spongiosa: composed of mixoid connective tissue
Fibrosa: composed of dense collagen with some elastic fibres

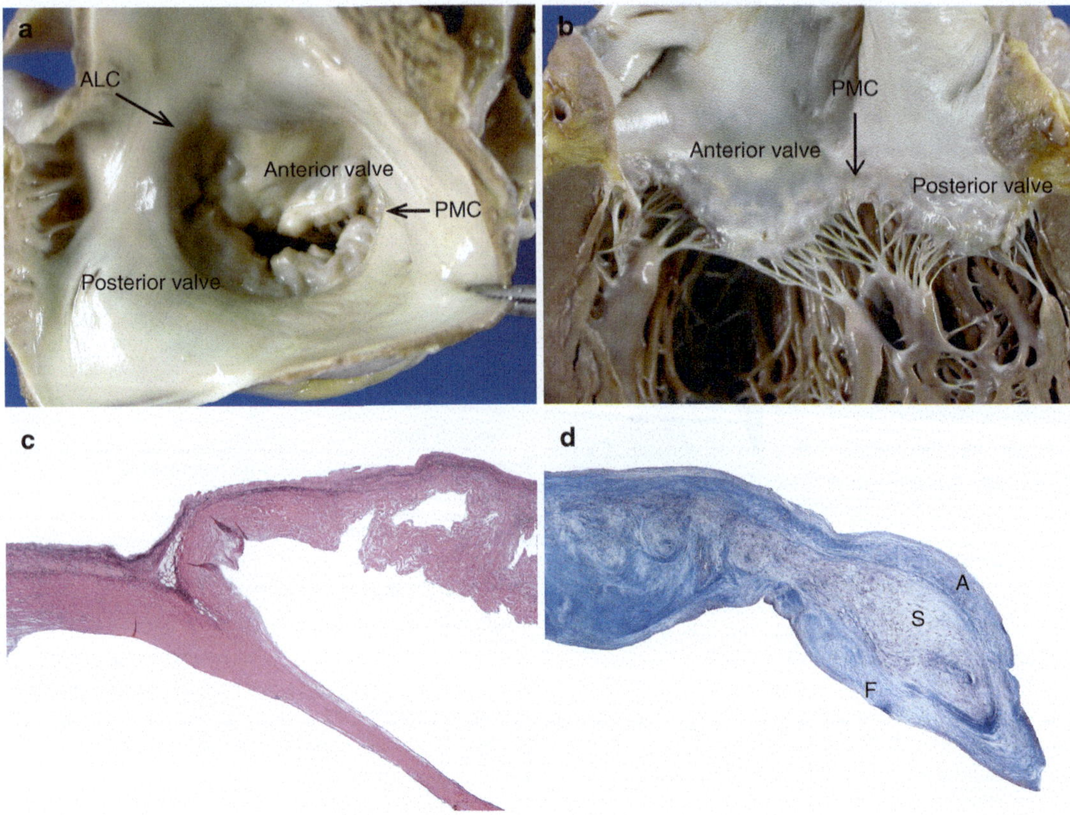

Fig. 7.20 *Normal mitral valve*. (**a**) Closed valve as seen from the left atrium. PMC: posteromedial commissure. ALC: anterolateral commissure. (**b**) Open valve. (**c**) Microscopic image of the posterior leaflet and *chordae tendinae* at low magnification (Weigert's stain) (**d**) Higher magnification of the edge of the posterior leaflet: *A* atrialis, *S* spongiosa, *F* fibrosa (Masson's trichrome)

7.3.1 Mitral Regurgitation

Mitral regurgitation can be primary when it is caused by a lesion of some of its components, or secondary if it is the result of other heart diseases (Table 7.4). It is important to differentiate the type of involvement in order to establish the most adequate surgical strategy (repair or replacement). Percutaneous endovascular repair of mitral regurgitation is currently under development as an alternative to surgical repair or replacement.

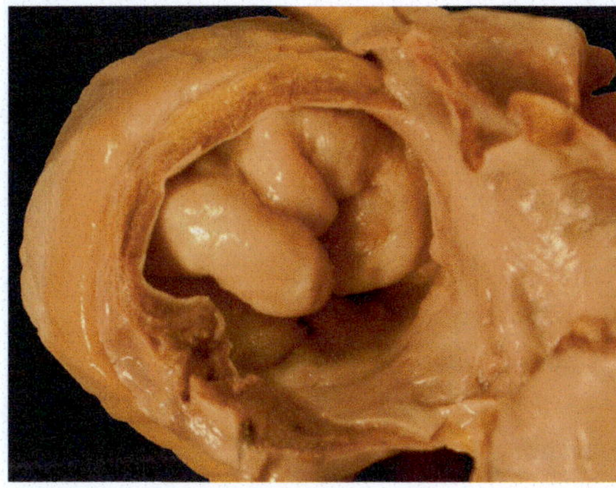

Fig. 7.21 *Mitral valve prolapse*. A 48-year-old woman who died suddenly at home. Mitral valve as seen from the atrium showing prolapse with billowing (hood-like form) of the posterior leaflet into the atrium

7.3.1.1 Mitral Valve Prolapse

Mitral valve prolapse (myxomatous degeneration of the mitral valve) is one of the most common forms of valvular heart disease and it is the most frequently encountered pathology in patients undergoing mitral valve surgery. Many patients are asymptomatic, and the anomaly is discovered incidentally during physical examination. In a minority of patients, it can manifest as mitral regurgitation, with angina-type chest pain, ventricular arrhythmia and even sudden death. Diagnosis is generally established by echocardiography.

In this pathology, one or both of the mitral leaflets billows over the atrial chamber forming a hood-like bulge (Figs. 7.21 and 7.22a). The affected leaflets are enlarged, redundant and rubbery. The posterior leaflet is the most frequently involved, especially its medial and central portions (Fig. 7.22b). Involvement of the anterior leaflet is only partial and proximal to the medial commissure. There also changes in the *chordae tendinae*, which tend to be elongated and thinned. As a general rule, macroscopic diagnosis is sufficient, unless infective endocarditis is suspected.

Histologically, the principal change is the deposition of myxomatous material in the *spongiosa* layer (Fig. 7.22c). Table 7.5 summarises the pathology of mitral valve prolapse.

Table 7.4 Causes of mitral regurgitation

Primary mitral regurgitation:
Mitral valve prolapse: the most common cause of mitral regurgitation in developed countries
Rheumatic disease: more common in underdeveloped countries. Usually manifests as mitral stenosis or double mitral lesion
Annular calcification: associated with aging. Does not cause severe mitral regurgitation
Infective or marantic endocarditis (associated with connective tissue disease)
Trauma that may produce rupture of the *chordae tendinae* and acute mitral regurgitation
Congenital anomalies, principally mitral cleft
Secondary to drugs: anorexigens, carbergoline, pergolide or ergotamine
Secondary mitral regurgitation:
Dilation and dysfunction of the left ventricle that causes stretching and distortion on the mitral valve annulus leading to centrally directed mitral regurgitation
Ischaemic heart disease, usually as a result of ischaemia or necrosis of the posterior papillary muscle, which can lead to acute mitral regurgitation
Hypertrophic obstructive cardiomyopathy, caused by the movement of the mitral valve anterior leaflet due to the "Venturi effect" caused by outflow tract obstruction

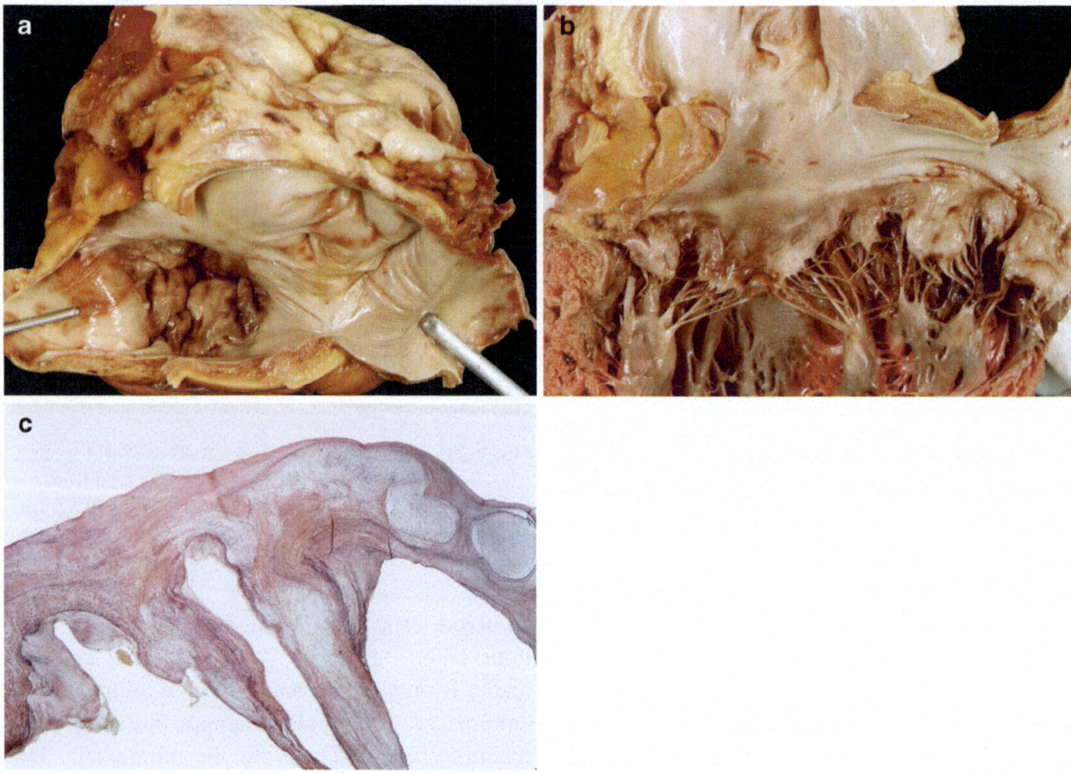

Fig. 7.22 *Mitral valve prolapse*. A 69-year-old male with severe mitral regurgitation secondary to mitral valve prolapse. (**a**) Marked billowing of the posterior leaflet that extends like a festoon into the atrium. (**b**) Appearance of the valve once opened. (**c**) Microscopic image in which an enlarged *spongiosa* in the leaflet and *chordae tendinae* can be appreciated due to mucopolysaccharide deposition (Weigert's stain). (Compare with Fig. 7.20c)

Table 7.5 Pathology of mitral valve prolapse

Macroscopic features:
Intercordal billowing of the affected mitral leaflet
Billowing of the leaflets into the left atrium
Elongation of the affected leaflets, which increases the valve area
Thickened, soft, whitish and rubbery leaflets (in the young)
Elongated, tortuous and thinned *chordae tendinae*
If secondary fibrosis is present, *chordae tendinae* are thickened
Friction lesions of the ventricular endocardium under the *chordae tendinae*
Abnormal *chordae tendinae*: anomalous insertions, fusions, etc.
Annular dilation in patients with severe mitral regurgitation
Microscopic features:
Thickening of the *spongiosa* layer with accumulation of glycosaminoglycans (proteoglycans) and invasion and interruption of the fibrous layer
Occasionally fibrino-platelet deposits in the surface above underlying fibrinoid lesion
Accumulation of proteoglycans at the point where the *chordae tendinae* insert into the ventricular surface of the leaflet, and around them
Histochemistry: Alcian blue, PAS stain diastase-resistant and colloidal iron positive for mucopolysaccharides

7.3.1.2 Mitral Annular Calcification

It is the most common valvular heart disease in women over 60 years of age (see Chap. 1). Usually, it is discovered incidentally during physical examination, although in some instances it can be very severe and lead to mitral regurgitation.

Morphology varies from whitish and hard focal nodules to large nodules that extend alongside the entire mitral valve annulus, between the valve leaflets and the ventricular wall (Figs. 7.23 and 7.24).

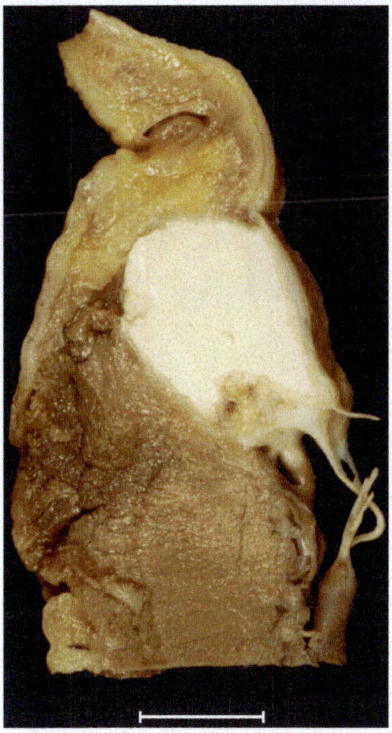

Fig. 7.23 *Mitral annular calcification.* Posterior-side section of the mitral annulus in an 83-year-old woman who died from stroke. Extensive calcification can be appreciated in the mitral annulus, with a pasty white appearance

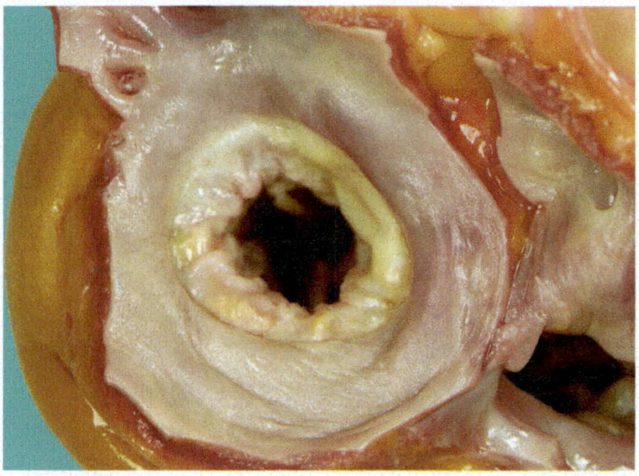

Fig. 7.24 *Circumferential calcification of the mitral annulus (atrial view).* A 73-year-old woman who was dead on arrival to hospital. Heart weight: 490 g with concentric left ventricular hypertrophy

7.3.2 Mitral Stenosis

Mitral stenosis is the most common valvular heart disease in underdeveloped countries. Symptomatology is secondary to the retrograde congestion it causes. Prolonged pulmonary congestion leads to progressive pulmonary hypertension, which ultimately provokes right ventricular pressure overload and systemic congestive symptoms. In cases of severe stenosis, marked atrial dilation may also arise, leading to an early onset of atrial arrhythmias and embolic complications.

7.3.2.1 Rheumatic Valvular Disease

Rheumatic heart disease is considered the most common cause of mitral stenosis, although its incidence has declined remarkably during the last decades. Valvular lesions, mainly mitral valve stenosis, are the major chronic complications of rheumatic fever. The aortic valve can also be affected in 25 % of cases.

Post-rheumatic inflammatory changes in the valve determine its degree of stenosis. Morphological features are summarised in Table 7.6.

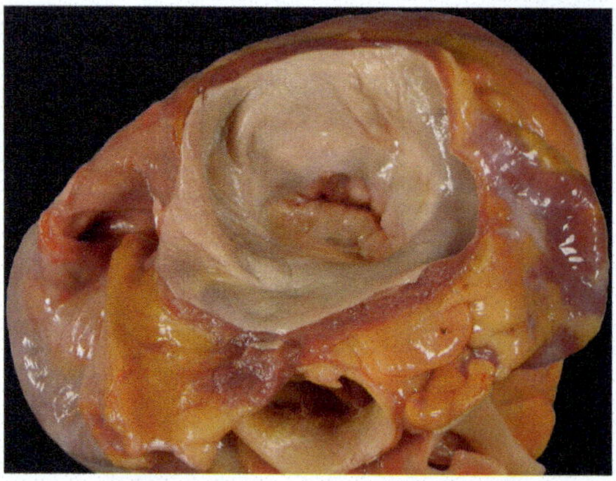

Fig. 7.25 *Rheumatic mitral stenosis (post-inflammatory).* Mitral valve as seen from the atrial side: fibrous leaflet thickening and commissural fusion can be appreciated. Dilatation of the left atrium. The aortic valve was also affected

Table 7.6 Pathology of rheumatic valve disease

Macroscopic features:
Acute phase:
Little verrucae along the lines of closure on the atrial side of the leaflet (on the ventricular side in the semilunar valves)
Chronic phase: (Figs. 7.25, 7.26 and 7.27)
Leaflet thickening and fibrosis
Commissural fusion
Shortening, thickening and fusion of the *chordae tendinae*
"Fish-mouth" appearance of the valve, as seen from the atrium
Leaflet calcification in the ventricular side
Fibrosis of the papillary muscles
Left atrial dilation. McCallum's plaques
Hypertrophy and dilation of the RV in advanced stages caused by pulmonary hypertension
Microscopic features:
Acute phase:
Verrucae formed by fibrin and platelet deposition over the inflamed leaflet, with fibrinoid necrosis of collagen tissue and a mononuclear inflammatory infiltrate with fibroblasts
Chronic phase:
Fibrosis, neovascularization, lymphocytic inflammatory infiltrate

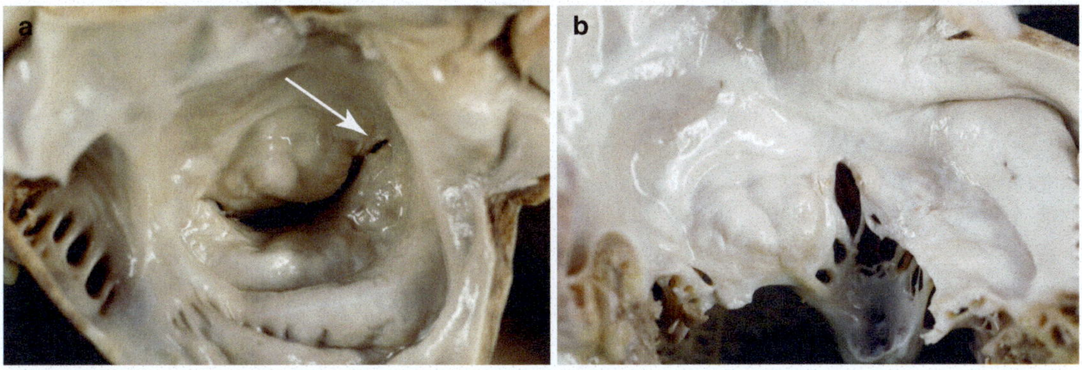

Fig. 7.26 *Mitral valvuloplasty due to rheumatic stenosis*. A 49-year-old woman who underwent mitral valvuloplasty 2 years before. (**a**) Mitral valve as seen from the left atrium. The valve has a "fish-mouth" appearance with whitish leaflets. Posteromedial commissurotomy can be appreciated (*arrow*). (**b**) The same valve opened. Thick and shortened *chordae tendinae* can be observed

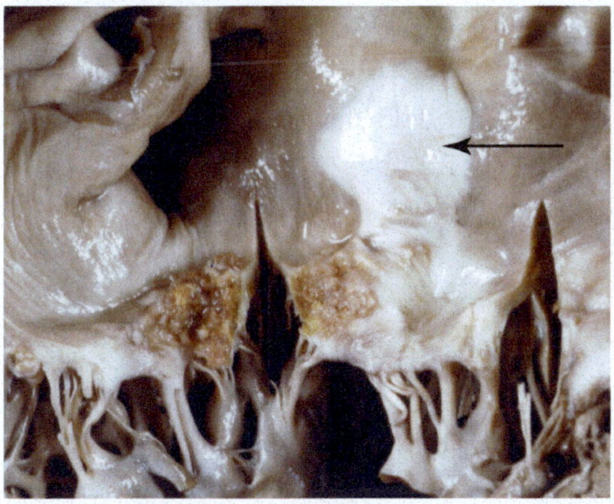

Fig. 7.27 *McCallum's plaque in left atrium*. A 61-year old-male with a rheumatic mitral valve: leaflet fibrosis and calcification, fibrosis and shortening of the *chordae tendinae*, and whitish plaque on the posterior side of the left atrium, above the annulus (jet lesion) (*arrow*)

7.4 Tricuspid Valve Pathology

The tricuspid valve consists of three leaflets (anterior: triangular or quadrangular in shape, septal, and posterior: with grooves); three commissures (anteroseptal, posteroseptal and anterosuperior); three papillary muscles (anterior: unique and large, septal: double or multiple, and posterior: formed by several small muscles); and between 17–36 *chordae tendinae* (Fig. 7.28) (see Chap. 1).

Tricuspid valve regurgitation generally has a functional origin caused by annular dilatation while the valve itself is unaffected. Primary valve disease is very rare. The most frequent causes of tricuspid valve regurgitation are summarised in Table 7.7.

Tricuspid valve stenosis is rare and exceptional as an isolated lesion. The most common cause is rheumatic disease and it is associated with lesions in other valves: mitral and aortic.

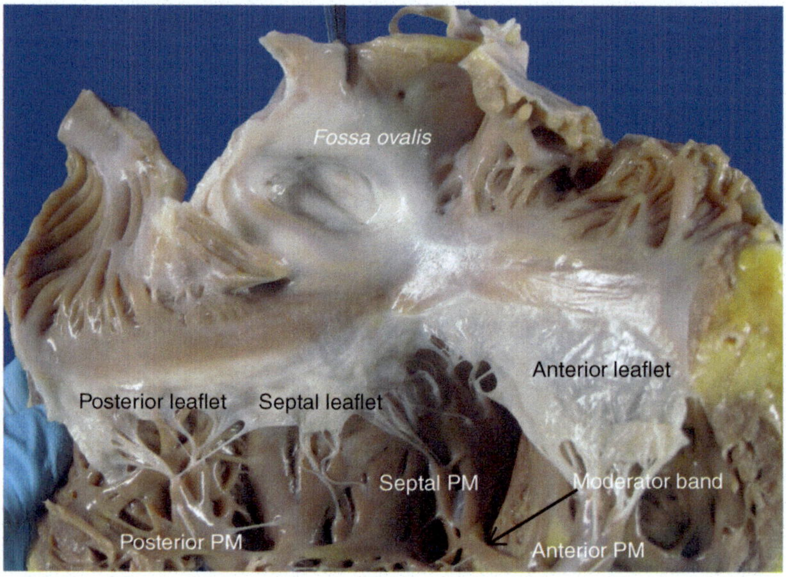

Fig. 7.28 *Opened normal tricuspid valve*

Table 7.7 Causes of tricuspid valve regurgitation

Valve perforation caused by a pacemaker electrode or when endomyocardial biopsy is performed
Thoracic trauma
Endocarditis
Rheumatic disease
Carcinoid syndrome
Ebstein's anomaly
Mixomatous degeneration with prolapse
Ischaemic heart disease
Connective tissue disease
Drugs

7.4.1 Ebstein's Anomaly

Ebstein's anomaly is a malformation in the structure and location of the tricuspid leaflets. The anterior leaflet originates in the annulus but is elongated and sail-shaped, and there is a downward displacement of the septal and posterior leaflets to the apex of the right ventricle (Fig. 7.29a). This displacement divides the right side of the heart into a right atrium, an atrialized portion of the right ventricle and the functional remnant of the right ventricle, which frequently leads to tricuspid regurgitation of varying degrees of severity. Furthermore, this anomaly tends to associate with interatrial shunt and ventricular septal defects (Fig. 7.29b).

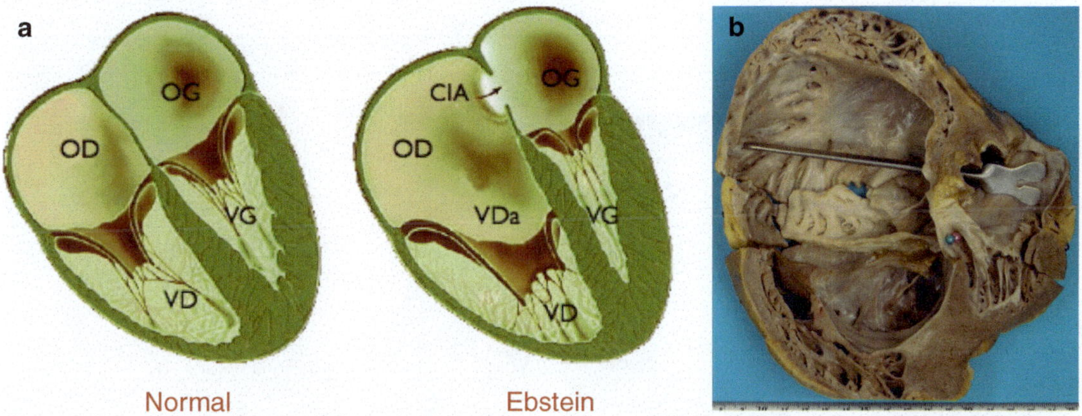

Fig. 7.29 *Ebstein's anomaly*. (**a**) Schematic illustration showing the downward displacement of the septal and posterior leaflets of the tricuspid valve. The superior portion of the RV is part of the atrium. (**b**) A 51-year-old male with a history of valvular disease since childhood, under anticoagulation and digoxin therapy. He suddenly lost consciousness and suffered ventricular fibrillation. Severe cardiomegaly (725 g) with dilatation and hypertrophy of the right atrium. The four-chamber view of the posterior wall of the heart shows interatrial communication due to *Patent foramen ovale* (probe) and low tricuspid valve position and a partly atrialized RV. The RA occupies two thirds of the heart whereas the RV, LA and LV are notably reduced

7.5 Pulmonary Valve Pathology

7.5.1 Quadricuspid Pulmonary Valve

It is a rare congenital anomaly and usually clinically silent. Its incidence in autopsy studies ranges from 0.0014 to 0.0033 %. The pulmonary valve has four cusps, three of equal size and a primitive accessory cusp (the most common) (Fig. 7.30A); two larger cusps next to two smaller ones (Fig. 7.30b); or four cusps of equal size.

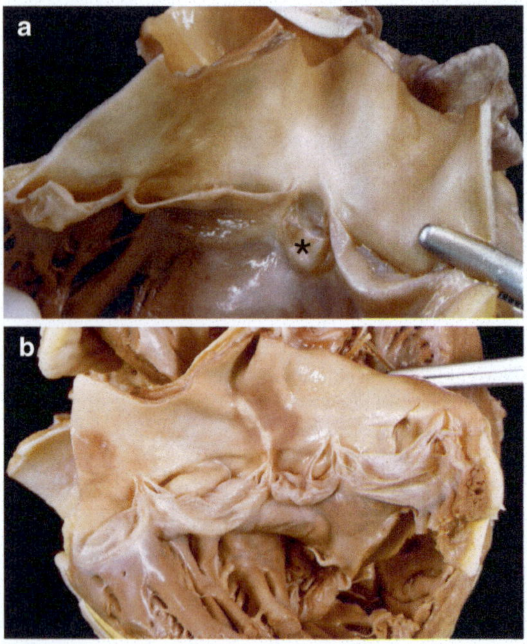

Fig. 7.30 *Quadricuspid pulmonary valve*. (**a**) A 19-year-old male who died in a car accident. Image shows a small primitive accessory cusp (*). (**b**) A 75-year-old male who died suddenly. The pulmonary valve is dilated with four cusps

7.5.2 Congenital Pulmonary Valve Stenosis

Pulmonary valve stenosis is usually congenital and may occur as an isolated defect or, more often, as part of a more complex abnormality (i.e. tetralogy of Fallot) (Figs. 7.31 and 7.32). Some mild forms exist that may be asymptomatic.

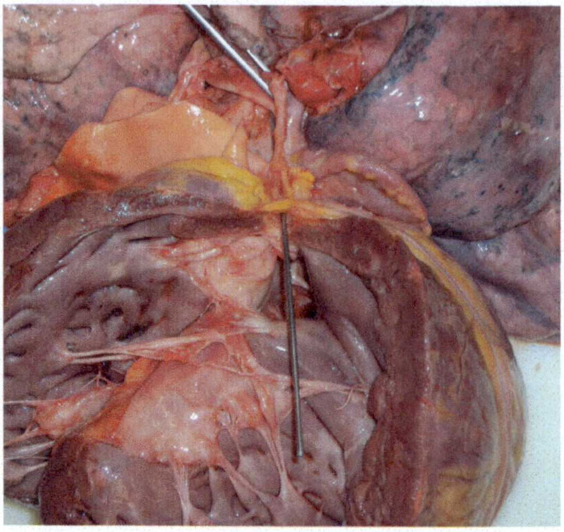

Fig. 7.31 *Pulmonary valve stenosis due to hypoplasia*. A 33-year-old male who died from congestive cardiac regurgitation due to extreme tetralogy of Fallot. Pulmonary stenosis that does not allow the passage of a probe, with severe hypoplasia of the pulmonary artery

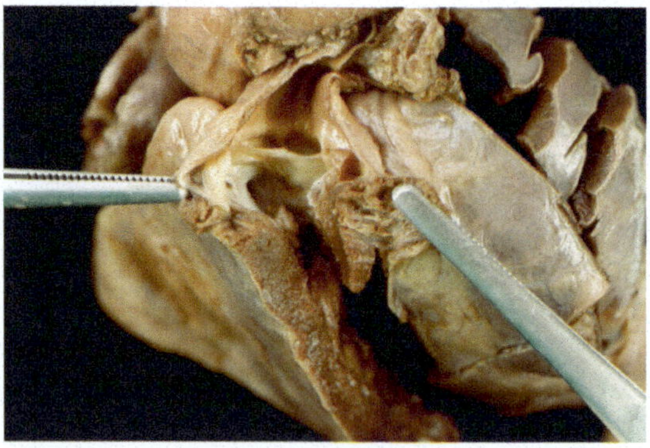

Fig. 7.32 *Pulmonary valve stenosis*. A 15-month-old baby boy with tetralogy of Fallot who died suddenly at home. Image shows a bicuspid pulmonary valve with stenosis and RV hypertrophy

7.6 Endocarditis

Endocarditis refers to an inflammation of the endocardium although the term is typically attributed to infection of the heart valves. According to the aetiology, two major types of endocarditis exist: infective and non-infective.

7.6.1 Infective Endocarditis

Infective endocarditis is an infection of the valvular endocardium by a microorganism. It can manifest as an abrupt (acute) form or more insidious (subacute).

This infection results in vegetations of different sizes usually located near the free edge of the valve. Vegetations are composed of platelets, fibrin and microorganisms (Figs. 7.33 and 7.34). A valvular lesion with necrosis and leaflet perforation is characteristic (Figs. 7.33, 7.34 and 7.35). Vegetations are friable masses that can fragment and embolize (Fig. 7.36). Table 7.8 describes the pathology of bacterial endocarditis.

Infective endocarditis can appear on previously normal valves or on predisposing valvulopathies such as rheumatic heart disease (nowadays with lower incidence), myxomatous degeneration of the mitral valve, and calcific aortic stenosis. Prosthetic heart valves account for 1–6 % of infective endocarditis, with a mortality of 50 %. When infections occur on bioprotheses, a destruction of the leaflets may follow, leading to perforations and valve regurgitation, septic emboli and myocarditis. Mechanical prosthesis-infection results in annular abscesses and paravalvular leaks.

Table 7.8 Pathology of infective endocarditis

Macroscopic features:
Friable vegetations near the free border that destroys the leaflets
Perforation of the valve leaflets (nearly pathognomonic)
Abscesses can form outside the annulus (ring abscess) and affect the neighbouring myocardium
Underlying valve disease except in people at high risk of bacteraemia (drug addicts)
Microscopic features:
Vegetation: masses of fibrin, platelets, neutrophils and bacteria
Chronic inflammation, neovascularization and destruction of tissue

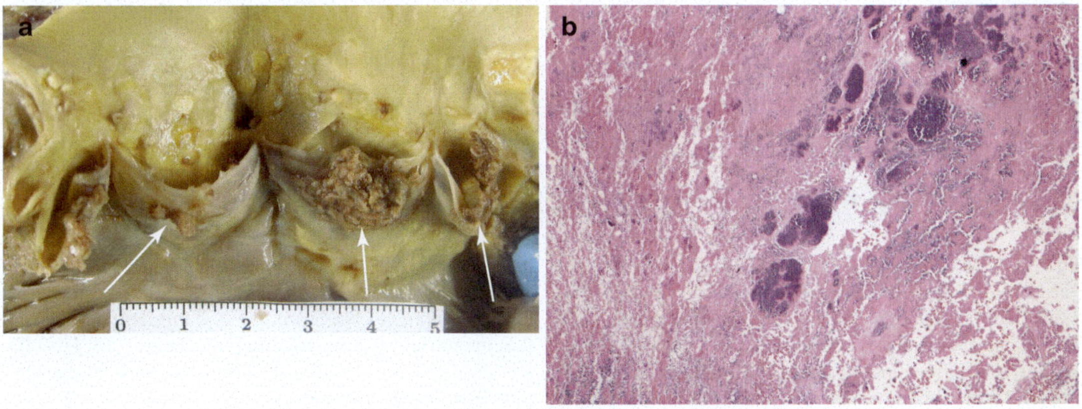

Fig. 7.33 *Acute infective endocarditis*. A 52-year-old male, diabetic, who died at hospital 4 days after being admitted with febrile syndrome caused by a urinary tract infection. (**a**) Aortic valve with vegetations (*arrows*). (**b**) Histological image of vegetation with fibrin and numerous bacterial colonies. Disperse micro abscesses were present in the myocardium and in the kidney

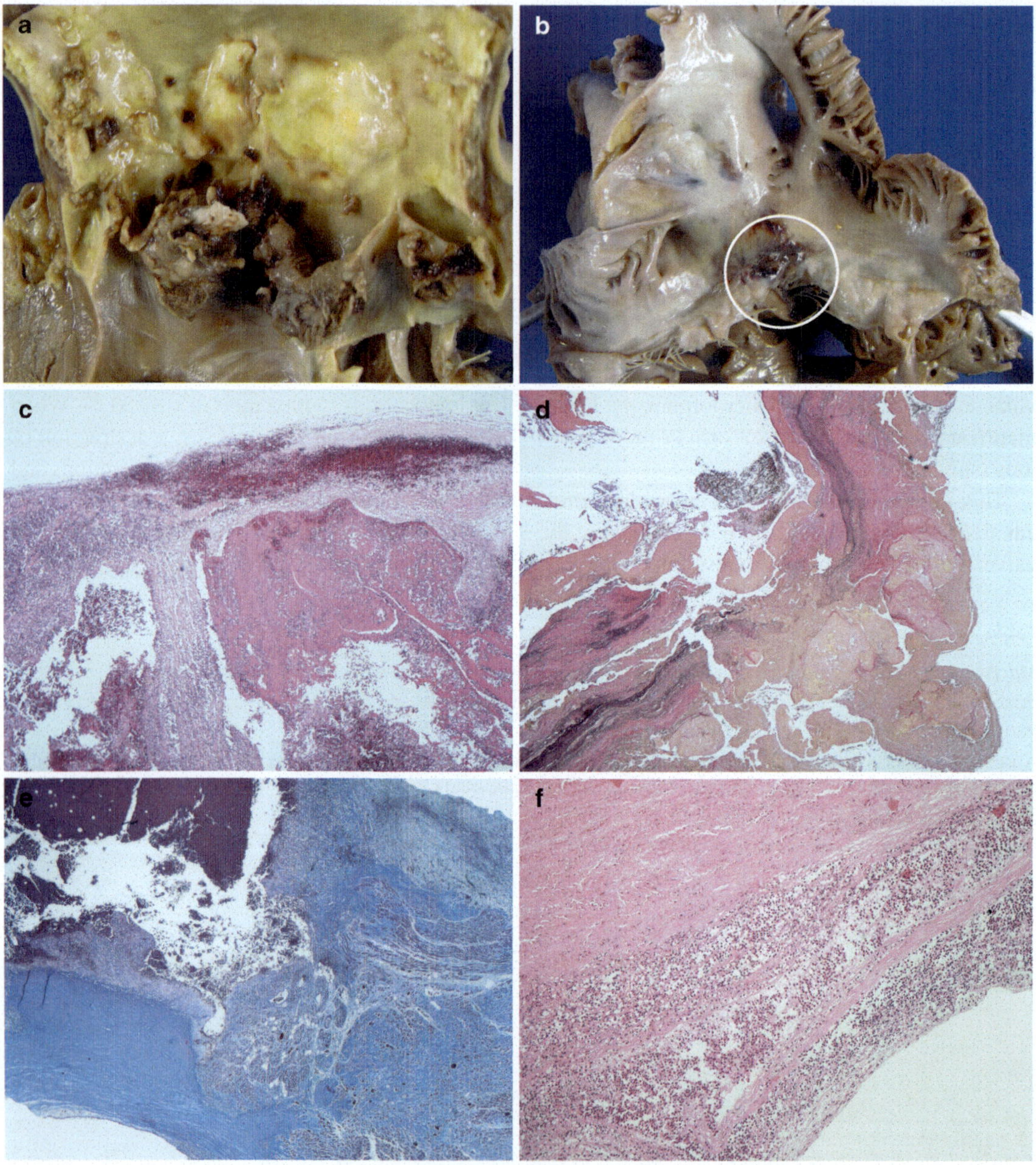

Fig. 7.34 *Infective endocarditis*. (**a**) Large vegetations adhered to the three cusps of the aortic valve. (**b**) Inflammatory infiltrate in the interatrial septum that reaches the membranous septum and the tricuspid valve (*circle*). (**c**) Microscopic image of one of the cusps where a severe acute inflammatory infiltrate with haemorrhage in the leaflet and *chordae tendinae* can be seen, and masses of fibrin and neutrophils under the leaflet. (**d**) Vegetations destroy the valve leaflets (Weigert's stain of elastic fibres). (**e**) Haemorrhage and purulent exudate in the interatrial septum affecting the bundle of His. (**f**) Detail of the aortic wall with extensive purulent exudate that forms microabscesses

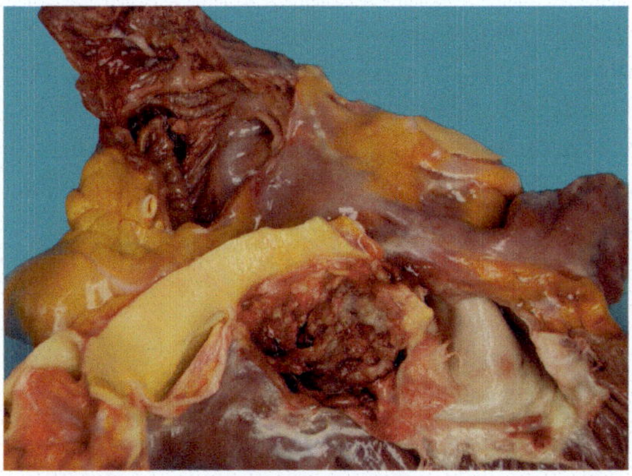

Fig. 7.35 *Acute bacterial endocarditis of the aortic valve*. The posterior leaflet is severely compromised and endocarditis is associated with destructive and ulcerative phenomena in the surrounding tissue that extends and infects the mitral valve (fresh tissue). A 52-year-old sub-Saharan black man who was admitted to hospital with heart failure, dying 4 days later

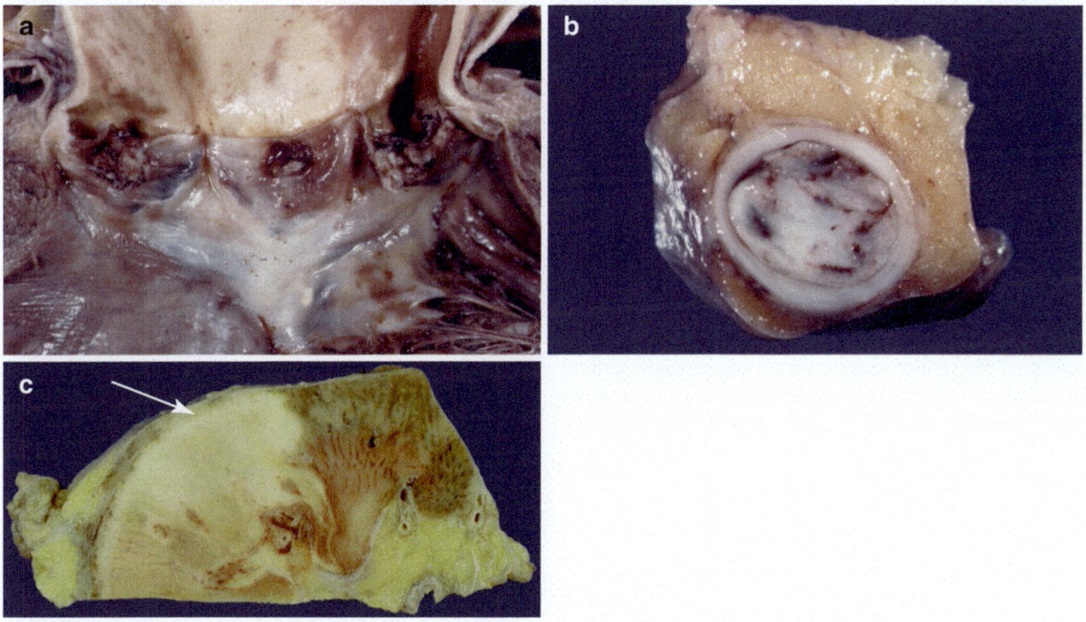

Fig. 7.36 *Acute infective endocarditis with embolization of vegetations*. A 55-year-old male who died suddenly in the street. (**a**) Large vegetations in the aortic valve cusps. (**b**) Section of the left main coronary with a vegetation fragment occluding the lumen. (**c**) Renal infarct in the same person due to a septic embolus (*arrow*)

7.6.2 Non-Infective Endocarditis (Non-Bacterial Thrombotic Endocarditis)

Non-infective endocarditis (also known as non-bacterial thrombotic endocarditis) usually occurs in patients with a wide variety of diseases associated with general debility: cancer, sepsis, disseminated intravascular coagulation, uraemia, burns or intracardiac catheters. The pathological findings are summarised in Table 7.9 and are shown in Fig. 7.37.

Table 7.9 Pathology of non-infective endocarditis

Macroscopic features:
Small vegetations (<5 mm) along the line of closure of the valves
May give rise to distal emboli
Microscopic features:
Fibrino-platelet material with no inflammatory cells

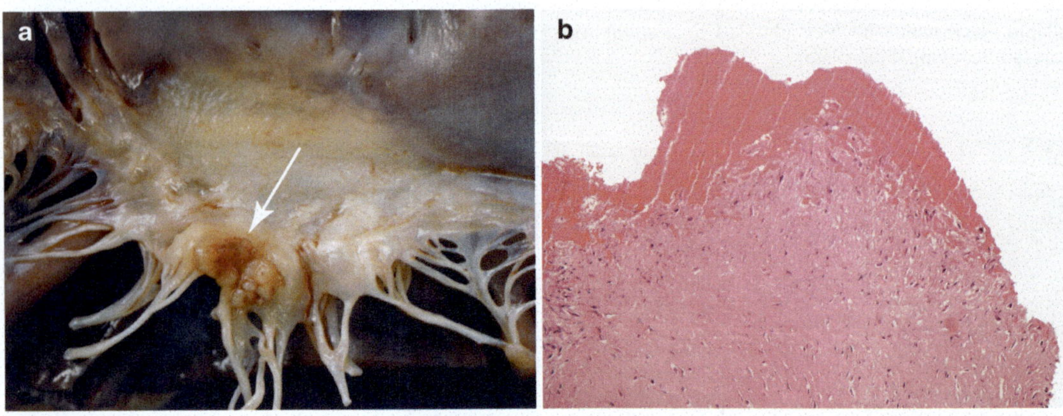

Fig. 7.37 *Non-infective endocarditis in the mitral valve*. (**a**) Vegetations along the line of closure of the anterior leaflet of the mitral valve can be appreciated, without destruction or erosion of the structure (*arrow*), in the heart of a 26-year-old man who died suddenly on a building site. Heart weight was 570 g and the LAD coronary artery presented a luminal stenosis of 75 % due to an atheroma plaque. (**b**) Microscopic image of the vegetation, showing fibrin deposition on the leaflet with no bacteria or inflammatory infiltrate

7.7 Pathology of Prosthetic Heart Valves

Replacement of diseased valves in symptomatic patients with prosthetic valves is a relatively common surgical intervention.

Prosthetic heart valves can be classified into two major types, biological and mechanical. The key difference is the material used to manufacture the occlusion system. Bioprosthetic valves are manufactured from animal tissue (heterografts) or from human cadavers (homografts). Heterografts can be derived either from porcine aortic valves (porcine valves, such as the Hancock or Carpentier Edwards valves) or, more frequently, from bovine pericardium (bovine valves, such as the Perimount or Trifecta valves). Mechanical valves are most commonly manufactured from pyrolytic carbon, usually coating a graphite substrate. These valves can be single tilting disk (Medtronic Hall, Omnicarbon, Björk-Shiley and Monostrut) or more frequently double tilting disk devices (St. Jude, ATS, and Carbomedics) (Figs. 7.38, 7.39, 7.40, 7.41 and 7.42).

Both types of valves have advantages and disadvantages. Biological valves have a limited duration but are less thrombogenic so anticoagulation therapy is not required. In contrast, mechanical valves have excellent durability but patients require chronic anticoagulation therapy. As a general rule, mechanical prosthetic valves are used in young people whereas in older patients, bioprosthetic valves are preferred. Complications of prosthetic valves are described in Table 7.10.

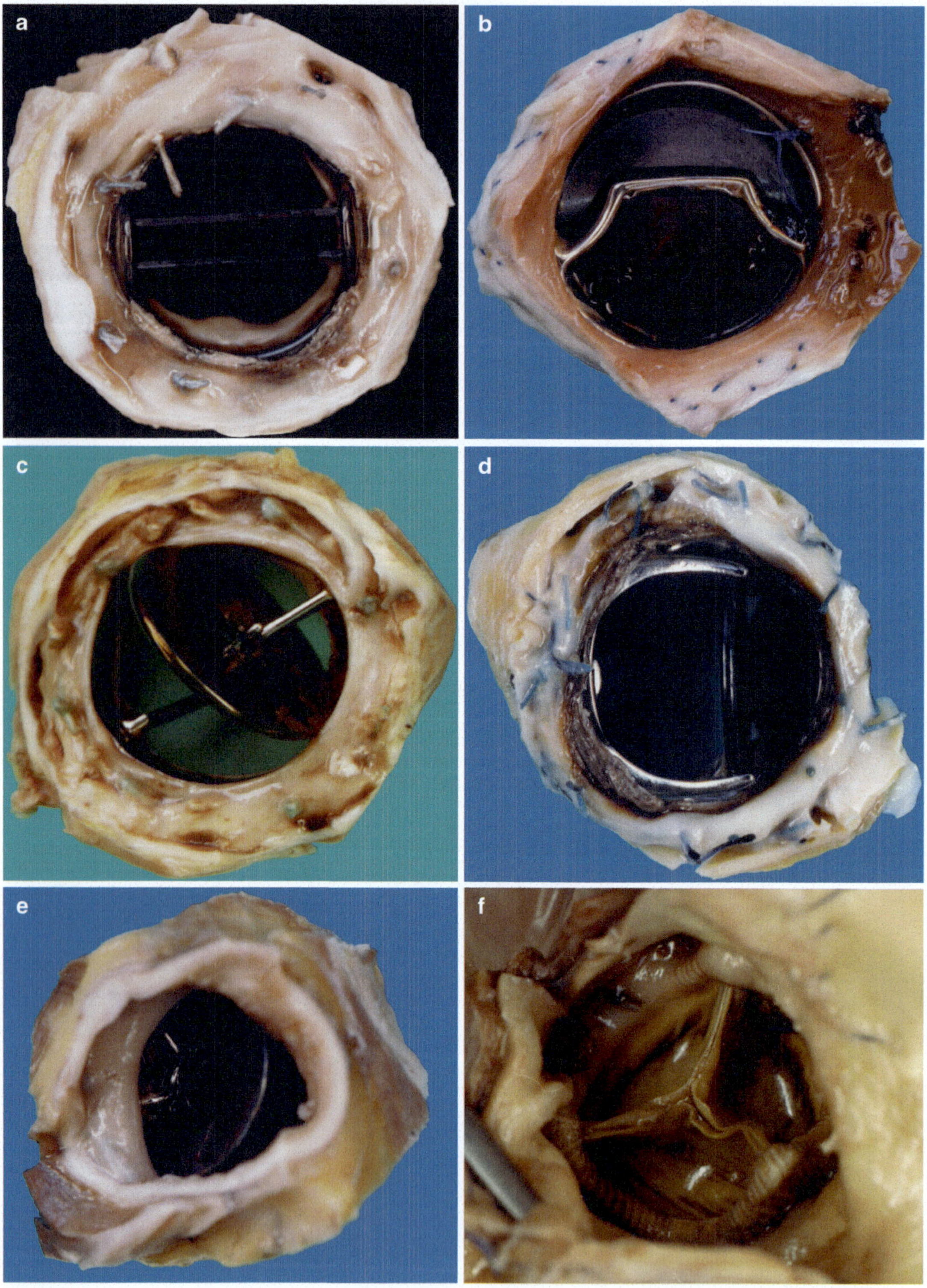

Fig. 7.38 *Types of prosthetic valves*. (**a**) Bileaflet prosthetic mechanical valve (St. Jude Medical). (**b**) Prosthetic valve with a single tilting disk (Björk-Shiley). (**c**) Mechanical valve with a single tilting disk (Medtronic-Hall). (**d**) Prosthetic valve with a single tilting disk (Lillehei-Kaster). (**e**) Prosthetic valve with a single tilting disk (Björk-Shiley). (**f**) Bioprosthetic valve

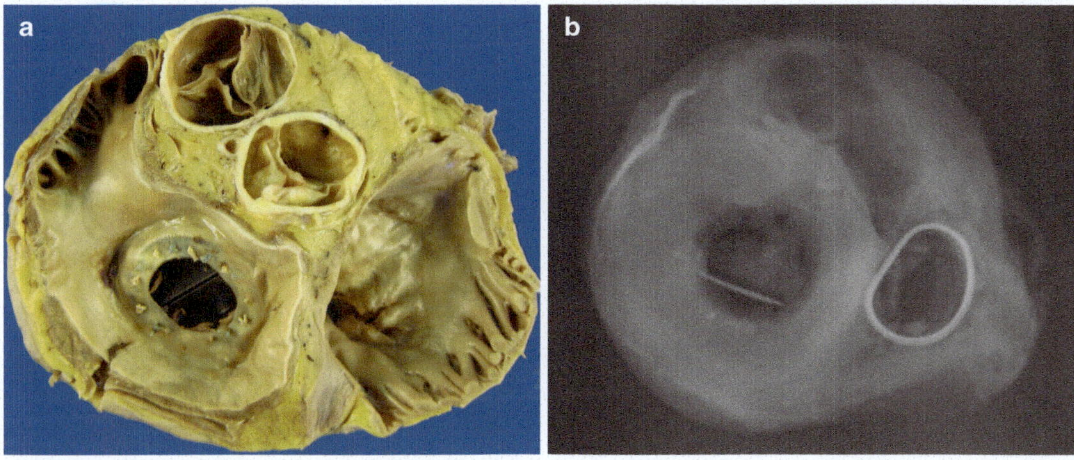

Fig. 7.39 *Bileaflet prosthetic mechanical valve*. (**a**) A 69-year-old woman who died suddenly at home. St. Jude Medical mechanical mitral valve. Heart weight was 788 g. (**b**) X-Ray image showing a prosthetic mechanical mitral valve similar to the one in (**a**) and annuloplasty of the tricuspid valve in a 70-year-old woman, performed a few months before her death

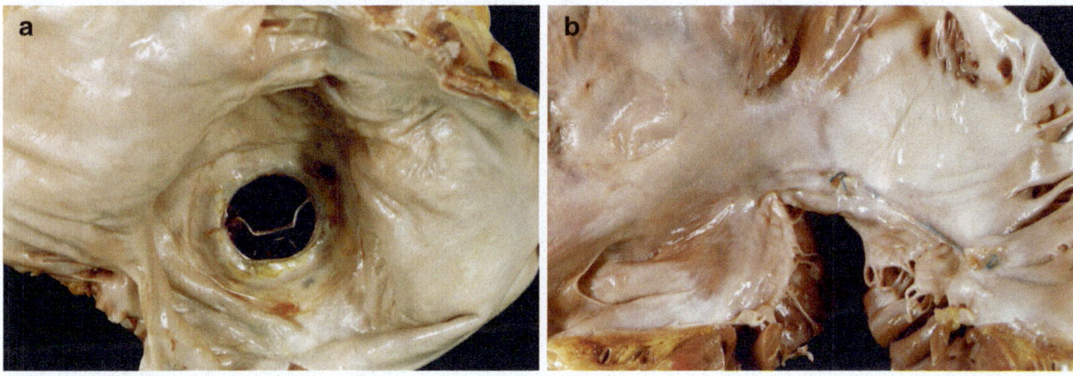

Fig. 7.40 *The single tilting-disk mechanical valve*. A 63-year-old male with previous valve replacement surgery who died in a traffic accident. Sudden death prior to the accident is suspected. Heart weight was 1.117 g and the LV was dilated (5.5 cm) with disperse scars. (**a**) Prosthetic valve with a single tilting disk (Björk-Shiley). Marked dilation of the left atrium can be appreciated. (**b**) Annuloplasty ring of 4.6 cm to restore the natural shape of the tricuspid valve

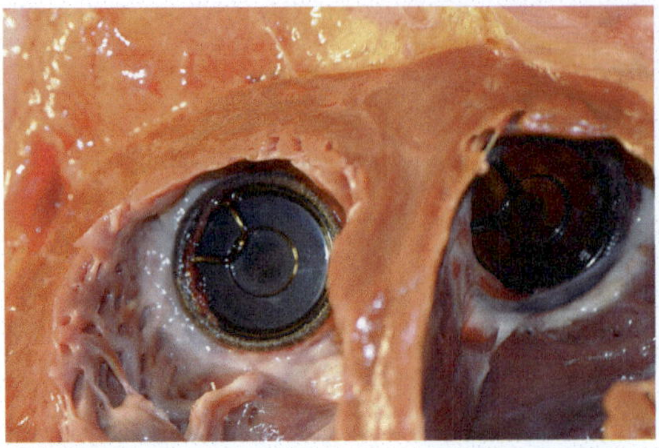

Fig. 7.41 *Double replacement of tricuspid and mitral valves with single tilting-disk mechanical prosthese*. View from the ventricles. The motion of the two valvular complexes was correct

Table 7.10 Complications derived from prosthetic valves

Biological and mechanical prostheses	Biological prostheses	Mechanical prostheses
Abscess in the annulus	Calcification	Thrombosis and embolization (Figs. 7.44, 7.45, 7.46 and 7.47)
Endocarditis	Perforation	Dysfunction in the valve movement
Paravalvular leak	Tear (Fig. 7.43)	
Pannus		
Ostia coronaria occlusion		

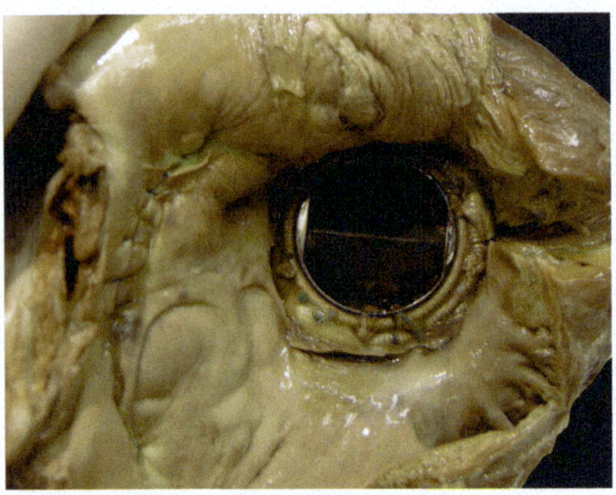

Fig. 7.42 *Mitral valve replacement with a bileaflet mechanical prosthesis*. Competent sutures. An adult woman who died from haemodynamic complications 10 days after undergoing cardiac surgery

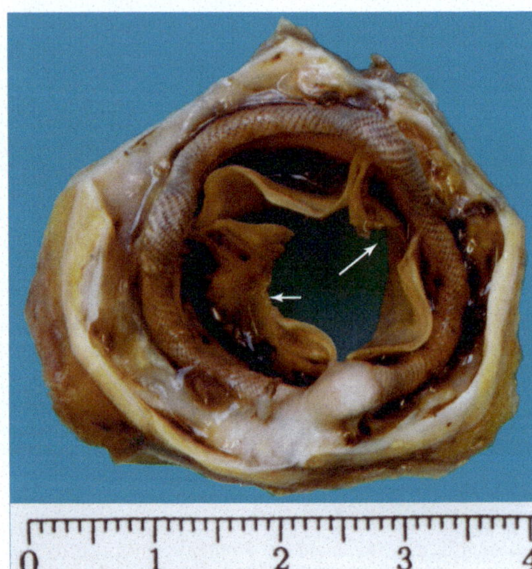

Fig. 7.43 *Tear in a biological prosthetic valve*. A 79-year-old male surgically intervened for a double aorto-coronary by-pass and replacement of the aortic valve with a bioprosthesis. He suddenly died on his way for a hospital check-up. Autopsy revealed a heart of 729 g, severe three-vessel coronary artery disease, scars from a postero-septal infarct, permeable double aorto-coronary by-pass and biological prosthesis in the aortic valve with tears in two of its leaflets (*arrows*). Sutures were competent

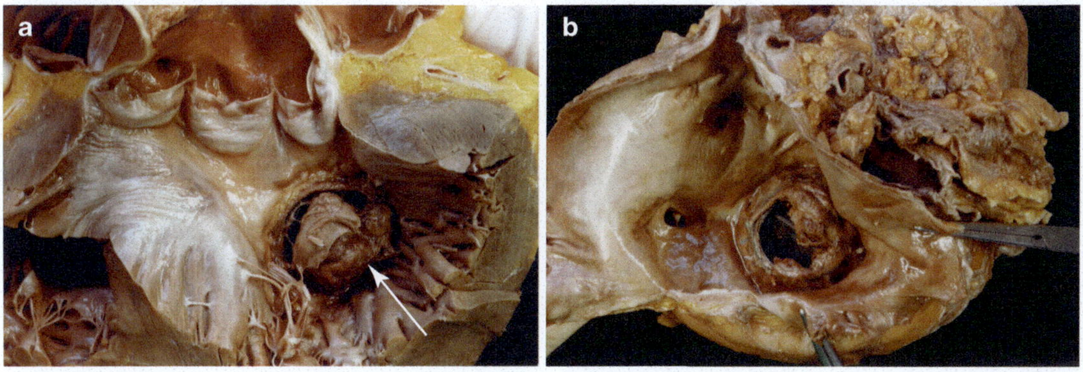

Fig. 7.44 *Thrombosis of a mechanical prosthetic valve.* A 38-year-old woman, with severe mitral regurgitation, LV dilation and pulmonary hypertension, surgically intervened for mitral valve replacement with a mechanical prosthesis. The postoperative period was normal. After 14 days she presented with severe hyponatremia, tachypnea and metabolic acidosis. Assisted ventilation was initiated but she developed right ventricular failure and died. Other diagnosis: right basal pneumonia, pulmonary hypertension and sepsis. (**a**) Large thrombus on the mitral prosthesis, in the LV outflow tract (*arrow*). (**b**) Thrombus on the atrial side of the prosthesis. The interventricular septum and the anterior wall of the RV showed foci of myocarditis

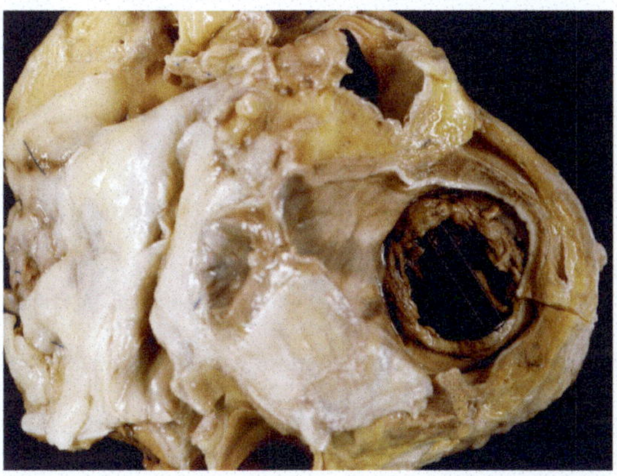

Fig. 7.45 *Thrombosis of a bileaflet mechanical prosthetic valve.* A 82-year-old male with brain trauma and residual hemiplegia due to an accident. He died 6 days after discharge. Autopsy revealed semi-solid material obstructing the glotis. Heart weight was 419 g, there was a stent implanted in the left circumflex coronary artery and a healed lateral transmural infarct in the LV. The mitral valve had been replaced by a mechanical prosthesis which presented annular thrombi that, under histology, were composed of fibrin and platelets

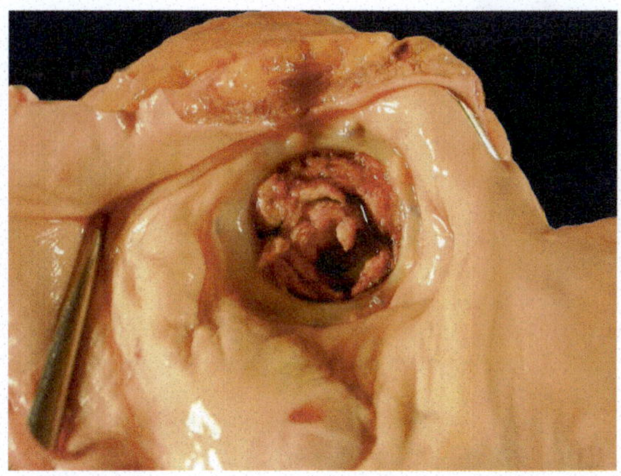

Fig. 7.46 *Thrombosis of a mechanical prosthesis.* Very large thrombosis that completely covers the surface of a single tilting mechanical prosthesis in the mitral valve (fresh tissue). A 38-year-old woman who died in the hospital and the relatives sued for alleged medical malpractice

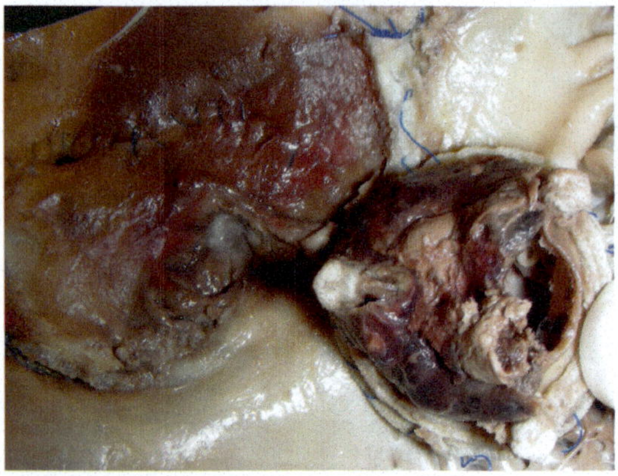

Fig. 7.47 *Thrombosis of a biological prosthesis*. An 18-year-old male surgically intervened for aortic valve replacement with a biological prosthesis. Thrombotic material on the bioprosthetic leaflets leading to luminal reduction greater than 60 %. The right coronary *ostium* was also compromised and histology revealed a recent subendocardial myocardial infarction affecting the interventricular septum and the free wall of the LV

Bibliography

Aboulhosn J, Child JS. Left ventricular outflow obstructions. Subaortic stenosis, bicuspid aortic valve, supravalvar aortic stenosis, and coarctation of the aorta. Circulation. 2006;114(22):2412–22.

Anders S, Said S, Schulz F, Püschel K. Mitral valve prolapse syndrome as cause of sudden death in young adults. Forensic Sci Int. 2007;171(2–3):127–30.

Barkhordarian R, Uemura H, Rigby ML, Sethia B, Shore D, Goebells A, et al. A retrospective review in 50 patients with subaortic stenosis and intact ventricular septum: 5-year surgical experience. Interact Cardiovasc Thorac Surg. 2007;6(1):35–8.

Boudoulas KD, Boudoulas H. Floppy mitral valve (FMV)/mitral valve prolapse (MVP) and the FMV/MVP syndrome: pathophysiologic mechanisms and pathogenesis of symptoms. Cardiology. 2013;126:69–80.

Boudoulas KD, Borer JS, Boudoulas H. Etiology of valvular heart disease in the 21st century. Cardiology. 2013;126:139–52.

Burke A, Tavora F. Cardiomyopathies and endocardial diseases. In: Practical cardiovascular pathology. Philadelphia: Wolters Kluwer Health/Lippincott Wiliams & Wilkins; 2011. p. 143–95.

Butany J, Ahluwalia MS, Munroe C, Fayet C, Ahn C, Blit P, et al. Mechanical heart valve protheses: identification and evaluation (erratum). Cardiovasc Pathol. 2003;12(6):322–44.

Castonguay MC, Burneo KD, Edwards WD, Baddour LM, Maleszewski JJ. Surgical pathology of native valve endocarditis in 310 specimens from 287 patients (1985–2004). Cardiovasc Pathol. 2013;22:19–27.

El Khoury G, Glineur D, Rubay J, Verhelst R, d'Acoz YD, Poncelet A, et al. Functional classification of aortic root/valve abnormalities and their correlation with etiologies and surgical procedures. Curr Opin Cardiol. 2005;20(2):115–21.

Fealey ME, Edwards WD, Miller DV, Maleszewski JJ. Unicommissural aortic valves: gross, histological, and immunohistochemical analysis of 52 cases (1978–2008). Cardiovasc Pathol. 2012;21(4):324–33.

Fernández B, Arqué JM, Durán AC, Such M, Sans-Coma V. Diferentes etiologías de las válvulas aórticas bicúspides: implicaciones genéticas, patológicas, clínicas y quirúrgicas. Cardiocore. 2010;45:68–71.

Otto CM. Calcific aortic stenosis — time to look more closely at the valve. N Engl J Med. 2008;359(13):1395–8.

Roberts WC. Morphologic aspects of cardiac valve dysfunction. Am Heart J. 1992;123(6):1610–32.

Roberts WC, Ko JM. Frequency by decades of unicuspid, bicuspid, and tricuspid aortic valves in adults having isolated aortic valve replacement for aortic stenosis, with or without associated aortic regurgitation. Circulation. 2005;111(7):920–5.

Schoen FJ. Cardiac valves and valvular pathology. Update on function, disease, repair, and replacement. Cardiovasc Pathol. 2005;14(4):189–94.

Siu SC, Silversides CK. Bicuspid aortic valve disease. J Am Coll Cardiol. 2010;55(25):2789–800.

Suarez-Mier MP, Aguilera B. Causas de muerte súbita asociada al deporte en España. Rev Esp Cardiol. 2002;55:347–58.

Suárez-Mier MP, Morentin B. Supravalvular aortic stenosis, Williams syndrome and sudden death. A case report. Forensic Sci Int. 1999;106:45–53.

Verma S, Siu SC. Aortic dilatation in patients with bicuspid aortic valve. N Engl J Med. 2014;370:1920–9.

Virmani R, Atkinson JB, Forman MB, Robinowitz M. Mitral valve prolapse. Hum Pathol. 1987;18(6):596–602.

Cardiomyopathies

8

Beatriz Aguilera, M. Paz Suárez-Mier,
Gonzalo Guzzo-Merello, and Pablo García-Pavía

Contents

8.1 Introduction

The term cardiomyopathy encompasses a heterogeneous group of disorders characterized by an intrinsic pathology of the cardiac muscle. Cardiomyopathies differ from other cardiac diseases in that they are not the result of other pathologies, such as pericardial diseases, hypertension, congenital abnormalities, valvulopathies or ischemic heart diseases. Diverse types of cardiomyopathies can be distinguished according to their morphology or aetiology. The major morphological patterns are: hypertrophic, dilated, restrictive, arrhythmogenic and non-compacted. Based on the aetiology, cardiomyopathies can be primary/secondary or familial/non-familial.

Several classifications exist (Table 8.1). The most accepted are those from the World Health Organisation (Richardson et al. 1996), the American Heart Association (Maron et al. 2006) and the European Society for Cardiology (Elliott et al. 2008). According to the WHO,

B. Aguilera (✉) • M.P. Suárez-Mier
Histopathology Service, National Institute of Toxicology and Forensic Sciences, Las Rozas (Madrid), Spain
e-mail: beatriz.aguileratapia@justicia.es; mariapaz.suarez@justicia.es

G. Guzzo-Merello
Heart Failure Unit, Cardiology Department, Hospital Universitario de Getafe, Getafe (Madrid), Spain

P. García-Pavía
Heart Failure and Inherited Cardiac Diseases Unit, Cardiology Department, Hospital Universitario Puerta de Hierro, Majadahonda (Madrid), Spain
e-mail: pablogpavia@yahoo.es

J.S. Lucena et al. (eds.), *Clinico-Pathological Atlas of Cardiovascular Diseases*,
DOI 10.1007/978-3-319-11146-9_8, © Springer International Publishing Switzerland 2015

Table 8.1 Classification of cardiomyopathies

1995 (WHO)	2006 (AHA)	2008 (ESC)
Idiopathic	*Primary*	*Familial/genetic*
DCM	*Genetic*	or
HCM	HCM	*Non-familial*
RCM	ACM	
ARVC	LVNC	HCM
Unclassified	Glycogen storage defects	DCM
	Conduction defects	RCM
Specific	Mitochondrial myopathies	ARVC
	Ion channel disorders	Unclassified
	Mixed:	
	DCM	
	RCM	
	Acquired:	
	Inflammatory (myocarditis)	
	Stress-provoked (Tako–Tsubo)	
	Peripartum	
	Tachycardia-induced	
	Infants of insulin-dependent diabetic mothers	
	Secondary	

DCM dilated cardiomyopathy, *HCM* hypertrophic cardiomyopathy, *RCM* restrictive cardiomyopathy, *ARVC* arrhythmogenic right ventricular cardiomyopathy, *ACM* arrhythmogenic cardiomyopathy, *LVNC* left ventricular non-compaction cardiomyopathy

cardiomyopathies are divided into two main groups depending on whether they are intrinsic diseases of the myocardium (cardiomyopathies) or are associated with a specific cardiac disease or are part of generalised systemic diseases (specific cardiomyopathies). The AHA classification distinguishes between primary and secondary cardiomyopathies based on whether the disorder involves predominantly the heart or is part of a multisystem syndrome/disease. Furthermore, the AHA classification includes channelopathies as a type of cardiomyopathy, in which there are molecular abnormalities without structural cardiac disease.

Finally, the ESC classification is based on the main morphological patterns but then makes distinctions between familial and non-familial entities. This classification does not include channelopathies as a type of cardiomyopathy.

Remarks
Hypertrophic and arrhythmogenic cardiomyopathies are the most common cause of sudden death associated with sports, among the young
More than one half of heart transplants are performed on patients affected by a cardiomyopathy

8.2 Hypertrophic Cardiomyopathy

Hypertrophic cardiomyopathy (HCM) is a heart muscle disease characterised by the presence of significant primary hypertrophy in the left ventricle (in adults ≥ 15 mm), which means that it is not caused by disorders leading to increased LV afterload (resistance to blood ejection). In clinical practice, conditions than can provoke compensatory or secondary ventricular hypertrophy are mainly hypertension and aortic stenosis.

HCM is a relatively common clinical entity with an incidence of 1 in every 500 individuals. Familial origin occurs in 50–60 % of the cases, presenting an autosomal dominant pattern of transmission in most of the cases, with incomplete penetrance.

Mutations in several genes have been identified as associated with HCM. The majority of the mutated genes encode sarcomeric proteins, which are proteins that form the contractile apparatus of the myocyte.

Clinically, it is important to highlight the great variability in symptomatology among these patients, not only in clinical presentation but also evolution and prognosis. Patients can report chest pain, dyspnoea on exertion, dizziness or syncope. The ECG usually shows a pattern of LV hypertrophy with increased voltage in the left precordial leads, I and aVL, and prominent and pathological Q waves, although deep and narrow ("dagger-like"). Alterations in the T waves and depression in the ST segment are also common. A definitive diagnosis is carried out with echocardiography by measuring the parietal thickness at different planes and projections, and discarding other clinical entities that could cause a similar degree of hypertrophy. In adults, HCM is diagnosed when the maximal wall thickness is greater or equal to 15 mm measured in end-diastole.

A small subgroup of patients have a particularly malignant course with episodes of sudden cardiac death (SCD) due to ventricular arrhythmias. In order to identify patients at higher risk, who would benefit from an automatic implantable cardiac defibrillator (ICD), a prognostic stratification should be done taking into account several risk factors associated with SD.

HCM can be considered obstructive or non-obstructive depending on the presence of obstruction of the left ventricular outflow tract (LVOTO). One third of patients with HCM have a significant obstruction (>30 mmHg) at rest and an additional third develops obstruction on exertion but not at rest. Based on the location at which hypertrophy is more manifest, HCM can be clinically divided in concentric, asymmetric with sigmoid septal contour (septal HCM), asymmetric with reverse pattern and apical forms. Gross and microscopic features are shown in Table 8.2.

Table 8.2 Pathology of HCM

Macroscopic features:
Myocardial hypertrophy (heart weight can exceed 500 g)
Predominance of left ventricular hypertrophy, although the right ventricle may also be involved
Based on the location, LV hypertrophy adopts several forms: Type I: hypertrophy of the anterior part of the ventricular septum, usually basal (Fig. 8.1) Type II: hypertrophy affecting the entire septum, without affecting the free wall of the LV (Fig. 8.2) Type III: hypertrophy of the septum and the anterior wall of the LV (Fig. 8.3) Type IV: hypertrophy anywhere in the ventricle, including the apex (Figs. 8.4 and 8.5)
Prominent trabeculae and papillary muscles
"Whorled" appearance of the myocardium
White scarring foci randomly distributed (do not follow the vessels as in scarring in infarcts)
Plaque of subaortic endocardial fibrosis due to the contact of the anterior mitral valve leaflet (it is not pathognomonic, it also may appear in aortic valve stenosis) (Fig. 8.6)
LV dilatation may be present at end-stages of the disease
Microscopic features: (Fig. 8.7)
Myocyte hypertrophy, with hyperchromatic nuclei
Myocyte disarray, with myocytes arranged at angles or adopting a whirling appearance or as fascicles organised perpendicularly
Myofibril disarray
Increased interstitial cellularity
Thickened intramyocardial coronary arteries due to hypertrophy, disarray in the medial layer and intimal fibrosis, which contributes to ischaemia and myocardial fibrosis
Evolution towards interstitial or replacement fibrosis
Foci of myocarditis in some cases

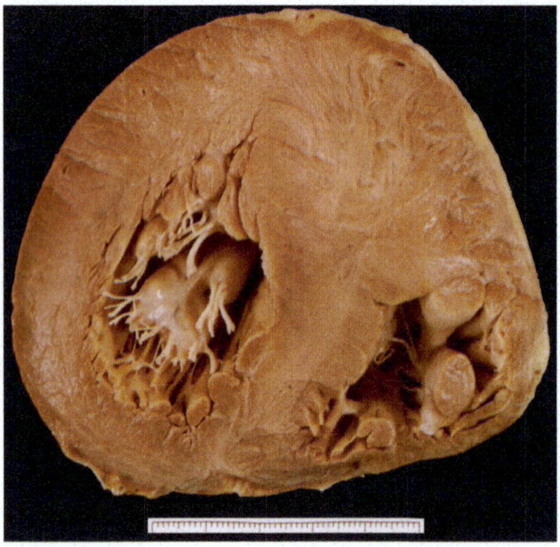

Fig. 8.1 *HCM type I.* Sudden death in a 22-year-old male. Heart weight was 505 g. Asymmetric HCM with major hypertrophy of the anterior portion of the ventricular septum (4 cm thickness) and whirling appearance of the myocardium

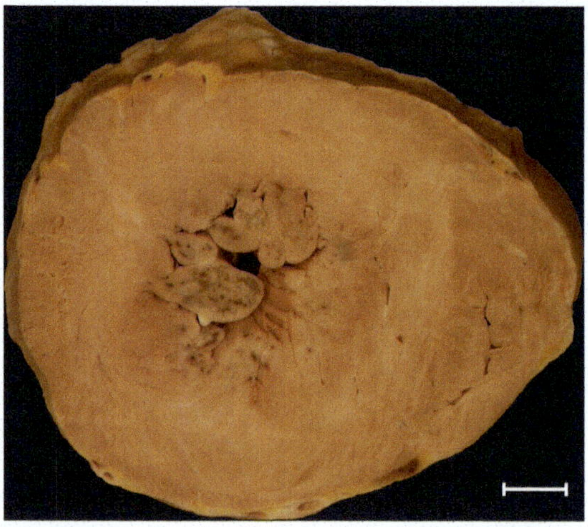

Fig. 8.2 *HCM type II.* A 27-year-old woman who died suddenly at home. Severe HCM with predominant involvement of the septum. Microinfarctions in papillary muscles and trabeculae can be observed, caused by small vessel disease (viewed microscopically)

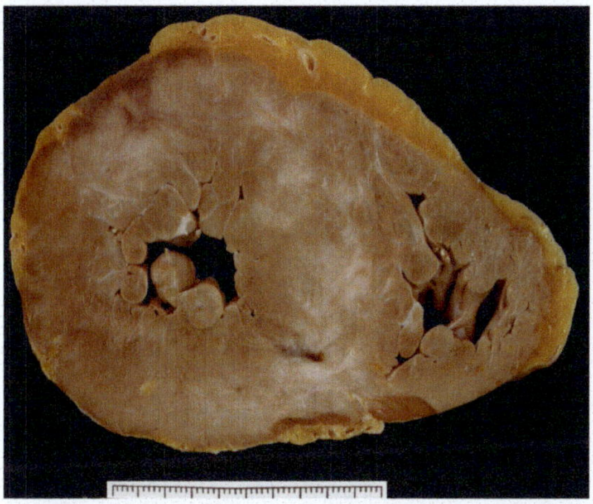

Fig. 8.3 *HCM type III*. A 24-year-old male who was diagnosed with obstructive HCM at the age of 8. He died suddenly at work while getting changed. Heart weight was 602 g and showed major hypertrophy of the ventricular septum and the anterior wall of the LV with extensive foci of fibrosis. Microscopic study revealed intense dysplasia in the intramural coronary arteries (Image reproduced with permission of Aguilera and Suárez-Mier 2003)

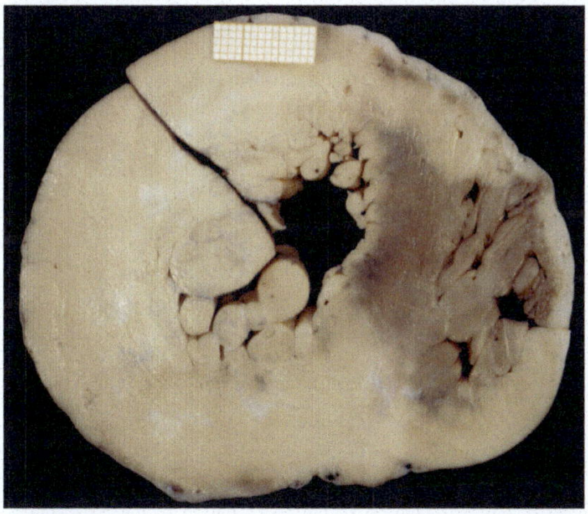

Fig. 8.4 *HCM type IV*. A 13-year-old boy who died suddenly at school during physical education class. Heart weight was 344 g and gross examination showed hypertrophy of the free wall of the LV (15 mm) with multiple scarring foci. Septum thickness was preserved (11 mm)

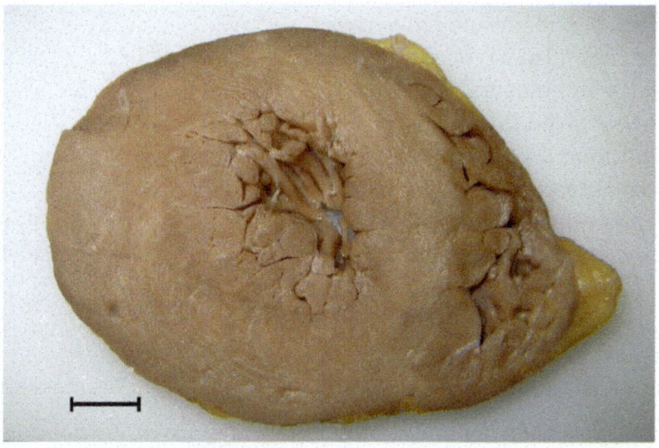

Fig. 8.5 *Apical HCM*. A 60-year-old male. Sports-related sudden death. Heart weight: 521 g. LV: 1.8 cm in posterior and lateral walls and 1.3 cm in anterior wall and septum, RV: 0.5 cm. Apical and posterolateral LV hypertrophy. 1 cm

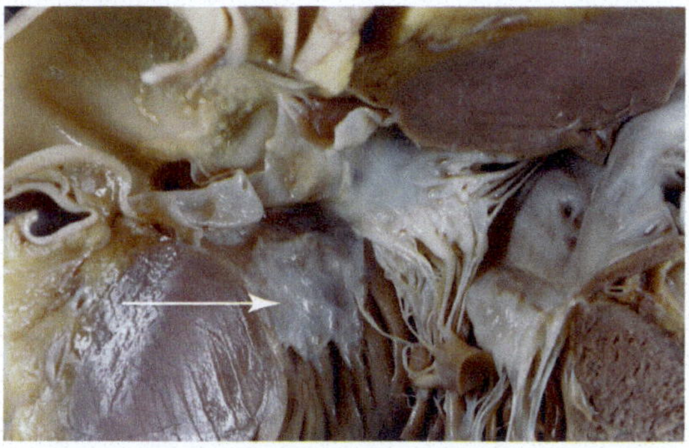

Fig. 8.6 *Subaortic endocardial fibrosis*. A 36-year-old male, diagnosed with moderate obstructive HCM. Subaortic endocardial fibrosis. Specular image of the anterior leaflet of the mitral valve (*arrow*)

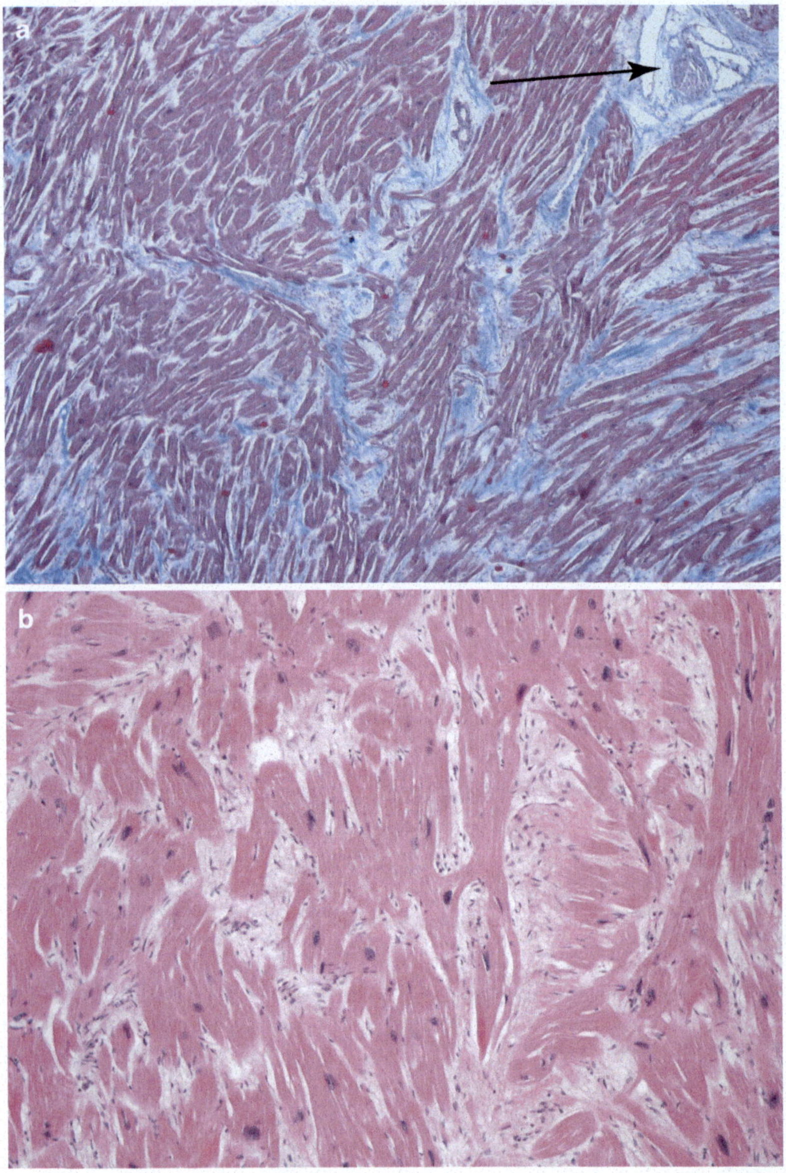

Fig. 8.7 *Microscopic features of HCM*. (**a**) Myocardium with architectural disarray, fibrosis and thickened intramural arteries (*arrow*). (**b**) Hypertrophic myocytes perpendicularly distributed and with hyperchromatic nuclei. (**c**) Intramural coronary artery dysplasia with thickened wall and medial layer alterations. (**d**) Patch of fibrosis. (**a**, **c**, **d**: Masson's trichrome)

Fig. 8.7 (continued)

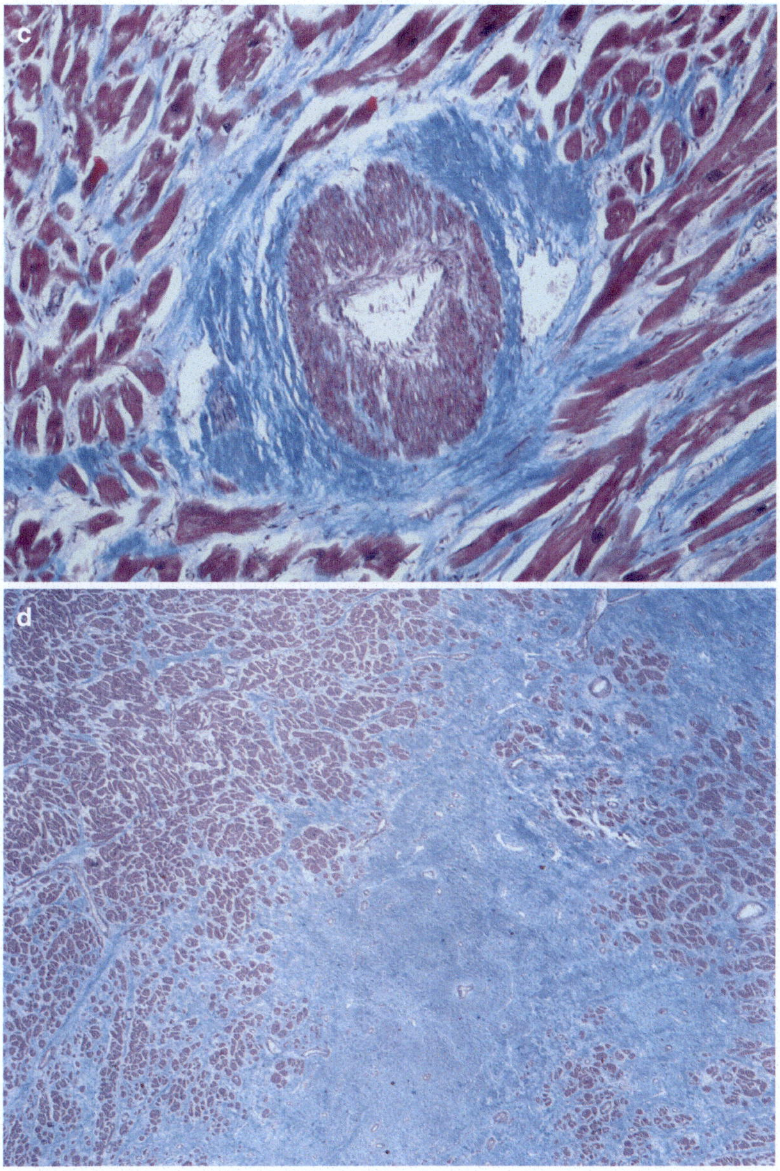

8.2.1 Idiopathic Left Ventricular Hypertrophy

Idiopathic left ventricular hypertrophy is characterised by concentric LV hypertrophy in the absence of hypertension, valve disease or aortic disease. Microscopically, myocyte disarray, characteristic of HCM, is not found (Fig. 8.8). Clinically, it is not a recognised entity and it is included within HCM. So far, limited molecular studies have been carried out in these patients and results are diverse: in some cases, mutations in genes encoding sarcomeric proteins (as in HCM) have been described, but in other cases no mutation was found. This form of LV hypertrophy has been associated with sports-related SCD and, from a pathologist's point of view, a differential diagnosis with the athletes' heart must be done.

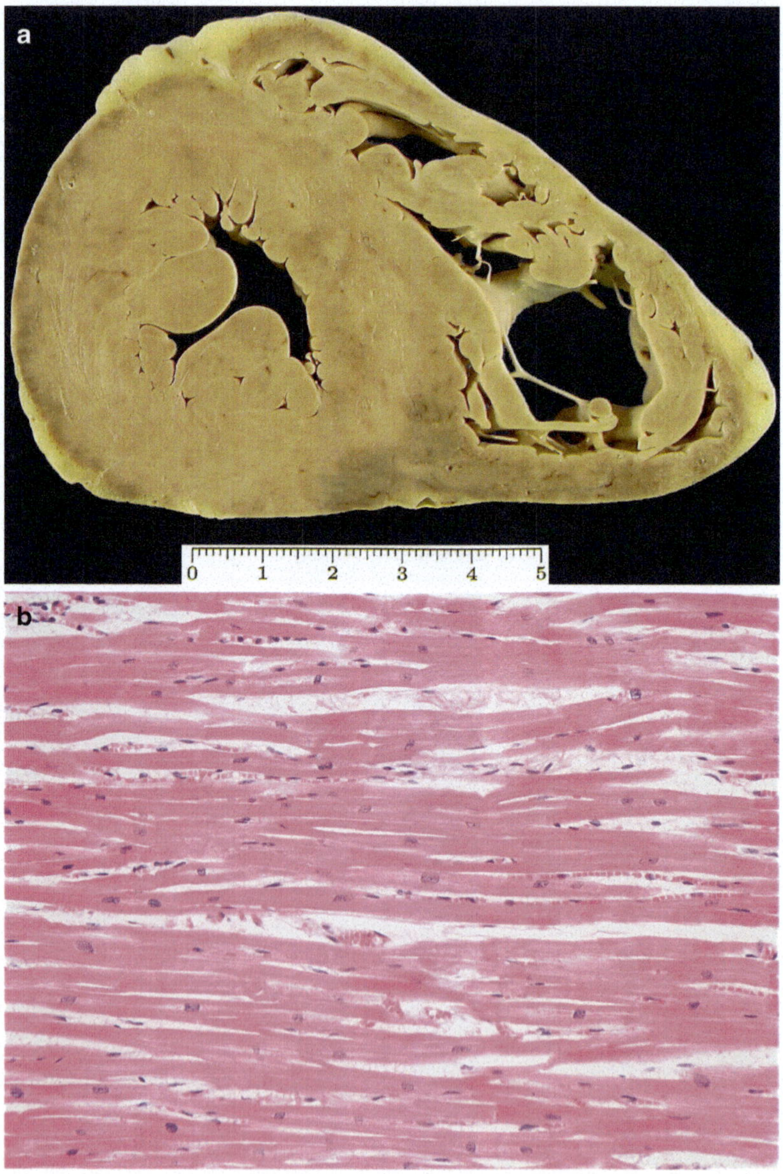

Fig. 8.8 *Idiopathic left ventricular hypertrophy*. A 15-year-old teenager suffered sudden death after cycling for 30 km. (**a**) Heart weight of 490 g and thickness of the LV walls of 15 mm. (**b**) Myocytes are hypertrophic and neatly arrayed. Cardiac family screening was negative

8.3 Arrhythmogenic Cardiomyopathy

Arrhythmogenic cardiomyopathy (ACM) is a condition characterised by fibrous or fibro-fatty replacement of the myocardium. ACM was initially described as arrhythmogenic right ventricular cardiomyopathy/dysplasia (ARVC/D), but during the last decade it has been demonstrated that the disease can affect both ventricles or even exclusively or predominantly the LV. A high density of ventricular arrhythmias, with varying degrees of complexity, is frequently found in patients with ACM. Characteristic features of ACM include the presence of ventricular wall thinning, associated with aneurysms preferentially located at the infundibular, subtricuspid or apical levels in the RV, or regional wall motion abnormalities in the LV.

A family history of the disease is present in approximately 50 % of patients, and inheritance shows an autosomal dominant pattern with variable expressivity and incomplete penetrance. Mutations in genes encoding desmosomal proteins constitute the most commonly found genetic defect in these patients. Desmosomes are molecular protein complexes that participate in the maintenance of cellular structure and cell-to-cell adhesions. Defects in these proteins can lead to separation between cells due to defective intercellular junctions, which can cause apoptosis and death of myocytes, and ultimately their replacement with fibrous and adipose tissue.

ACM usually debuts in adolescents and before the fourth decade of life. Clinical presentation can be very variable, from progressive heart failure to a mainly arrhythmogenic disease with cardiac arrest as the first manifestation. Ventricular arrhythmias are very common, hence the name of the disease, and can vary from isolated ventricular extrasystoles to sustained ventricular tachycardia (VT) or ventricular fibrillation (VF).

ACM must be suspected in individuals suffering frequent and more or less complex ventricular arrhythmias, in patients that have had a cardiac arrest, and in those with RV dilatation non-attributab. to pulmonary disease or to a systemic-pulmonary shunt. Diagnosis may be difficult, and in the clinical practice ACM is diagnosed when the patient fulfils several major and/or minor criteria. These criteria are established depending on the degree of sensitivity, specificity and predictive value of the regional wall motion abnormalities identified through imaging techniques such as echocardiography or MRI, through ECG abnormalities, ventricular arrhythmias, familial background/genetic, and histopathological studies.

Treatment of ACM is focused on the prevention of SCD and the management of heart failure. Competitive physical exercise is contraindicated. Pharmacological treatment is initiated in patients with ventricular arrhythmias or progressive ventricular dysfunction. An implantable cardiac defibrillator could be an option for patients that have suffered a cardiac arrest, syncope, or badly-tolerated ventricular arrhythmias, and in patients with severe left or right ventricular dysfunction, as a bridge to heart transplantation. Radiofrequency ablation may diminish the occurrence and complexity of the arrhythmias suffered by these patients, although this is a transient solution.

Three forms of ACM exist depending on the ventricle involved:

1. Right or classical form, which can ultimately involve also the LV (Fig. 8.9).
2. Biventricular, simultaneously compromising both ventricles (Figs. 8.10, 8.11 and 8.12).
3. Left form, in which the LV lesion is predominant (Fig. 8.11).

The pathologic characteristics of ACM are shown in Table 8.3.

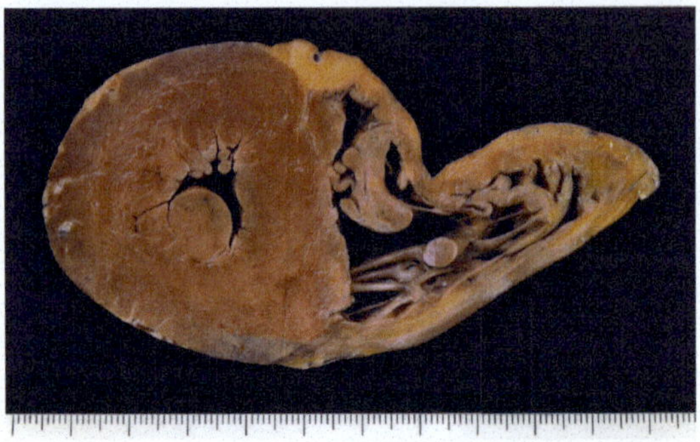

Fig. 8.9 *Arrhythmogenic right ventricular cardiomyopathy (ARVC)*. A 29-year-old male who died suddenly while playing tennis. The RV is dilated with a thinned and yellow wall. No alterations were found in the LV

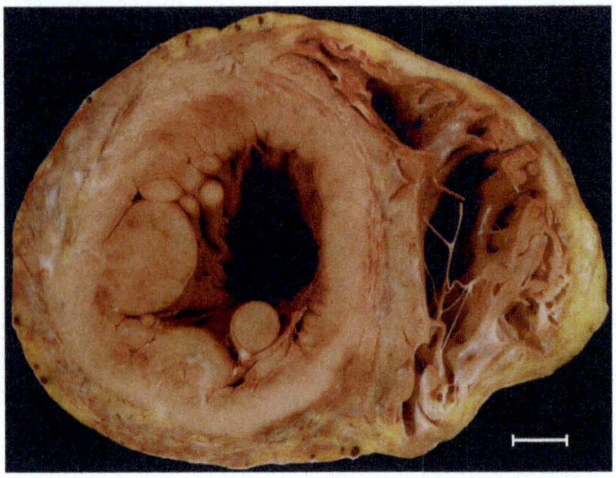

Fig. 8.10 *Arrhythmogenic biventricular cardiomyopathy*. A 23-year-old male who died while out having a drink (alcohol). Heart weight: 483 g. Involvement of the subepicardial half of the myocardium, the free wall of the LV, and the middle zone of the ventricular septum, adopting a circumferential pattern. The RV is dilated, with only the trabecular myocardium preserved (Image reproduced with permission of Aguilera and Suárez-Mier 2003)

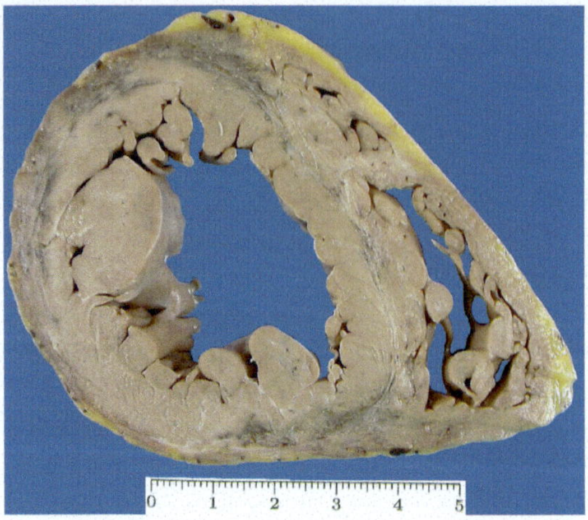

Fig. 8.11 *Arrhythmogenic left ventricular cardiomyopathy (ALVC)*. A 37-year-old woman, 12 weeks pregnant, who died suddenly while watching TV. Family history of premature SD of her father and two of her sisters (no autopsy was performed). ACM with left predominance and with a circumferential pattern. The affected myocardium appears darker and depressed due to the fixative solution

Fig. 8.12 *Arrhythmogenic biventricular cardiomyopathy.* A 38-year-old male suffered SD whilst playing paddle tennis. (**a**) Four-chambered section of the heart (460 g). Lesion is very extended in the RV; in the LV, the lesion is more focal. (**b**) Panoramic view of the RV wall where transmural fibro-fatty replacement of the myocardium can be observed. Only the trabeculae are preserved (Masson's trichrome). (**c**) Large subepicardial replacement of the free wall of the LV (Masson's trichrome)

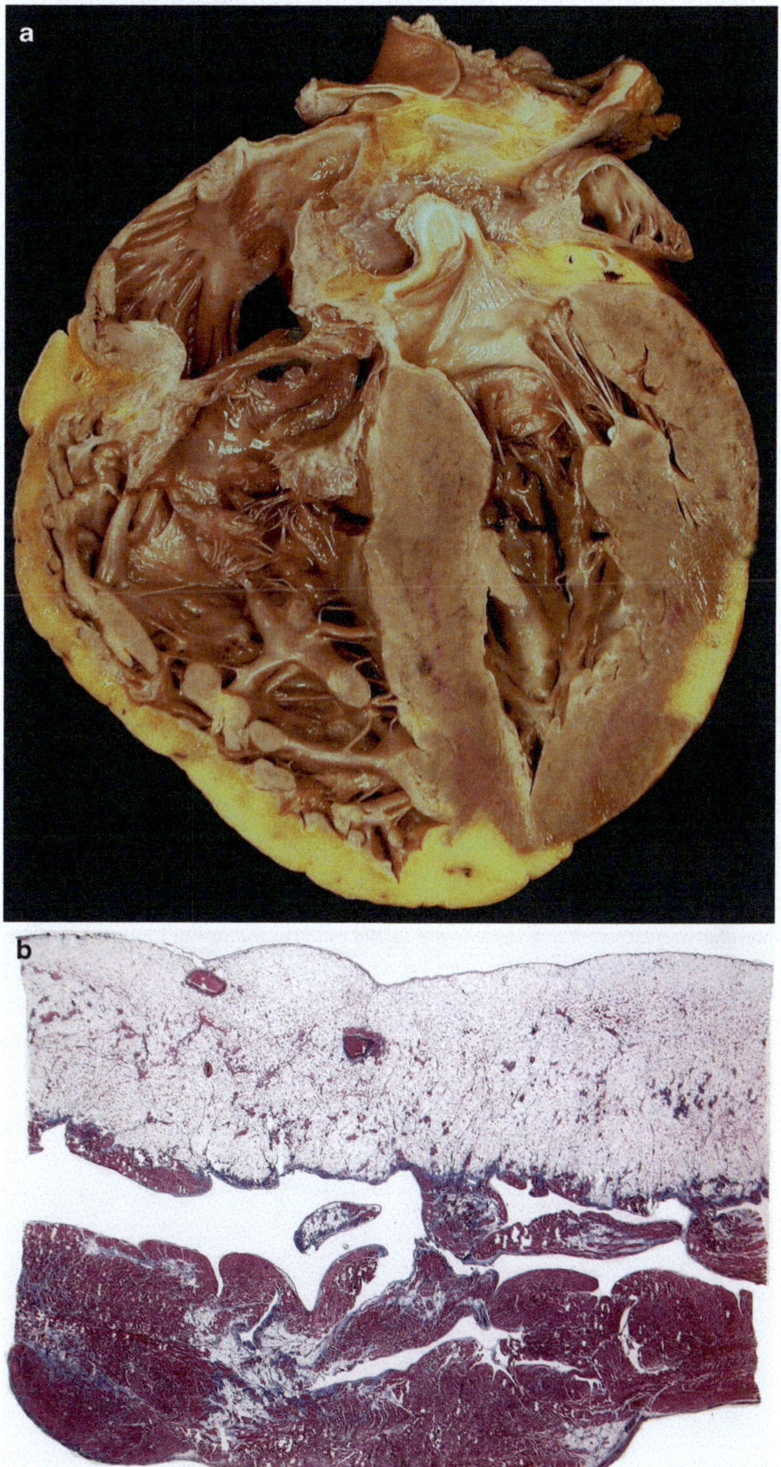

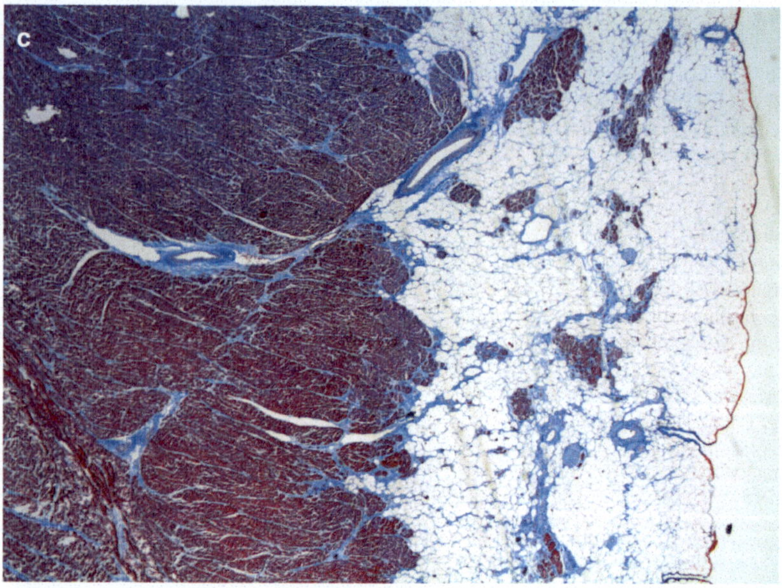

Fig. 8.12 (continued)

Table 8.3 Pathology of ACM

Macroscopic features:
Increased heart weight
Fibrous or fibro-fatty replacement of myocardial tissue that initiates in the subepicardial ventricular region (in the RV only the trabecular myocardium is preserved)
Patchy or diffuse myocardial lesion
Circumferential involvement of the LV with an impaired ventricular septum
Ventricular dilatation with formation of aneurysms, especially in the RV
Microscopic features:
Fibro-fatty replacement in a variable fibrous/fat ratio (in contrast to Lipomatous RV where the replacement is only fatty) (Figs. 8.12 and 8.13)
Alterations in myocyte morphology with apoptosis and cytoplasmic vacuolation (Fig. 8.14)
Frequent foci of inflammatory cell infiltrates

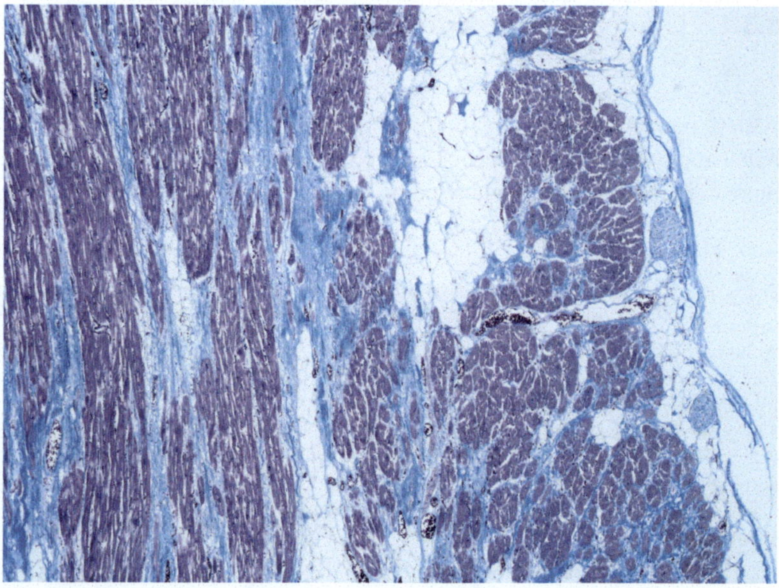

Fig. 8.13 Subepicardial region of the lateral wall of the LV in a patient with ALVC where myocytes have been replaced with fibro-fatty tissue (Masson's trichrome)

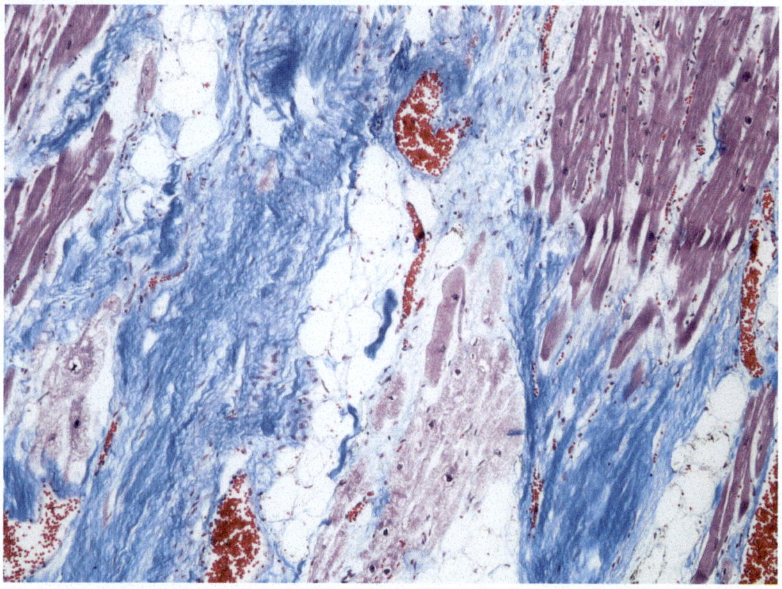

Fig. 8.14 Degenerative changes in the myocardium. Myocytes show pyknotic nuclei and pale cytoplasm with vacuoles (Masson's trichrome)

8.3.1 Lipomatous Hypertrophy of the Right Ventricle

ARVC must be differentiated from lipomatous hypertrophy of the right ventricle, which is much more common and is characterised by thickening and adipose infiltration of the ventricular wall, disrupting the myocytes (Fig. 8.15). Fibrous tissue and ventricular dilatation are absent.

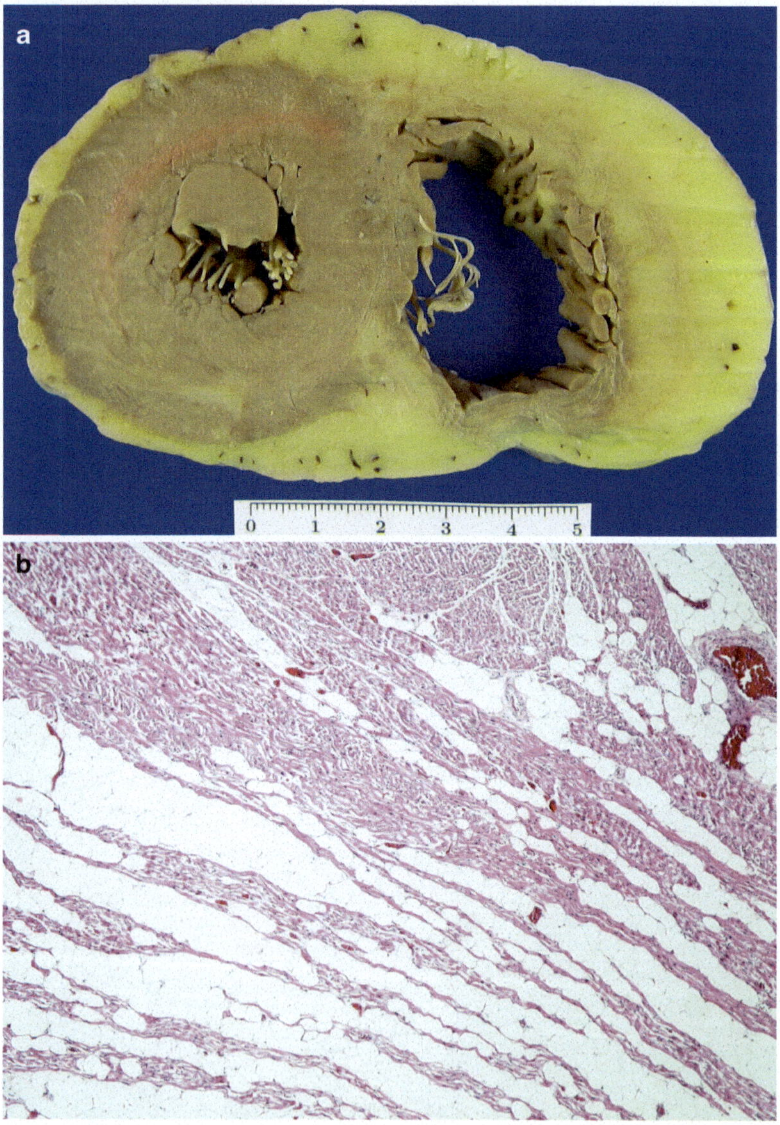

Fig. 8.15 *Lipomatous hypertrophy of the RV.* A 59-year-old woman who died from decompensated liver cirrhosis. Heart weight: 430 g. (**a**) Thickness of the RV wall is 10 mm. Significant adipose infiltration can be observed in the RV wall and subendocardium. (**b**) Microscopic image showing adipose tissue infiltrating the ventricular wall isolating thin muscle fascicles, which causes wall thickening

8.4 Dilated Cardiomyopathy

The term dilated cardiomyopathy (DCM) refers to dilatation of the left ventricular chamber that can affect only the LV or may also involve the RV, causing progressive contractile dysfunction. DCM occurs due to the presence of a diffuse myocardial lesion that weakens the heart's contractile capacity, resulting in a decrease in the volume of blood ejected in each heartbeat.

The residual volume that remains in the ventricle tends to produce an increase in the diameter and pressure of the ventricles. The initial compensatory response of the heart is hypertrophy and a progressive increase in sarcomere length, in an attempt to improve contractility. As a result, the heart gradually enlarges and adopts a more spherical shape.

DCM can be caused by a great variety of factors, the most common being coronary artery disease (CAD) and significant valvular disease, which must always be excluded by the use of imaging techniques and, when appropriate, coronary angiography. A possible history of hypertension, myocarditis, alcoholism or exposure to toxic, medical or chemotherapeutic drugs should always be assessed, in addition to the presence of associated systemic diseases (e.g. muscular dystrophy).

It has been suggested that between 25 and 40 % of patients with DCM have a hereditary basis and a diverse group of gene mutations has been implicated in DCM. For this reason, DCM is considered a mixed form of cardiomyopathy in the AHA classification (2006).

Clinically, DCM manifests with progressive dyspnoea and loss of exertional capacity due to dysfunction of the LV; and with systemic venous congestion, increase in venous pressure and oedema in the lower limbs, due to RV dysfunction. Ventricular arrhythmias may also appear and could lead to syncope and SCD, being the first manifestation of the disease.

Pharmacological treatment with beta-blockers such as carvedilol, metoprolol, bisoprolol and nebivolol, with inhibitors of the renin-angiotensin system, and with anti-aldosterone therapy has proven to be effective and to improve the prognosis of patients with DCM.

Cardiac resynchronization therapy (CRT) using a pacing device has been demonstrated to improve prognosis in patients with dilated cardiomyopathy who, despite medical treatment, present symptoms of severe ventricular dysfunction, with an ejection fraction below 35 % and a QRS in the ECG greater than 120 ms in duration. If the ejection fraction falls below 30–35 %, a defibrillator should be implanted in order to treat life-threatening ventricular arrhythmias.

Cardiac transplantation is the only definitive treatment for patients with end-stage disease (advanced heart failure), in which therapeutic treatment no longer works, and without contraindications.

Table 8.4 summarizes the main pathological features of DCM.

Table 8.4 Pathology of DCM (Figs. 8.16, 8.17 and 8.18)

Macroscopic features:
Increased heart mass (over 500 g in adults)
Globoid shape
Dilatation of all chambers with wall thinning and flattened trabeculae (Ratio between wall thickness and diameter of the chamber is reduced)
Ventricular septum can be straight with large RV dilatation
Appearance of the myocardium is normal or with variable scattered areas of fibrosis
Endocardial fibrosis and mural thrombi are usually present
Microscopic features:
Minimal myocardial alterations may be present
Myocytes of variable size, multinucleated or degenerated
Interstitial or replacement fibrosis

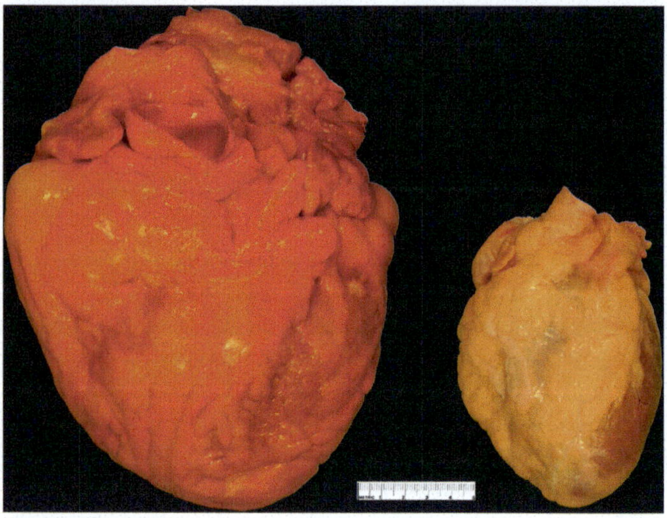

Fig. 8.16 *Comparative image between a heart with DCM and a normal heart*. Heart with DCM (830 g) from a 65-year-old male who died suddenly at home. Comparison with a normal heart (277 g) from a 23-year-old male who died in a traffic accident (*right*)

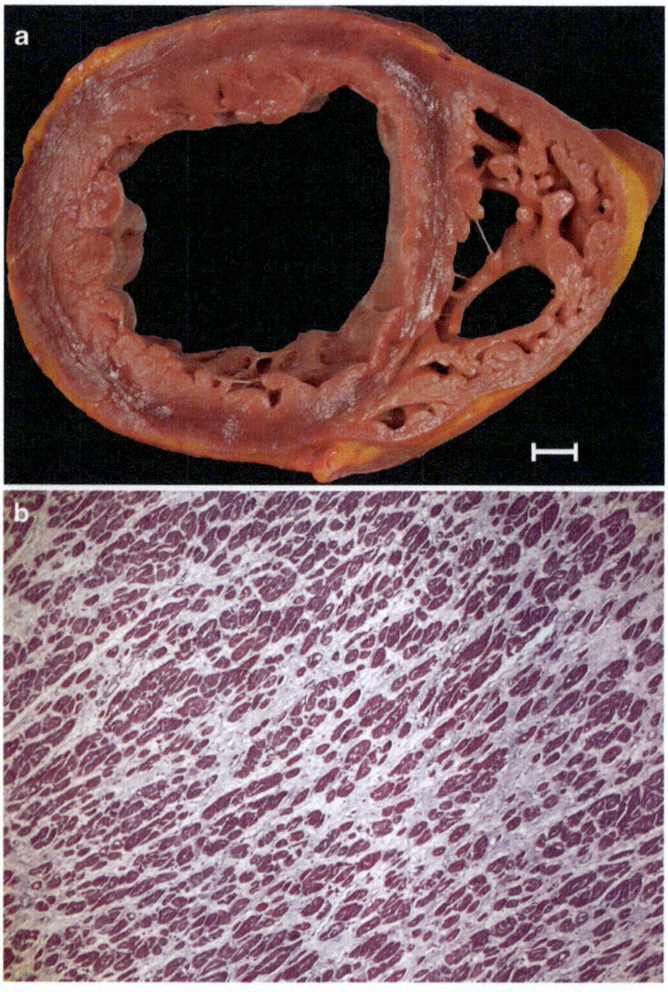

Fig. 8.17 *DCM*. A 53-year-old male diagnosed with DCM, with an implanted automatic defibrillator. He died suddenly in the street. (**a**) Cardiomegaly (735 g) with increased diameter of the ventricular chambers and ventricular septal displacement towards the right. (**b**) Abundant interstitial fibrosis (Masson's trichrome)

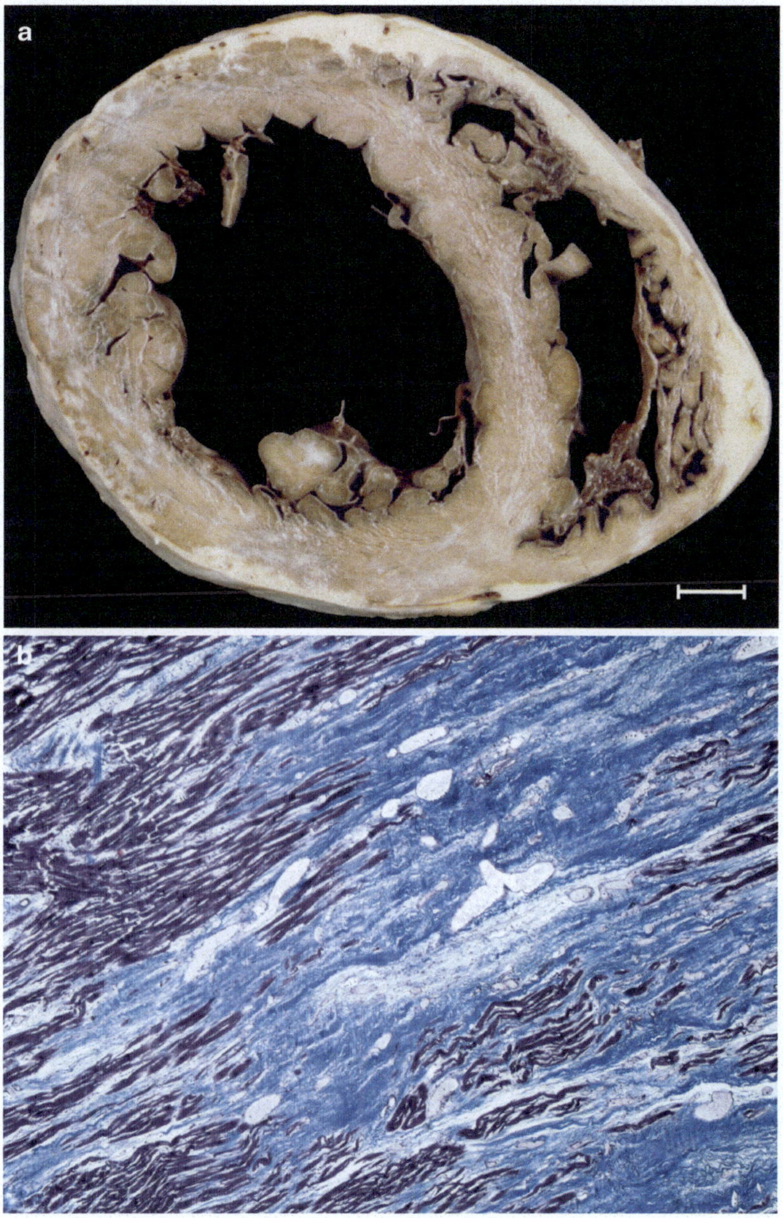

Fig. 8.18 *DCM*. A 16-year-old male diagnosed with advanced Duchenne muscular dystrophy and infectious mononucleosis 4 weeks before suddenly dying. Heart weight: 752 g. (**a**) Marked dilatation of the LV (5.5 cm of diameter) with wall thinning (10 mm) and extensive myocardial fibrosis. (**b**) Large areas of replacement fibrosis of the myocardium. Necrotic foci with lymphocytic inflammatory infiltrates could be observed in other areas (Masson's trichrome)

8.5 Restrictive Cardiomyopathy

Restrictive cardiomyopathy (RCM) is a cardiac muscle disease characterized by severe alterations in myocardial compliance, resulting in a disproportionate increase in the end-diastolic ventricular pressure and an impaired ventricular filling during diastole. Although dilatation of the LV may occur, the diameters are usually normal or slightly reduced.

The aetiology of RCM can be primary, familial or acquired, or secondary to systemic diseases such as amyloidosis, sarcoidosis, carcinoid syndrome or scleroderma.

No specific treatment exists for patients affected by RCM, apart from diuretics to release congestion and drugs to control atrial arrhythmias that usually occur.

8.5.1 Amyloidosis

Amyloidosis is the term used to describe a group of diseases wherein there is deposition of extracellular proteins that stain positive for Congo Red and appear green under a polarizing microscope. Deposition of this material on the heart (cardiac amyloidosis) usually provokes an increase in cardiac mass and a clinical presentation that resembles RCM. In the heart, amyloid deposition occurs on the atrial and ventricular myocardium, and also on the valves, conduction system and small vessels. Among all the proteins that can result in amyloid deposits, the main proteins that are deposited in the heart are: light-chain immunoglobulins (AL amyloidosis), serum amyloid A (SAA amyloidosis), mutated transthyretin protein (ATTR familial amyloidosis), and non-mutated transthyretin (senile systemic amyloidosis SSA). There are also other less common forms of cardiac familial amyloidosis (AApoAI and AFib).

Diagnosis requires great clinical suspicion. Amyloidosis tends to manifest as congestive heart failure. Since the increase in the cardiac mass is due to amyloid deposition and not due to myocyte hypertrophy, amyloidosis should be suspected when imaging techniques show an increased wall thickness and reduced QRS voltages on ECG.

Treatment of cardiac amyloidosis is twofold: On the one hand, therapy for cardiac symptoms (mainly the use of diuretics to relieve congestive symptomatology) and, on the other hand, treatment of the underlying disease. Thus, in AL amyloidosis it is necessary to treat the underlying blood dyscrasia, in SAA amyloidosis to reduce infection/inflammation, whereas in ATTR familial amyloidosis the only option, until recently, was liver transplantation to avoid the production of the mutated protein. No treatment exists for senile systemic amyloidosis since the accumulated protein is wild-type, non-mutated, transthyretin.

The pathologic characteristics of amyloidosis are shown in Table 8.5.

Table 8.5 Pathology of cardiac amyloidosis

Macroscopic features:
Dilated atria with normal-sized ventricles
Left ventricular hypertrophy (can be misdiagnosed as classical HCM) or right ventricular hypertrophy
Waxy appearance of the myocardium and elastic texture (Fig. 8.19)
Yellow stippling in the atrial endocardium (Fig. 8.20)
Microscopic features:
Deposits of eosinophilic material in interstitium, valves, conduction system and in the medial layer of the intramyocardial coronary arteries
Deposits appear red after Congo Red staining (Fig. 8.19b) and show an apple-green birefringence when viewed under polarized light
Deposits can be nodular, perifibrilar or mixed

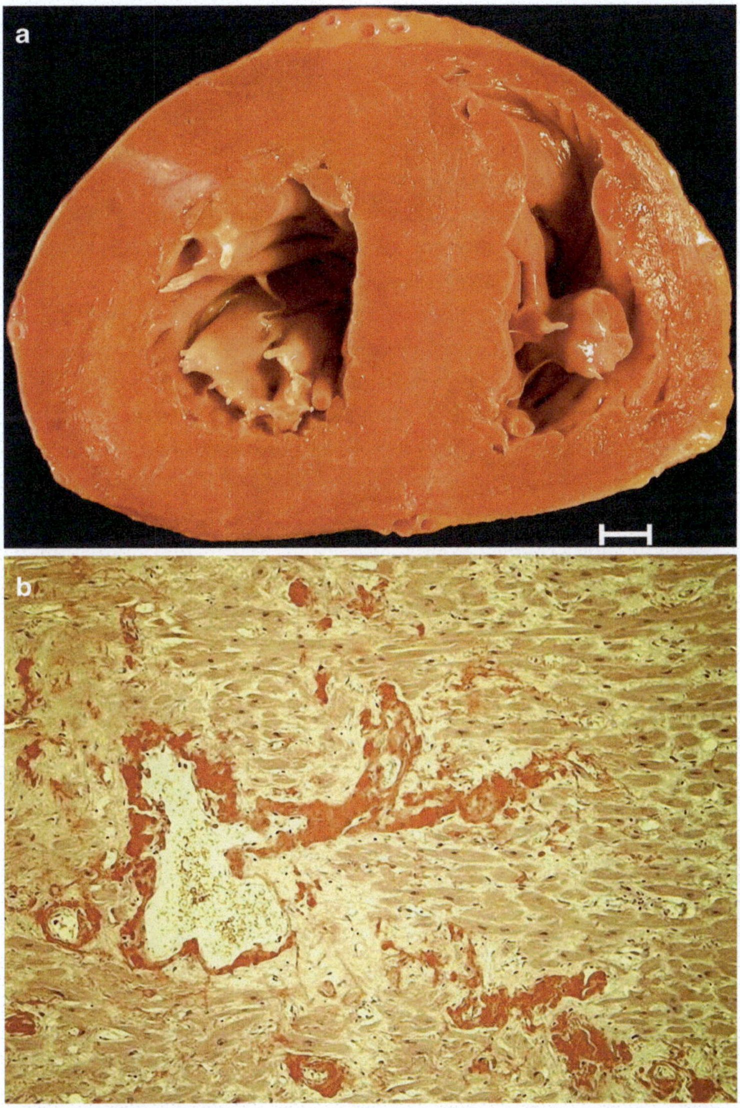

Fig. 8.19 *Cardiac amyloidosis*. A 73-year-old male with restrictive heart failure who died during a pacemaker implantation procedure. (**a**) Cardiomegaly (720 g) with a more consistent myocardium, rubbery in appearance. (**b**) Congo Red staining reveals amyloid deposition on the vessel's walls and the interstitium

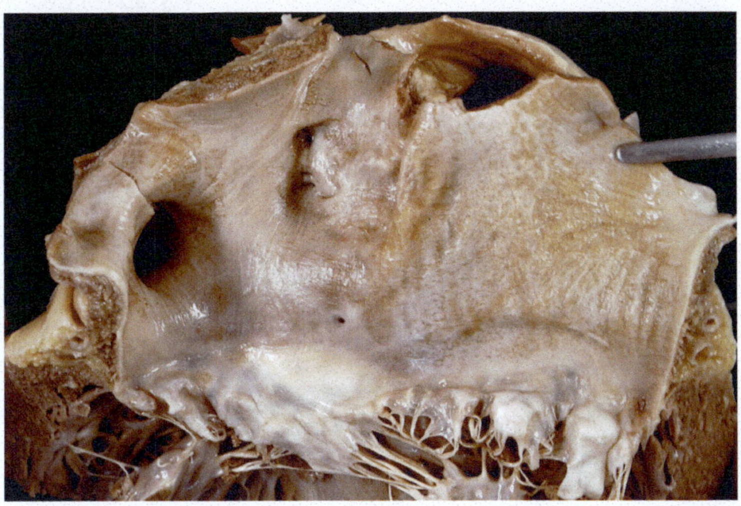

Fig. 8.20 *Senile cardiac amyloidosis*. Glistening, partially translucent (amber-like) amyloid deposits in the left atrial endocardium

8.6 Left Ventricular Non-Compaction Cardiomyopathy

Left ventricular non-compaction cardiomyopathy (LVNC) is a condition characterized by excessive trabeculations and deep recesses within the myocardium. Its origin is thought to be a genetic defect in the process of myocardial compaction during embryogenesis with persistence of embryonal sinusoids. Since a certain degree of trabeculation can exist in a healthy heart and there is not an exact definition of what it is considered normal, it is necessary to be cautious when diagnosing LVNC. The criterion currently used to diagnose this condition is the numerical ratio between the non-compacted subendocardial myocardium and the normal compacted myocardium. A ratio greater than 2, as measured by echocardiography in end-systole, is considered indicative of LVNC. Presence of deep perfused intertrabecular recesses must also be determined.

It is frequent for LVNC to run in families and inheritance usually follows an autosomal dominant or X-linked pattern. In many cases where hyper-trabeculation is prominent, mutations in genes encoding sarcomeric proteins and other proteins also affected in HCM or DCM can be found, suggesting some overlap between these conditions. LVNC can manifest as an isolated entity or can be associated with other cardiomyopathies and several congenital cardiopathies.

The disease can produce left ventricular systolic dysfunction, thromboembolism, arrhythmias and sudden death.

Table 8.6 summarises the pathology of LVNC.

Table 8.6 Pathology of LVNC

The apical and medial thirds of the heart are affected (Figs. 8.21 and 8.22)
Absence of properly-formed papillary muscles
Histologically, deep intertrabecular recesses from the endocardium towards the epicardium. Depth >50 % of the LV wall thickness (Fig. 8.22b)
Polypoid or anastomosing trabeculae

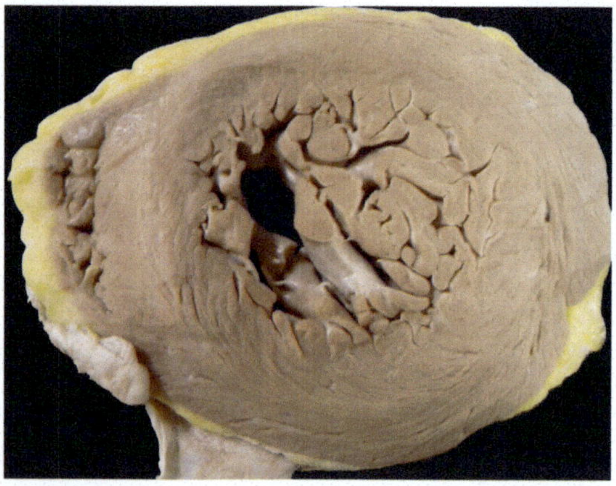

Fig. 8.21 *LVNC*. A 54-year-old woman who died suddenly while sleeping. Cardiomegaly with left ventricular hypertrophy (Heart weight: 670 g). A thick internal and non-compacted layer with hypertrabeculation and deep recesses can be observed in the LV myocardium

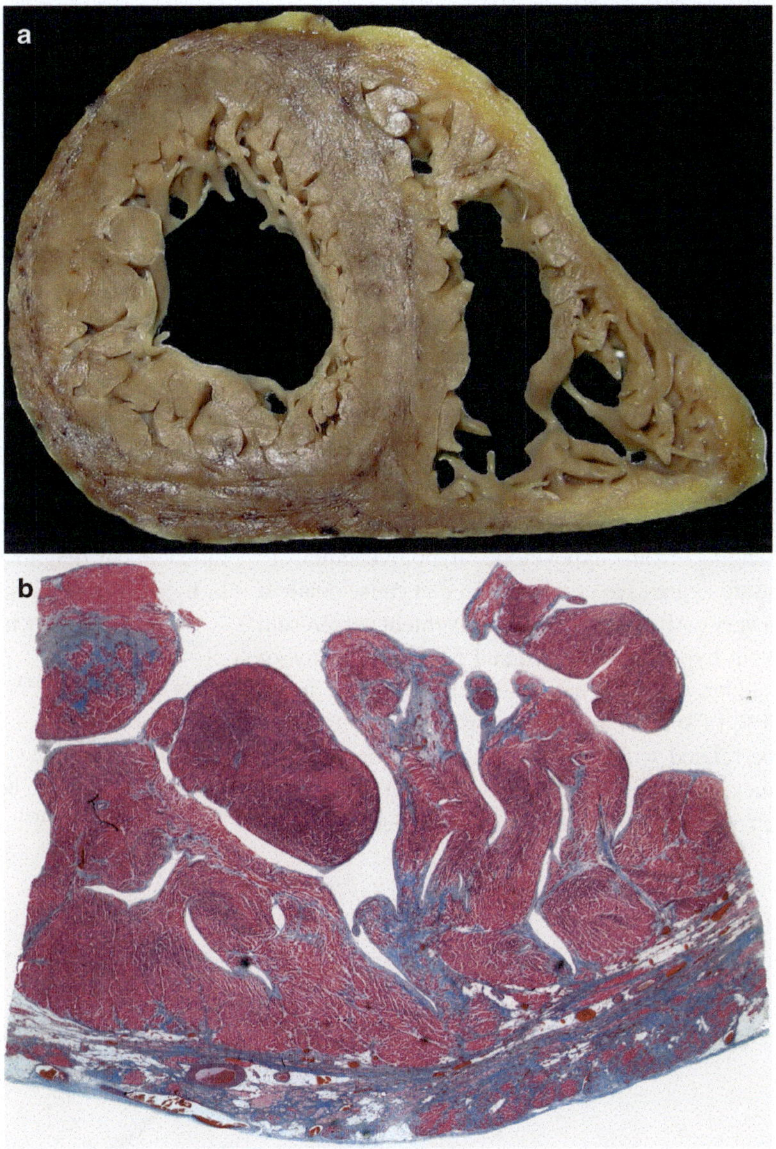

Fig. 8.22 *LVNC associated with ACM*. A 14-year-old male who died suddenly at school during physical education class. (**a**) Heart weight of 292 g with a dilated RV and fibro-fatty infiltration of the anterior and posterior walls. Left ventricle with hypertrabeculation involving more than one half of the ventricular thickness. (**b**) LV myocardium with endocardial hypertrabeculation and prominent subendocardial fibroelastosis. The scarce band of compacted tissue shows severe fibro-fatty infiltration with degenerative myocyte changes characteristic of ALVC (Masson's trichrome)

8.7 Cardiac Pathology in Systemic Diseases

8.7.1 Defects in Mitochondrial Fatty-Acid Beta-Oxidation

Defects in mitochondrial fatty-acid beta-oxidation comprise a group of inherited diseases with an autosomal recessive inheritance pattern, characterized by a premature onset in the newborn or in early childhood. Although incidence may seem low, its global prevalence could be underestimated.

Symptoms and prognosis are very diverse and depend on the type of enzymatic defect. Thus, symptoms can fall anywhere on the spectrum, ranging from hypoketotic hypoglycaemia, to more aggressive episodes that can cause death at a very early age. Cardiac involvement can present with hypertrophic or dilated cardiomyopathy, or sudden death. It has been estimated that up to 5 % of sudden deaths occurring in infants could be related with this type of genetic disorder; in these cases, the most common defect is medium-chain acyl-CoA dehydrogenase deficiency, often known as MCAD deficiency, where situations of metabolic crisis can be triggered by viral infection, fasting or vaccination.

Clinically, a fatty acid oxidation disorder should be suspected in individuals with cardiomyopathies who present with hypoketotic hypoglycaemia, ketonuria, lethargy, muscle weakness or hypotony, peripheral neuropathy or pigmentary retinopathy. Analytical determinations show increased levels of creatine phosphokinase (CPK) and transaminases, or typical alterations consistent with rhabdomyolysis or myoglobinuria. Levels of total and free carnitine, butirylcarnitine, octanoylcarnitine and acylcarnitine help to delineate the type of enzymatic defect. A definitive diagnosis is done by *in vitro* confirmation of biochemical diagnosis using fibroblast cultures, or by genetic studies.

Pathologic characteristics are shown in Table 8.7.

Table 8.7 Pathology of the fatty-acid beta-oxidation disorder (Fig. 8.23)

Cardiomegaly
Lipid deposits in liver, heart, skeletal and smooth muscle, and renal tubuli

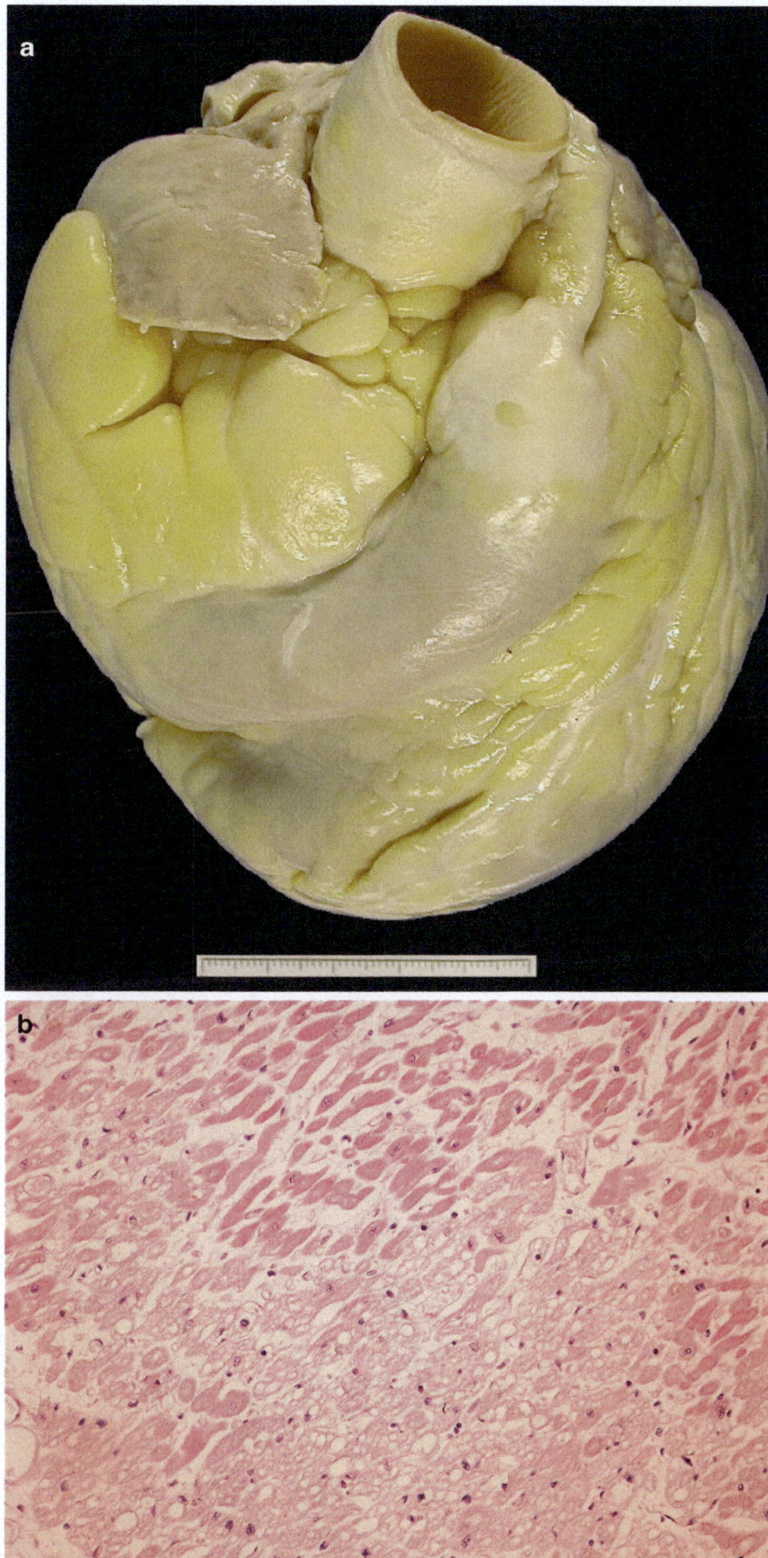

Fig. 8.23 A 35-year-old woman, with unknown medical history, who died from a haemorrhage after an abortion. (**a**) Increased heart weight (390 g) with fatty appearance. (**b**, **c**) Myocardium with extensive myocytolysis. (**d**, **e**) Lipid deposits (Nile blue-positive) in myocardium and renal tubuli; findings suggestive of a defect in fatty-acid beta-oxidation

Fig. 8.23 (continued)

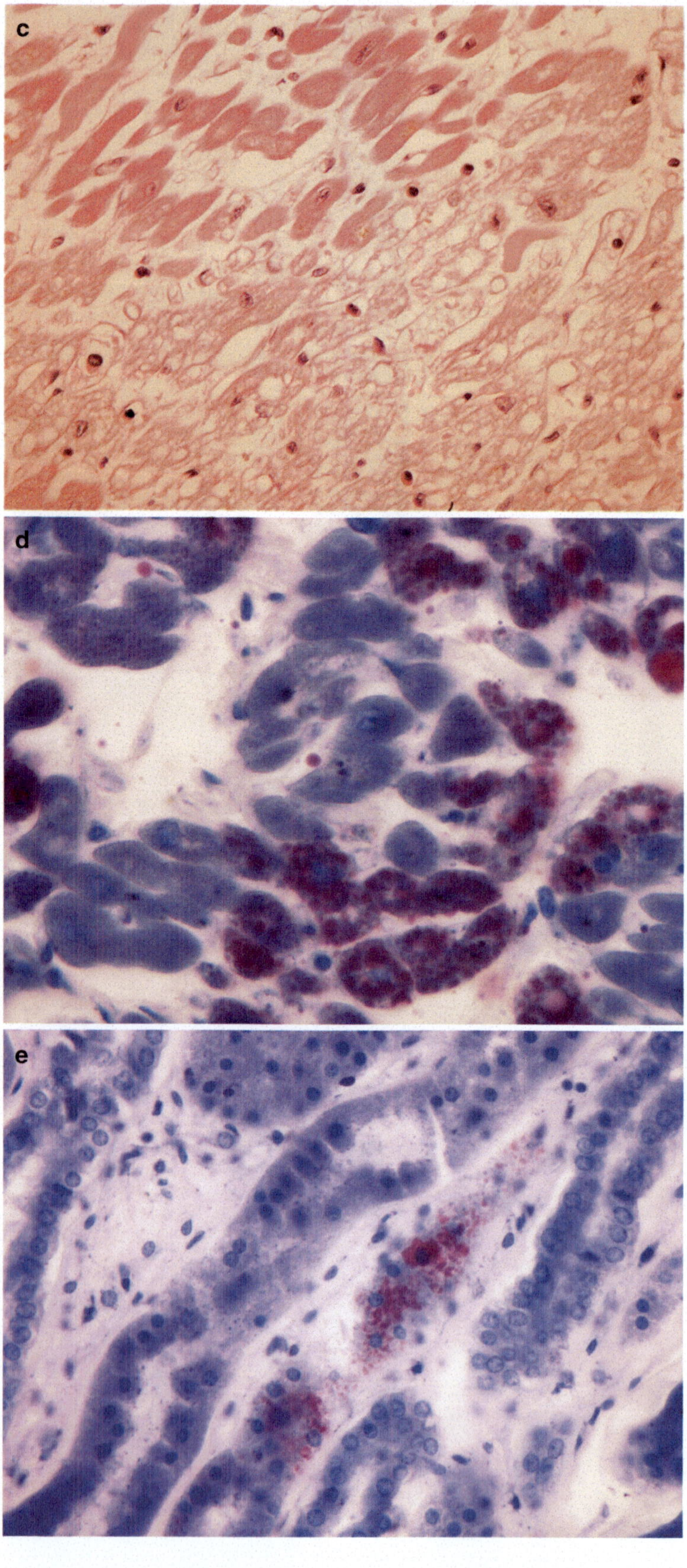

8.7.2 Anderson-Fabry Disease

Anderson-Fabry disease is a rare X-linked disorder caused by mutations of the *GLA* gene encoding the alpha-galactosidase A enzyme. The ensuing enzyme deficiency results in intracellular accumulation of glycosphingolipids and progressive failure in the function of the tissues where deposition occurs, mainly the central nervous system, kidney and heart.

Concentric left ventricular hypertrophy is the most common cardiac manifestation, although asymmetric hypertrophy may also occur. Glycosphingolipid accumulation on the heart valves can lead to mild valve insufficiency in 15 % of the cases. Clinically, Anderson-Fabry disease should be suspected in patients with LV hypertrophy without hypertension or valvular disease, and in those also presenting with peripheral neuropathy, mainly acroparesthesia, deafness, tinnitus, postprandial discomfort with alterations in intestinal motility, and cutaneous angiokeratomas.

Diagnosis in males is established by enzymatic determination of alpha-galactosidase A activity or by genetic testing. In females, the disease is diagnosed by genetic testing as the enzymatic activity could be normal. Treatment consists of enzyme replacement therapy.

Table 8.8 summarises the main pathological findings in Anderson-Fabry disease.

Table 8.8 Pathology of Anderson-Fabry disease (Fig. 8.24)

Cardiac hypertrophy generally concentric, but can also present with septal asymmetry similar to HCM
Vacuolated myocytes diffusely distributed in both ventricles
It can involve the AV node and bundle of His
Myocytic hypertrophy and fibrosis, which can be extensive
Areas of disarray may be seen around the scars
Electron microscopy: intracytoplasmic concentric lamellar bodies

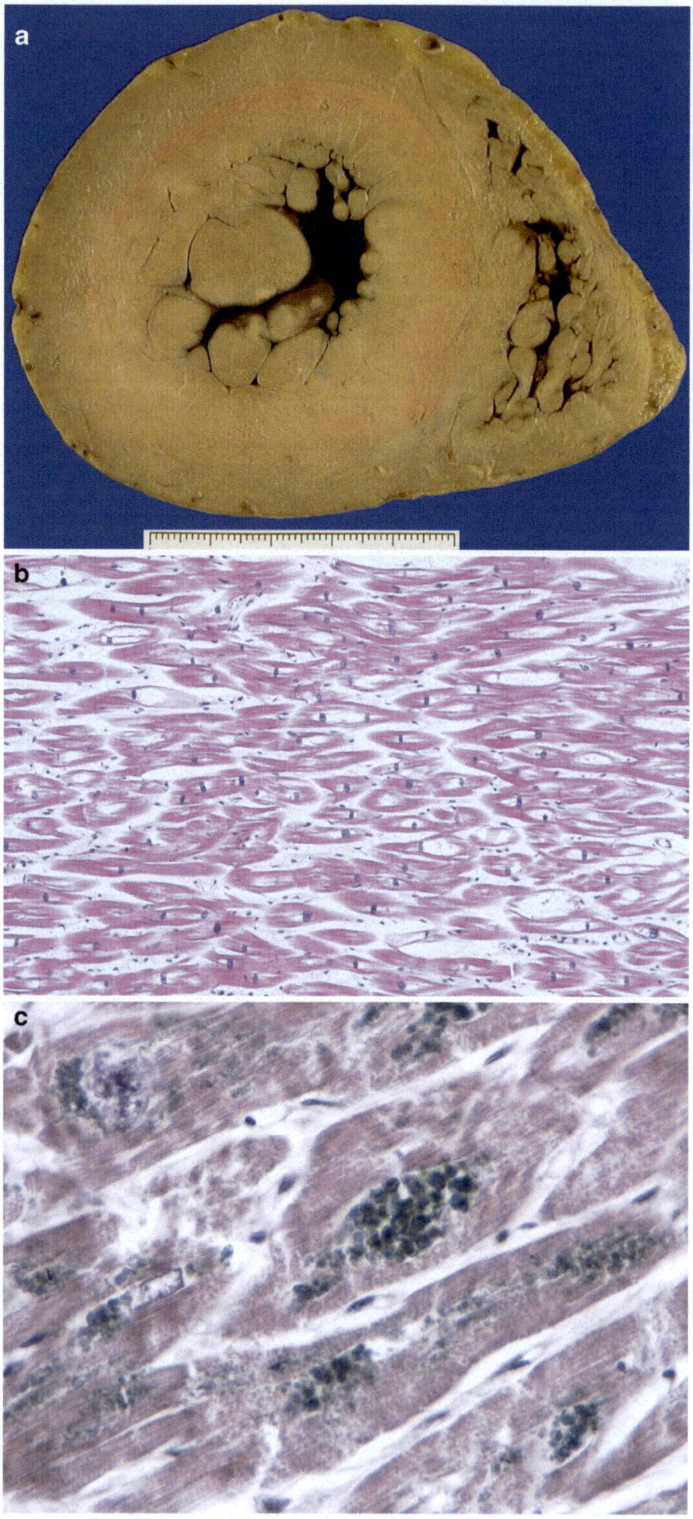

Fig. 8.24 *Anderson-Fabry disease*. A 54-year-old male clinically diagnosed with HCM. (**a**) Cardiomegaly (705 g) with concentric LV hypertrophy (20 mm). (**b**) Vacuolated myocytes and (**c**) Lipid deposits in the myocardium suggestive of Anderson-Fabry disease (Post-fixation of formalin-fixed tissue with osmium tetroxide). *Post-mortem* genetic studies identified a mutation in the *GLA* gene (alpha-galactosidase A) (*Courtesy of Mª Soledad Sánchez de León. INTCF, Madrid*)

8.7.3 Mitochondrial Cardiomyopathies

Mitochondrial diseases result from abnormalities in the mitochondrion, the essential organelle that provides energy in the form of ATP through the oxidative phosphorylation system. Alterations in any of the steps involved in this process lead to mitochondrial dysfunctions. They are caused by genetic defects exhibiting variable expressivity. Mitochondria are distributed among all the cells in an organism, except for red blood cells, and hence mitochondrial diseases can affect any organ from an early age. However, those tissues with more energy demand such as muscle and central nervous system are preferentially affected.

The most common clinical presentations, from a cardiovascular point of view, are ventricular hypertrophy with or without LV outflow tract obstruction, and dilated cardiomyopathy. Clinically, the disease is very heterogeneous ranging from asymptomatic patients to those with progressive heart failure. Mitochondrial disease manifests early in childhood and must be suspected in patients with ventricular hypertrophy or dilatation associated with fragile or weak skeletal muscle, deafness, epilepsy, ataxia or encephalopathy. Analytical determinations show increased plasma levels of CPK and lactic acid.

Table 8.9 summarises the main pathological findings.

Table 8.9 Pathology of mitochondrial cardiomyopathy

Dilated ventricles as in classical DCM (Fig. 8.25a)
Giant mitochondria can be seen in the cytoplasm with an optical microscope at high magnification (Fig. 8.25b)
Electron microscopy shows abnormal mitochondria: alterations in the mitochondrial crests, giant mitochondria, electron dense inclusions and calcifications

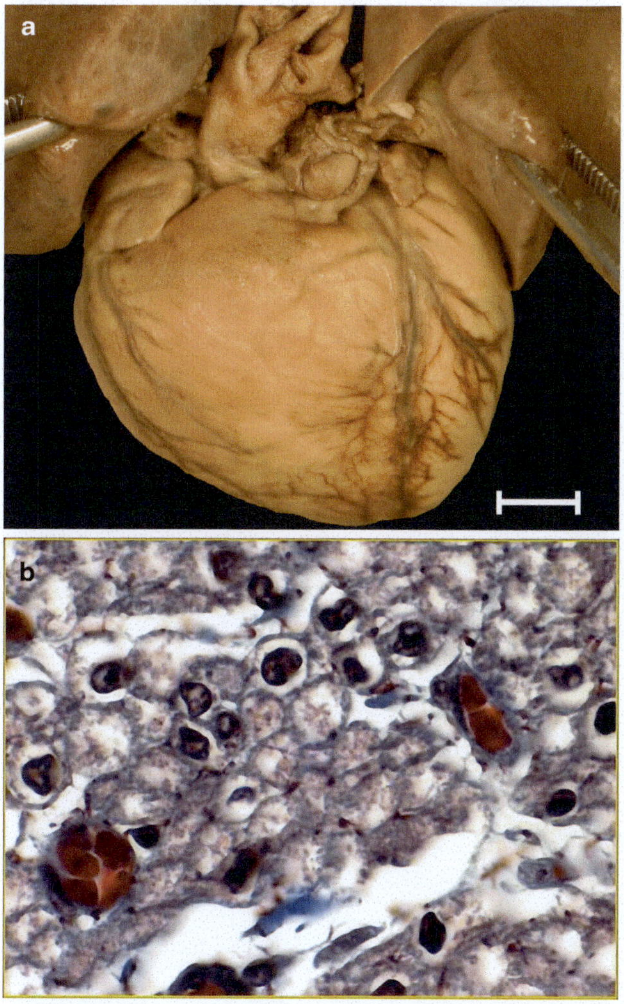

Fig. 8.25 *Mitochondrial cardiomyopathy*. A 4-day-old baby girl who died from heart failure. Her brother had died suddenly at 7 months the previous year. (**a**) Cardiomegaly (120 g) with a globoid-shaped heart. (**b**) Myocytes with intense intracytoplasmic pinkish stippling corresponding to giant mitochondria (Masson's trichrome)

8.8 Endocardial Fibroelastosis

Idiopathic endocardial fibroelastosis is a disease characterized by a thickening of the endocardium that mainly affects the ventricles. The disease causes heart failure in infants and children and can debut as sudden infant death. Its aetiology is not completely known, although it has been demonstrated that it could be related with viral infections causing parotitis, and with genetic defects.

The genetic defect is known as Barth syndrome (BTHS), also referred to as 3-methylglutaconic aciduria type II. BHTS is an extremely rare X-linked disorder caused by mutations in the tafazzin gene. As the inheritance pattern is X-linked recessive type, only males are affected. The genetic defect causes alterations in the remodelling of cardiolipin in mitochondrial membranes, leading to a dysfunction in mitochondrial respiration chain. The disease manifests early after birth, either suddenly or progressively. Clinically, it must be suspected in cases of heart failure or dilated cardiomyopathy associated with myopathy or muscle weakness, intermittent neutropenia with immunosuppression and infections, or with hypocholesterolaemia. Diagnosis can be established by analytical determination of cardiolipin in lymphocytes or by the detection of 3-methylglutaconic acid.

The pathologic characteristics are shown in Table 8.10.

Table 8.10 Pathology of endocardial fibroelastosis

Macroscopic features:
Dilated form (similar to DCM in the adult):
Globoid-shaped heart
Dilated LV with endocardial thickening mainly in the LV outflow tract (Fig. 8.26a)
Heart valves can be involved
Contracted form:
Normal LV with dilatation of the right chambers and less endocardial fibrosis
Microscopic features:
Endocardial and subendocardial thickening caused by fibroelastic tissue (Fig. 8.26a)
Myocardium is preserved, with fibrotic patches or focal ischaemia
Macro or microscopic mural thrombi may be present

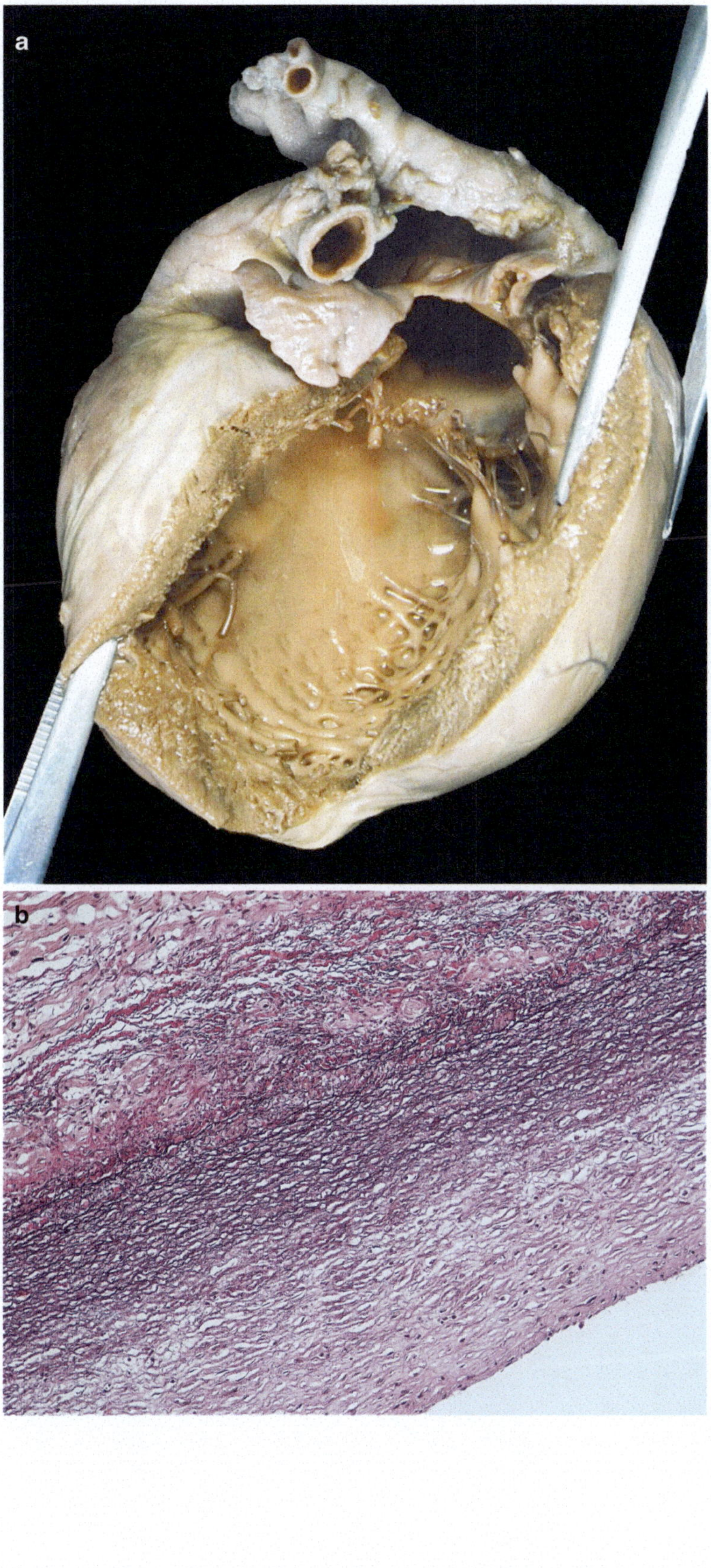

Fig. 8.26 *Endocardial fibroelastosis*. Sudden death in a 5-month-old baby girl. (**a**) Heart with cardiomegaly and rounded shape. Heart weight was 98 g (normal: 36 ± 5 g), LV was dilated and the endocardium was thick and pearly white, with flattened trabeculae. (**b**) Microscopic image showing a thickened LV endocardium due to elastic fibres and collagen (Weigert's stain). (**c**) View of the septum where fibrosis of the left bundle branches of the conduction system can be observed (Masson's trichrome)

Fig. 8.26 (continued)

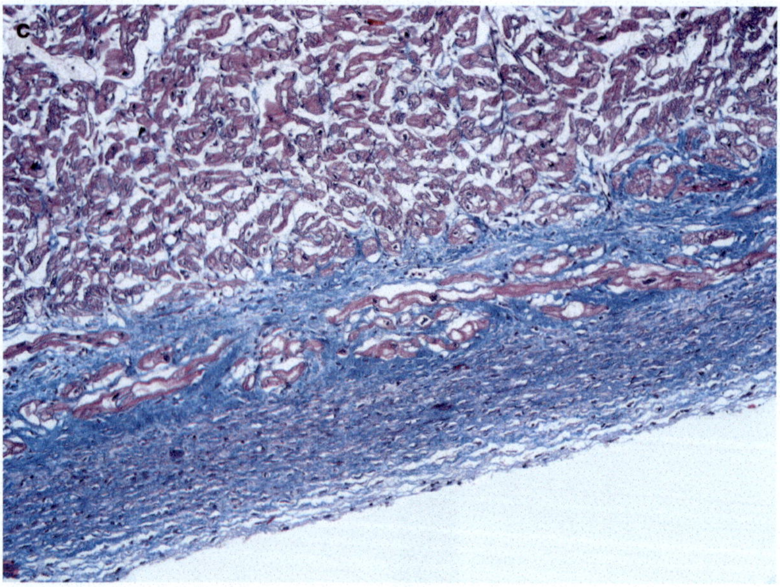

Bibliography

Aguilera B, Suárez-Mier MP. Muerte súbita cardiaca. Revisión. REA: E J Autopsy. 2003;1:21–34.

Aguilera B, Suárez Mier MP, Morentin B. Miocardiopatía arritmogénica como causa de muerte súbita en España. Presentación de 21 casos. Rev Esp Cardiol. 1999;52:656–62.

Asimaki A, Tandri H, Huang H, Halushka MK, Gautam S, Basso C, et al. A new diagnostic test for arrhythmogenic right ventricular cardiomyopathy. N Engl J Med. 2009;360:1075–84.

Basso C, Thiene G. Adipositas cordis, fatty infiltration of the right ventricle, and arrhythmogenic right ventricular cardiomyopathy. Just a matter of fat? Cardiovasc Pathol. 2005;14(1):37–41.

Basso C, Corrado D, Marcus FI, Nava A, Thiene G. Arrhythmogenic right ventricular cardiomyopathy. Lancet. 2009;373(9671):1289–300.

Burke A, Tavora F. Cardiomyopathies and endocardial diseases. In: Practical cardiovascular pathology. Philadelphia: Wolters Kluwer Health/Lippincott Williams & Wilkins; 2011. p. 143–95.

Burke A, Mont E, Kutys R, Virmani R. Left ventricular non–compaction: a pathological study of 14 cases. Hum Pathol. 2005;36:403–11.

Elliott P, Andersson B, Arbustini E, et al. Classification of the cardiomyopathies: a position statement from the European Society of Cardiology Working Group on Myocardial and Pericardial Diseases. Eur Heart J. 2008;29:270–6.

Garcia-Pavia P, Syrris P, Salas C, Evans A, Mirelis JG, Cobo-Marcos M, et al. Desmosomal protein gene mutations in patients with idiopathic dilated cardiomyopathy undergoing cardiac transplantation: a clinicopathological study. Heart. 2011a;97:1744–52.

Garcia-Pavia P, Tome-Esteban MT, Rapezzi C. Amyloidosis. Also a cardiac disease. Rev Esp Cardiol. 2011b;64:797–808.

Hershberger RE, Siegfried JD. Update 2011: clinical and genetic issues in familial dilated cardiomyopathy. J Am Coll Cardiol. 2011;57:1641–9.

Jeffries JL, Towbin JA. Dilated cardiomyopathy. Lancet. 2010;375:752–62.

Marcus FI, McKenna WJ, Sherrill D, et al. Diagnosis of arrhythmogenic right ventricular cardiomyopathy/ dysplasia: proposed modification of the Task Force Criteria. Eur Heart J. 2010;31:806–14.

Maron BJ, Towbin JA, Thiene G, Antzelevitch C, Corrado D, Arnett D, et al. Contemporary definitions and classification of the cardiomyopathies: an American Heart Association Scientific Statement from the Council on Clinical Cardiology, Heart Failure and Transplantation Committee; Quality of Care and Outcomes Research and Functional Genomics and Translational Biology Interdisciplinary Working Groups; and Council on Epidemiology and Prevention. Circulation. 2006;113: 1807–16.

Nihoyannopoulos P, Dawson D. Restrictive cardiomyopathies. Eur J Echocardiogr. 2009;10:iii23–33.

Noorman M, Hakim S, Asimaki A, Vreeker A, van Rijen HVM, Marcel AG, et al. Reduced plakoglobin immunoreactivity in arrhythmogenic cardiomyopathy: methodological considerations. Cardiovasc Pathol. 2013;22:314–8.

O'Mahony C, Elliott P. Anderson-Fabry disease and the heart. Prog Cardiovasc Dis. 2010;52:326–35.

Oeschlin E, Jenni R. Left ventricular non-compaction revisited: a distinct phenotype with genetic heterogeneity? Eur Heart J. 2011;32:1446–56.

Quarta G, Muir A, Pantazis A, et al. Familial evaluation in arrhythmogenic right ventricular cardiomyopathy: impact of genetics and revised task force criteria. Circulation. 2011;123(23):2701–9.

Richardson P, McKenna W, Bristow M, Maisch B, Mautner B, O'Connell J, et al. Report of the 1995 World Health Organization/International Society and Federation of Cardiology Task Force on the Definition and Classification of cardiomyopathies. Circulation. 1996;93:841–2.

Romero J, Mejia-Lopez E, Manrique C, Lucariello R. Arrhythmogenic Right Ventricular Cardiomyopathy (ARVC/D): a systematic literature review. Clin Med Insights: Cardiol. 2013;7:97–114.

Saguner AM, Brunckhorst C, Duru F. Arrhythmogenic ventricular cardiomyopathy: a paradigm shift from right to biventricular disease. World J Cardiol. 2014; 6(4):154–74.

Seki A, Patel S, Ashraf S, Pernees G, Fishbein MC. Primary endocardial fibroelastosis: an underappreciated cause of cardiomyopathy in children. Cardiovasc Pathol. 2013;22:345–50.

Sen Chowdry S, Syrris P, Prasad S, et al. Left-dominant arrhythmogenic cardiomyopathy: an under-recognized clinical entity. J Am Coll Cardiol. 2008;52:2175–87.

Soor GS, Luk A, Ahn E, Abraham JR, Woo A, Ralph-Edwards A, Butano J. Hypertrophic cardiomyopathy: current understanding and treatment objectives. J Clin Pathol. 2009;62:226–35.

Suárez Mier MP, Aguilera B, Mosquera RM, de Leon MS S. Pathology of sudden death during recreational sports in Spain. Forensic Sci Int. 2013;226(1–3): 188–96.

Tuttolomondo A, Pecoraro R, Simonetta I, Miceli S, Pinto A, Licata G. Anderson-Fabry disease: a multiorgan disease. Curr Pharm Des. 2013;19(33):5974–96.

Watkins H, Ashrafian H, Redwood C. Inherited cardiomyopathies. N Engl J Med. 2011;364:1643–56.

Wybraniec M, Mizia-Stec K, Krzych L. Stress cardiomyopathy: yet another type of neurocardiogenic injury. 'stress cardiomyopathy'. Cardiovasc Pathol. 2014;23: 113–20.

Myocarditis

9

Benito Morentin, Beatriz Aguilera,
María Dolores García-Cosío Carmena,
and Luis A. Alonso-Pulpón

Contents

B. Morentin (✉)
Forensic Pathology Department,
Institute of Legal Medicine, Bizkaia, Spain

Forensic Medicine, Basque Country University,
Bizkaia, Spain
e-mail: morentin.b@aju.ej-gv.es

B. Aguilera
Histopathology Service, National Institute
of Toxicology and Forensic Sciences,
Las Rozas (Madrid), Spain

M.D. García-Cosío Carmena
Heart Failure and Transplant Unit, Department
of Cardiology, Hospital Santa Creu i Sant Pau,
Barcelona, Spain

L.A. Alonso-Pulpón
Cardiology Department, Hospital Universitario
Puerta de Hierro, Majadahonda (Madrid), Spain

Universidad Autónoma de Madrid, Madrid, Spain
e-mail: luispulpon@secardiologia.es

9.1 Introduction

Myocarditis is a disease characterized by inflammation of the heart muscle that is not secondary to ischemia. It can be caused by the action of several infectious (viruses, bacteria, fungi, protozoa, etc.) and non-infectious (drugs, toxins, autoimmune diseases, systemic diseases) agents. On many occasions, the underlying cause of myocarditis cannot be adequately established (Table 9.1). In developed countries, viral infections are the most common cause of myocarditis, with enteroviruses accounting for the majority of the cases.

The inflammatory process, which occurs during infections, can be produced by direct pathogen invasion, toxin production or through immunological mediators. Medication can induce myocardial damage either directly by a toxic effect or can be mediated by hypersensitivity reactions.

J.S. Lucena et al. (eds.), *Clinico-Pathological Atlas of Cardiovascular Diseases*,
DOI 10.1007/978-3-319-11146-9_9, © Springer International Publishing Switzerland 2015

Table 9.1 Causes of myocarditis

Causes	Remarks
Infections:	
Virus: adenovirus, enterovirus (Coxsackie), echovirus, herpes simplex virus, hepatitis C virus, human immunodeficiency virus, parvovirus B19	In the majority of cases diagnosed at autopsy, it is not possible to identify the pathogen and myocarditis is attributed to viral infection
Bacteria: *Staphylococcus, Streptoccocus* sp., *Legionella, Mycobacterium* sp., *Mycoplasma pneumoniae*	Endomyocardial biopsy has low sensitivity since inflammation is usually patchy
Fungi: *Aspergillus, Candida, Cryptococcus*	Young adults constitute the age group more commonly affected
Protozoa: *Trypanosoma cruzi, Toxoplasma gondii*	Myocarditis is a significant cause of sudden death in infants and children
Rickettsias: *Coxiella burnetti, Rickettsia typhi*	
Immunological syndromes:	
Churg-Strauss syndrome, inflammatory bowel disease, giant cell myocarditis, hypereosinophilic syndrome, Wegener's granulomatosis, systemic lupus erythematosus, rheumatoid arthritis, sarcoidosis	
Hypersensitivity reactions:	
Penicillin, ampicillin, cephalosporins, tetracyclines, digoxin, tricyclic antidepressants, diuretics, clozapine	
Venoms from scorpion or viper	
Toxic reactions to therapeutic and illicit drugs:	
Cocaine, amphetamines, anthracyclines, phenytoin, ethanol	

Symptomatology in the majority of patients involves dyspnoea, chest pain or palpitations, and can be part of the following clinical entities:

- Associated with pericarditis.
- Recently diagnosed heart failure.
- Acute coronary syndrome without coronary artery disease.
- Syncope and ventricular arrhythmias.
- Dilated cardiomyopathy with chronic heart failure.

To confirm a diagnosis of myocarditis, it is necessary to analyse the cardiac tissue. The Dallas criteria were proposed in 1987 and provided a histopathological categorization by which the diagnosis of myocarditis could be established from endomyocardial biopsies. According to these criteria, myocarditis requires an interstitial inflammatory infiltrate and associated myocyte necrosis and/or degeneration. When these criteria are applied, a diagnosis of myocarditis is established in 25 % of the biopsied patients, or in up to two thirds of patients if more than five samples are taken. This low sensitivity is due to the fact that the inflammatory involvement is often patchy and can be missed, and also due to technical limitations. Currently, an endomyocardial biopsy is indicated in situations of a recently diagnosed heart failure (less than 3 months) with ventricular dysfunction that provokes a haemodynamic compromise or does not respond adequately to treatment. This biopsy should be carried out in experienced hospitals to avoid complications, which arise in around 1 % of the cases.

The wide spectrum of possible aetiologies, mechanisms of pathogenicity, variability of clinical presentation and the absence of a systematic histological examination, results in a poor understanding of the disease and limits progress in this field.

It is interesting to study the clinicopathological correlation in the most frequent forms of myocarditis, which are the following:

- Fulminant myocarditis: typically occurs 2 weeks after a viral infection. It manifests with a very severe decrease in cardiac function that threatens life, although a complete recovery is possible if the patient survives the acute period. Cardiac chambers appear with thickened walls, but are not dilated. Histology reveals abundant diffuse and active inflammatory infiltrates compatible with a diagnosis of myocarditis.
- Acute myocarditis: Progression to heart failure is not as quick and serious as in fulminant myocarditis, although there is a greater risk of incomplete recovery of cardiac function. Myocardial tissue may show limited criteria for the diagnosis of myocarditis with few isolated foci of active inflammation.
- Chronic active myocarditis: Clinical presentation is slower but clinical and histological recurrences are observed, with fibrosis and

ventricular dysfunction in the medium-long term. Biopsy may show limited criteria for the diagnosis of myocarditis with few isolated foci of active inflammation.

- Chronic persistent myocarditis: Biopsy may show limited criteria for the diagnosis of myocarditis with few isolated foci of active inflammation, but the patient suffers neither loss of cardiac function nor symptoms of heart failure.

Advances in several techniques have modified the perspectives of this disease and, undoubtedly, will improve its clinical approach:

- The use of immunohistochemistry and molecular biology techniques (the viral genome has been isolated using the polymerase chain reaction) that allow the identification of viruses in the myocardium even in instances where the Dallas criteria are not met.
- Cardiac Magnetic Resonance (CMR) provides accurate location and quantification of the inflammatory process, oedema and myocardial fibrosis. This imaging technique permits the identification of patients with myocarditis in up to 30 % of the cases that present with chest pain but without lesions in the coronary arteries. Currently, CMR is not as accessible as echocardiography; hence, it is not yet routinely used.

9.2 Lymphocytic Myocarditis

Lymphocytic myocarditis is caused mainly by viral infections. The most common viruses are enteroviruses, predominantly *Coxsackie* A and B. It is a common cause of sudden death in children and the young, accounting for 5–25 % of deaths depending on the series. In the cases of sudden death, patients usually reported a previous viral infection and it is not uncommon for sudden death to be the first clinical manifestation of myocarditis.

The heart may appear normal or dilated under gross examination. Occasionally, the heart has a more flaccid consistency and the appearance of the myocardium is patchy or mottled (Fig. 9.1a).

Lesions are usually focal, although sometimes there is a diffuse myocardial involvement. In acute viral myocarditis, histology shows myocyte necrosis, interstitial oedema and inflammatory infiltrates composed of mononuclear cells (mainly lymphocytes, with variable amounts of macrophages and plasma cells) (Fig. 9.1b). At autopsy, diffuse infiltrates can usually appear in several sections of the myocardium (Fig. 9.2), although sometimes the inflammatory involvement is more focal. In infantile acute viral myocarditis, in contrast to what occurs in other age groups, the inflammatory infiltration of neutrophils prevails, as in cases of widespread myocytic necrosis, although in the latter a polymorphic cellular infiltrate can be observed.

Newborn infants are at high risk of developing severe infections caused by enteroviruses, which can even be lethal. Neonatal enterovirus myocarditis is a rare but serious infection that usually manifests with signs and symptoms of congestive heart failure, and it is associated with high mortality. As previously discussed, in the early stages of the disease, the inflammatory infiltrate is mainly composed of polymorphonuclear leukocytes (neutrophils) (Fig. 9.3). In some instances, myocarditis can be fulminant with clinical features of congestive heart failure or cardiogenic shock that can result in death in less than 24 h (Fig. 9.4).

At *post-mortem* studies on persons who have died from natural or violent causes, it is not uncommon to find isolated foci of lymphocytic infiltrates in the interstitium, with minimal or absent myocyte necrosis or degeneration (Fig. 9.5). The significance of these foci of focal myocarditis is uncertain, although they could be a contributing factor to death in some cases. However, there is no consensus regarding the degree of extension of myocarditis required to justify death in the absence of other causes.

The use of immunohistochemistry techniques (Fig. 9.6) has demonstrated that the majority of cells are T lymphocytes, while B lymphocytes and macrophages are present in lower numbers.

During the resolution period in lymphocytic myocarditis, necrotic myocytes are replaced with fibroblasts and collagen, which will finally create fibrous tissue, randomly distributed. Cases of sudden death have been reported in the subacute (Figs. 9.7 and 9.8) or the chronic (Figs. 9.9 and 9.10) phases, in which it is possible to find focal areas of necrosis and inflammation next to areas of replacement fibrosis (chronic myocarditis).

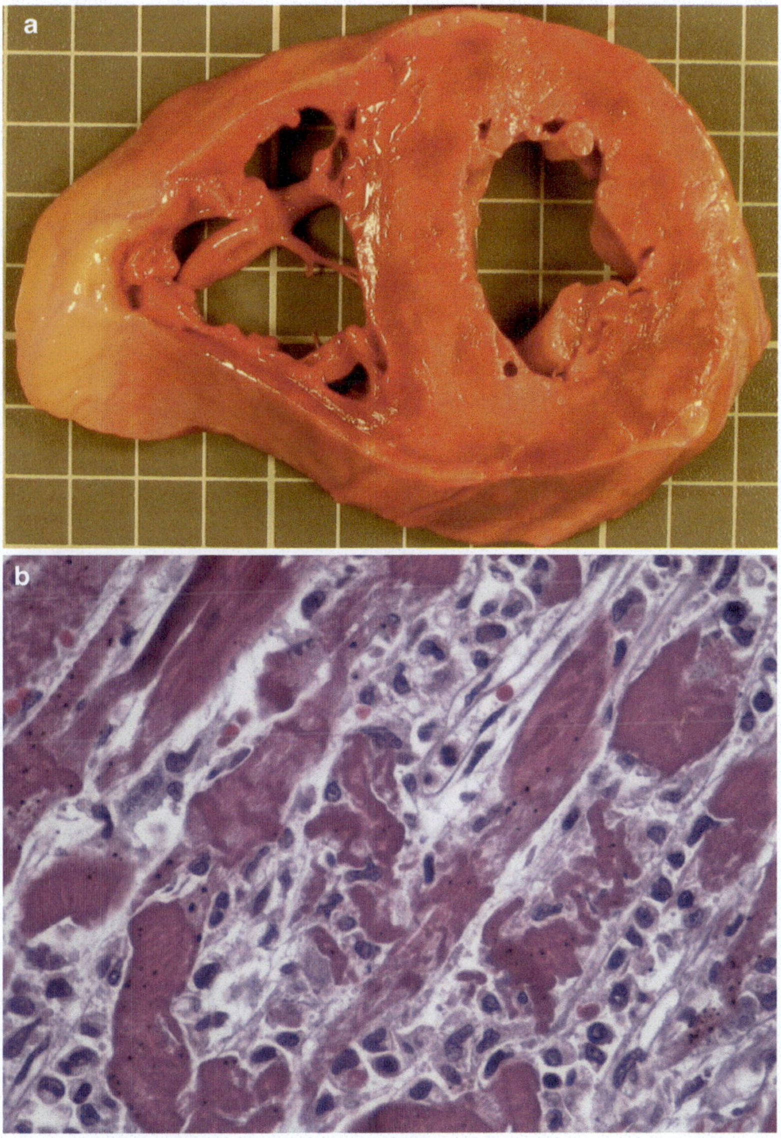

Fig. 9.1 A 24-year-old woman who died suddenly. (**a**) Transverse section of the heart in which the patchy appearance of the myocardium can be appreciated in the LV. (**b**) Histological image showing myocyte necrosis and inflammatory infiltrate, mainly composed of mononuclear cells (Lucena J, et al. (2007). *Reproduced with permission from Cuadernos de Medicina Forense*)

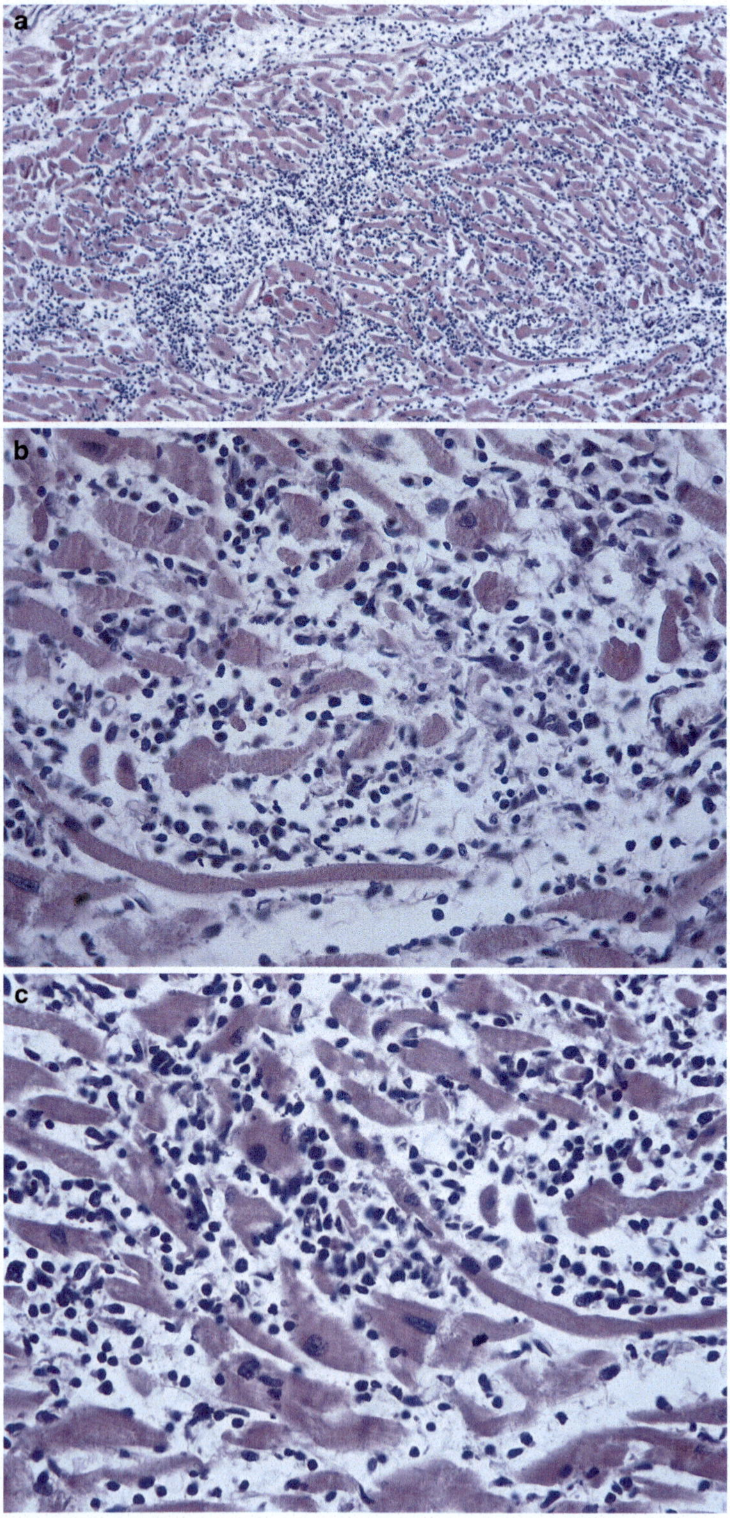

Fig. 9.2 Sudden death in a 21-year-old male, with no previous medical history, presenting symptoms of a viral syndrome 2 days before death. (**a**) Microscopic study revealed acute myocarditis with a widespread lymphocytic infiltrate and associated necrosis. (**b**, **c**) Detailed images showing focal myocyte necrosis and a mononuclear inflammatory cell infiltrate. PCR analysis from paraffin-embedded myocardial samples allowed the identification of Coxsackievirus B as the causal infectious agent

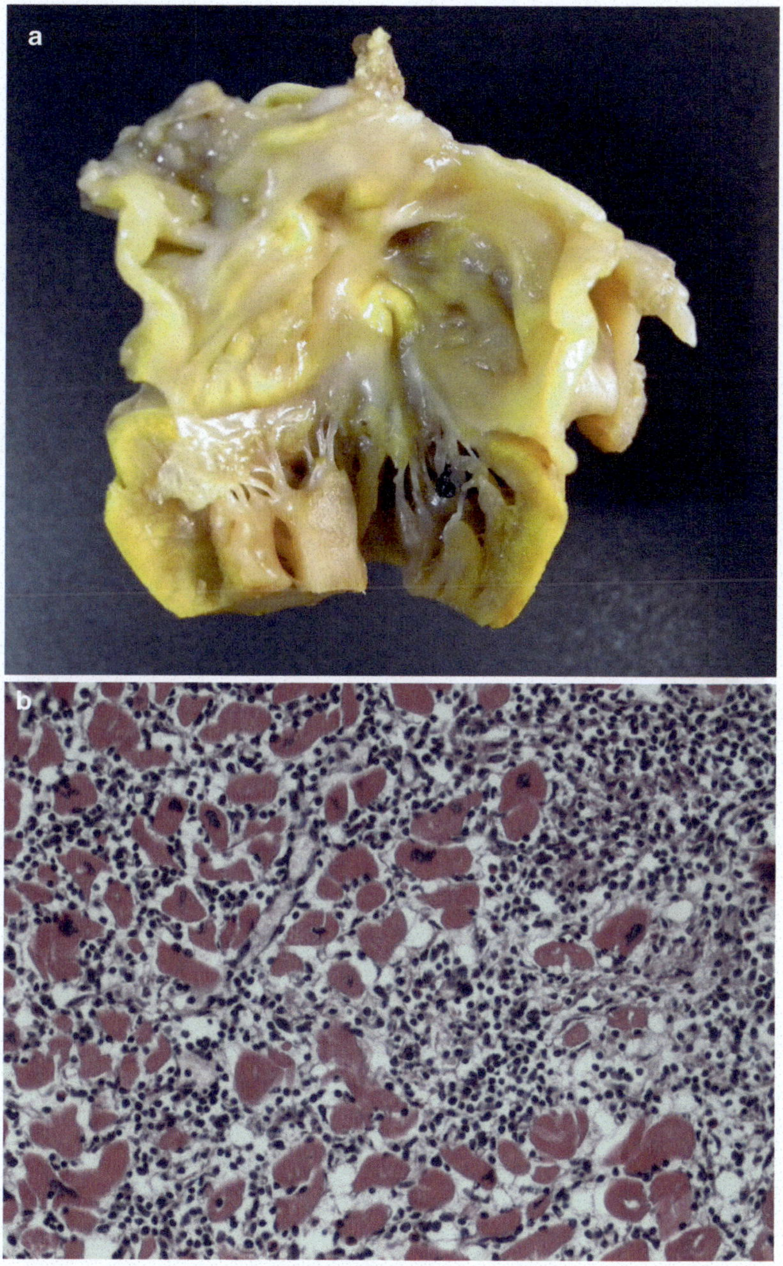

Fig. 9.3 Newborn-term infant with a normal neonatal period. Ten days after birth, the paediatrician observed hypotony, hepatomegaly and extreme bradycardia. He was taken to the casualty department, dying soon after. Necropsy studies showed acute myocarditis, more extended in the LV. (**a**) Gross examination of the heart showed a yellow-tan colour and necrotic appearance. (**b**) Photomicrograph showing a pleomorphic inflammatory infiltrate mainly composed of polymorphonuclear leukocytes, and widespread myocyte necrosis. Meningoencephalitis was also diagnosed. PCR determination identified presence of Coxsackievirus B in spleen and myocardial tissue samples taken at autopsy

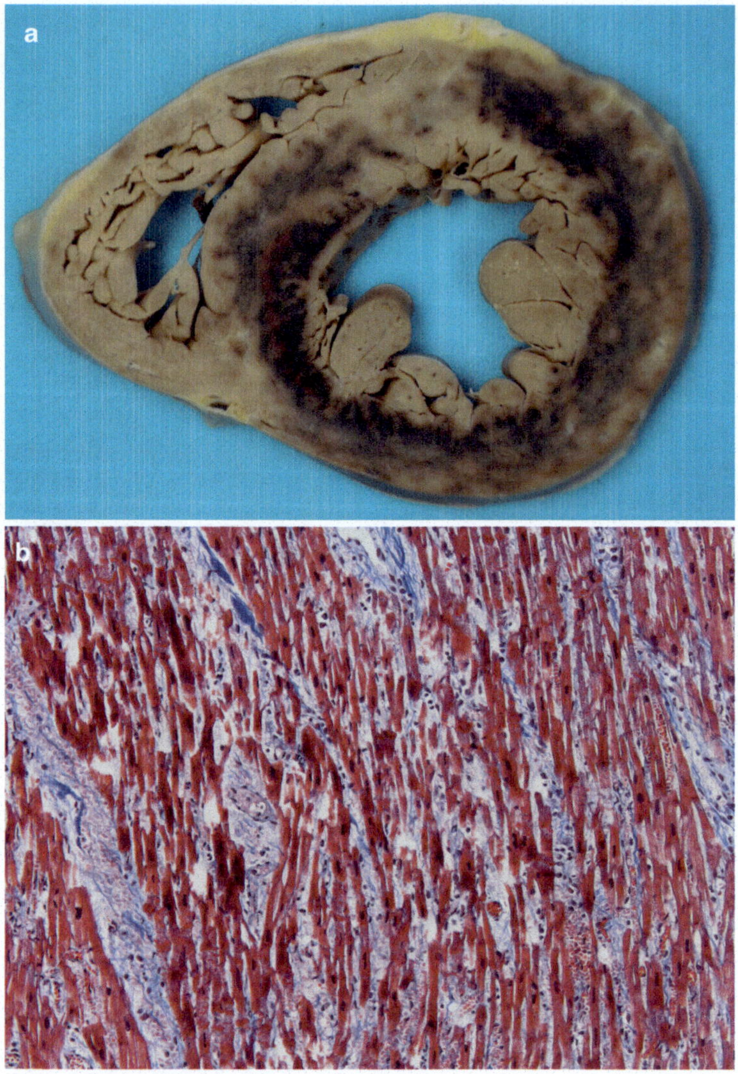

Fig. 9.4 A 6-year-old girl, with a recent episode of pharyngotonsillitis, who died in 10 h as a consequence of a cardiogenic shock with multiple organ failure. (**a**) Transverse section of the heart showing a diffuse patchy appearance of the LV myocardium, as well as marked circumferential haemorrhagic infiltration, probably due to the intensive treatment received. Microbiological analysis identified Influenza B virus, Parainfluenza virus type 4 and Rhinovirus in the pharyngeal exudate. (**b**) Microscopic image showing myocyte necrosis and inflammatory infiltrate, mainly lymphocytic, along with interstitial haemorrhage (Masson's trichrome)

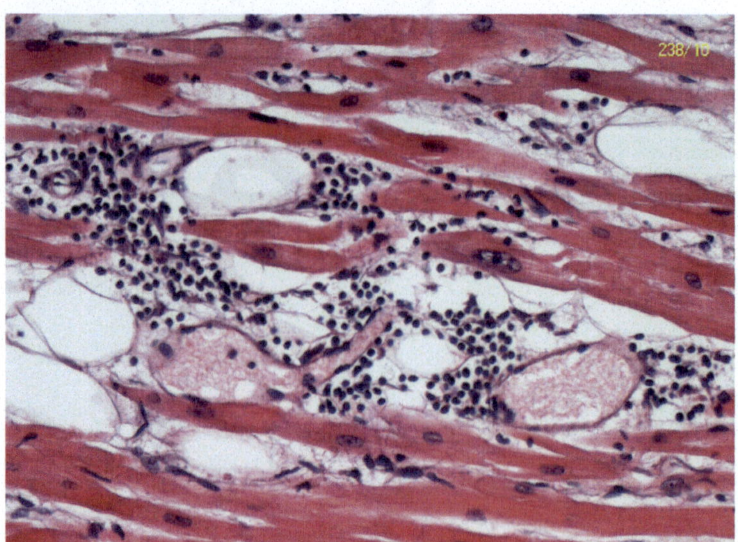

Fig. 9.5 Lymphocytic infiltrates in the interstitium were observed in the myocardium of a person who died in a traffic accident

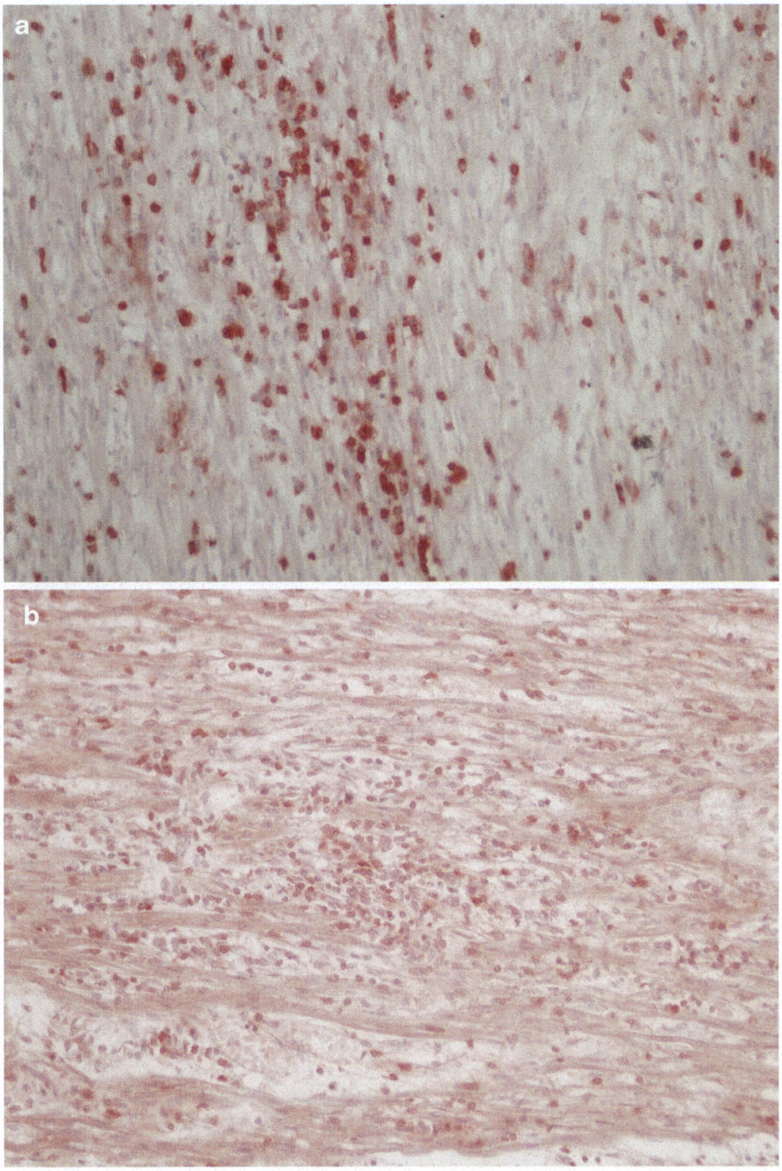

Fig. 9.6 A case of myocarditis showing inflammatory cells that stain positive for CD45 (common leukocyte antigen) (**a**) and CD3 immunostaining (T lymphocytes) (**b**)

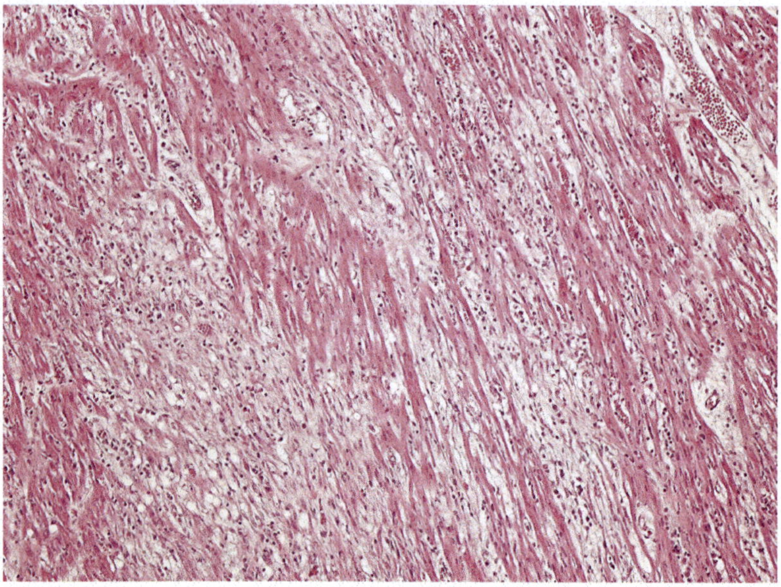

Fig. 9.7 Image shows a transmural involvement of the free wall of the LV with replacement fibrosis and mononuclear cell infiltration

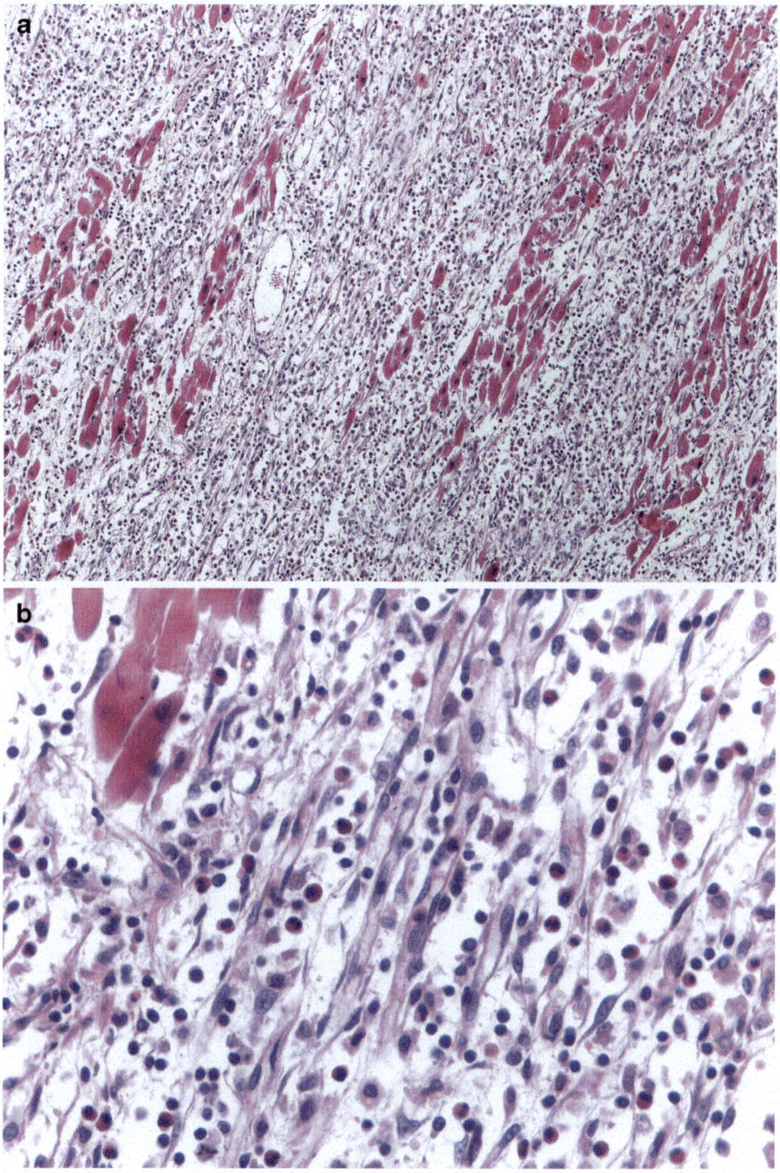

Fig. 9.8 Subacute myocarditis. (**a**) Replacement of necrotic myocytes with fibroblasts and lymphocytic infiltrate can be observed. (**b**) Image at higher magnification showing some eosinophils among the inflammatory infiltrate

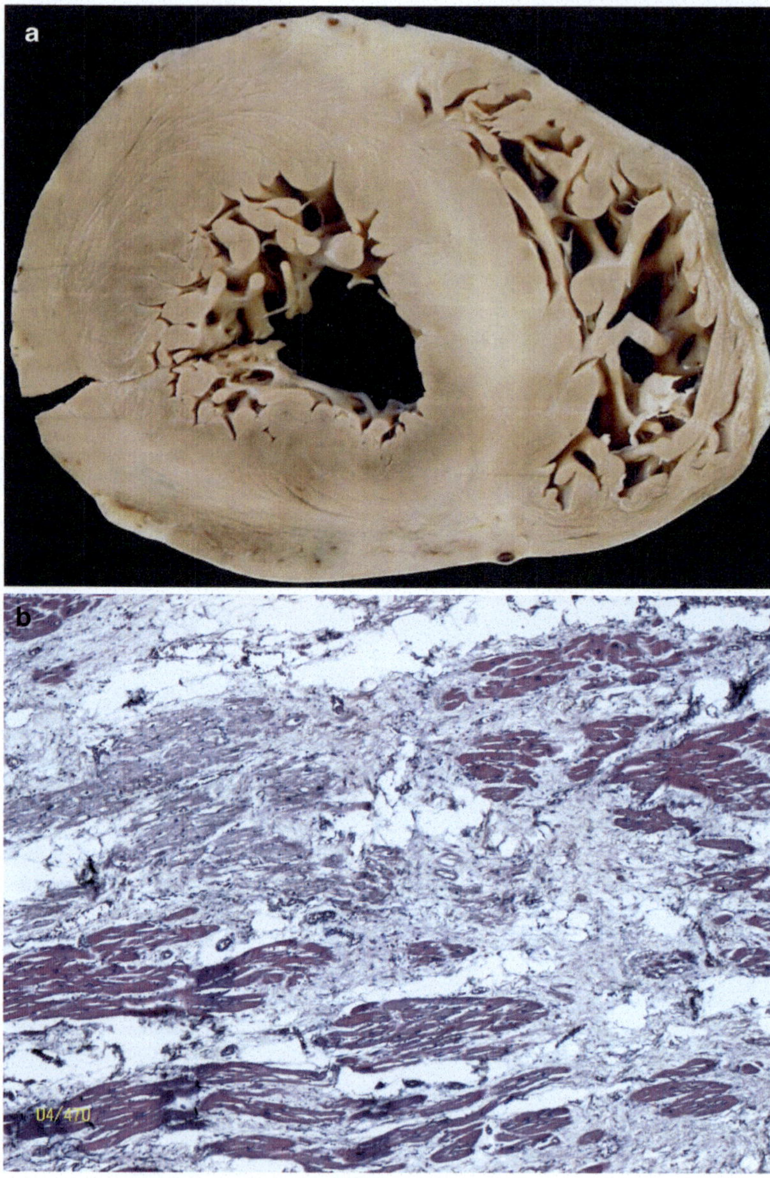

Fig. 9.9 Chronic myocarditis. (**a**) Macroscopic section of the heart showing large whitish areas of scarring in septum and free wall of the LV. Coronary arteries were permeable. This type of fibrosis distribution that does not follow the coronary flow is suggestive of chronic myocarditis. (**b**) Image shows an area of replacement fibrosis in the absence of inflammatory infiltrate. Lesions were not predominantly subendocardial (typical of ischaemic lesions)

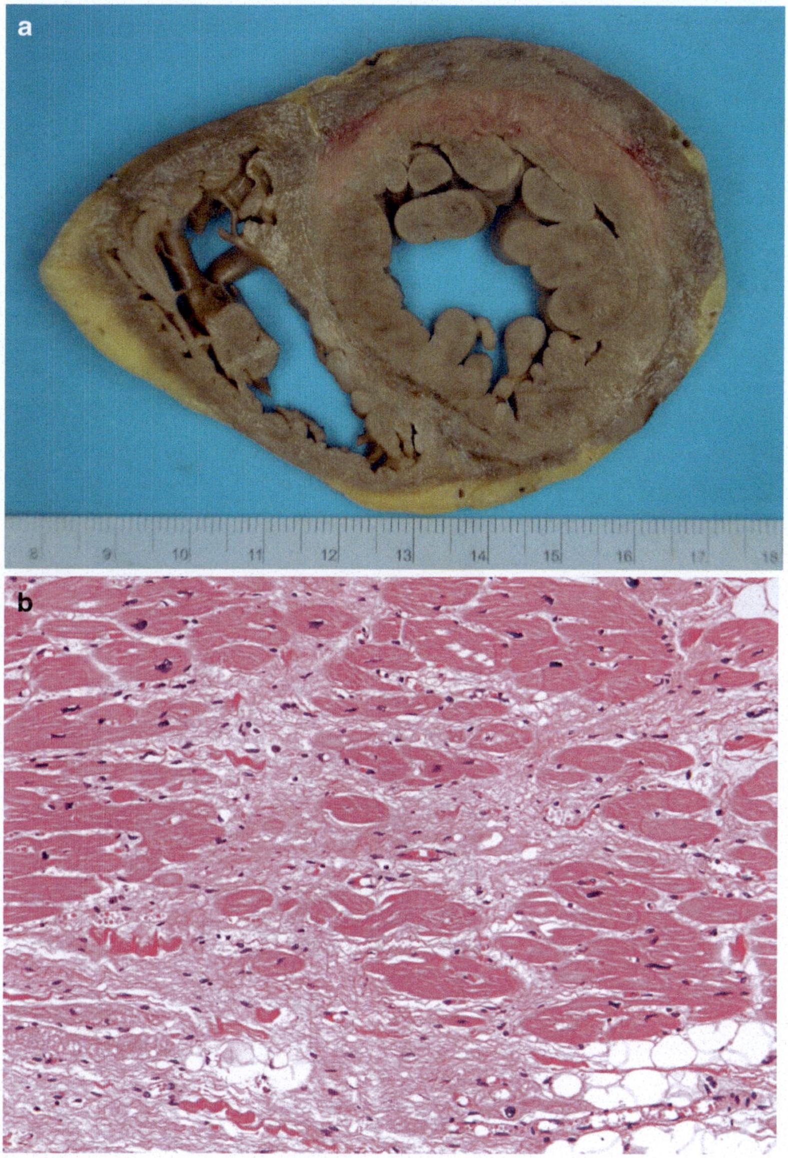

Fig. 9.10 Sudden death in a 19-year-old woman as a consequence of chronic lymphocytic myocarditis. (**a**) Transverse section of the heart in which areas of fibrosis and necrosis can be appreciated in the LV subepicardium. (**b**) Histological image showing areas of patchy fibrosis. Microbiological analysis identified respiratory syncytial virus type A

9.3 Bacterial Myocarditis

Inflammation of the heart muscle caused by bacterial infection is not commonly encountered in the daily clinical practice. Bacterial myocarditis is characterized by myocardial necrosis with inflammatory infiltrates mainly composed of neutrophils, and microabscess formation, sometimes visible macroscopically (Fig. 9.11), which contains bacterial colonies (Fig. 9.12). This type of myocarditis can appear as the progression of a pneumonic process or in the context of overwhelming sepsis (Figs. 9.13 and 9.14).

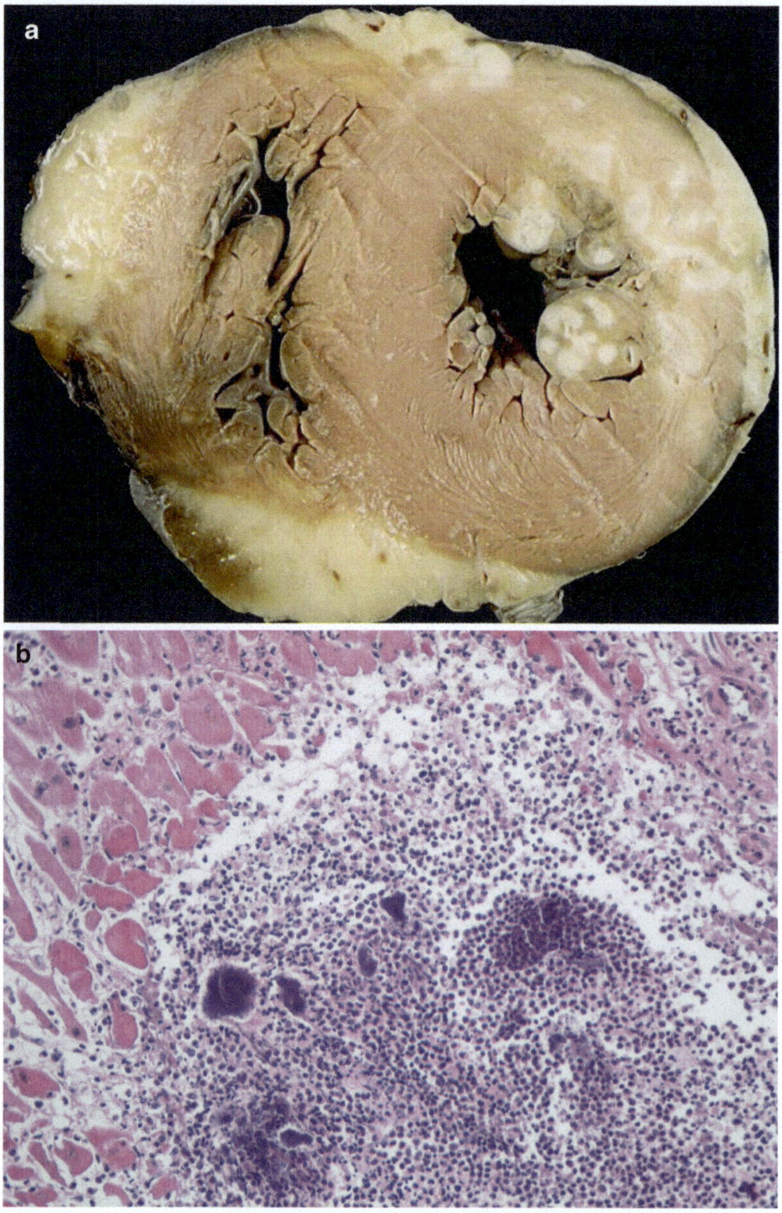

Fig. 9.11 *Bacterial myocarditis*. An adult male with clinical antecedents of insulin-dependent diabetes mellitus and diabetic neuropathy. (**a**) Transverse section of the heart. In the free wall of the LV several white-yellowish foci can be appreciated in the myocardium, which are irregular in shape and size. (**b**) Microscopic image showing a myocardial abscess containing bacterial colonies and a neutrophil exudate

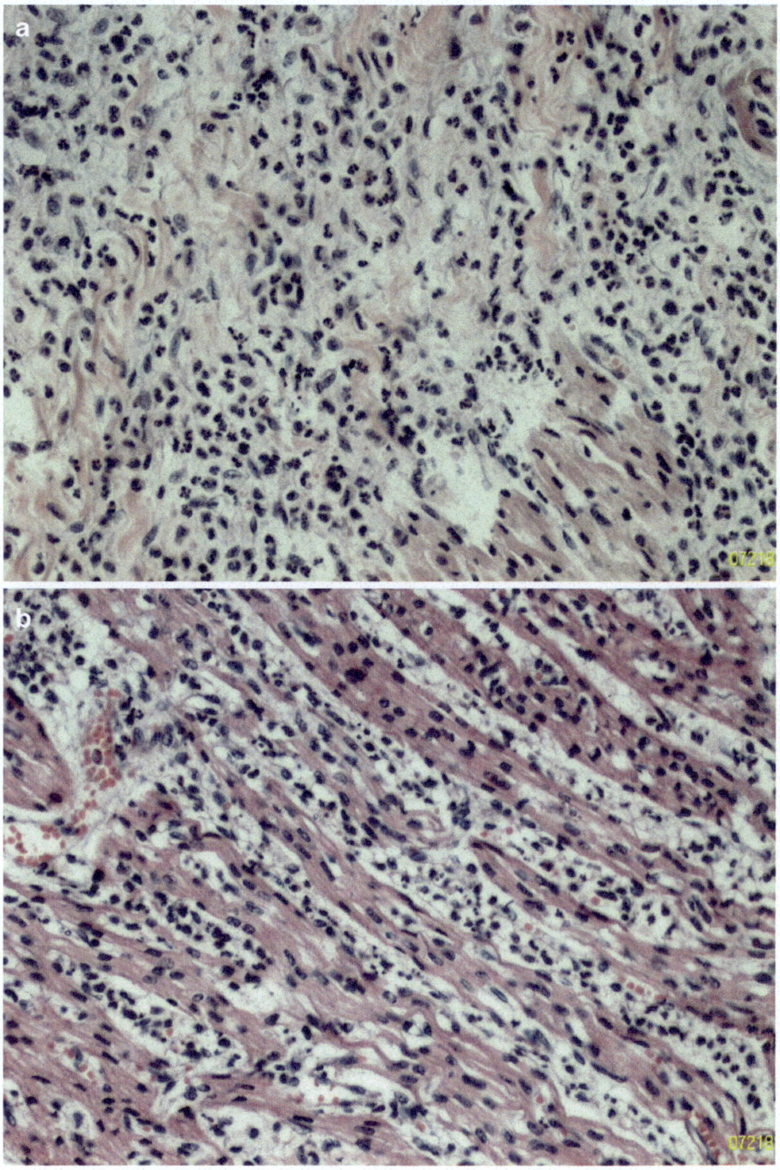

Fig. 9.12 Sudden death in a 25-year-old woman who died of fulminant sepsis. Microbiological analysis was positive for *Neisseria meningitidis* (serogroup C). (**a**, **b**) Extensive neutrophil infiltration of the myocardium forming microabscesses

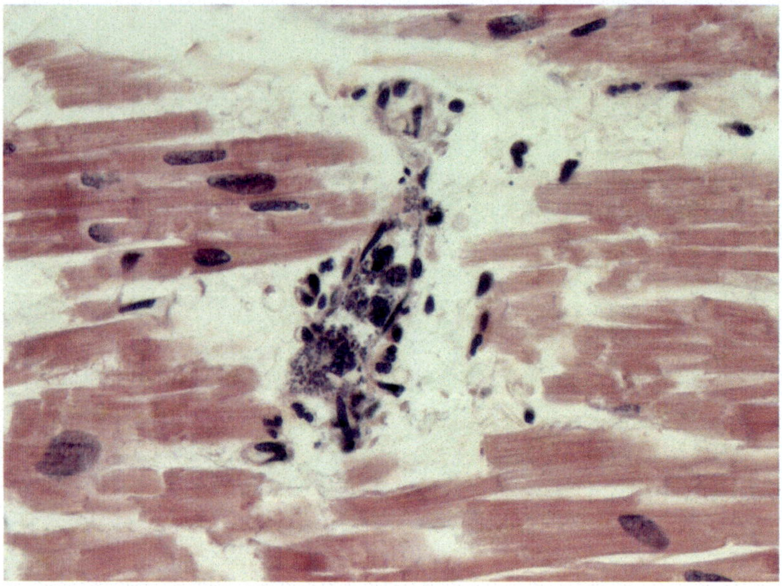

Fig. 9.13 Another case of fulminant meningococcemia where neutrophils and bacteria (cocci) can be observed inside a capillary vessel

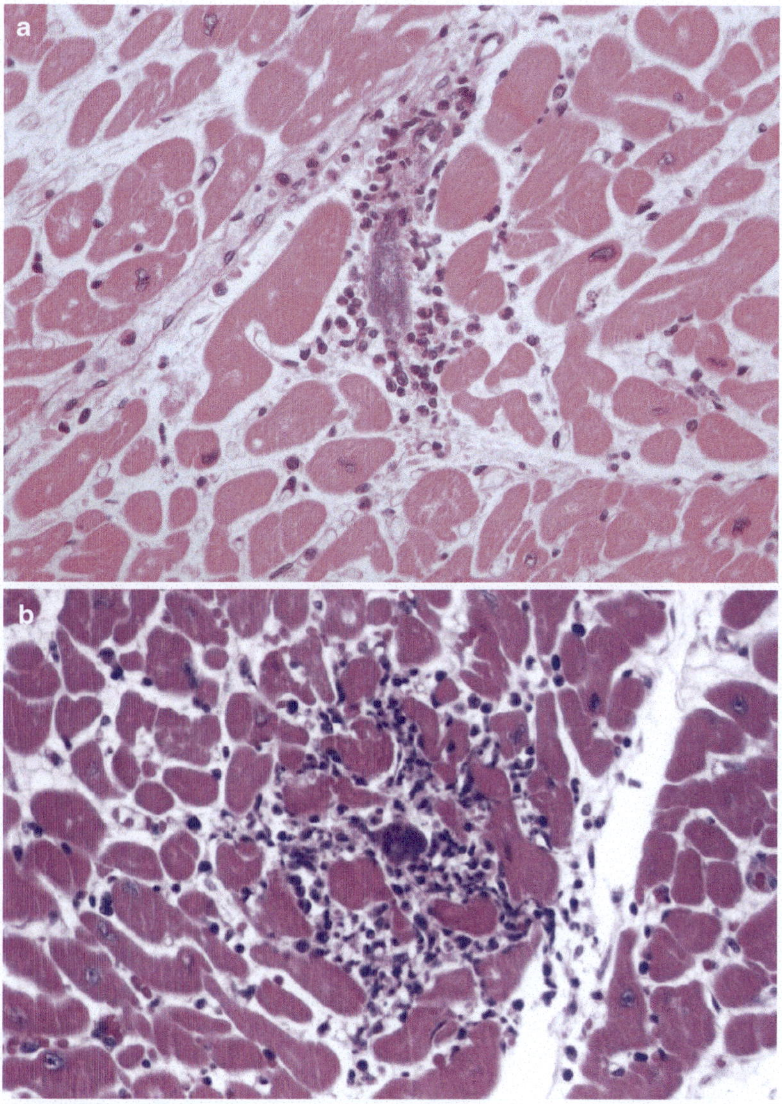

Fig. 9.14 A 33-year-old male with prodromic signs of fever and cutaneous exanthema. Microbiological studies resulted in isolation of *Neisseria meningitidis* serogroup B from blood, lung, kidney and brain. (**a**) Histology revealed foci of perivascular myocarditis around capillary vessels with bacterial colonies in their lumen. (**b**) Focus of myocyte necrosis associated to inflammatory infiltrate

9.4 Myocarditis Caused by Other Infectious Agents

9.4.1 Myocarditis Caused by *Toxoplasma gondii*

Toxoplasma gondii is an obligate intracellular parasitic protozoan. In immunocompromised individuals, infection can be very serious and occasionally fatal, constituting an important cause for morbi-mortality in patients infected with the human immunodeficiency virus (HIV). Typically, it produces abscesses in the brain, although in some instances it can also affect other organs, including the heart (Fig. 9.15).

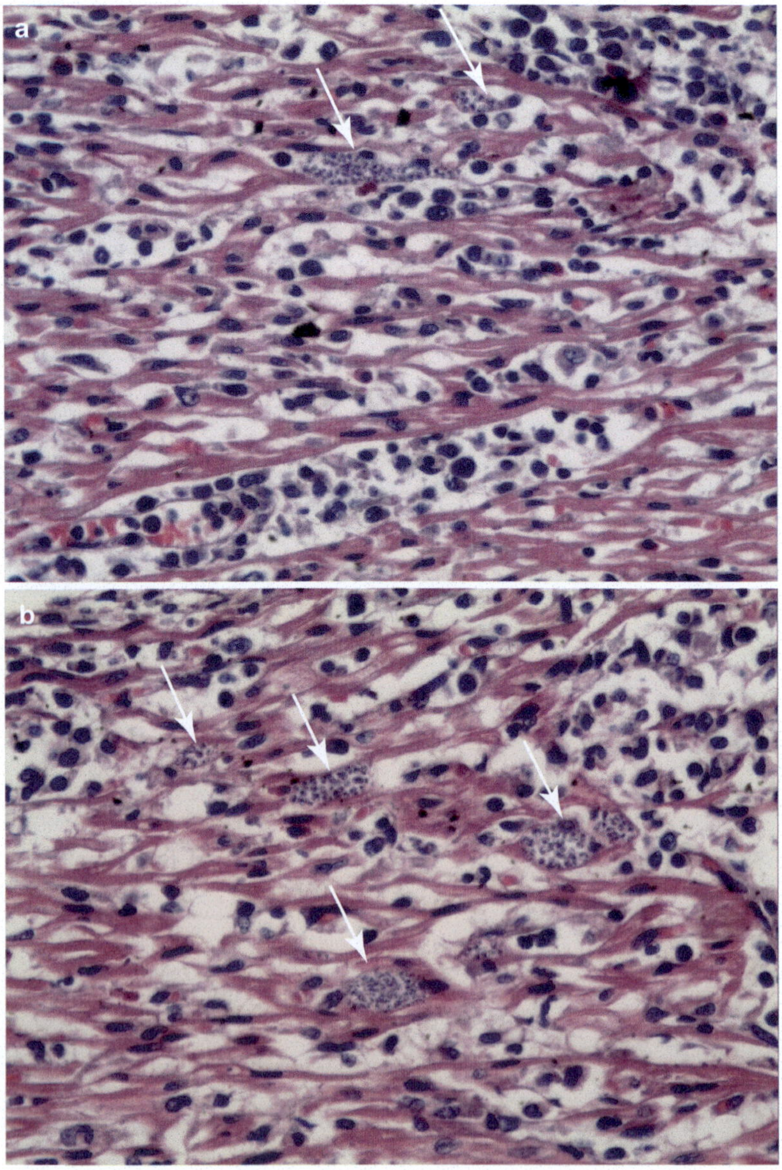

Fig. 9.15 Myocarditis caused by *T. gondii* in an 8-month-old girl. (**a**, **b**) Toxoplasma cysts (*arrows*) can be observed inside the myocytes, alongside inflammatory cells in the interstitium (*Courtesy of Dr. Carlos de Bernard. Institute of Legal Medicine and Forensic Sciences of Panama*)

9.4.2 Chagas Disease

Chagas disease is a tropical parasitic disease caused by the protozoan *Trypanosoma cruzi*, endemic in South America. However, due to population migratory movements, the disease is now extending to other countries.

Chagas disease occurs in two phases: an acute stage (it generally goes unnoticed since it manifests flu-like symptoms), and a chronic stage. Cardiac involvement is very different in each stage. In the acute phase, pericarditis and pericardial effusion can occur, with little inflammation of the heart muscle. However, in the chronic phase, patients can develop dilated cardiomyopathy with biventricular dysfunction. Chagas disease must be suspected in patients with dilated cardiomyopathy who have lived in endemic regions, and an analysis of parasite-specific antibodies in serum should be performed.

On gross examination, the heart affected by cardiomyopathy related to Chagas disease is similar to any other case of dilated cardiomyopathy although the presence of an apical aneurysm is characteristic of the disease (Fig. 9.16). Microscopically, lymphocytic myocarditis is also typical of the disease, although its intensity can vary. Indeed, in some cases, myocarditis is not detected in the *post-mortem* histological examination of the heart. However, in other instances, it is possible to identify the amastigote form of the parasite within the myocardium. Interstitial fibrosis is very common, usually surrounding isolated groups of cardiomyocytes (perimysial fibrosis) (Fig. 9.17).

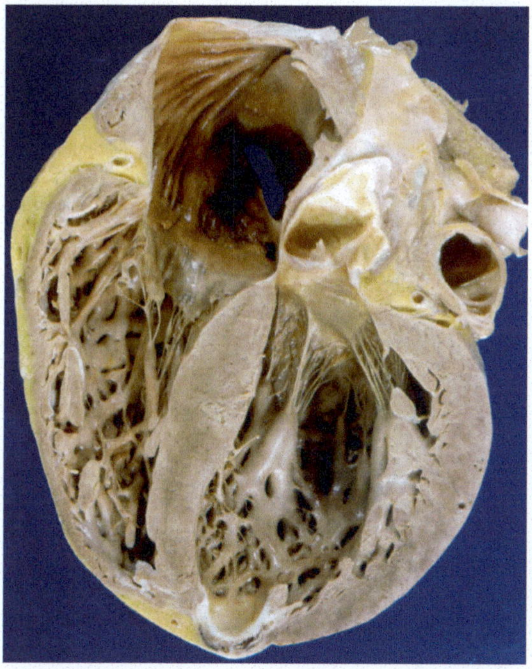

Fig. 9.16 A 57-year-old male with chronic chagasic cardiomyopathy: gross examination of the heart after longitudinal section showing the characteristic apical aneurysm and myocardial fibrosis in both ventricles (*Courtesy of Dr. Luiz A. Benvenuti. Pathology Laboratory. Heart Institute. School of Medicine. University of Sao Paulo, Brazil*)

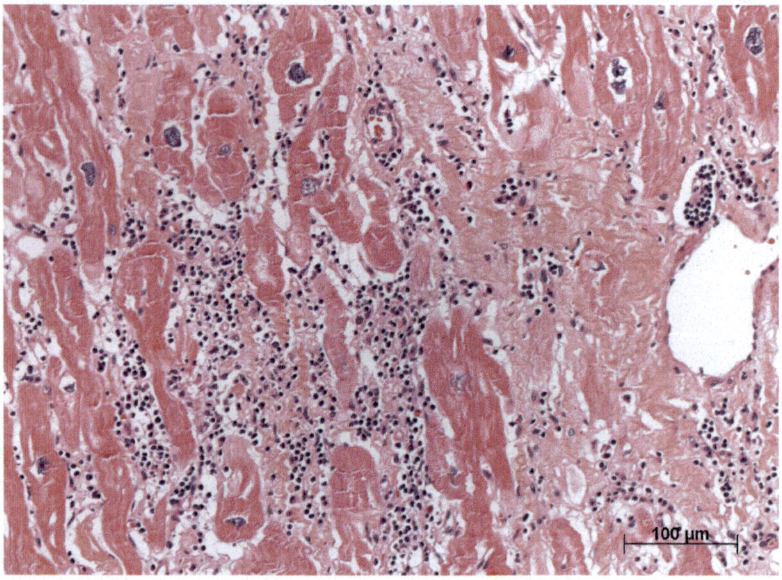

Fig. 9.17 A 35-year-old male with chronic Chagas cardiomyopathy was referred for heart transplantation. Histology of the explanted heart showed lymphocytic myocarditis, interstitial fibrosis and cardiomyocyte hypertrophy (*Courtesy of Dr. Luiz A. Benvenuti. Pathology Laboratory. Heart Institute. School of Medicine. University of Sao Paulo, Brazil*)

9.4.3 Myocarditis Caused by *Cryptococcus neoformans*

This is a fungal infection transmitted by inhalation. The degree of symptomatology largely depends on the immunological state of the patient and ranges from a non-specific mild pneumonitis to the formation of abscesses in the brain in individuals with a weakened immune system, such as those infected with HIV in the AIDS phase. After the initial infection, *C. neoformans* can remain latent for some time and only produce symptomatology in situations where the hosts' defences decline. The response of an organism to *C. neoformans* infection depends on the host's immunological status and, in cases of severe immunosuppression, there is no inflammatory response (Fig. 9.18).

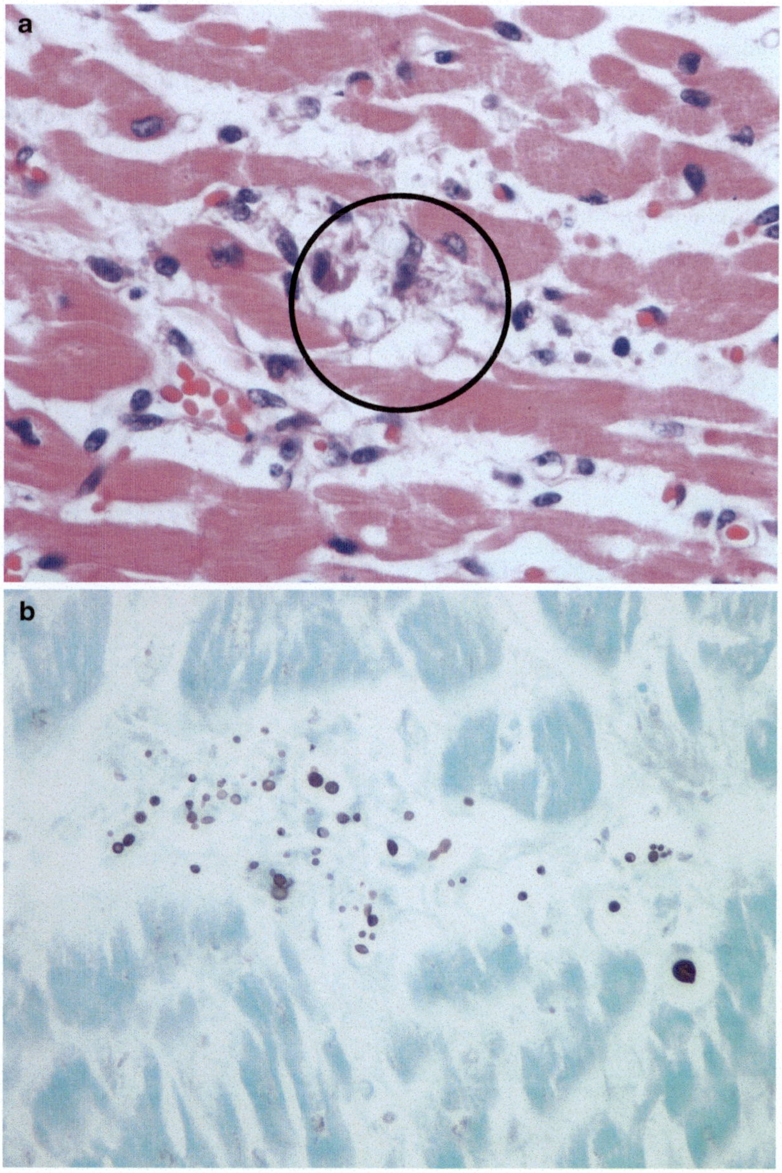

Fig. 9.18 A 40-year-old woman with HIV died shortly after being admitted to hospital with severe hypoglycaemia and metabolic alkalosis. (**a**) Histology revealed the presence of large numbers of Cryptococci in the myocardium (*circle*). (**b**) Grocott staining allows the identification of the several morphologies adopted by the fungus (*Courtesy of Rosario Mosquera. Histopathology Service. National Institute of Toxicology and Forensic Sciences. Madrid*)

9.5 Giant Cell Myocarditis

This is a rare type of myocarditis. The cause is not known but the presence of concomitant autoimmune processes and its response to immunosuppressive treatment suggest an autoimmune origin. Clinically, it manifests with subacute congestive heart failure of unexplained aetiology and recent onset, left ventricular dilatation, ventricular arrhythmias and atrioventricular blocks. Clinical suspicion of giant cell myocarditis is one of the reasons that justify an endomyocardial biopsy, since the therapeutic approach would change if diagnosis is confirmed. This variant carries a poor prognosis and quickly progresses to refractory congestive heart failure. It is commonly fatal unless heart transplantation can be performed; if this is the case, morphological study of the explanted heart allows confirmation of diagnosis. Recurrences have been described in the transplanted heart; hence, for some time, the disease was questioned as a contraindication for heart transplantation. Currently, given that recurrences are uncommon and it is understood that immunosuppression is part of the treatment, patients with giant cell myocarditis can be candidates to receive a heart transplant.

This type of myocarditis has similar macroscopic alterations to lymphocytic myocarditis, with hearts appearing normal or with mottled or whitish areas. Histology (Fig. 9.19) reveals the presence of a characteristic inflammatory infiltrate consisting of multinucleated giant cells intermingled with macrophages, lymphocytes, eosinophils and plasma cells. Necrosis of myocardial fibres is usually widespread, although some cases of focal involvement have been described. The inflammatory infiltrate does not extend towards the epicardial adipose tissue. The differential diagnosis in the endomyocardial biopsy mainly includes sarcoidosis.

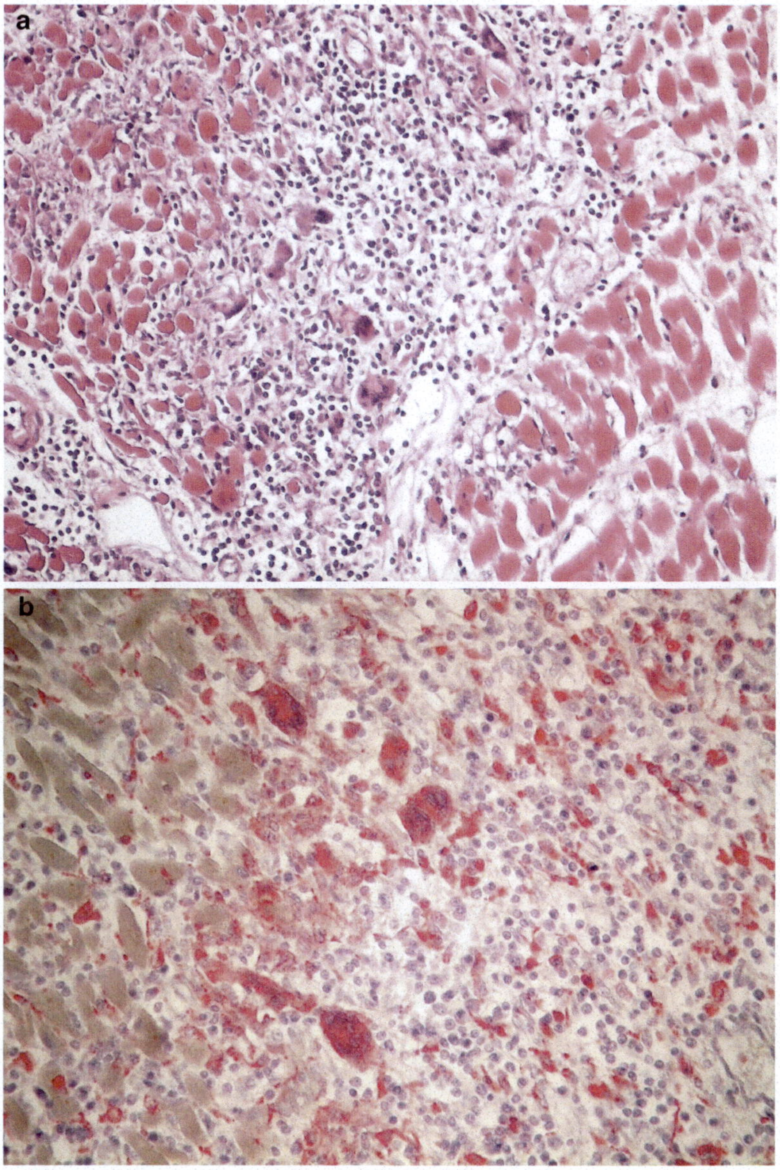

Fig. 9.19 Giant cell myocarditis. (**a**) Presence of multinucleated giant cells and inflammatory mononuclear cell infiltrate associated with myocardial necrosis. (**b**) Immunohistochemistry for CD68 was positive in giant cells and macrophages

9.6 Hypersensitivity Myocarditis

Several drugs and medications can produce myocardial inflammation either by direct toxic mechanisms or mediated by hypersensitivity reactions, which is usually associated with eosinophilic myocarditis. The clinical spectrum is broad: in some instances, the disease is asymptomatic but in other cases, there could be a precipitous onset of heart failure, arrhythmias, or even sudden death.

Hypersensitivity myocarditis occurs several days or weeks after exposure to medication and usually resolves spontaneously once drug administration is discontinued.

At autopsy, the myocardium is frequently normal in appearance, although in some cases the heart may have pale or dark areas (Fig. 9.20a). Histologically the myocardium shows patchy areas of interstitial inflammatory infiltrate, mainly composed of eosinophils and, to a lesser extent, of inflammatory mononuclear cells (lymphocytes, macrophages) (Fig. 9.20b). This infiltrate tends to locate in perivascular or large interstitial spaces. There is little necrosis of myocytes and fibrosis is not usually present.

Acute necrotizing eosinophilic myocarditis is a fulminant form of myocarditis with giant cells, which clinically manifests with acute congestive heart failure and an increase in cardiac enzymes. In contrast to hypersensitivity myocarditis, it shows widespread myocyte necrosis (Fig. 9.21).

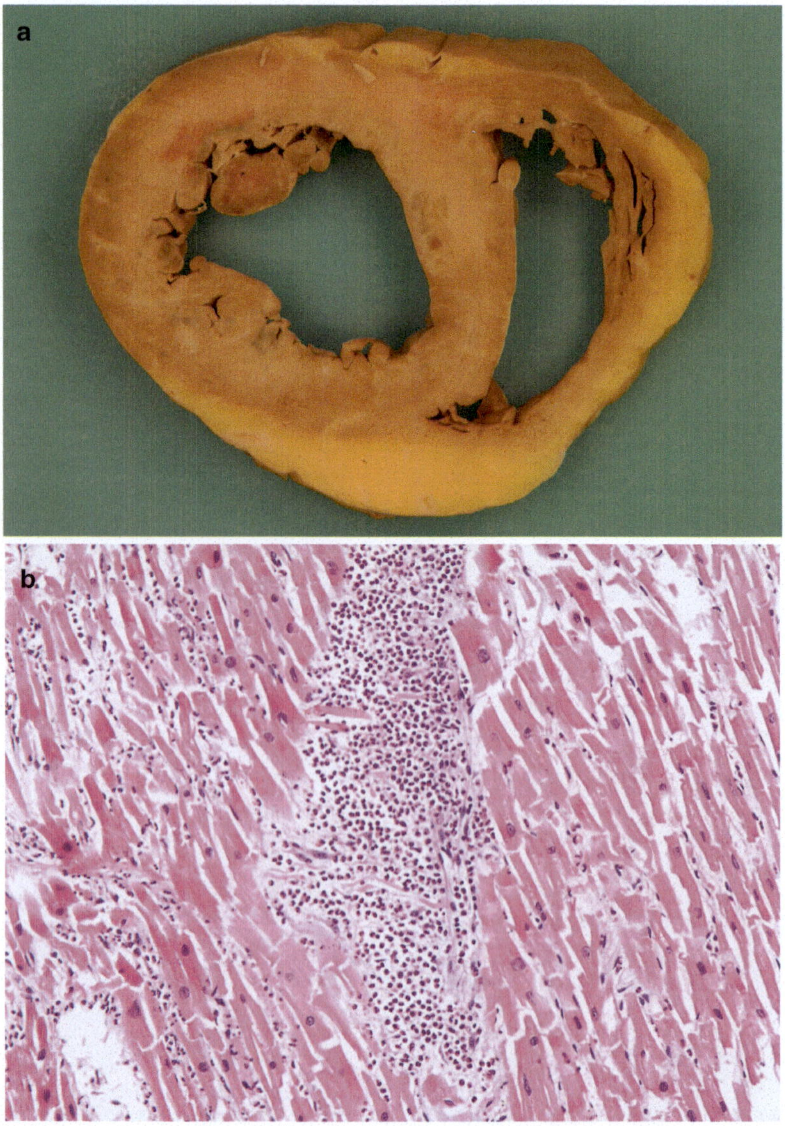

Fig. 9.20 Eosinophilic myocarditis in a 50-year-old woman who was found dead at home. She had a clinical history of asthma, and fever the week before death. Heart weight, 405 g. (**a**) A section of the heart shows greyish mottled areas in the lateral wall of the LV and the septum. Microbiological analysis for viruses was negative. (**b**) Histology showed interstitial eosinophil-rich inflammatory infiltrates

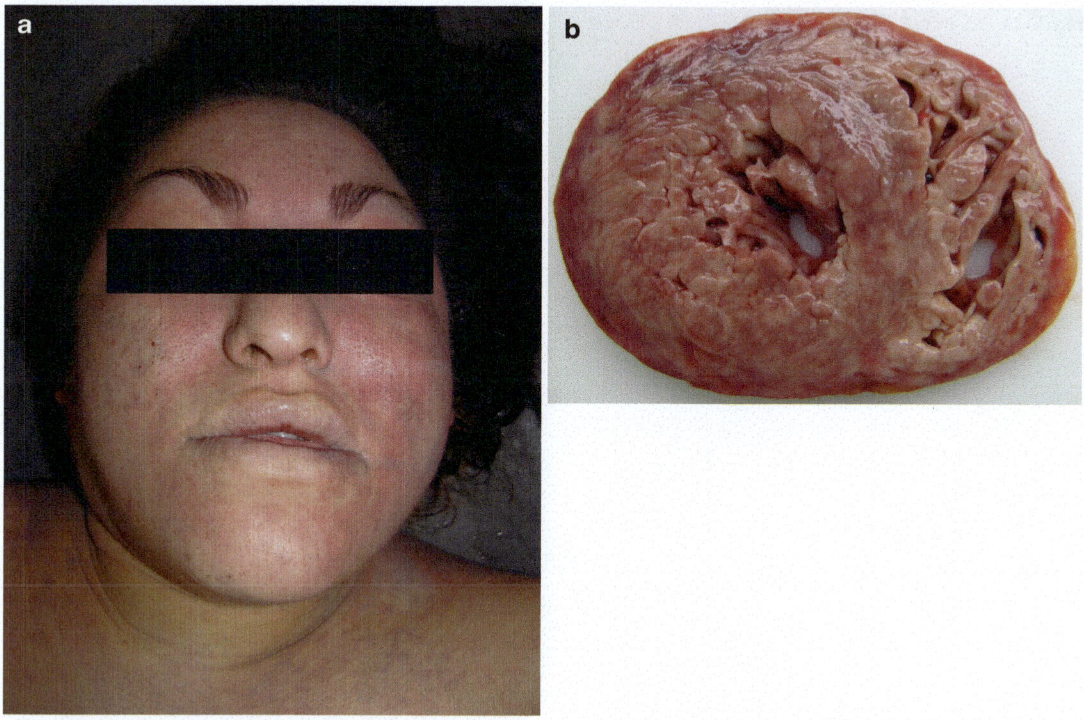

Fig. 9.21 Acute necrotizing eosinophilic myocarditis. A 17-year-old woman with acute tonsillitis who was prescribed amoxicillin. She died 36 h after the initial dose. (**a**) Autopsy revealed extensive erythema and facial oedema. (**b**) Gross examination of the heart showed an oedematous myocardium with a mottled appearance. (**c**) Histology revealed a dense inflammatory infiltrate mainly composed of eosinophils, with widespread myocyte necrosis and abundant giant cells. (**d**) Higher magnification of the lesion showing broad necrosis and myocardial inflammation (*Courtesy of Dr. Soledad Martínez Latrach. Medico Legal Service. Santiago de Chile*)

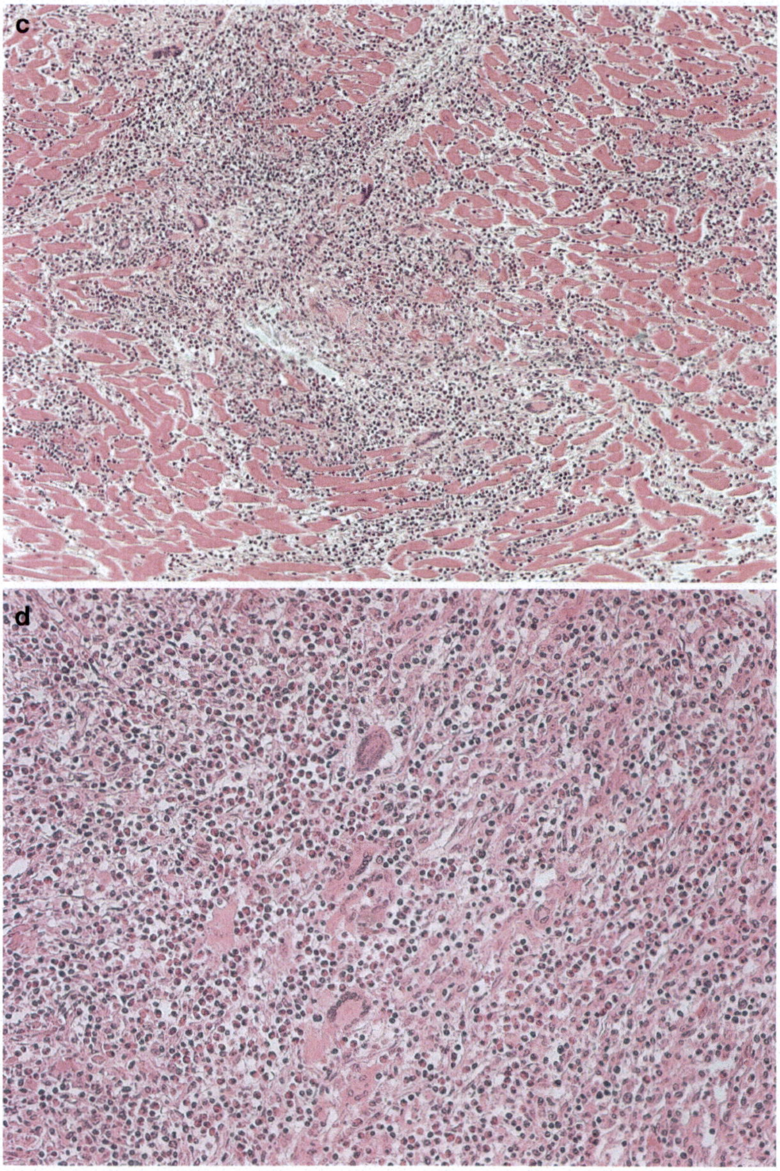

Fig. 9.21 (continued)

9.7 Eosinophilic Myocarditis

Eosinophilic myocarditis is the term used to describe the syndrome in which myocardial inflammation is associated with persistent peripheral blood eosinophilia. Differential diagnosis should include hypersensitivity myocarditis and parasitosis with eosinophilic inflammatory infiltrates. Eosinophilic myocarditis usually involves both ventricles. The initial stages of the disease are characterized by marked eosinophilic infiltration and widespread necrosis of myocytes (Fig. 9.22).

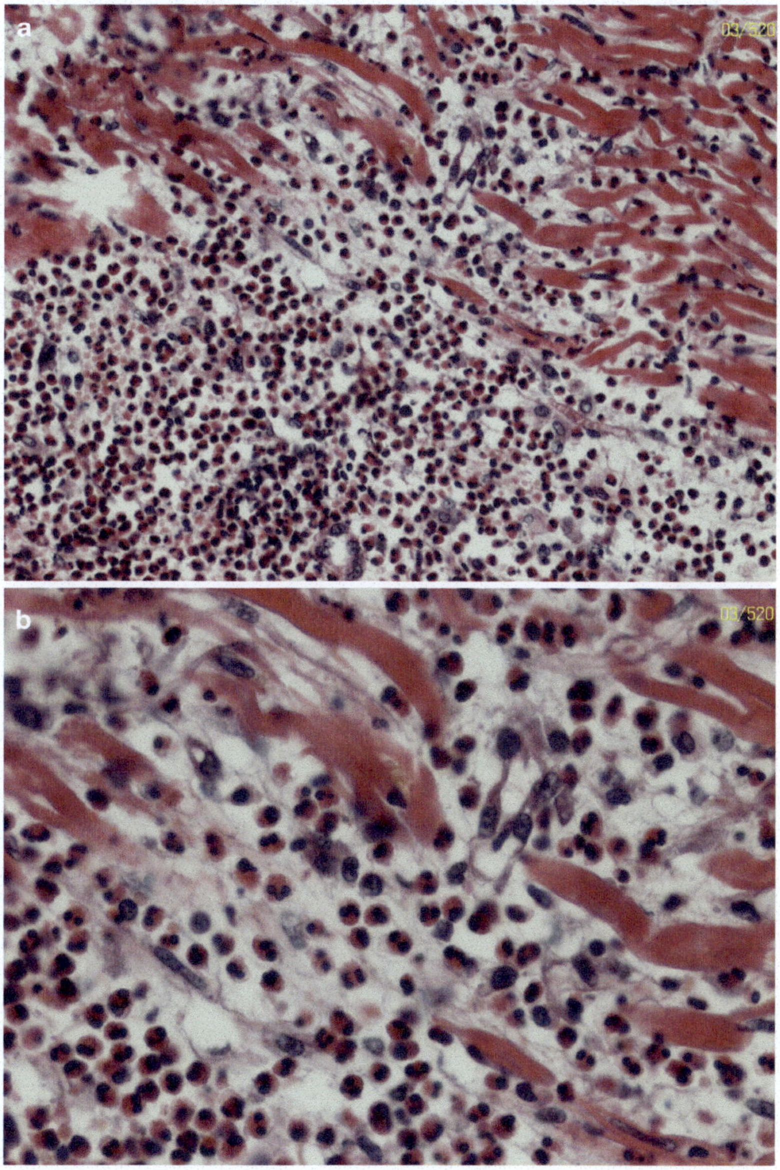

Fig. 9.22 *Fulminant eosinophilic myocarditis/Hypereosinophilic syndrome*. A 50-year-old male was admitted to hospital with anorexia and weight loss. Analytical determinations showed eosinophilia. He died suddenly a few hours after admission. Autopsy revealed multiple organ involvement: heart, lungs, liver, kidney, bone marrow, etc. (**a**, **b**) show extensive inflammatory infiltrate composed of eosinophils, with necrosis of myocytes that are better appreciated in the higher magnification image

9.8 Cardiac Sarcoidosis

Sarcoidosis is a systemic inflammatory disease of unknown aetiology that manifests as non-caseating granulomas in several organs. The most common clinical presentation is pulmonary sarcoidosis, with bilateral hilar lymphadenopathy or reticular opacities that are often casual findings. Clinical features and diagnosis are very broad and vary with the degree of generalised inflammation (it causes tiredness and exercise intolerance) and the organ involved (arthritis, dermatologic lesions or ocular involvement).

Cardiac involvement occurs in 5–10 % of the patients. While in the majority of cases of cardiac sarcoidosis there is also systemic involvement, the heart can be the first organ where the disease manifests. Clinical presentation in cardiac sarcoidosis also varies with the extent and location of the lesions, and includes atrioventricular conduction abnormalities, ventricular arrhythmias and congestive heart failure, among others. Diagnosis requires imaging techniques and endomyocardial biopsy. Delayed gadolinium-enhanced MRI has allowed a breakthrough in the confirmation of diagnosis. Since the disease is focal, a negative result from endomyocardial biopsy does not exclude a positive diagnosis.

Cardiac lesions mainly involve the left ventricle and basal septum. Lesions can be visible on gross examination and appear as nodules or firm whitish areas. The typical non-caseating granulomas can be observed on histological studies (Fig. 9.23a). With time, granulomas are surrounded or replaced by fibrous scar tissue (Fig. 9.23b). While the presence of this type of granuloma is suggestive of sarcoidosis, other causes of granulomatous inflammation such as mycotic infections and tuberculosis need to be ruled-out by the use of specific staining techniques. Nevertheless, the epicardial and coronary artery involvement in the cases of cardiac sarcoidosis facilitates the differential diagnosis with giant cell myocarditis.

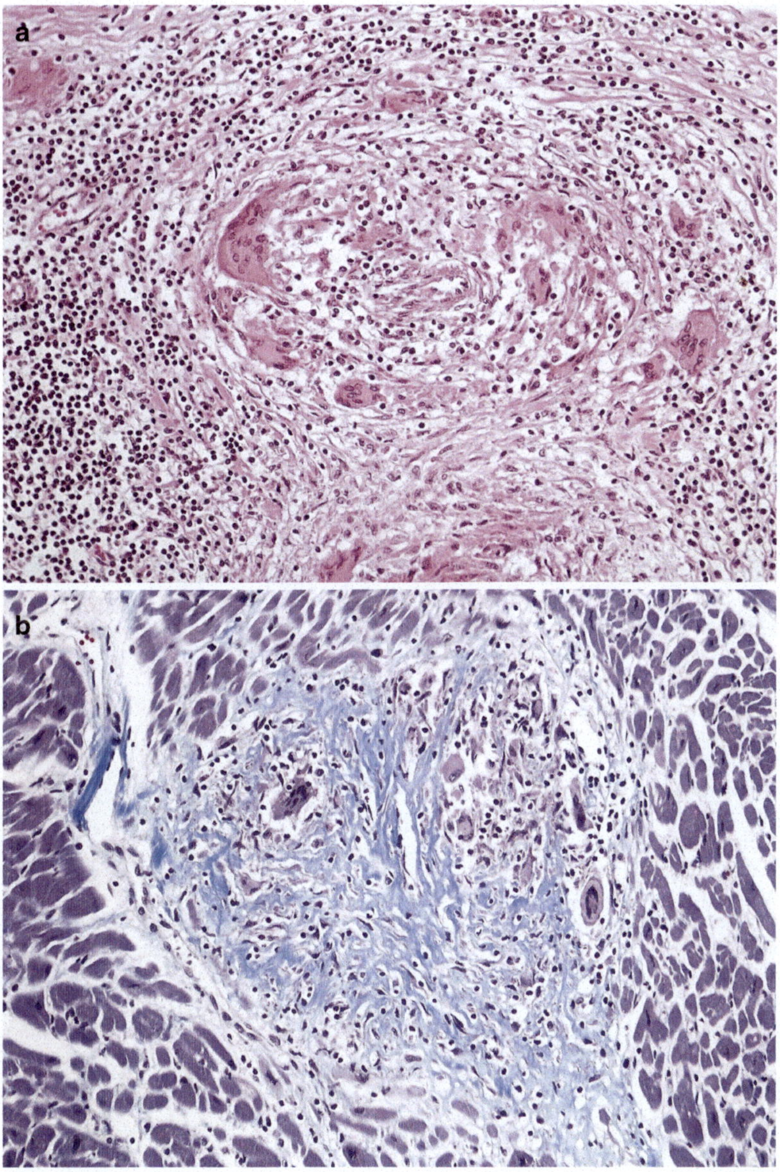

Fig. 9.23 A 51-year-old male, alcoholic. Heart weight, 717 g. Microscopic studies revealed pulmonary, lymph nodes, splenic and hepatic sarcoidosis. (**a**) Sarcoid granuloma in the AV node of the conduction system. (**b**) Evolution to fibrosis can be appreciated (Masson's trichrome)

9.9 Rheumatic Myocarditis

Rheumatic heart disease typically occurs in underdeveloped or developing countries. Although the pathogeny of the disease is not well established, it is accepted that rheumatic myocarditis is a late complication of pharyngitis caused by group A-haemolytic streptococci and actually, treatment with antibiotics considerably diminishes its incidence.

During acute rheumatic fever, any one of the three layers of the heart can be affected (pancarditis). In the myocardium, the typical lesion present is the Aschoff body that appears disperse in the interstitial connective tissue, predominantly perivascular. Aschoff bodies are fibroinflammatory lesions composed of mononuclear inflammatory cells (lymphocytes, macrophages and plasma cells), with focal fibrinoid necrosis (Fig. 9.24).

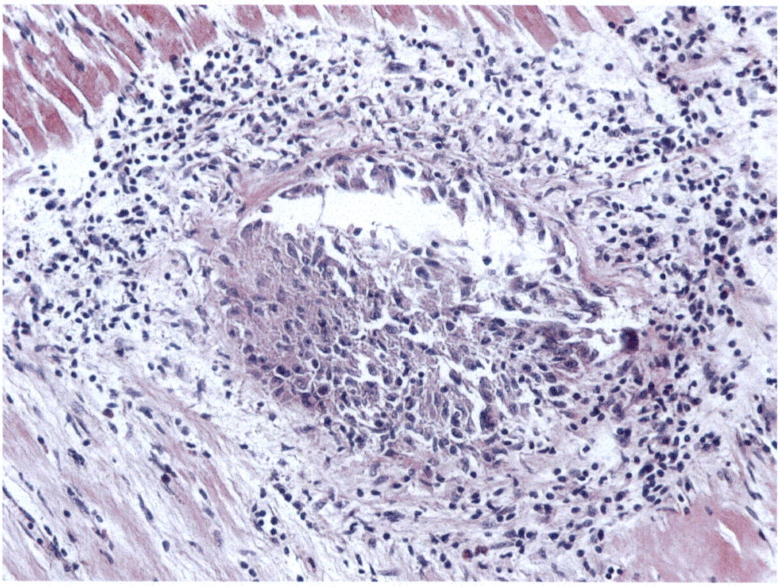

Fig. 9.24 A 76-year-old woman with a clinical history of deep venous thrombosis and hypertension. Presence of fibrosis in the perivascular interstitium can be appreciated in the interventricular septum and in the LV, with numerous foci of inflammatory infiltrates composed of lymphocytes, histiocytes and scarce leukocytes (neutrophils with some eosinophils), that form granulomas with fibrinoid debris in its interior (Aschoff nodules)

Bibliography

Abdel-Aty H, Boye P, Zagrosek A, et al. Diagnostic performance of cardiovascular magnetic resonance in patients with suspected acute myocarditis: comparison of different approaches. J Am Coll Cardiol. 2005; 45:1815–22.

Amabile N, Fraisse A, Bouvenot J, Chetaille P, Ovaert C. Outcome of acute fulminant myocarditis in children. Heart. 2006;92:1269–73.

Aretz HT, Billingham ME, Edwards WD, Factor SM, Fallon JT, Fenoglio Jr JJ, et al. Myocarditis. A histopathologic definition and classification. Am J Cardiovasc Pathol. 1987;1:3–14.

Basso C, Calabrese F, Angelini A, Carturan E, Thiene G. Classification and histological, immunohistochemical, and molecular diagnosis of inflammatory myocardial disease. Heart Fail Rev. 2013;18:673–81.

Blauwet LA, Cooper LT. Myocarditis. Prog Cardiovasc Dis. 2010;52:274–88.

Blauwet LA, Cooper LT. Idiopathic giant cell myocarditis and cardiac sarcoidosis. Heart Fail Rev. 2013; 18(6):733–46.

Caforio ALO, Pankuweit S, Arbustini E, Basso C, Gimeno-Blanes J, Felix SB, et al. Current state of knowledge on aetiology, diagnosis, management, and therapy of myocarditis: a position statement of the European Society of Cardiology Working Group on Myocardial and Pericardial Diseases. Eur Heart J. 2013;34:2636–48.

Calabrese F, Rigo E, Milanesi O, Boffa GM, Angelini A, Valente M, Thiene G. Molecular diagnosis of myocarditis and dilated cardiomyopathy in children; clinicopathologic features and prognostic implications. Diagn Mol Pathol. 2002;11:212–21.

Casali MB, Lazzaro A, Gentile G, Blandino A, Robnchi E, Zoja R. Forensic grading of myocarditis: an experimental contribution to the distinction between lethal miocarditis and incidental myocarditis. Forensic Sci Int. 2012;223:78–86.

Cooper LT. Myocarditis. N Engl J Med. 2009;360: 1526–38.

Cooper LT, Baughman KL, Feldman AM, et al. The role of endomyocardial biopsy in the management of cardiovascular disease: a scientific statement from the American Heart Association, the American College of Cardiology, and the European Society of Cardiology. Circulation. 2007;116:2216–33.

Felker GM, Boehmer JP, Hruban RH, Hutchins GM, Kasper EK, Baughman KL, Hare JM. Echocardiographic findings in fulminant and acute myocarditis. J Am Coll Cardiol. 2000;36:227–32.

Gaaloul I, Riabi S, Harrath R, Evans M, Salem NH, Mlayeh S, Huber S, Aouni M. Sudden unexpected death related to enterovirus myocarditis: histopathology, immunohistochemistry and molecular pathology diagnosis at post-mortem. BMC Infect Dis. 2012; 12:212.

Gupta S, Markham DW, Drazner MH, Mammen PP. Fulminant myocarditis. Nat Clin Pract Cardiovasc Med. 2008;5:693–706.

Kindermann I, Kindermann M, Kandolf R, et al. Predictors of outcome in patients with suspected myocarditis. Circulation. 2008;118:639–48.

Kindermann I, Barth C, Mahfoud F, Ukena C, Lenski M, Yilmaz A, et al. Update on myocarditis. J Am Coll Cardiol. 2012;59:779–92.

Lieberman EB, Hutchins GM, Herskowitz A, Rose NR, Baughman KL. Clinicopathologic description of myocarditis. J Am Coll Cardiol. 1991;18:1617–26.

Lucena J, et al. Miocarditis aguda y muerte súbita en jóvenes. Presentación de un caso. Cuad Med Forense. 2007;13(48–49):191–6.

Magnani JM, William G. Myocarditis: current trends in diagnosis and treatment. Circulation. 2006;113: 876–90.

Nielsen TS, Nyengaard JR, Møller J, Banner J, Nielsen LP, Thorngren Baandrup UT. Quantitative diagnosis of lymphocytic myocarditis in forensic medicine. Forensic Sci Int. 2014;238:9–15.

Schultz JC, Hilliard AA, Cooper Jr LT, Rihal CS. Diagnosis and treatment of viral myocarditis. Mayo Clin Proc. 2009;84:1001–9.

Zhang M, Tavora F, Zhang Y, Ripple M, Fowler D, Li L, Zhao Z, Burke A. The role of focal myocardial inflammation in sudden unexpected cardiac and noncardiac deaths – a clinicopathological study. Int J Leg Med. 2013;127:131–8.

Pathology of the Cardiac Conduction System

10

M Paz Suárez-Mier and Ignacio Fernández-Lozano

Contents

M.P. Suárez-Mier (✉)
Histopathology Service, National Institute of Toxicology and Forensic Sciences, Las Rozas (Madrid), Spain
e-mail: mariapaz.suarez@justicia.es

I. Fernández-Lozano
Arrhythmia Section, Cardiology Department, Hospital Universitario Puerta de Hierro, Majadahonda (Madrid), Spain

Universidad Autónoma de Madrid, Madrid, Spain

10.1 Introduction

The cardiac conduction system (CCS) is constituted by specialised myocardial cells responsible for the initiation of the cardiac impulse and its fast propagation throughout the myocardium, in a regular and orderly manner, causing the heart to contract efficiently and send blood to the great arteries. The CCS consists of the *sinus node* (discovered by Keith and Flack in 1906), which is the normal pacemaker of the heart, the *internodal pathways*, and the *atrioventricular* (AV) *system* including the atrioventricular node (described by Tawara in 1906), the bundle of His (described by Wilhelm His in 1893), and the ventricular branches (described by Purkinje in 1845).

Compared to other cardiac structures, little work has been published regarding the CCS, and this is due mainly to two reasons: the tedious methods of examination used by the first anatomists and pathologists devoted to the study of the CCS; and the lack of correlation between clinical manifestations and pathological findings. However, the CCS has been broadly studied in relation to sudden death, especially in the 1970s (Thomas James) and the 1980s (Saroja Bharati and Maurice Lev), and recently due to the development of radiofrequency ablation techniques to correct

J.S. Lucena et al. (eds.), *Clinico-Pathological Atlas of Cardiovascular Diseases*,
DOI 10.1007/978-3-319-11146-9_10, © Springer International Publishing Switzerland 2015

arrhythmias initiated by accessory pathways, and atrial or nodal tachycardia.

In this chapter we will describe, based on a simplified method of study, the normal histology of the CCS, several anatomical variations, age-related alterations, and lesions encountered in forensic autopsies, mainly associated to heart block and sudden death.

10.2 Anatomy of the Cardiac Conduction System

Sinus Node The sinus node or sinoatrial (SA) node is a superficial structure located just below the epicardium and close to the endocardium, at the lateral junction of the right atrial appendage and the entrance of the superior vena cava in the smooth slope of the *crista terminalis*. It is situated subepicardially, and arranges as a wedge with its base opposed to the *crista terminalis*. It generally extends towards the orifice of the inferior vena cava, although in 10 % of hearts the node forms a "horseshoe" arrangement around the orifice of the superior vena cava.

Internodal Pathways They do not have proper morphology, but are preferential tracts for the transmission of the sinus node impulse to the AV node by virtue of the disposition of the atrial muscle bands that surround the natural orifices. Classically, three internodal tracts have been described: the posterior tract or Thorel's bundle that passes through the *crista terminalis*; the middle tract or Wenckebach's bundle, behind the superior vena cava; and, the anterior tract or Bachman's bundle that runs from the anterior margin of the sinus node toward the interatrial septum and divides in two fascicles, one that travels posteriorly around the aortic root to the left atrium, and the other that travels through the interatrial septum to the AV node.

Atrioventricular Node It is located above the tricuspid valve annulus, anterior to the *ostium* of the coronary sinus at the base of the triangle of Koch (Fig. 10.1).

Bundle of His It is the continuation of the distal part of the AV node, and is immersed in the fibrous skeleton. It travels forward over the crest of the interventricular septum finishing ahead of the membranous septum, at the vertex of the triangle of Koch (Fig. 10.1).

Left Bundle Branches They arise from the bundle of His in its forward course and are distributed as a continuous "fan" of cells along the left ventricular septal surface.

Right Bundle Branch It is the direct continuation of the bundle of His after the origin of the last of the left bundle branches.

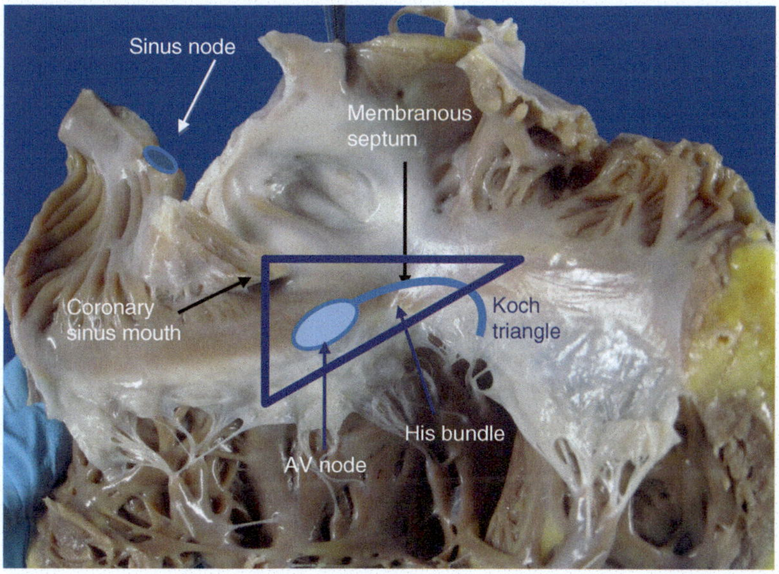

Fig. 10.1 Anatomy of the AV node and bundle of His in relation to the *ostium* of the coronary sinus, the septal leaflet of the tricuspid valve and the membranous septum. The limits of the triangle of Koch are defined by these structures and by the tendon of Todaro, a subendocardial fibrous cord that constitutes the superior side of the triangle

10.3 Method of Study

Sinus Node In order to study the SA node, the right atrium must be opened with scissors, from the entrance of the inferior vena cava towards the atrial appendage (following the dotted line shown in Fig. 10.2a). Next, the atrial wall is opened between both venae cavae, but sparing the anatomical location of the SA node. The *crista terminalis* is removed as a whole and is cross-sectioned every 3–4 mm to obtain 5 or 6 sections (Fig. 10.3).

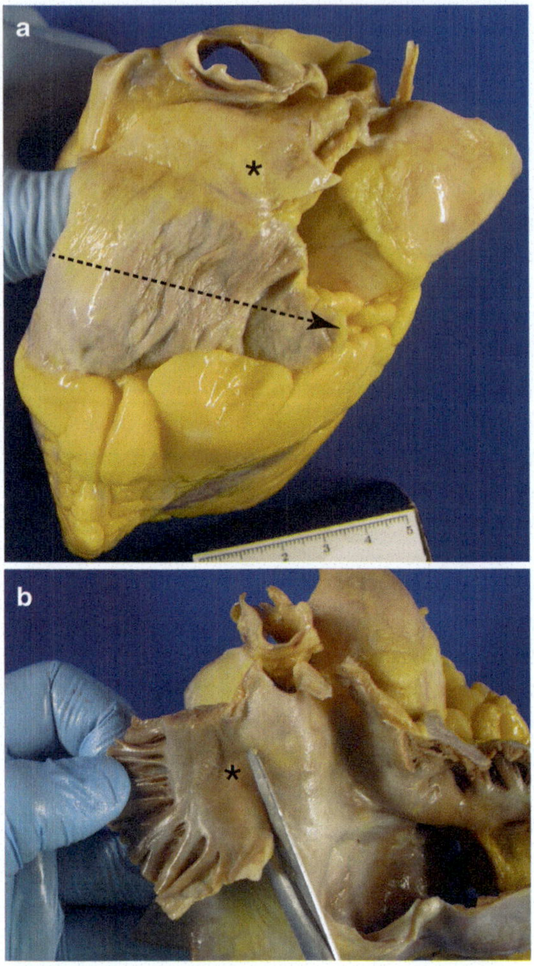

Fig. 10.2 Location of the SA node (*): (**a**) as seen from the heart's surface and (**b**) from the internal side of the right atrium

Atrioventricular System A block of tissue is obtained containing the atrioventricular junction at the level of the insertion of the septal leaflet of the tricuspid valve, from the entrance of the coronary sinus to the septum past the membranous septum (Figs. 10.4 and 10.5a). This block should include 1 cm of the interatrial septum above the tricuspid valve annulus and around 2 cm of the interventricular septum below (Fig. 10.5). This block of tissue is cut every 2–3 mm, obtaining between 6 and 8 sections that are conventionally processed and embedded in paraffin (Figs. 10.5b and 10.6).

Each of the sections obtained is automatically processed according to standard methods and the slides are stained with haematoxylin-eosin and Masson's trichrome. This is the simplest method that allows visualisation of all the elements of the CCS although it has many limitations, especially to identify anomalous pathways (see below). The optimal method (only possible in research laboratories) is the manual embedding of the tissue in paraffin and the serial sectioning of the entire block, obtaining more than 1,000 slides for every conduction system. An intermediate approach consists of obtaining serial sections of the blocks that have been automatically processed, and staining every 15–20 slides.

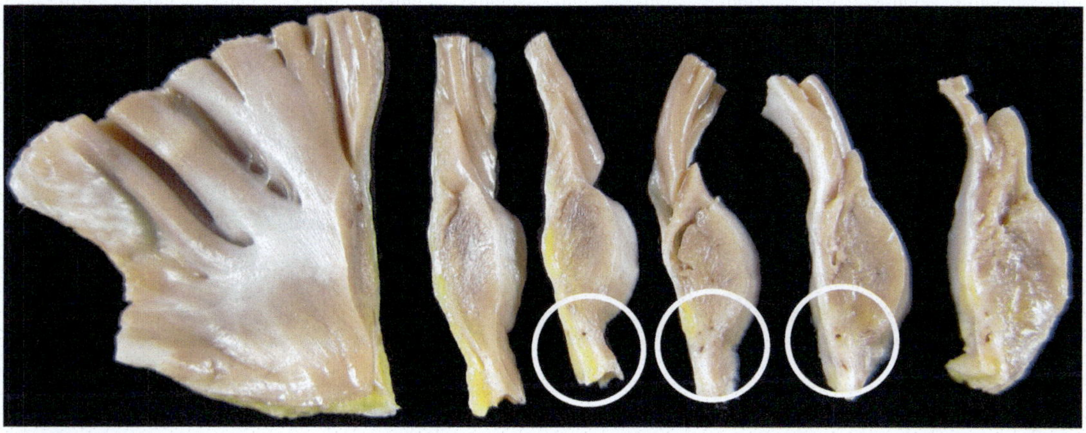

Fig. 10.3 Serial sectioning of the *crista terminalis*. Gross examination allows visualisation of the SA nodal artery and, on some occasions, of the surrounding whitish nodal tissue (*circled*)

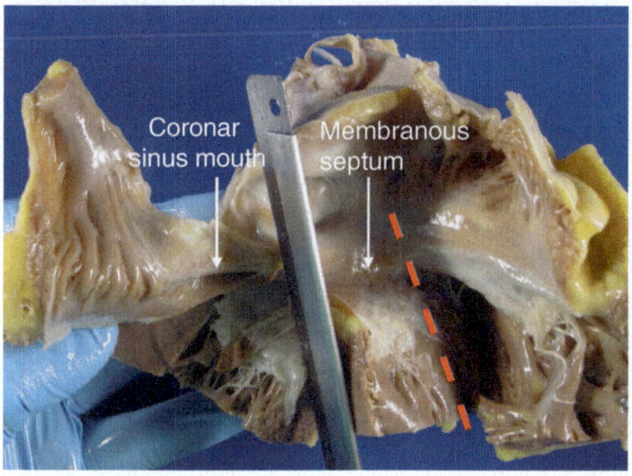

Fig. 10.4 Selection of the block of tissue to study the atrioventricular system. The knife blade points to the excision past the *ostium* of the coronary sinus. The *discontinuous line* indicates the direction of the excision, parallel to the previous one, in the anterior part of the septum

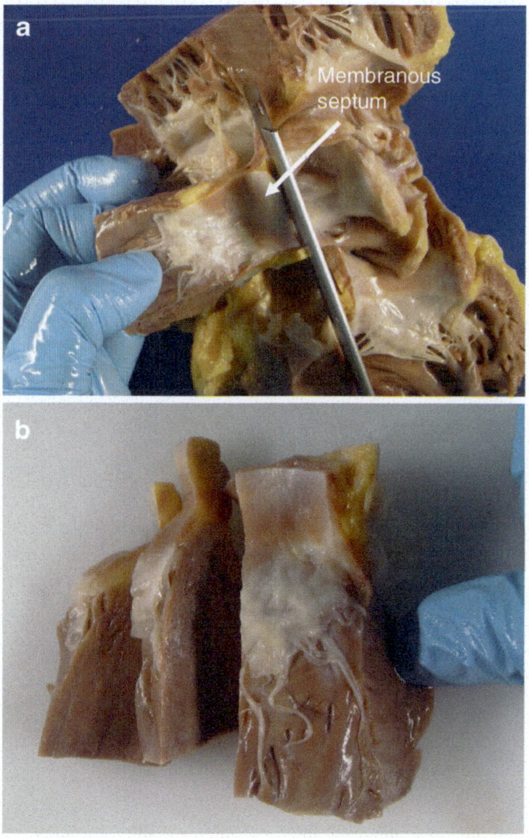

Fig. 10.5 (**a**) Superior excision of the block, 1 cm above the tricuspid valve annulus. (**b**) Further sectioning of the block in sections 2–3 mm in width

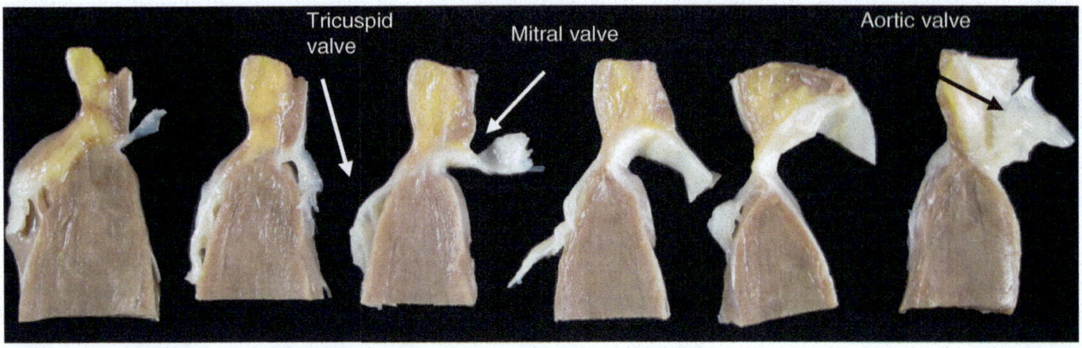

Fig. 10.6 Image showing the different sections obtained after serial sectioning of the block

10.4 Histology of the Cardiac Conduction System

10.4.1 The Normal Cardiac Conduction System

Sinus Node The SA node is composed of fusiform cells, with a diameter smaller than those of the atrial cells, and the interstitium contains large amounts of elastic tissue and collagen, which increase with age. The nodal tissue is typically crossed by the SA nodal artery, which can appear as a single artery or can branch into two or more arteries, and frequently contains many nervous tracts and neurovegetative ganglions, especially in the periphery (Fig. 10.7a).

Atrioventricular (AV) Node It is an ovoid structure with a maximum length of 5–8 mm. It can be divided into a poorly-defined portion, composed of fascicles arranged parallel to one another, and a portion that corresponds to the proper compact node that sits above the central fibrous body (Fig. 10.7b).

Bundle of His It gradually arises from the AV node, which begins penetrating the central fibrous body up to a point where all the conduction cells are completely included in it (this region is termed the *penetrating bundle*) (Fig. 10.8a). This part is very short and from the His bundle exit conduction fibres forming a cascade down the left ventricular septal surface (the left branches). This is the so-called *branching bundle*, situated beneath the commissure of the right and non-coronary aortic sigmoid cusps, below the membranous septum, adopting a triangular shape (Fig. 10.9b).

Left Bundle Branches Left bundle branches are multiple and originate from the bundle of His as a continuous "fan" of cells. The left bundle branches primarily travel subendocardially along the left ventricular septum, and an interstitium of lax connective tissue separates the left bundle branches from the contractile myocardium. The cells vary in shape in comparison with the ventricular cells and can be compact or vacuolated. These are the so-called Purkinje cells (Figs. 10.8b and 10.9).

Bifurcation It is the end of the His bundle, from where the last of the left bundle branches and the right bundle branch emerge.

Right Bundle Branch It is the continuation of the bundle of His once all the left branches have arisen (Fig. 10.9b). Moving down the right side of the interventricular septum, it travels intramyocardially near its origin, whereas in the medium and lower thirds of the septum it becomes a subendocardial structure. Histologically, it is similar to the ventricular myocardium and it commonly contains one blood vessel and adipose tissue.

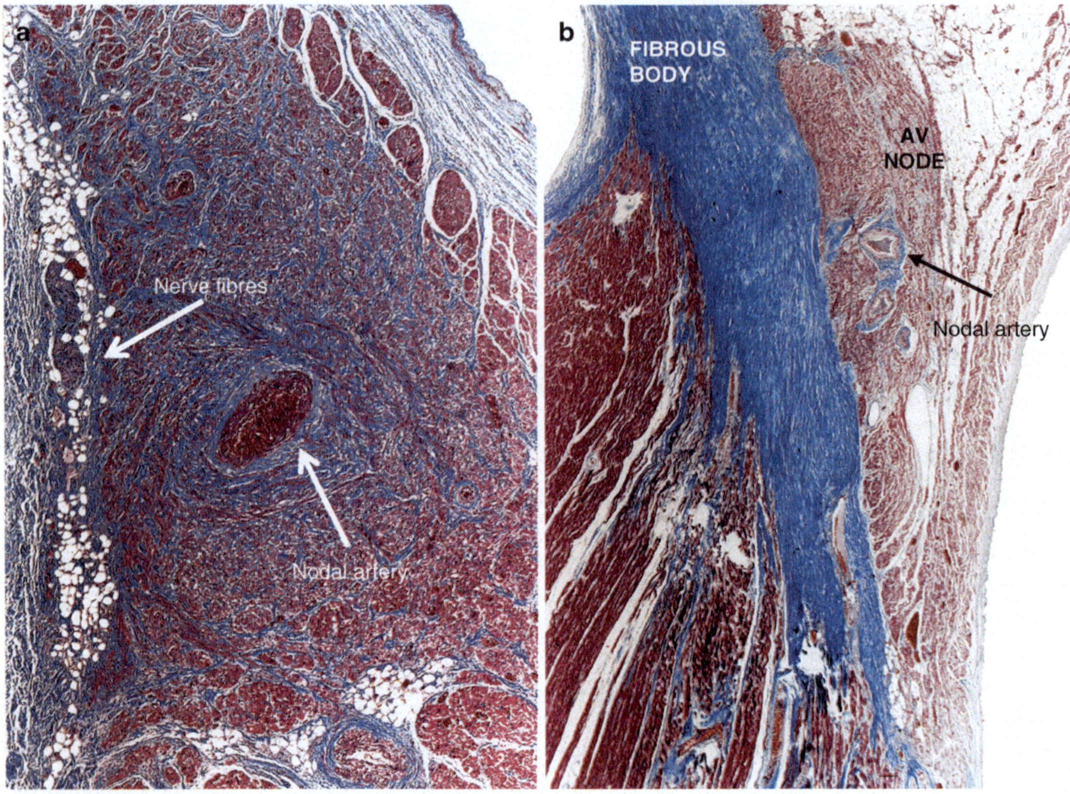

Fig. 10.7 (**a**) Normal sinus node. (**b**) Position of the AV node above the central fibrous body

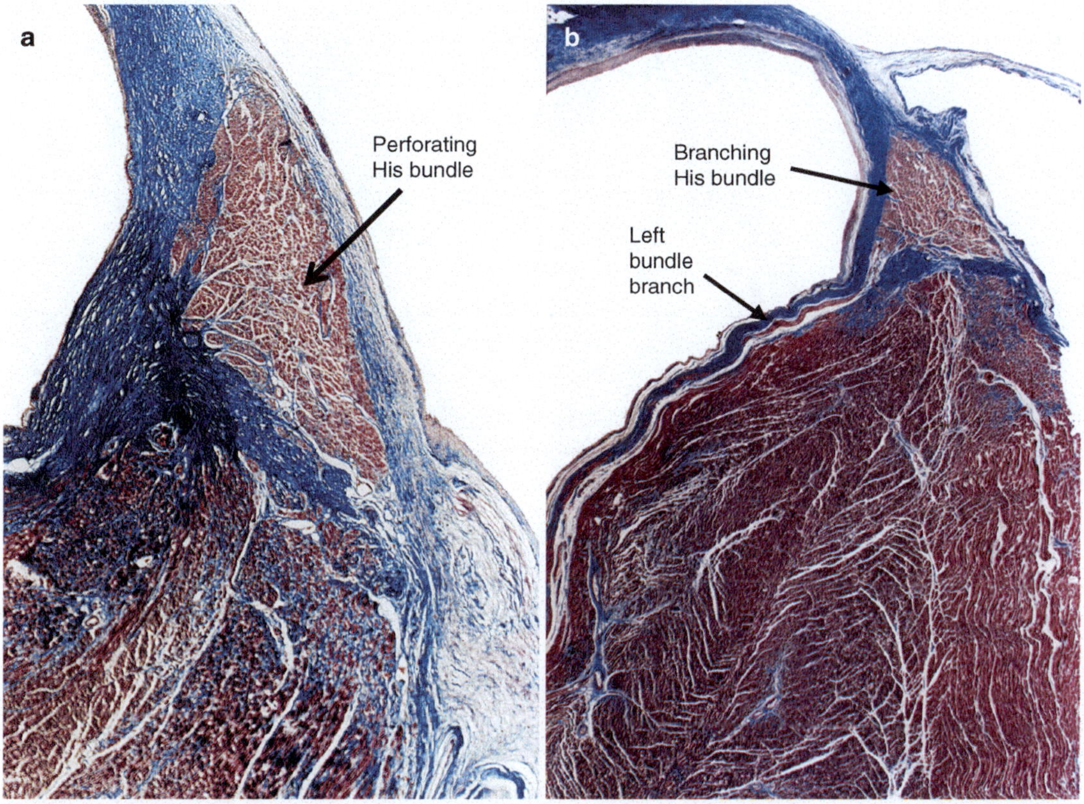

Fig. 10.8 (**a**) Penetrating portion of the bundle of His within the central fibrous body. (**b**) Branching portion of the bundle of His situated at the crest of the interventricular septum, and one of the left branches travelling down the left side of the septum

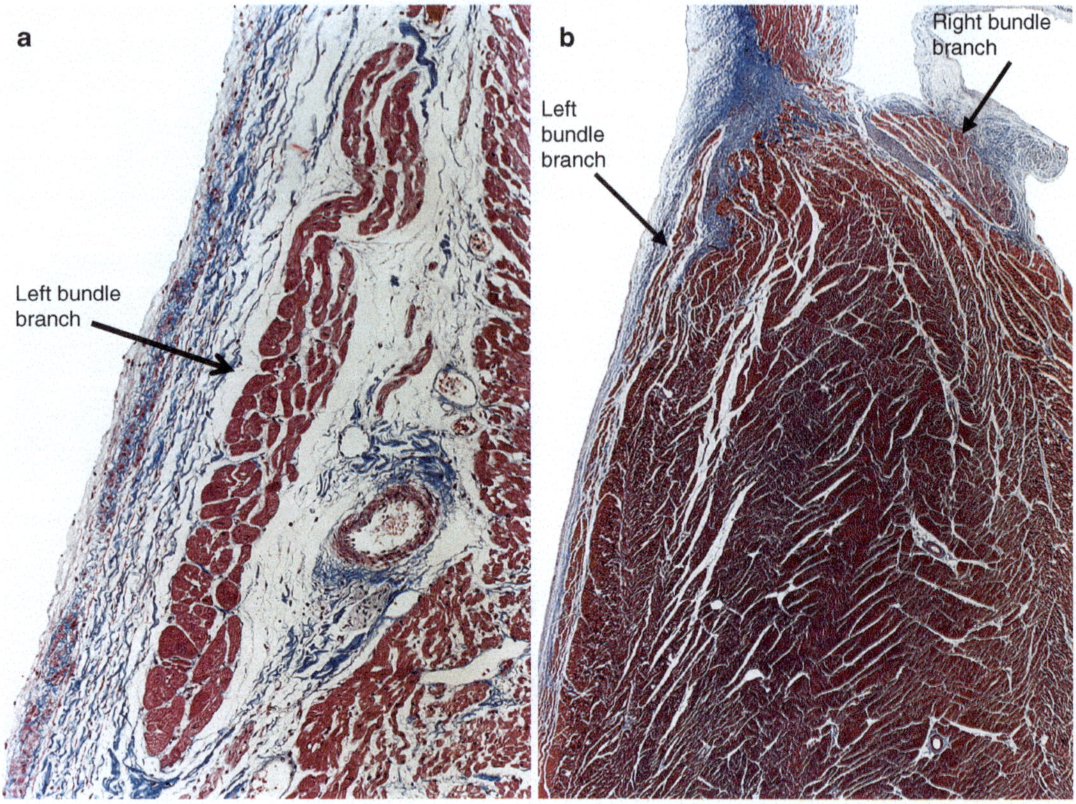

Fig. 10.9 (**a**) Detail of the left bundle branch under the endocardium. (**b**) Image showing the emergence of the last of the left bundle branches and the initial portion of the right bundle branch (Masson's trichrome)

10.4.2 Normal Variations

The study of hundreds of hearts from persons who died due to non-cardiac causes has allowed the recognition of several patterns of morphology, which, despite the fact that they do not follow the previously described anatomy, have no functional consequences in the majority of cases. The most common variations are as follows:

10.4.2.1 Persistent Foetal Dispersion of the AV Node and His Bundle Fragmentation

In many hearts, the conduction system does not form a compact structure but rather forms islands or tongues of conduction tissue, adopting an "archipelago" type morphology. This variation for the AV node is known as persistent foetal dispersion since this is the typical morphology of the AV node during normal development (Fig. 10.10a). The septation of the His bundle is characterised by fragmentation of the main bundle forming loops (Fig. 10.10b). For some time, these findings in adults were considered the cause of death in cases of sudden cardiac death where no other structural alterations were present; however, it was later established that these were normal variations of the CCS.

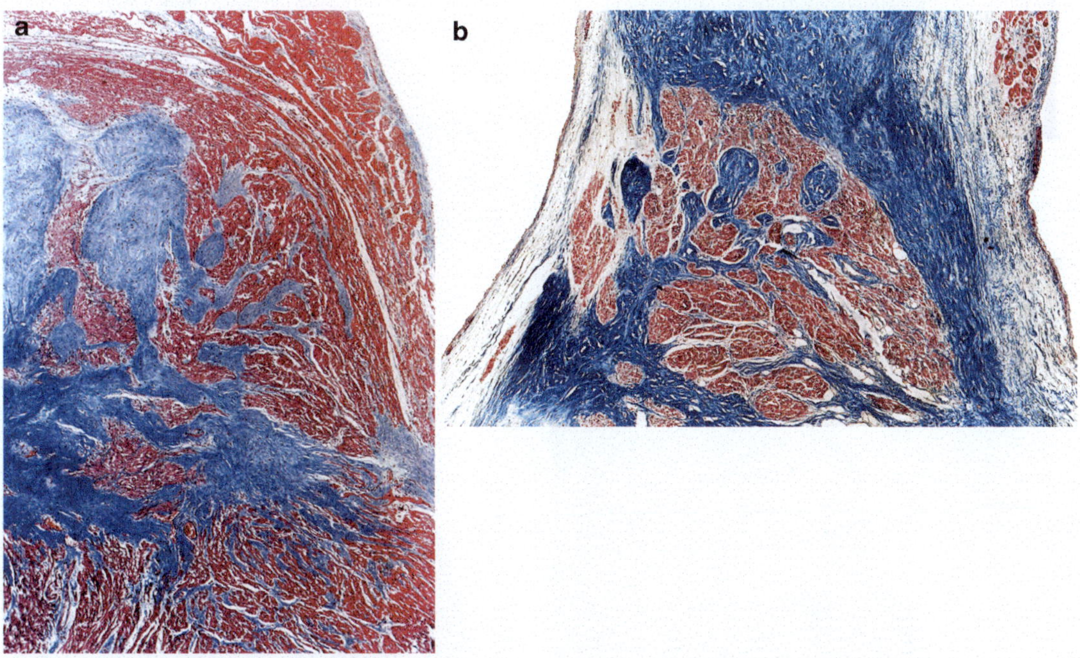

Fig. 10.10 (**a**) A 2-year-old boy who died suddenly from an endodermal sinus tumour in the mediastinum. The AV node penetrates the central fibrous body as tongues of conduction tissue. (**b**) A 48-year-old male who died from severe brain injury after falling from a height of 3 m. The His bundle presents a polylobulated morphology (Masson's trichrome) *(Image reproduced with permission of* Suárez-Mier and Gamallo 1998)

10.4.2.2 Variations in the Position of the His Bundle

It is relatively common for the bundle of His, especially its branching portion, not to be located as shown in Fig. 10.8, astride the interventricular septum, but rather in different positions, as illustrated in Fig. 10.11. Positions shown in C and D are very uncommon and their functional consequences are uncertain.

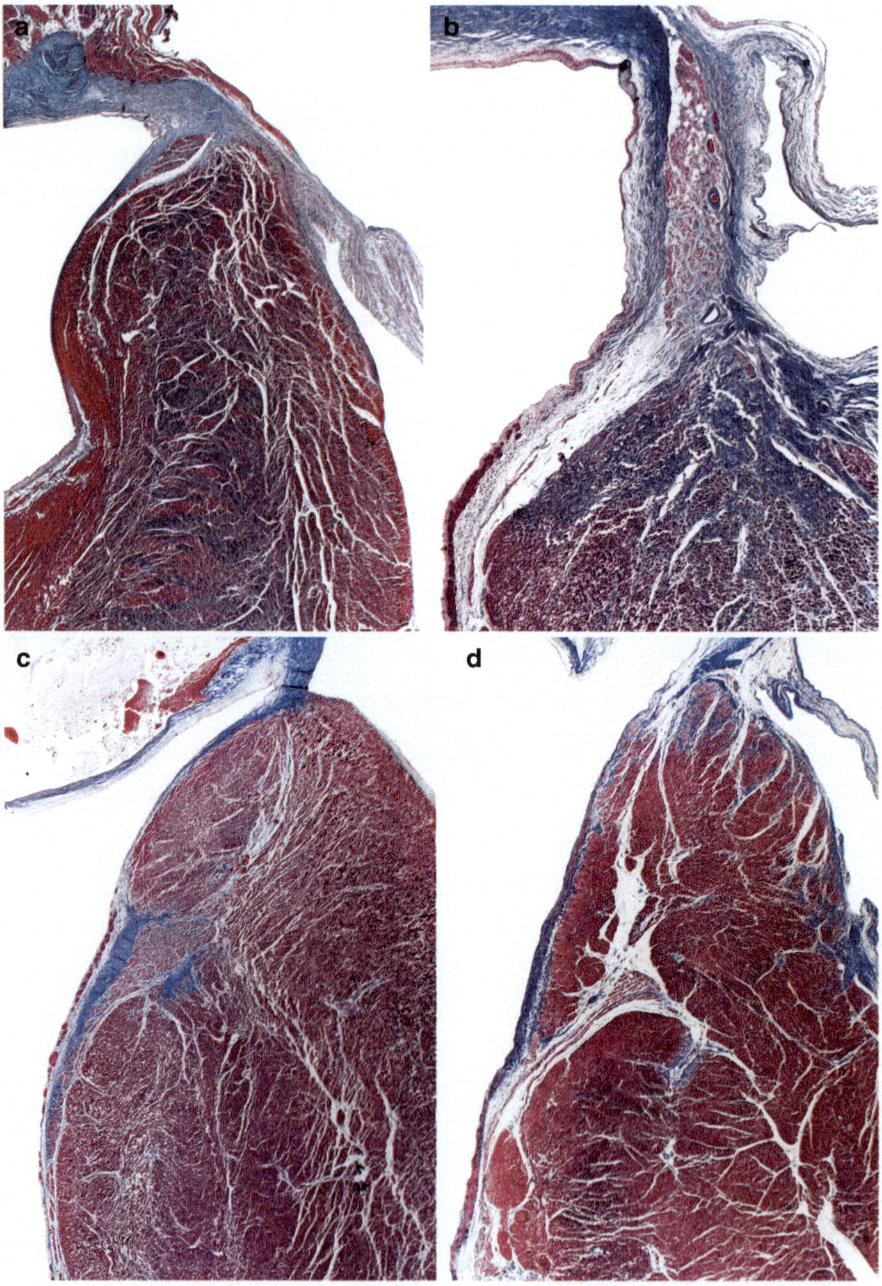

Fig. 10.11 (**a**) A 21-year-old male who died suddenly with no known medical history. Heart weight, 403 g. Bundle of His situated above the septum, displaced to the left. (**b**) An 81-year-old woman with no cardiac history. The His bundle is situated in the membranous septum. (**c**) A 32-year-old male. Work-related accident involving crushing. The bundle of His is in a much-lowered position, at the left side of the septum. (**d**) A 55-year-old male. His heart showed a healed anterolateral infarct with 90 % luminal stenosis in the left coronary trunk. The bundle of His is much lowered down and intramural (Masson's trichrome)

10.4.3 Age-Related Changes

10.4.3.1 Adipose Tissue Infiltration

Adipose tissue infiltration is a degenerative alteration of the CCS that appears around the age of 50 years and involves all the structures of the conduction system, with the nodes being the most affected structure followed by the His bundle bifurcation (Fig. 10.12).

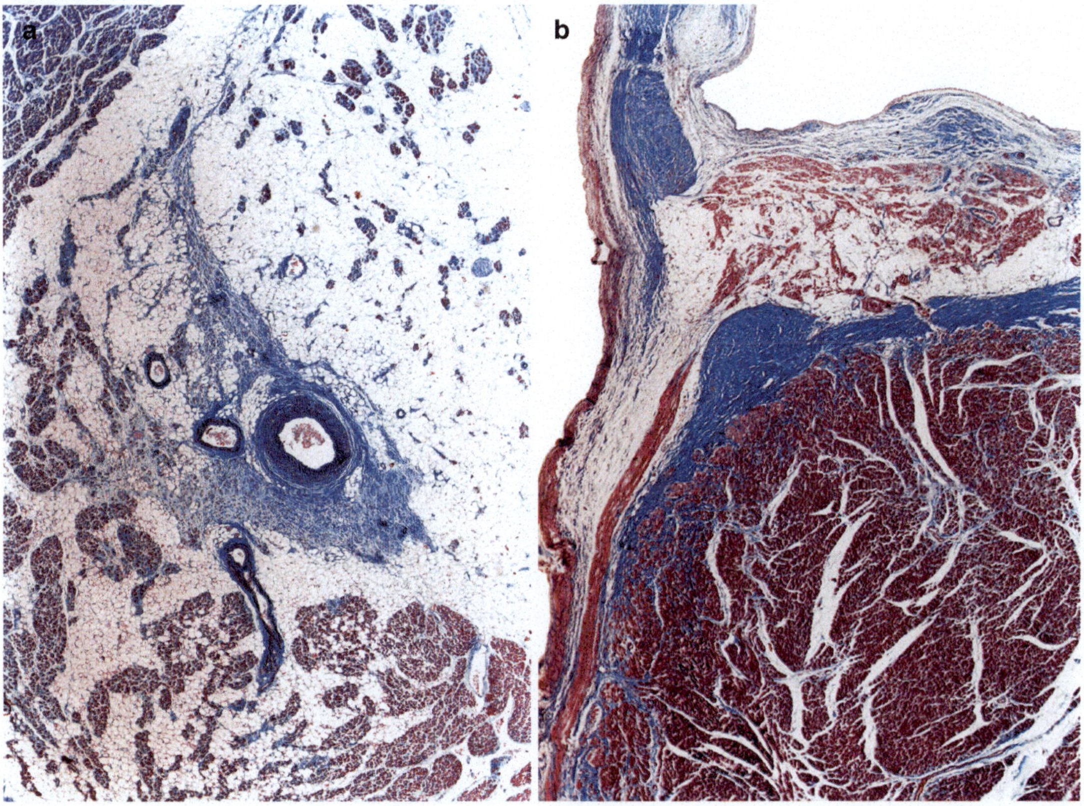

Fig. 10.12 (**a**) Marked adipose tissue infiltration in the sinus node, in a 68-year-old male with a history of chronic bronchitis. (**b**) Adipose tissue infiltration in the bifurcation of the His bundle in a 59-year-old male who died suddenly from ischaemic heart disease (Masson's trichrome)

10.4.3.2 Fibrosis

Fibrosis of the CCS is an age-related alteration, although it has also been associated with sick sinus syndrome, congenital block, collagen diseases and chronic atrioventricular block (see below) (Fig. 10.13).

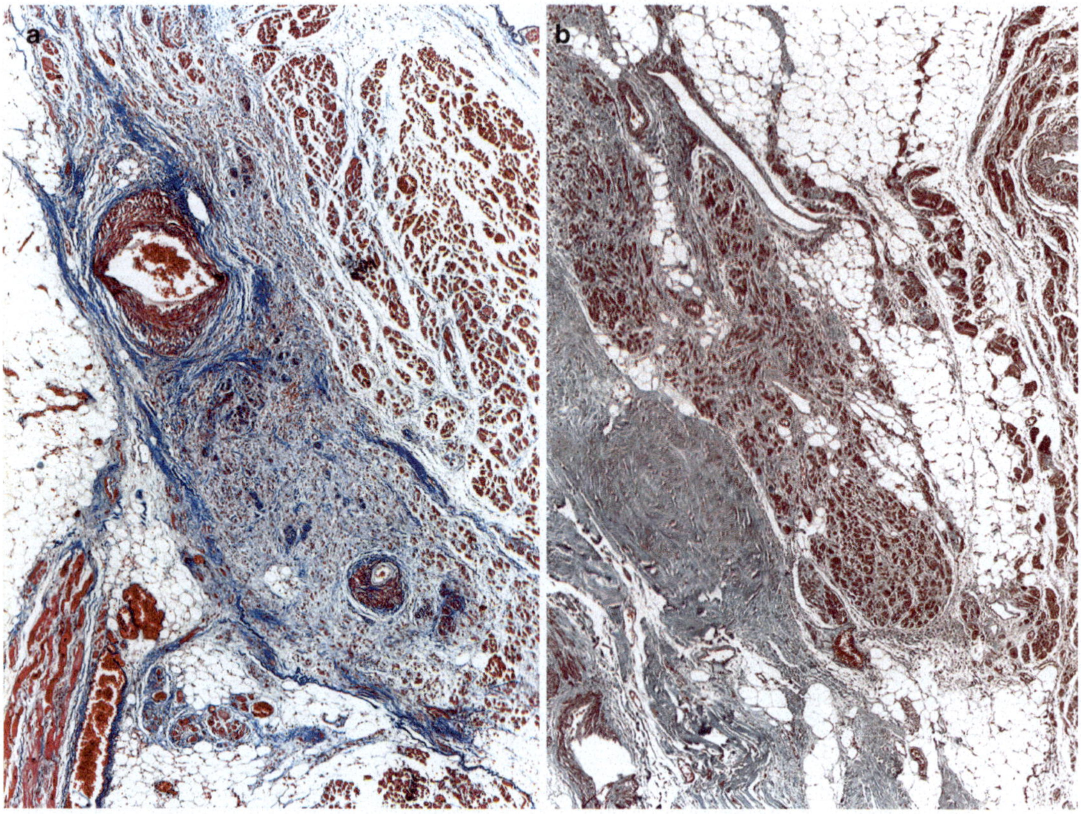

Fig. 10.13 (**a**) Fibrosis of the sinus node in a 90-year-old woman. (**b**) AV node with fibrosis and mild adipose tissue infiltration in a 52-year-old male carrying a pacemaker (Masson's trichrome)

10.4.4 Haemorrhage Secondary to Resuscitation Manoeuvres

In older literature dealing with sudden cardiac death in infants, haemorrhage in the CCS was thought to have a hypoxic origin, associated with respiratory problems, and was considered a possible cause of death. However, haemorrhage in the CCS is a lesion commonly seen in individuals who suffered sudden death and received cardiopulmonary resuscitation, as a result of the compression of the heart between the sternum and the spine during cardiac massage (Fig. 10.14).

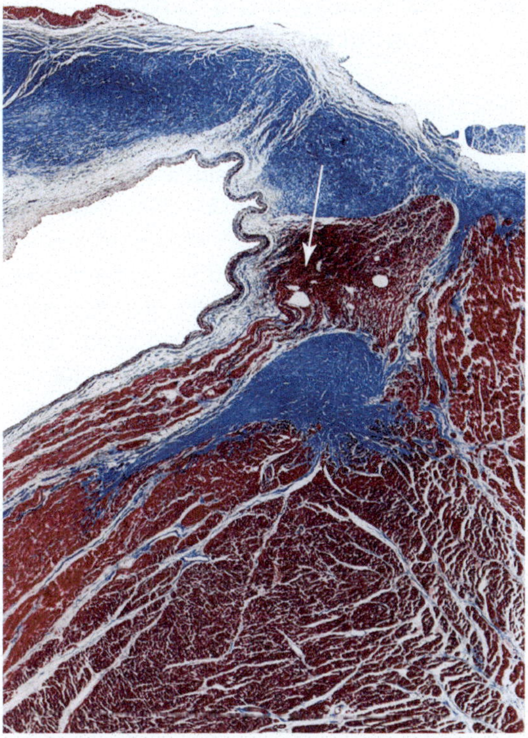

Fig. 10.14 Haemorrhage in the bundle of His (*arrow*) (Masson's trichrome)

10.5 Pathology of the Cardiac Conduction System

Clinical and histologic studies of the human conduction system do not always correlate, and alterations from what it is considered the normal morphology of the CCS are in many cases not associated with functional defects; moreover, alterations in the cardiac rhythm that have been clinically demonstrated even by electrophysiological studies, may not correspond to any morphological lesions. This is especially true for the pre-excitation syndromes, where identification of the anomalous conduction tracts between the atria and ventricles is extremely difficult. In this section we will review some pathologic conditions localised to the conduction system and others that, despite being non-specific, can manifest as conduction blocks and/or sudden death.

10.5.1 Sick Sinus Syndrome

The sinus node is the area of the heart that under normal conditions controls the cardiac rhythm. Sick sinus syndrome is the name given to a disorder where the sinus node does not function properly. It is generally caused by the aging of the node and/or surrounding atrial cells. Histologically, there is a replacement of the myocardial cells by fibrous tissue (Fig. 10.13a). Dysfunction of the sinus node can manifest as prolonged sinus pauses, symptomatic sinus bradycardia, sinus arrests, sinoatrial block, asystole following cardioversion or the presence of chronotropic incompetence. Sick sinus syndrome can be clinically associated with alterations in the atrioventricular conduction system or with paroxysmal supraventricular tachycardia, known as bradycardia-tachycardia syndrome.

Mortality in patients affected by sick sinus syndrome is significant and is not always due to cardiac causes. In the MOST study, by Flaker et al. 2003, from a total of 2,010 patients, 404 (20 %) died after a medium follow-up period of 33 months. The cause of death was cardiac in 35 % of the cases, non-cardiac in 49 %, and unknown in the remaining 16 %. The main predictive factors of death were old age, male gender, weight, and presence of a previous AMI or dilated cardiomyopathy. Sick sinus syndrome has been associated with sudden death only in anecdotal cases, although the difficulty in demonstrating this correlation makes for the possibility that the association is underestimated.

Treatment involves permanent pacemaker implantation once the manifestations of the disease are associated with symptomatology.

10.5.2 Atrioventricular Block

Atrioventricular block is defined as the delay or interruption of the electric impulse between the atria and ventricles of the heart, caused by an anatomical or functional alteration in the CCS. AV block can be classified as first-degree block if the electrical impulse moves more slowly than normal through the AV node but all the auricular impulses reach the ventricles; second-degree block, when some of the electrical signals do not reach the ventricles; and third-degree or complete AV block, when no atrial stimulus reach the ventricle. In general, the severity of the symptoms increases with the degree of the block. Presyncopal episodes, syncope or malignant ventricular arrhythmias secondary to the low heart rate can occur in complete heart block associated with slow escape rhythms or asystole. The association of AV block with sudden cardiac death is better documented than in the sick sinus syndrome, although its real relevance is unknown.

Treatment implies removal of the underlying cause by changing certain medications, managing cardiac ischaemia and, when necessary, implanting a permanent pacemaker.

Some disorders that affect AV conduction are associated with specific pathological findings. The most significant of these will be described below.

10.5.2.1 Progressive Cardiac Conduction Defect

Progressive Cardiac Conduction Defect (PCCD), also called Lenègre or Lev disease, represents one of the major causes of pacemaker implantation. It is characterised by progressive alteration of cardiac conduction through the atrioventricular system which manifests with right or left bundle branch block and widening of QRS complexes on the ECG. It can evolve to complete atrioventricular block, syncope and sudden death.

PCCD has been considered a primary degenerative disease, but the existence of familial cases and its clinical presentation during childhood suggests a genetic component, at least in some cases. Several mutations in the SCN5A sodium channel gene have been described in children with AV block. Mutations result in the loss of function of the sodium channel similar to that found in Brugada syndrome, which leads to a progressive deterioration in the electrical atrioventricular conduction associated with dilated cardiomyopathy and fibrosis. The pathological findings associated to PCCD are shown in Table 10.1.

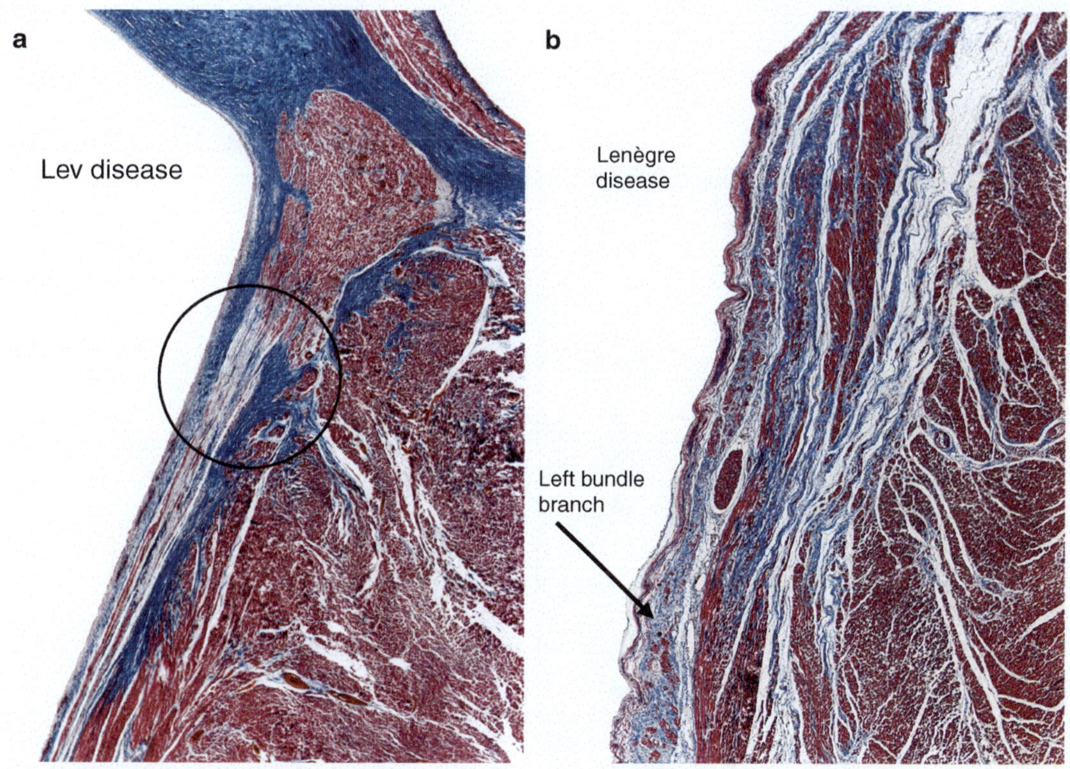

Fig. 10.15 (**a**) Loss of fibres in the proximal portion of a left bundle branch. (**b**) Fibrosis affecting the entire length of the left bundle branch (Masson's trichrome)

Table 10.1 Pathology of the progressive cardiac conduction defect

Lev disease: loss of continuity between the branching bundle of His and the majority of the left bundle branch fascicles. Conduction fibres vanish, leaving "ghost" structures (Fig. 10.15a)
Lenègre disease: loss of conduction fibres in the right bundle branch and in the middle and distal portions of the left bundle branch fascicles, with replacement fibrosis (Fig. 10.15b)

10.5.2.2 Calcific Atrioventricular Block

The conduction system is interrupted by calcium deposits that are visible on X-rays, echography or by gross examination during autopsy. Given the proximity of the penetrating bundle of His to the mitral valve annulus, membranous septum and aortic valve, nodular calcification in any of these structures can interrupt the conduction system and provoke third-degree or complete AV blocks (Box 10.1).

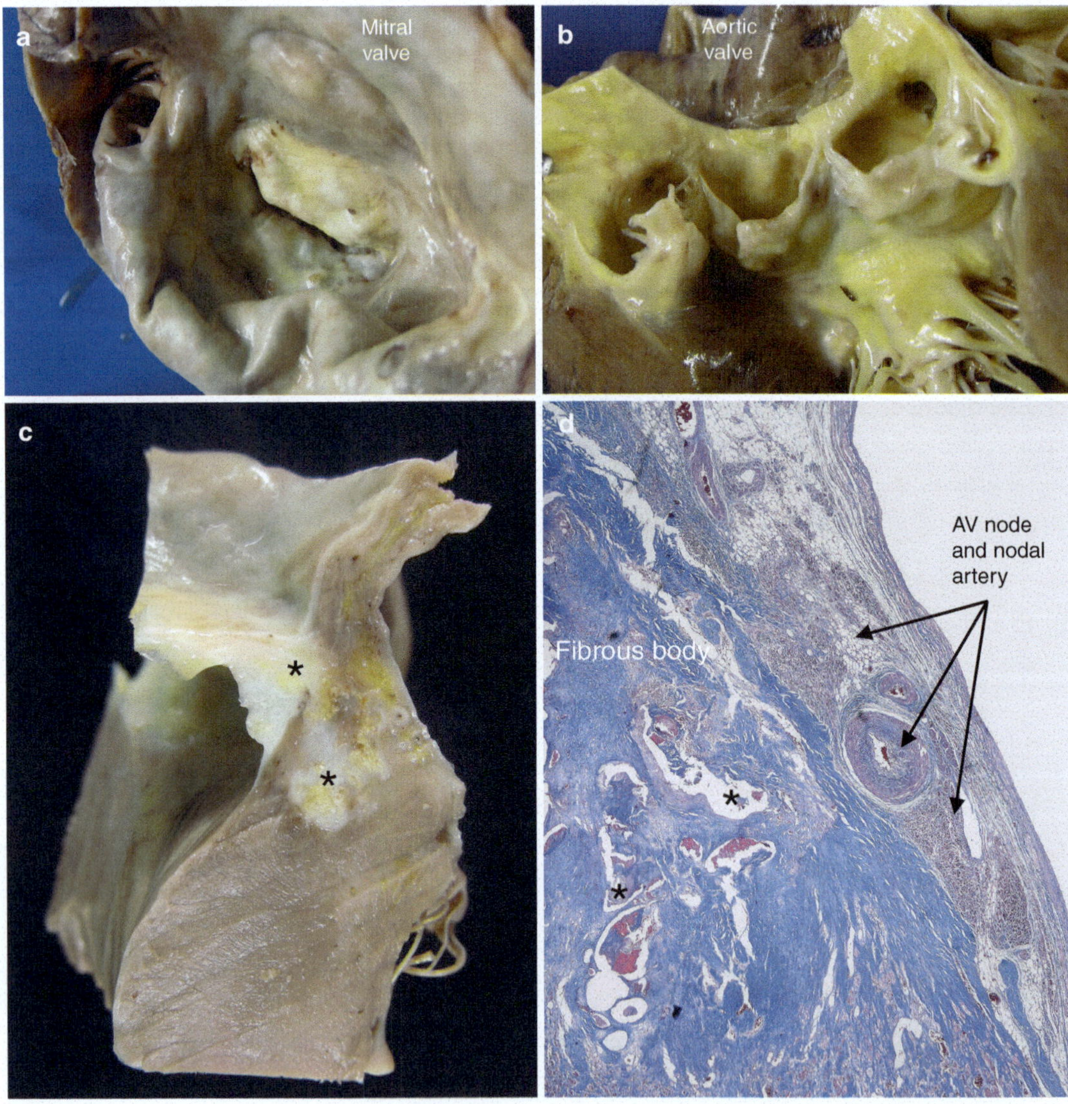

Fig. 10.16 A 77-year-old male with an implanted pacemaker. Heart weight, 615 g. (**a**, **b**) Calcific mitral and aortic valve disease, probably rheumatic in origin. (**c**) Calcification of the higher portion of the interventricular septum and of the central fibrous body (*). (**d**) Microscopic image of the AV node showing extensive calcification of the central fibrous body (*) (Masson's trichrome)

Box 10.1 Pathology of the Calcific AV Block

The pathology of the calcific AV block consists of large calcified nodules in the annulus and anterior leaflet of the mitral valve, which frequently extend to the base of the membranous septum and to the superior portion of the interventricular septum (Fig. 10.16).

10.5.2.3 Atrioventricular Node Tumour

The atrioventricular node tumour locates at the base of the interatrial septum, compromising the AV node, although in some exceptional situations it can locate outside the node. The origin of the tumour is endodermic and has been given several names depending on the different histogenetic theories: mesothelioma, benign congenital polycystic tumour of the AV node, endodermic heterotopia of the AV node and tumour of the AV nodal region. These tumours can be asymptomatic throughout life or clinically manifest with AV blocks or sudden death, which can occur even if a pacemaker has been implanted. Table 10.2 summarises the pathological characteristics of AV nodes tumours.

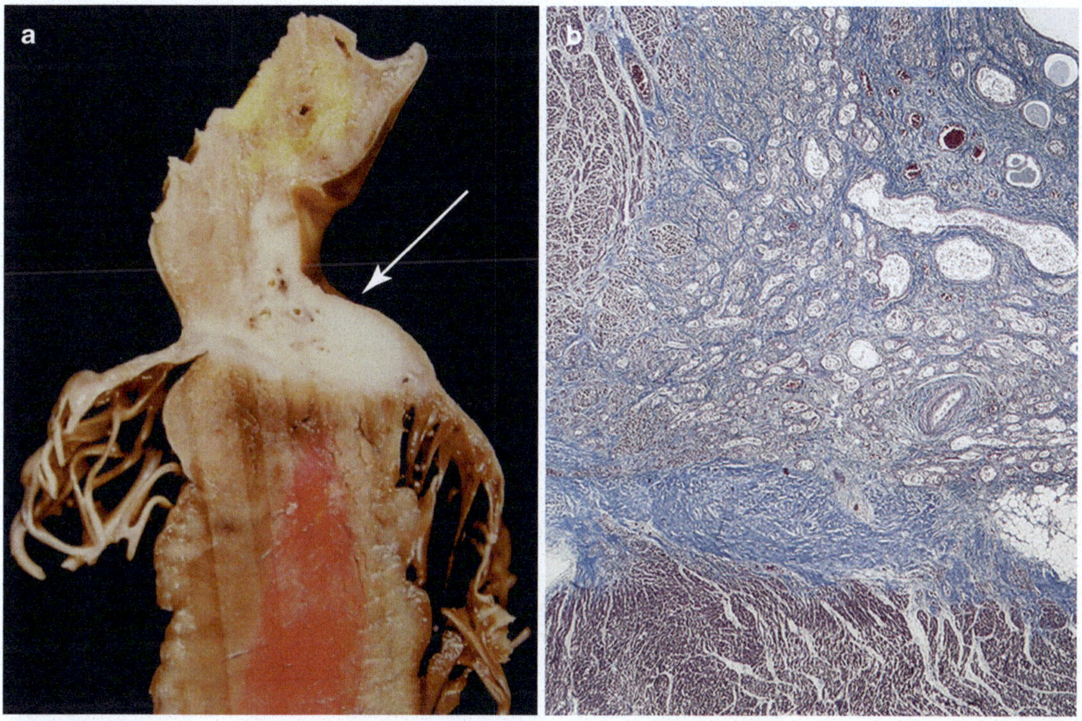

Fig. 10.17 A 33-year-old woman, with a history of congenital bradycardia and depression. She died suddenly in bed. (**a**) Gross examination of the section corresponding to the AV node, where a whitish tissue containing multiple cysts can be observed (*arrow*). (**b**) Histology shows cysts that are irregular in size and shape (some of them contain secretion material) and small nests of epithelial cells immersed in a fibrous stroma

Table 10.2 Pathology of the AV node tumour

Macroscopic features:
Whitish lesion of spongy appearance in the area of the AV node (Fig. 10.17a)
Microscopic features:
It is formed by irregular cords, solid nests of cuboidal, transitional or squamous cells, or by glandular structures or cysts in a fibrous stroma. Cysts are tortuous in shape, lined by a flattened epithelium and contain PAS-positive material (Fig. 10.17b)
Immunohistochemistry: positive for CEA, epithelial membrane antigen, cytokeratin and B72.3
Ultrastructure: the tumours present microvilli and monofilaments

10.5.2.4 Infiltrative Diseases

Sarcoidosis

Sarcoidosis is a systemic granulomatous disease of unknown aetiology that affects all population and ethnic groups, mainly adults between 20 and 40 years. It is a type of autoimmune disease that can affect any organ in the body but mostly the lungs, intrathoracic lymph nodes, eyes, skin and liver.

The heart is involved in 20–30 % of the patients, although symptomatology only appears in 5 % of those. Cardiac sarcoidosis presentation may range from asymptomatic alterations in the electric conduction to ventricular arrhythmias, constituting a cause (extremely uncommon) of sudden death. The pathological features of the cardiac sarcoidosis are shown in Table 10.3 (see also Chap. 9).

Table 10.3 Pathology of cardiac sarcoidosis

Macroscopic features (Fig. 10.18):
All cardiac layers can be involved, although the myocardium is the structure most commonly affected
In descending order: basal portion of the interventricular septum → free wall of the LV → free wall of the RV → atria → pericardium → endocardium
Well-defined whitish lesions with scar-like appearance and non-ischaemic distribution
Microscopic features (Fig. 10.19):
Non-necrotizing epithelioid granulomas with/without giant cells
Young lesions contain numerous lymphocytes whereas advanced lesions are densely collagenised with few granulomas and little chronic inflammation
It can be differentiated from giant cell myocarditis by the absence of myocyte necrosis and by its extension towards the epicardial adipose tissue

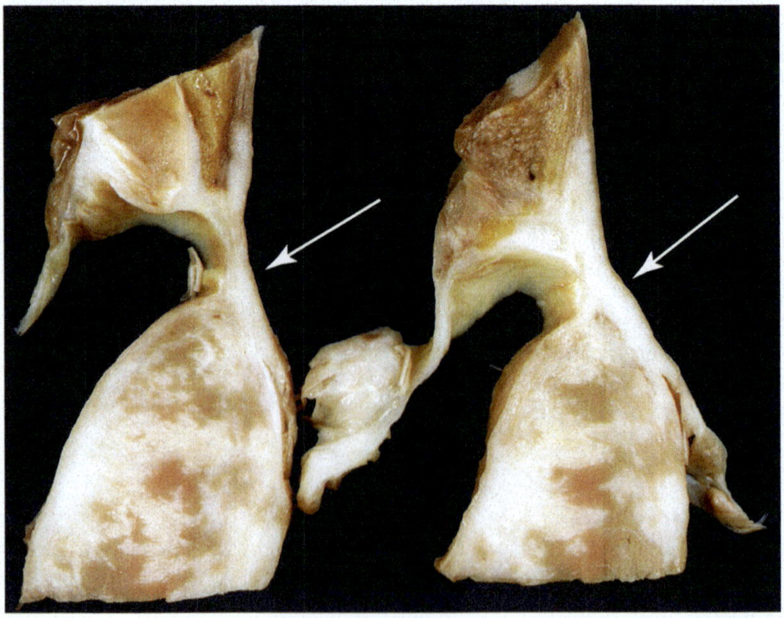

Fig. 10.18 A 61-year-old woman, with a history of third degree AV block who died 3 days after pacemaker implantation. Gross examination showed extensive replacement of the myocardium with whitish tissue, in the posterior wall of the LV and, especially in the interatrial and interventricular septum. Image shows two of the blocks obtained to study the CCS, where whitish tissue infiltration can be appreciated in the septum and AV node and His bundle regions (*arrows*)

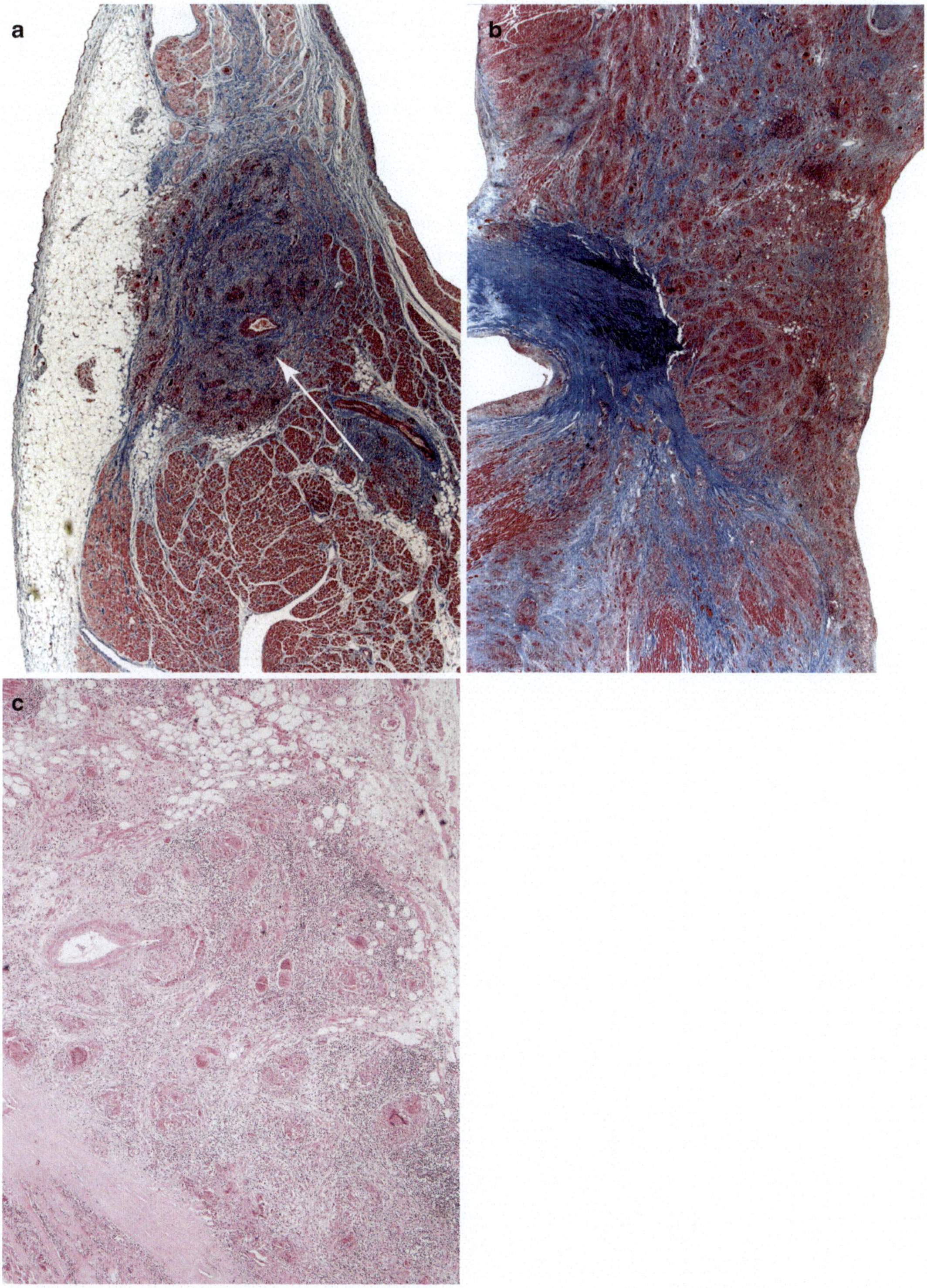

Fig. 10.19 Histology of the same case. (**a**) Image of the sinus node showing massive infiltration composed of sarcoid granulomas, sparing the nodal artery (*arrow*) (**b**) Diffuse infiltration of the interatrial septum, AV node region and central fibrous body. (**c**) Detail of the AV node infiltration with sarcoid granulomas. (**a**, **b**) (Masson's trichrome)

Amyloidosis

The term amyloidosis refers to a group of diseases of diverse aetiology and variable prognosis and treatment that all share a common feature: all are caused by the extracellular deposition of abnormal proteins known as amyloid. This insoluble protein material, resistant to proteolysis, was given the name amyloid by Virchow because it stained positive with iodine-staining techniques, similar to starch (*amylum* in Latin).

Currently, more than 20 proteins are known to be related with this disease. The systemic form of amyloidosis, where amyloid material is deposited on several organs, blood vessels and connective tissue, is usually fatal and causes approximately 1/1,000 of the deaths in developed countries.

Cardiac amyloidosis may or may not be associated with systemic involvement and can manifest as restrictive or hypertrophic cardiomyopathy (see Chap. 8). Atrial fibrillation and AV block are common and in some instances they are associated with amyloid deposits in the atria, nodes, His bundle and branches (Fig. 10.20).

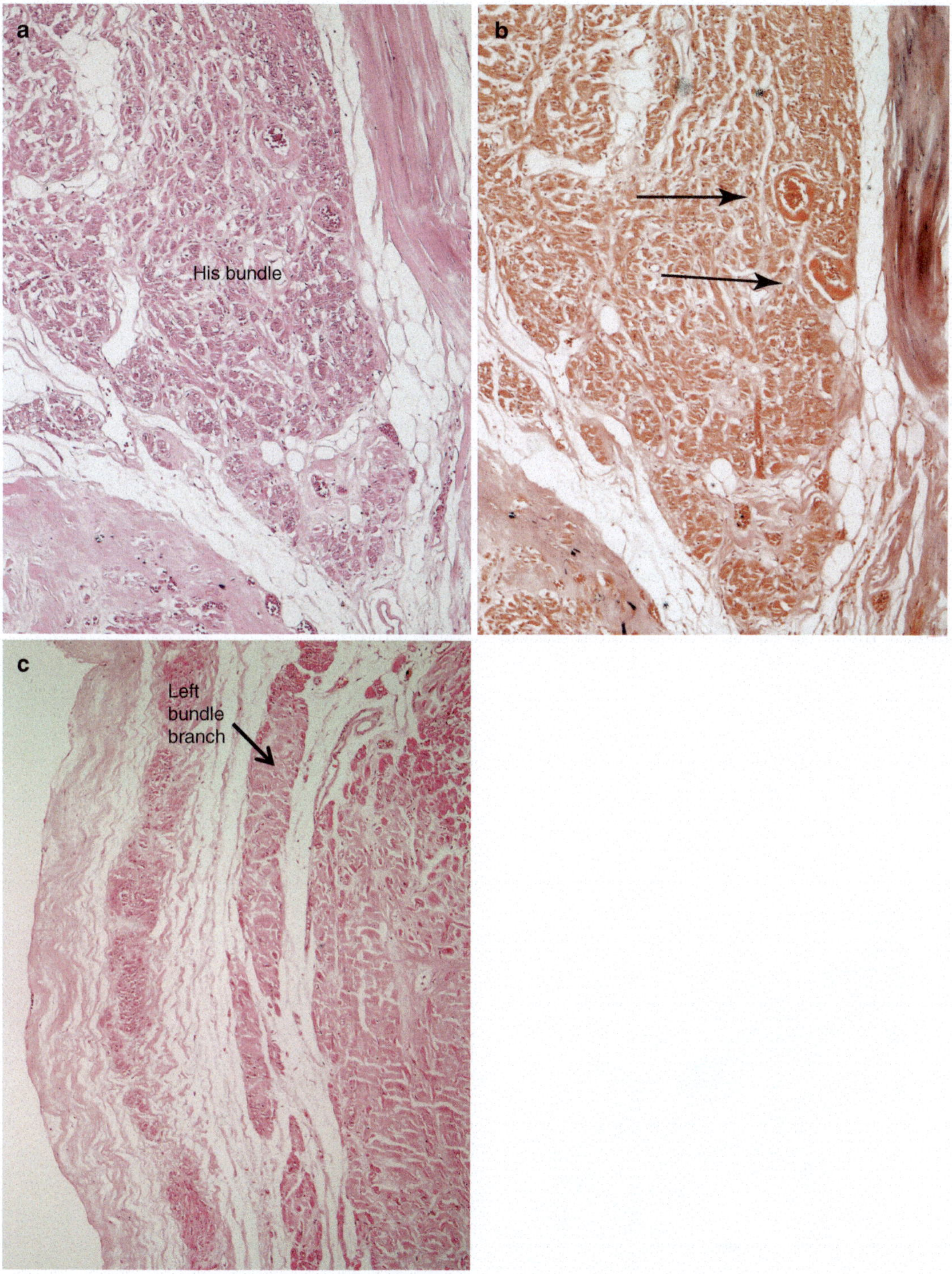

Fig. 10.20 An-85 year-old woman who died suddenly without known medical history. Autopsy revealed amyloid deposits in the arteries of several organs and in the myocardial interstitium. (**a**) Eosinophilic hyaline deposition in the arteries of the His bundle. (**b**) Amyloid deposit stained positive with Congo Red (*arrows*). (**c**) Diffuse infiltration in the left bundle branch

10.5.3 Sudden Death

The correlation between alterations in the CCS and sudden cardiac death, especially in cases with a structurally-normal heart, was broadly studied in the 1970s, chiefly by Thomas N. James, and in the 1980s by Saroja Bharati and Maurice Lev. These and other authors described many alterations associated with sudden death, but later work by other groups demonstrated that these alterations were normal variations with no functional relevance. The development of molecular cardiology since the 1990s, and the discovery of genetic alterations associated with channelopathies, have corroborated these findings.

Currently, only a few specific alterations in the conduction system are considered responsible for sudden death: significant stenosis of the nodal arteries, tumour of the AV node (previously mentioned), tumour of the Purkinje cells (histiocytoid cardiomyopathy), and pre-excitation syndromes. Other clinical entities affecting the myocardium, such as myocarditis, amyloidosis, sarcoidosis (already mentioned) etc., can also compromise the CCS and precipitate sudden death.

Alterations in the CCS have also been described in sudden infant deaths and the intrauterus unexplained death that occurs during the last weeks of gestation.

10.5.3.1 Stenosis of the Nodal Arteries

It is a lesion characterised by fibromuscular hyperplasia of the intimal and/or medial layers that causes a luminal stenosis >75 % (Fig. 10.21). The term dysplasia is used when there is mucopolysaccharide deposition with destruction of the internal elastic lamina. Ischemia of the AV node can cause sudden death before any lesions can be seen (necrosis or fibrosis). Physical exercise, or stress secondary to cocaine consumption are among the triggering factors. Stenosis of the nodal arteries has been associated with hypertrophic cardiomyopathy and with mitral valve prolapse.

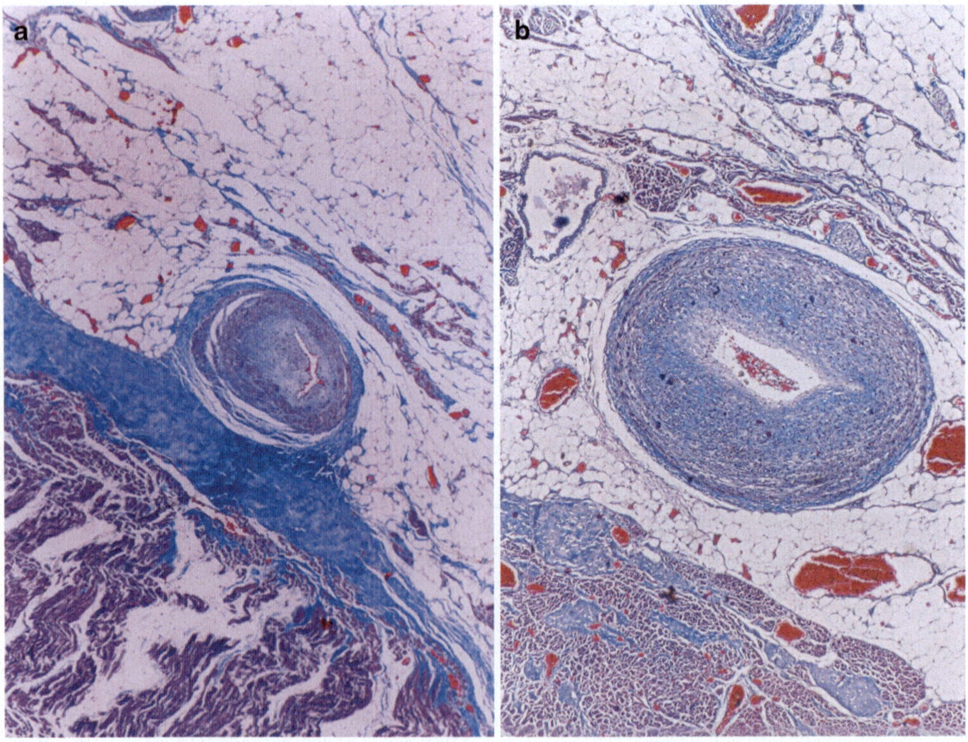

Fig. 10.21 (**a**) A 24-year-old male. Sudden death whilst working in a water-bottling plant. Toxicology was negative. Heart weight, 375 g. Unique finding: stenosis of the AV node artery (Masson's trichrome). (**b**) A 39-year-old male. Sudden death at home. He had been out the night before. Low levels of cocaine and its metabolite were found in blood. Epicardial coronaries were patent. AV node artery: fibrointimal hyperplasia with luminal stenosis of around 75 % (Masson's trichrome)

10.5.3.2 Purkinje Cell Tumour (Histiocytoid Cardiomyopathy)

The Purkinje cell tumour or histiocytoid cardiomyopathy is a multicentric hamartoma-like aggregation of cardiac myocytes with features similar to those of oncocytes. The tumour tends to affect girls younger than 4 years old, and is characterised by severe and untreatable arrhythmias and preexcitation syndromes that can lead to heart failure or sudden death. On some occasions, cardiac symptomatology is preceded by a pseudoviral infection with vomiting and diarrhoea or respiratory symptomatology. Without surgical intervention, death occurs within hours or months.

In the past, this lesion has been considered a tumour of Purkinje cells, although more recently it is thought to be a congenital or acquired metabolic disorder that affects the mitochondria of certain myocytes, mainly Purkinje cells. Immunohistochemistry studies have demonstrated an alteration in proteins of the cytoskeleton. The morphologic characteristics are described in Table 10.4.

Table 10.4 Pathology of the Purkinje cell tumour (histiocytoid cardiomyopathy)

Macroscopic features:
Small nodules, of 1–4 mm, yellowish in colour, and usually located in the subendocardium of the LV, although they can also locate anywhere (valves, myocardium, subepicardium)
Microscopic features (Fig. 10.22):
Large tumoural cells, polygonal in shape, with ample and clear cytoplasm that includes large granules composed of abnormal mitochondria
Immunohistochemistry: positive for desmin and myoglobin
Disorganised myofibrils resulting in abnormal striations

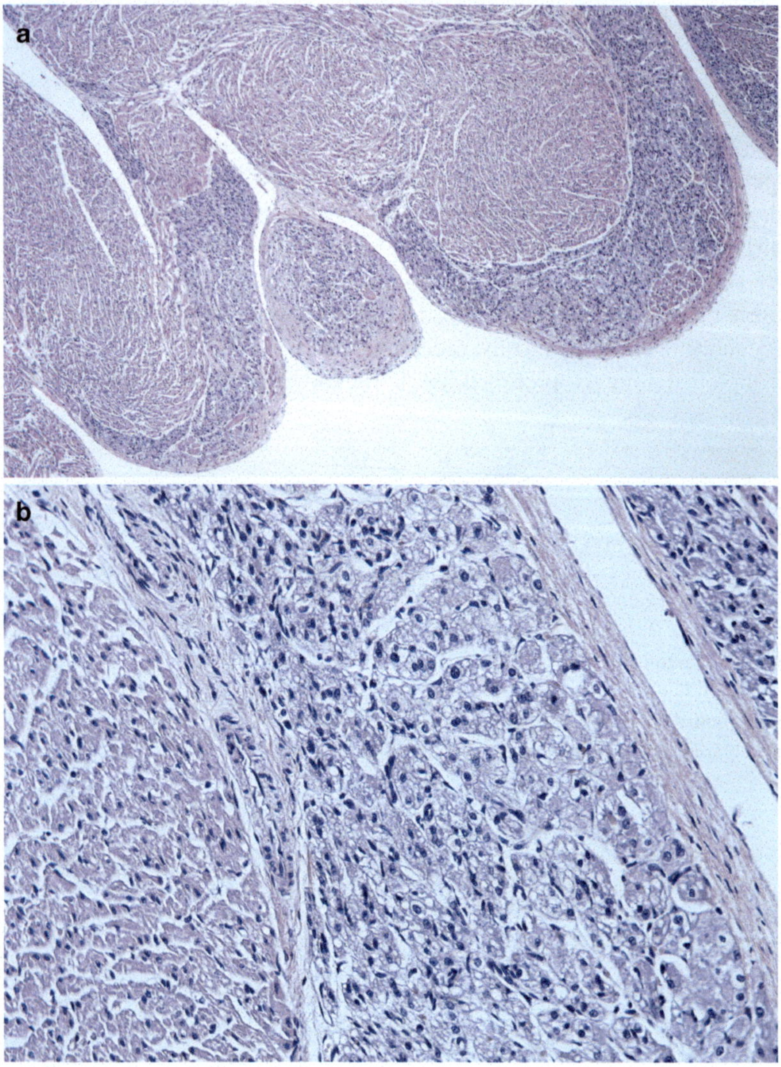

Fig. 10.22 Sudden cot death in a 4-month-old female. Multiple subendocardial plaques composed of polygonal cells with clear cytoplasm: (**a**, **b**). Same image at low and medium magnification, respectively (*Courtesy of Dr. Stephen D. Cohle. Spectrum Health East, Department of Pathology, Grand Rapids, Michigan (USA)*)

10.5.4 Pre-excitation Syndrome

Pre-excitation syndrome is a condition where part or all of the ventricular myocardium becomes depolarised prematurely because of an abnormal conduction of the impulse outside the CCS. This is due to the presence of accessory pathways from the atria to the ventricular chambers of the heart, which provokes a short circuit (by-pass) in the normal transmission of the electrical impulse. The arrhythmias associated with these syndromes are due to the retrograde transmission of the impulse from the ventricles to the atria through the normal or the accessory pathways, creating a circular movement.

From an anatomical point of view, it has been traditionally considered that the accessory pathways could be located at any point in the annuli of the atrioventricular valves, or could emerge from the conduction system (Fig. 10.23). The most important accessory pathways are: (a) Accessory atrioventricular connections, located in the annuli of the AV valves (Kent fibres); (b) Atrioventricular connections, situated in the posterior and anterior area of the septum; (c) Nodo-ventricular connections, which connect the atrioventricular node to the ventricular myocardium (Variety of Mahaim fibres); (d) Atriofascicular connections, which connect the atria with the penetrating portion of the His bundle (James fibres); (e) Fasciculoventricular connections, which connect the His bundle to the ventricular myocardium (Variety of Mahaim fibres) (Fig. 10.24); (f) Intranodal by-pass tracts, which are pathways within the atrioventricular node that by-pass the delay-producing core. This last type of connection has never been anatomically demonstrated, but electrophysiological studies suggest it should exist.

The morphological identification of these accessory pathways is very difficult without a previous electrophysiological mapping that indicates the location. Without mapping information, it would be necessary to study thousands of sections to include the entire structure of the CCS and the valve annuli. Strictly speaking, from a histopathological point of view, there should be *continuity* between the accessory pathway and the contractile myocardium, and not mere *contiguity*. Furthermore, discovery of abnormal connections does not necessarily mean they are functionally active, unless there is a compatible electrophysiological study.

Many of these accessory pathways are associated with alterations in the fibrous skeleton of the heart. This skeleton is composed of the central fibrous body, the membranous septum and the fibrous rings of the atrioventricular valves (Fig. 10.25). In the left side of the heart, the atrial and ventricular myocardium are separated by a firm collagen ring from which the mitral valve leaflets emerge. This fibrous structure completely surrounds the mitral annulus and is thicker in the right and left margins of the mitro-aortic fibrous continuity, forming the fibrous trigons. The right fibrous trigon is in continuation with the tricuspid valve ring through the central fibrous body, which in turn is in continuity with the membranous septum. Posterior to the fibrous body, in the area where the coronary sinus enters the right atrium, the septal atrioventricular plane is rarely a fibrous structure but the inferior wall of the coronary sinus is separated from the ventricular myocardium by a plane of adipose tissue. It is through this plane that the AV nodal artery travels forward to enter the node. Figures 10.26, 10.27, 10.28 and 10.29 show several examples of alterations in the central fibrous body.

Accessory pathways are also associated with Ebstein's anomaly, characterised by a low displacement of the septal and posterior tricuspid valve leaflets (see Chap. 7), which would explain its association with Wolf-Parkinson-White syndrome and other pre-excitation syndromes. In some cases, only the septal leaflet is involved and the AV node generally is elongated in shape in the direction of the valve insertion point (Fig. 10.30). It is at this point where defects in the central fibrous body can occur, allowing crossing of accessory pathways.

Development of modern electrophysiology has questioned the location of some of the anomalous connections described by the morphologists. Thus, it has been observed that Mahaim fibres are commonly located along the anterolateral margin of the tricuspid annulus and apparently originate in a structure similar to the node, and insert in three possible locations: (a) in distal branches of the right bundle branch (long atriofasciculars); (b) in the myocardium, close to the annulus (short AV connections); and (c) in the ventricular myocardium near the apex (long AV connections). These accessory pathways share similarities with those described by Kent in

1914, which other authors such as Davis, Anderson and Becker (1983) considered as a remnant node of specialised embryonic tissue preferentially located in the right atrioventricular junction, in particular in the anterolateral margin.

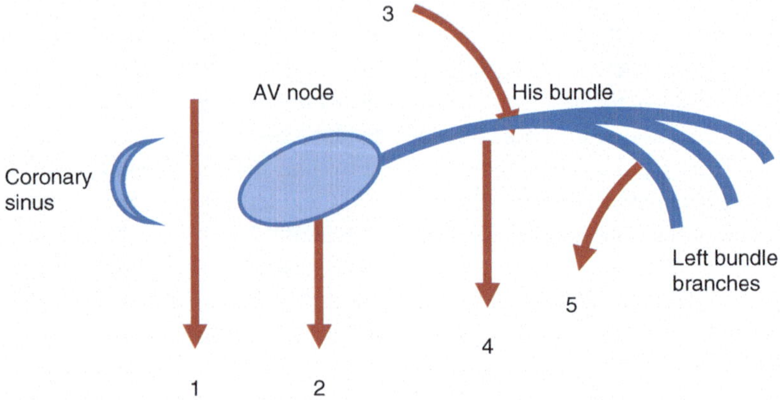

Fig. 10.23 Schematic representation of the accessory pathways in the CCS, according to histological findings. (*1*) Atrioventricular posteroseptal connection (posterior to AV node). (*2*) Nodo-ventricular connection (Mahaim fibres). (*3*) Atriofascicular connection (James fibres). (*4*) Fasciculoventricular connection (Mahaim fibres). (*5*) Connection between branches and ventricular septum (Mahaim fibres). Kent fibres are located in the annuli of the AV valves

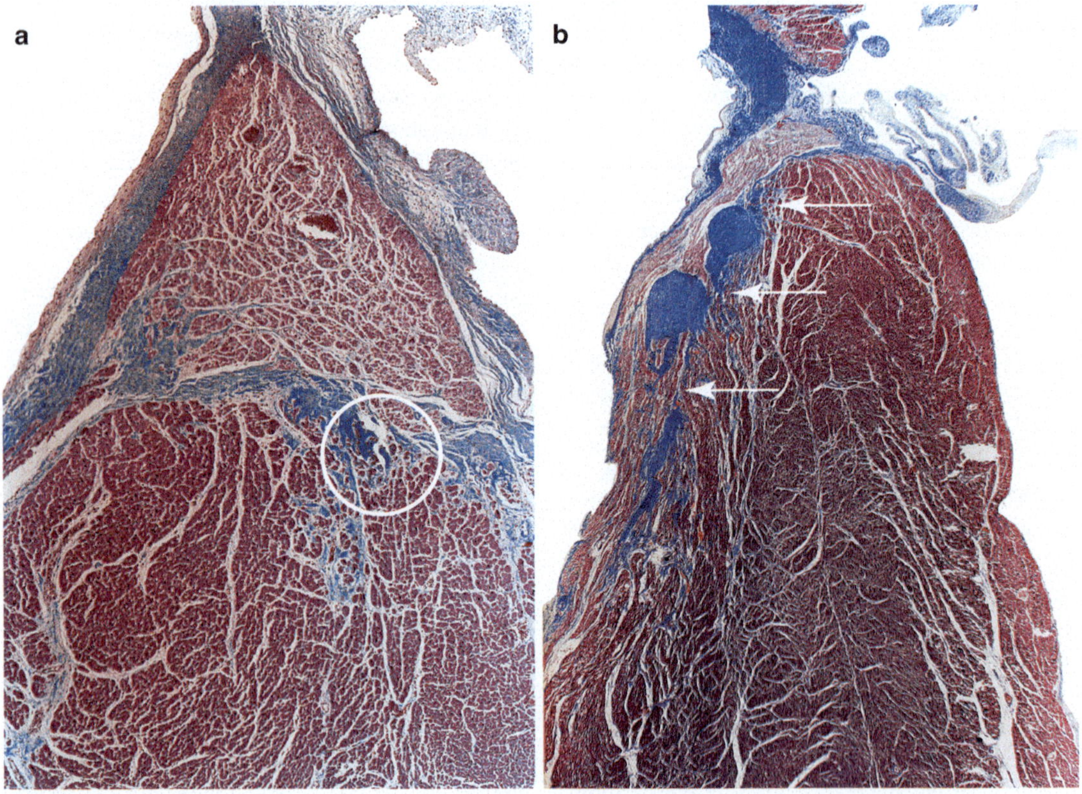

Fig. 10.24 (**a**) A 29-year-old woman. All the CCS structures were enlarged and there was a prolongation of the His bundle towards the interventricular septum (Mahaim fibres) (*circle*). (**b**) A 26-year-old male with a history of myotonic dystrophy. Accessory connections from the left bundle branch to the septum (*arrows*) (Masson's trichrome)

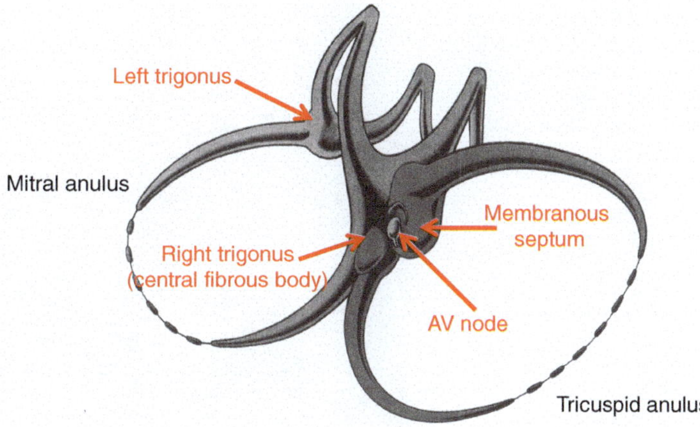

Fig. 10.25 Fibrous skeleton of the heart

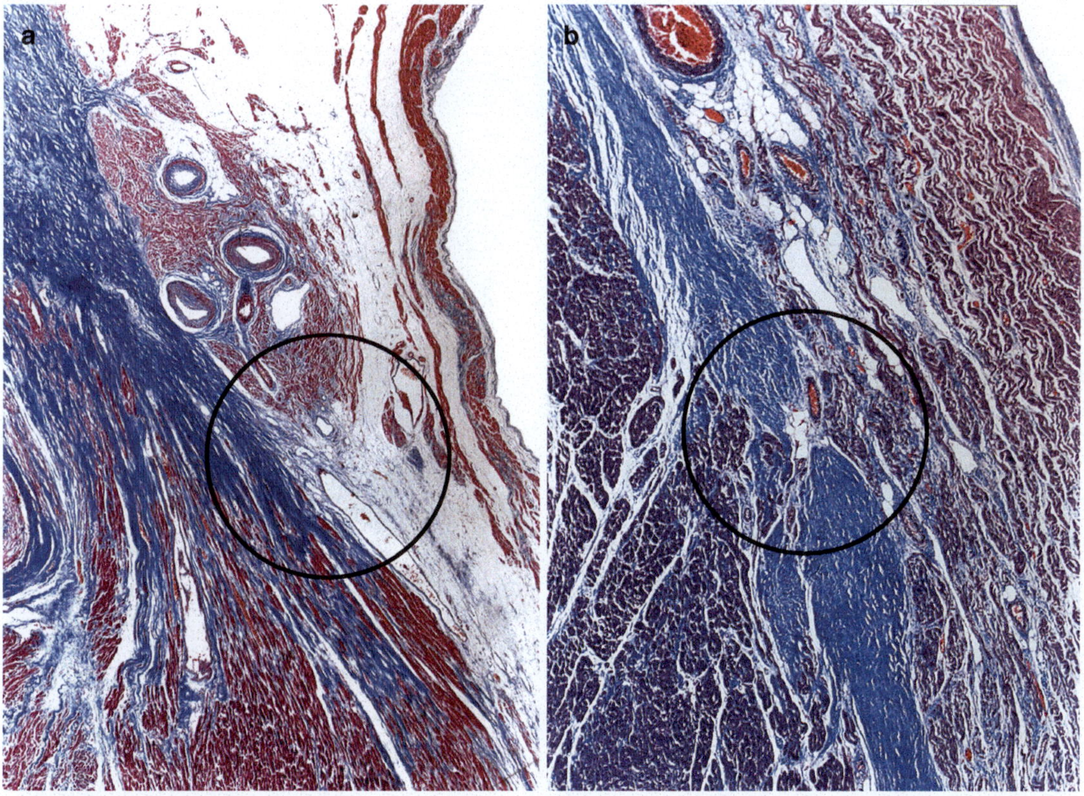

Fig. 10.26 (**a**, **b**) Defects in the central fibrous body are shown; through which accessory pathways can travel (*circles*) (Masson's trichrome)

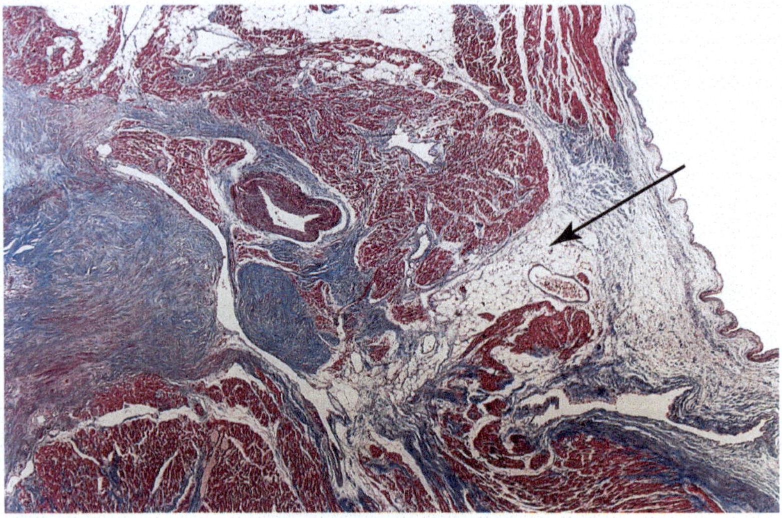

Fig. 10.27 Large defect in the fibrous skeleton (*arrow*) in the heart of a 55-year-old male who died suddenly. Heart weight, 534 g (Masson's trichrome)

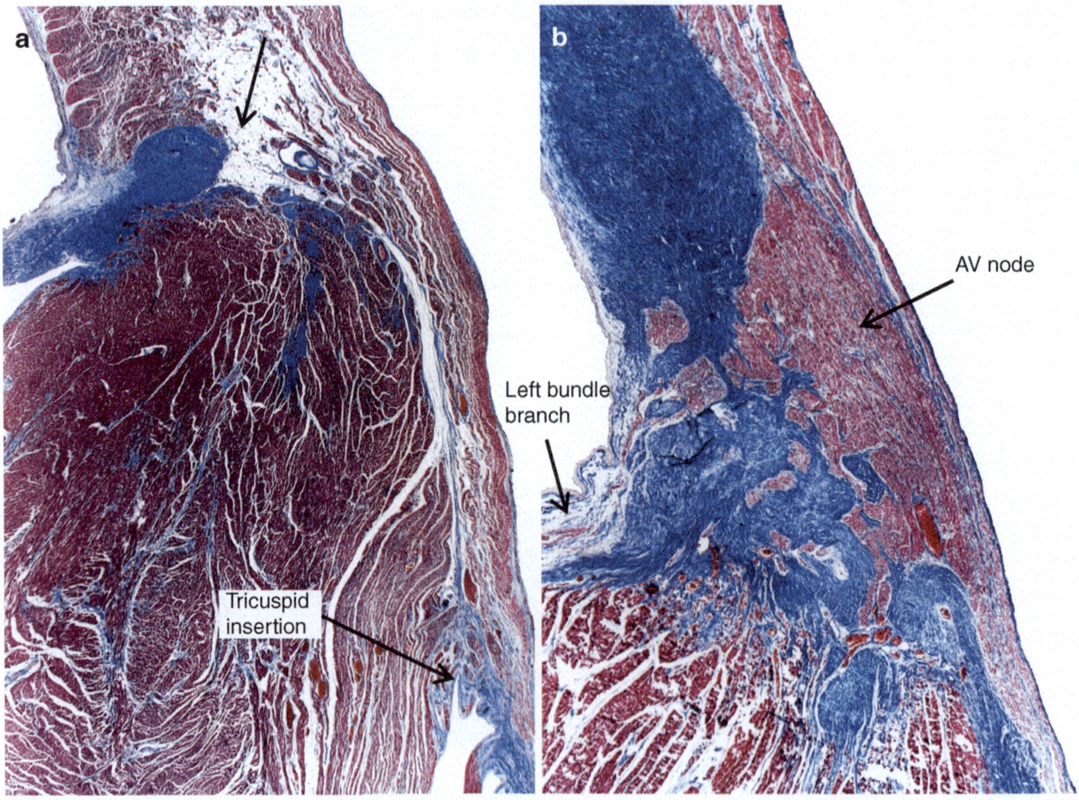

Fig. 10.28 A 26-year-old male with a history of myotonic dystrophy. (**a**) The fibrous skeleton is incomplete near the insertion of the mitral valve (*arrow*) and the insertion point of the septal leaflet of the tricuspid valve is slightly lower down. (**b**) AV node foetal dispersion with early origin of the left bundle branches. There were also anomalous connections (same case as in Fig. 10.24) (Masson's trichrome)

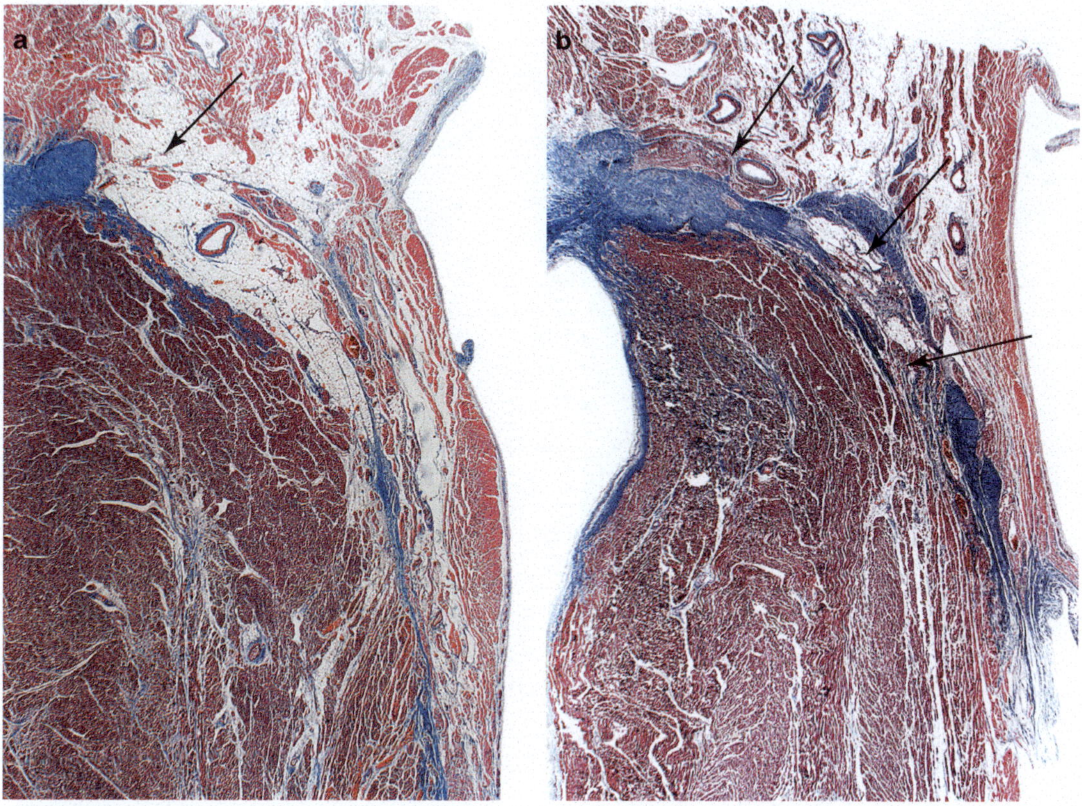

Fig. 10.29 A 30-year-old male with a history of epilepsy. Multiple anomalies in the CCS were found. (**a**) Incomplete fibrous skeleton in the superior portion (*arrow*) with the nodal artery surrounded by adipose tissue. (**b**) Discontinuous fibrous skeleton associated with a tripartite AV node (*arrows*) (Masson's trichrome)

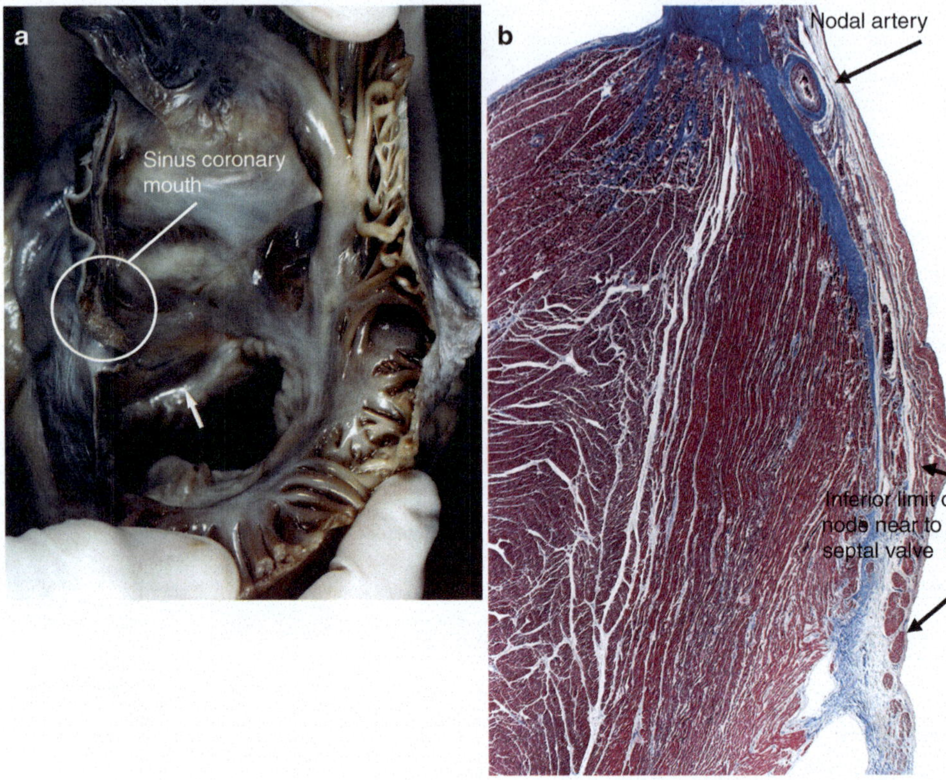

Fig. 10.30 A 23-year-old male who died in a traffic accident. (**a**) Apical displacement of the insertion point of the septal leaflet of the tricuspid valve (*arrow*), as seen from the right atrium. (**b**) Largely elongated AV node that ends near the valve's insertion point. At this level, the central fibrous body is extremely thin (Masson's trichrome)

Bibliography

Anderson RH, Boyett MR, Dobrzynski H, Moorman AF. The anatomy of the conduction system: implications for the clinical cardiologist. J Cardiovasc Transl Res. 2013;6(2):187–96.

Benditt DG, Lü F. Atriofascicular pathways: fuzzy nomenclature or merely wishful thinking? J Cardiovasc Electrophysiol. 2006;17(3):261–5.

Bharati S, Lev M. The cardiac conduction system in unexplained sudden death. Mount Kisco: Futura Publishing Company; 1990.

Burke A, Tavora F. Giant cell myocarditis and cardiac sarcoidosis. In: Practical cardiovascular pathology: an atlas. Philadelphia: Wolters Kluwer Health/Lippincott Williams & Wilkins; 2011. p. 258–61.

Cabrera JA, Sánchez-Quintana D. Cardiac anatomy: what the electrophysiologist needs to know. Heart. 2013;99(6):417–31.

Cohle SD, Lie JT. Pathologic changes of the cardiac conduction tissue in sudden unexpected death. A review. Pathol Annu. 1991;26(2):33–57.

Cohle SD, Suárez-Mier MP, Aguilera B. Sudden death resulting from lesions of the cardiac conduction system. Am J Forensic Med Pathol. 2002;23(1):83–9.

Davies MJ, Mann J. The cardiovascular system part B: acquired diseases of the heart. Systemic pathology, vol. 10. 3rd ed. Edinburgh: Churchill Livingstone; 1995.

Flaker G, Greenspon A, Tardiff B, Schron E, Goldman L, Hellkamp A, et al. Death in patients with permanent pacemakers for sick sinus syndrome. Am Heart J. 2003;146(5):887–93.

Gulino SP. Examination of the cardiac conduction system. Forensic application in cases of sudden cardiac death. Am J Forensic Med Pathol. 2003;24(3):227–38.

Hudson REB. The conducting system: anatomy, histology and pathology in acquired heart disease. In: Silver MD, editor. Cardiovascular pathology. New York: Churchill Livingstone; 1991. p. 1367–427.

James TN. The connecting pathways between the sinus node and the A-V node and between the right and the left atrium in the human heart. Am Heart J. 1963;66:498–508.

James TN. Normal variations and pathologic changes in structure of the cardiac conduction system and their functional significance. J Am Coll Cardiol. 1985; 5(6 Suppl):71B–8.

Law KB, Feng T, Nair V, Cusimano RJ, Butany J. Cystic tumor of the atrioventricular node: rare antemortem diagnosis. Cardiovasc Pathol. 2012;21(2):120–7.

Michaud K, Romain N, Taroni F, Horisberger B, Mangin P. Evaluation of a simplified method of the conduction system analysis in 110 forensic cases. Forensic Sci Int. 2002;130(1):13–24.

Monzón F, Aguilera B, Martínez Peñuela JM y, Oliva H. Heterotopia endodérmica poliquística del nódulo aurículo-ventricular. Med Clin (Barc). 1995;104:257–61.

Ottaviani G, Matturri L. Histopathology of the cardiac conduction system in sudden intrauterine unexplained death. Cardiovasc Pathol. 2008;17(3):146–55.

Patel J, Sheppard MN. Cystic tumour of the atrioventricular node: three cases of sudden death. Int J Legal Med. 2011;125(1):139–42.

Sánchez-Quintana D, Ho SY. Anatomía de los nodos cardíacos y del sistema de conducción específico aurículoventricular. Artículo de revisión. Rev Esp Cardiol. 2003;56:1085–92.

Sánchez-Quintana D, Pizarro G, López-Mínguez JR, Ho SY, Cabrera JA. Standardized review of atrial anatomy for cardiac electrophysiologists. J Cardiovasc Transl Res. 2013;6(2):124–44.

Suárez-Mier MP, Fernández-Simón L, Gamallo C. Pathologic changes of the cardiac conduction tissue in sudden cardiac death. Am J Forensic Med Pathol. 1995;16(3):193–202.

Suárez-Mier MP, Gamallo C. AV node fetal dispersion and His bundle fragmentation of the cardiac conduction system in sudden cardiac death. J Am Coll Cardiol. 1998;32(7):1885–90.

Suárez-Mier MP, Aguilera B. Histopathology of the conduction system in sudden infant death. Forensic Sci Int. 1998;93(2–3):143–54.

Suarez-Mier MP, Sánchez-de-León MS, Cohle SD. An unusual site for the AV node tumour: report of two cases. Cardiovasc Pathol. 1999;8(6):325–8.

Temple IP, Inada S, Dobrzynski H, Boyett MR. Connexins and the atrioventricular node. Heart Rhythm. 2013;10(2):297–304.

Zack F, Kutter G, Blaas V, Rodewald AK, Büttner A. Fibromuscular dysplasia of cardiac conduction system arteries in traumatic and nonnatural sudden death victims aged 0 to 40 years: a histological analysis of 100 cases. Cardiovasc Pathol. 2014;23:12–6.

Cardiovascular Pathology Secondary to Alcohol Consumption, Drugs of Abuse and Medication

11

Joaquín S. Lucena, Pedro M. Garamendi, Gonzalo Guzzo-Merello, and Pablo García-Pavía

Contents

J.S. Lucena (✉)
Department of Forensic Pathology, Institute of Legal Medicine, Seville, Spain

Forensic Medicine, University of Cadiz, Cadiz, Spain
e-mail: joaquin.lucena@uca.es

P.M. Garamendi
Department of Forensic Pathology, Institute of Legal Medicine, Huelva, Spain

G. Guzzo-Merello
Heart Failure Unit, Cardiology Department, Hospital Universitario de Getafe, Getafe (Madrid), Spain

P. García-Pavía
Heart Failure and Inherited Cardiac Diseases Unit, Cardiology Department, Hospital Universitario Puerta de Hierro, Majadahonda (Madrid), Spain

11.1 Introduction

A large number of chemical compounds, including drugs of abuse and certain medications, have proven harmful effects on different components of the cardiovascular system, especially on the myocardium; hence, some authours use the term chemical cardiomyopathy to refer to these negative effects.

Some of the chemical compounds, such as ethanol, have a direct toxic effect on the cardiomyocyte structure and function. Others are sympathomimetic drugs and mimic the effects of sympathetic activation on the heart and circulation, as occurs with cocaine and synthetic amphetamines; while chemotherapeutic drugs, especially anthracyclines, induce cardiomyocyte oxidative stress.

J.S. Lucena et al. (eds.), *Clinico-Pathological Atlas of Cardiovascular Diseases*,
DOI 10.1007/978-3-319-11146-9_11, © Springer International Publishing Switzerland 2015

11.2 Alcohol

For many decades, chronic and excessive alcohol consumption has been considered one the main causes of dilated cardiomyopathy (DCM) in developed countries.

Diagnosis of alcohol-induced cardiomyopathy is based on the existence of DCM, without cause for justification, in a patient with a history of excessive alcohol intake, which is, according to the most significant studies, superior to 80 g/day during at least 5 years. Although this is a diagnosis of presumption and exclusion, there is consensus among the major cardiology societies about considering alcoholic cardiomyopathy as a specific type of cardiomyopathy. The exact mechanism by which alcohol induces the onset of cardiomyopathy has not yet been fully elucidated. Although some authors defend hypertrophy as the initial physiopathological event, in general it is acknowledged that loss of myocytes is the cause of both structural alterations and left ventricular dysfunction.

Mechanisms that would promote cellular damage mediated by ethanol and its metabolite acetaldehyde include direct toxicity on the myocytes, oxidative stress, and structural changes in the contractile proteins, disruption of the function of intracellular organelles such as mitochondria and sarcoplasmic reticulum, involved in cellular energy metabolism, alterations in calcium homeostasis, direct neurohormonal stimulation and even pro-apoptotic events.

Echocardiography studies on patients with alcohol-induced cardiomyopathy showed that in the initial stages of the disease, there is progressive increase in ventricular mass and diameters associated with ventricular dysfunction, largely diastolic. As the disease progresses, in a subgroup of patients, probably due to individual susceptibility, decreased contractility or systolic dysfunction appears originating what is known as **alcoholic cardiomyopathy**. Given that alcohol consumption constitutes a major public health problem in developed countries, **alcoholic cardiomyopathy** is, as mentioned before, one of the main causes of DCM and could be responsible for 15–40 % of DCM cases.

From a pathological point of view, alcoholic cardiomyopathy is indistinguishable from other types of DCM. Gross examination shows increased heart weight with biventricular dilatation, preferentially in the left ventricle. Thickness of the LV walls can be normal or slightly reduced (Fig. 11.1a, b). In approximately 50 % of the cases, mural thrombi can be observed adhered to the LV endocardium, although they can also appear in the other cardiac chambers. Thrombi can detach causing distal embolisms. It is common to find normal or mild atheromatosis in the coronary arteries. Microscopically, hypertrophic cardiomyocytes and diffuse interstitial fibrosis are observed, which are more or less extensive depending on the degree of disease progression (Fig. 11.1c).

Classical histochemical and electron microscopy studies performed in the 60s demonstrated that no clear qualitative differences existed between patients with dilated cardiomyopathy and excess alcohol intake and those with idiopathic DCM with no alcohol abuse. These authours found several degrees of fibrosis, fatty acid accumulation, inflammatory infiltrates, enlarged and jagged nuclei, and different types of structural alterations in the sarcoplasmic reticulum and myofilaments, in addition to disproportions in the type and composition of the different sarcomeric proteins. The most characteristic feature was the increase in the number and size of the mitochondria, which also presented structural alterations with reduction or absence of mitochondrial crests.

Clinically, alcoholic cardiomyopathy is also indistinguishable from idiopathic DCM, and is characterised by the classical symptoms of heart failure, with dyspnoea, exertion intolerance and congestive signs that can be associated with hepatic pathology more commonly than in DCM patients. Prognosis is controversial, although it appears similar to that of idiopathic DCM once the patient stops or significantly reduces alcohol intake. Those patients in which alcohol abuse persists suffer the most severe forms of DCM and clearly have a worse prognosis.

Medical treatment of alcoholic cardiomyopathy overlaps with that of DCM and is based on the use of diuretics and vasodilators to relieve the symptoms. Also, beta-blockers, angiotensin converting enzyme (ACE) inhibitors or angiotensin II receptor blockers (ARBs) and anti-aldosterone treatment are used to improve prognosis.

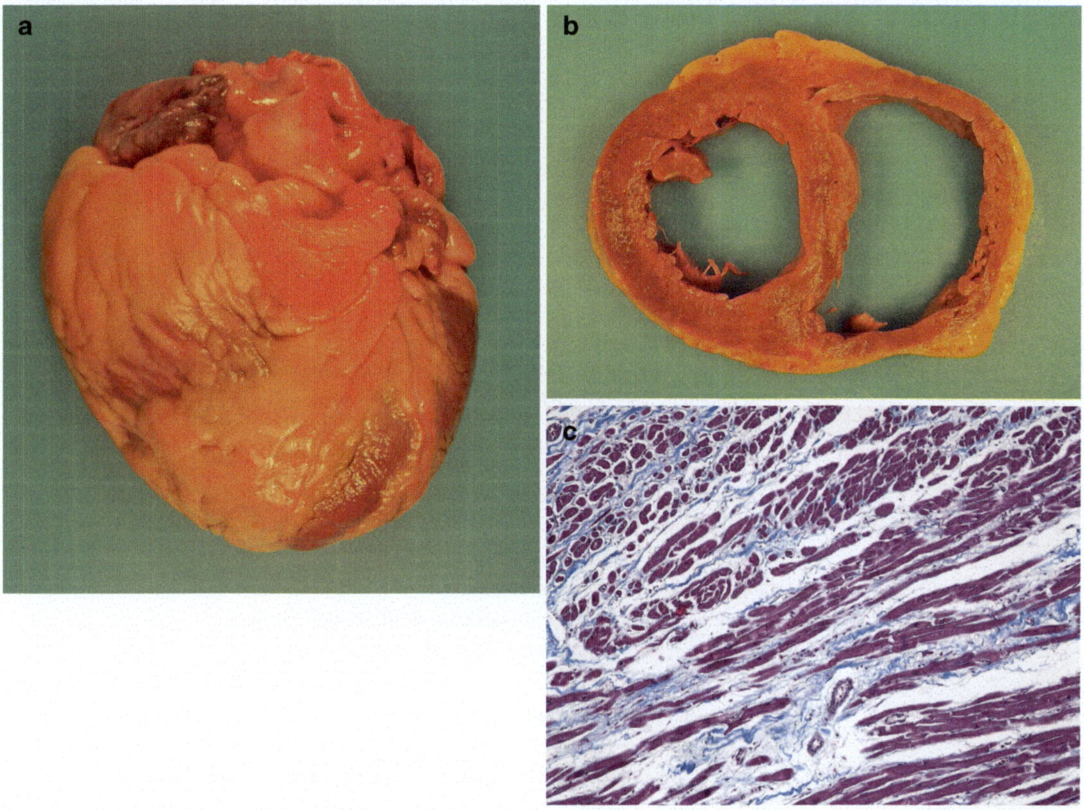

Fig. 11.1 *Alcoholic cardiomyopathy*. A 54-year-old male, alcoholic, who died suddenly. (**a**) Heart of 760 g with globoid appearance. (**b**) Dilatation of both ventricular chambers (RV chamber: 5 cm, LV chamber: 4.5 cm). Ventricular wall thickness is 5 mm in the RV and 12 mm in the septum and free wall of the LV. (**c**) Hypertrophic cardiomyocytes and interstitial fibrosis (Masson's trichrome)

11.3 Cocaine

Several studies have demonstrated that cocaine consumption has major adverse cardiovascular effects due to its actions on both coronary arteries and myocardium. Furthermore, cocaine is a pro-arrhythmic substance due to its effects on Na^+ and K^+ channels (Table 11.1).

These effects, in isolation or in combination, make cocaine consumption a risk factor for sudden cardiac death (Fig. 11.2).

Table 11.1 Cocaine-related cardiovascular pathology

Coronary arteries
Acute pathology:
Coronary artery spasm (mediated by α_1-adrenergic receptors)
Ischaemia, especially when associated with severe atheromatosis
Chronic pathology:
Accelerated atherosclerosis and thrombosis
Intramyocardial small vessel disease
Myocardium
Left ventricular hypertrophy due to:
Elevated systolic blood pressure
Direct stimulation of the myocardiac α-adrenergic receptors
Replacement fibrosis secondary to ischaemia due to:
Vasoconstriction
Accelerated atherosclerosis
Arrhythmias due to
Loss of function of Na^+ channels (Brugada-like electrocardiographic pattern)
Loss of function of HERG K^+ channels (Bradycardia, Long QT syndrome (LQTS))

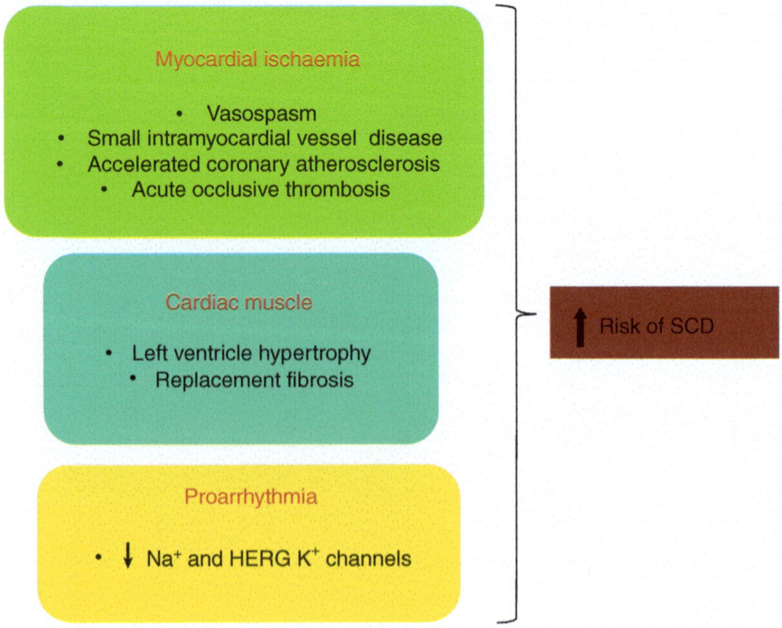

Fig. 11.2 Mechanisms through which cocaine increases the risk of sudden cardiac death (SCD)

11.3.1 Accelerated Atherosclerosis

It has been suggested that repeated administration of cocaine leads to endothelial damage, which favours the development of a premature and accelerated form of atherosclerosis. Lucena et al. (2010) reported severe coronary artery atherosclerosis in 76.2 % of the autopsies performed in cases of SCD associated with cocaine abuse, and atherosclerosis was the main cause of death in 28.6 % of the cases (Figs. 11.3 and 11.4). Cocaine has also been associated with intravascular thrombosis in the coronary and peripheral arteries (Figs. 11.5 and 11.6). Evidence suggests that cocaine activates platelet aggregation and stimulates the secretion of thrombogenic substances from the vascular endothelium.

Coronary artery atherosclerosis, with or without added thrombosis, results in acute or chronic myocardial ischaemia (Fig. 11.7).

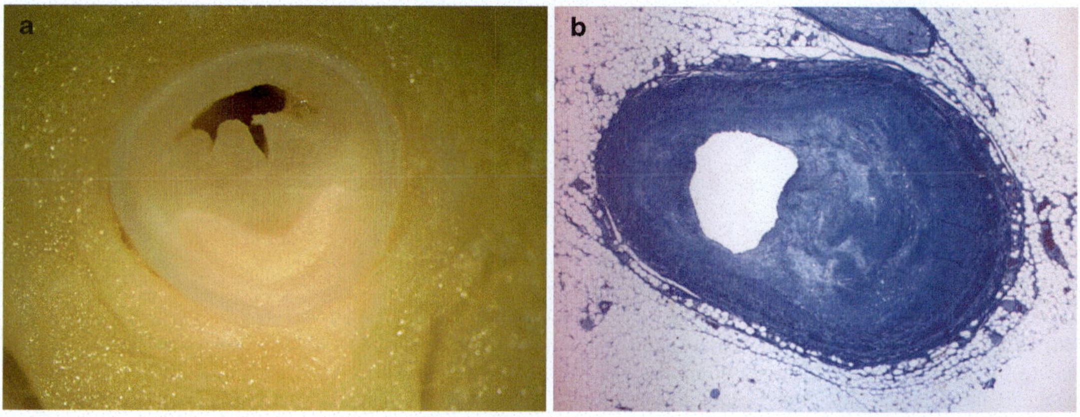

Fig. 11.3 A 21-year-old male, who committed suicide by shooting himself in the head after a weekend of consuming large amounts of cocaine. (**a**) Severe atherosclerosis in the proximal third of the LAD coronary artery. (**b**) Microscopic image showing atherosclerosis

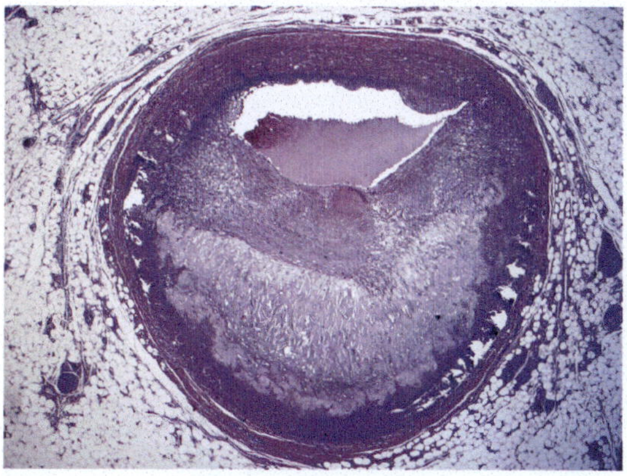

Fig. 11.4 A 28-year-old male, with a history of type I diabetes, who suffered cardiac arrest after smoking free-base cocaine. Severe atherosclerosis in the proximal third of the LAD coronary artery, with a large lipid core

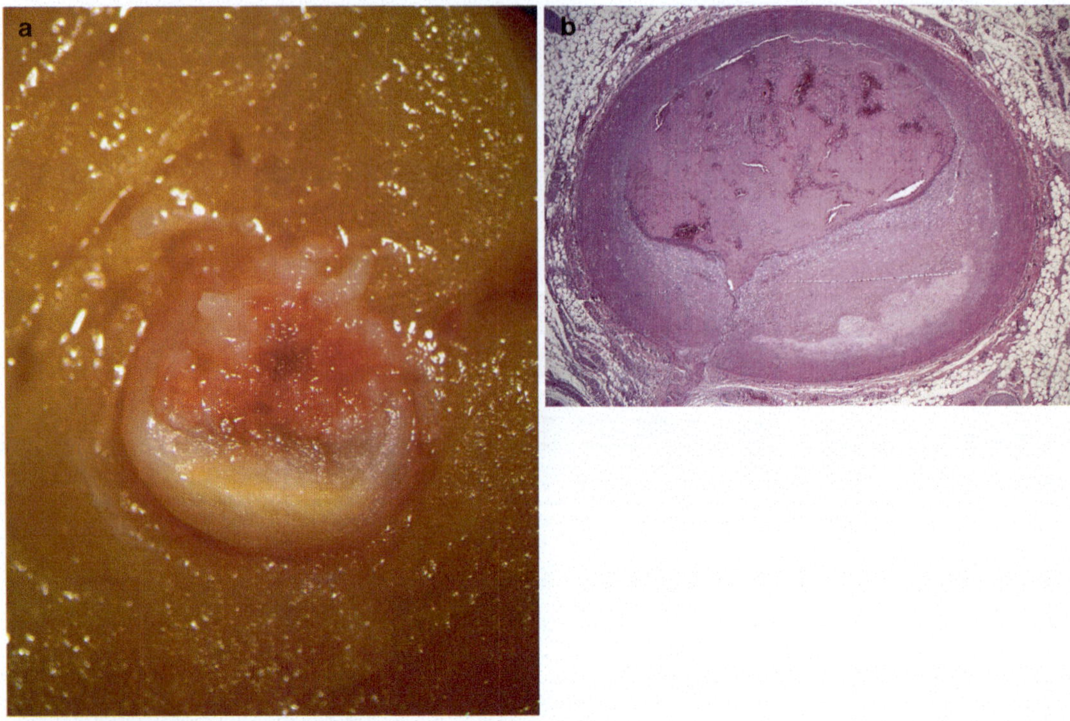

Fig. 11.5 A 28-year-old male who died suddenly in a pub. (**a**) Acute thrombosis superimposed to an atheroma plaque, in the proximal third of the LAD coronary artery. (**b**) Acute occlusive fibrin platelet thrombosis due to endothelial erosion superimposed to a fibrocellular plaque

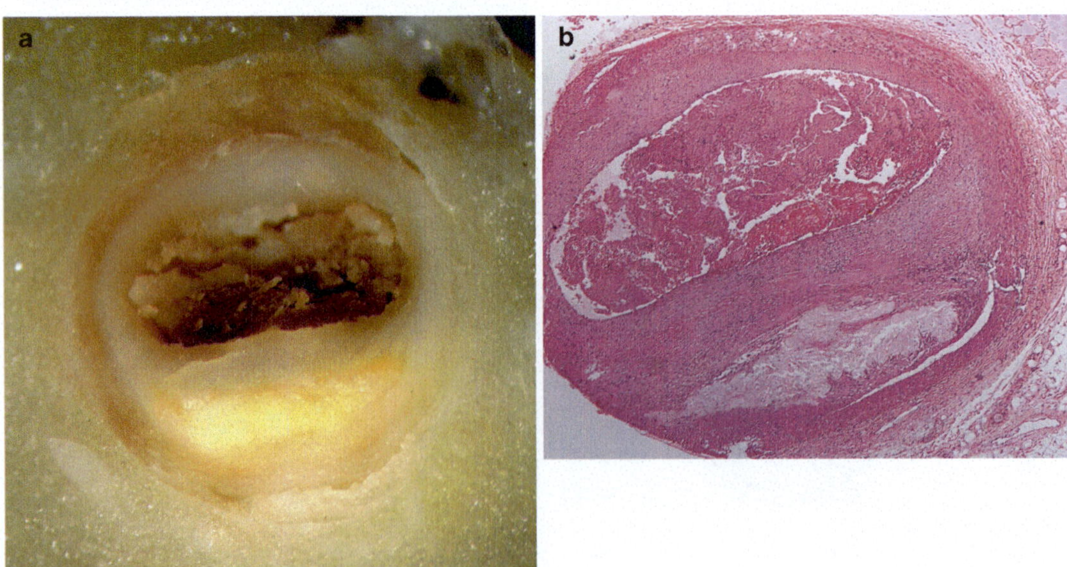

Fig. 11.6 A 33-year-old male, with a history of cocaine consumption, who died suddenly at home. (**a**) Acute thrombosis superimposed to an atheroma plaque, in the proximal portion of the LAD coronary artery. (**b**) Histological image showing the fibroatheromatous plaque with a small lipid core and acute thrombosis due to plaque erosion

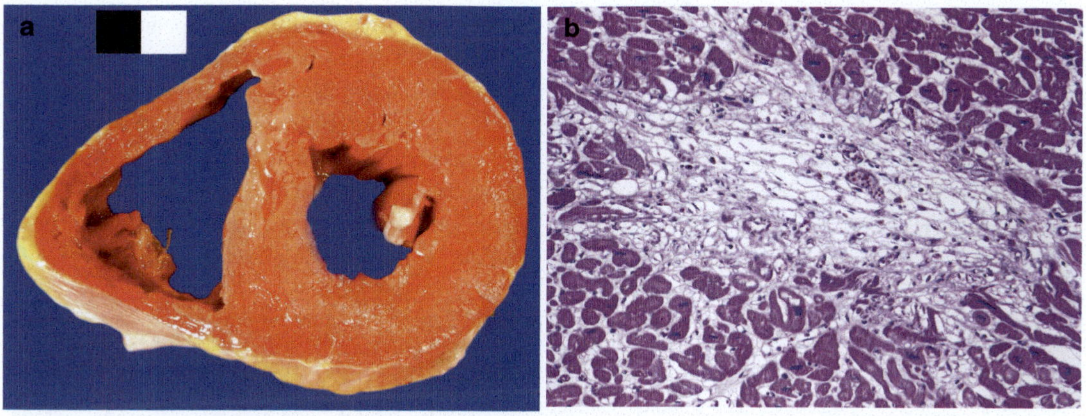

Fig. 11.7 A 28-year-old male, cocaine user, who died suddenly at home. (**a**) A recent myocardial infarction can be appreciated in the ventricular septum. (**b**) Microscopic image of the same case showing myocyte necrosis and granulation tissue (about 10–14 days old)

11.3.2 Intramyocardial Small Vessel Disease

In recent years, it has been demonstrated that myocardial perfusion can be diminished, even in the absence of atherosclerosis, in the epicardial coronary arteries. This could be explained by a thickening of the medial layer affecting small intramyocardial vessels that appears in cocaine consumers. This finding is considered a small vessel disease characterised by fibromuscular hypertrophy of the arterioles with medial hyperplasia and endothelial oedema (Fig. 11.8). Hypertrophy in the medial layer of the intramyocardial small vessel could result from sustained vasoconstriction due to the chronic effect of cocaine.

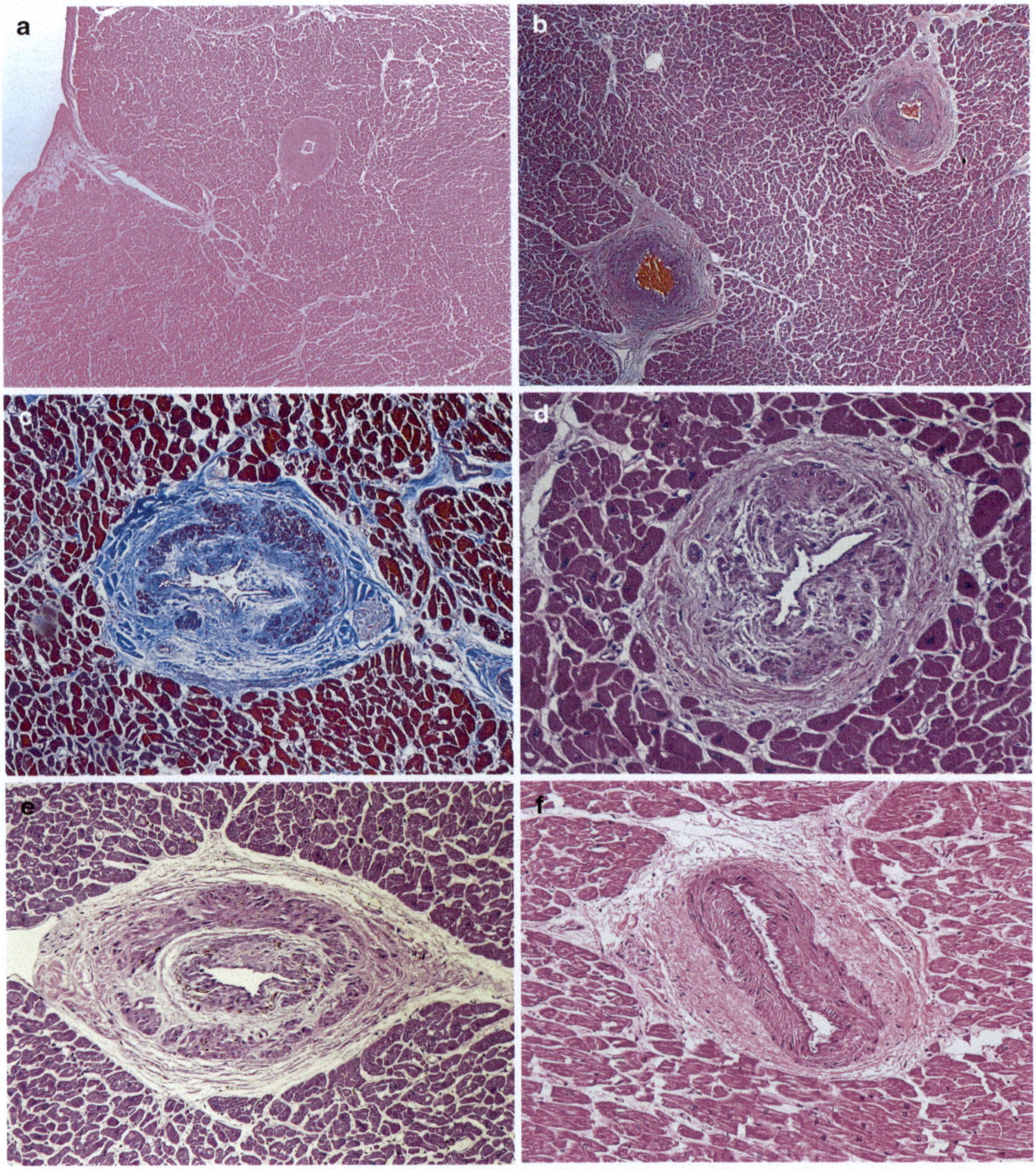

Fig. 11.8 Intramyocardial small vessel disease in cocaine consumers, from 19 to 34 years, who died suddenly. (**a**–**f**) Significant periadventitial fibrosis and hyperplasia of the medial layer can be appreciated, which notably reduces the vascular lumen (Masson's trichrome).

11.3.3 Cardiomegaly with Left Ventricular Hypertrophy

Cardiac hypertrophy *per se* increases the risk of SCD, even in the absence of cocaine consumption. Increased heart weight is a demonstrated consequence of chronic cocaine consumption, which can in turn favour the development of myocardial ischaemia (Fig. 11.9). Contradictory data exist in the literature regarding the heart weight in cocaine addicts. Some studies report an increase in heart weight whereas other studies show no differences. The most likely cause of LV hypertrophy could be the transient elevation of blood pressure that occurs after cocaine consumption. Another possible mechanism could be a direct stimulation of cardiac α-adrenergic receptors by cocaine.

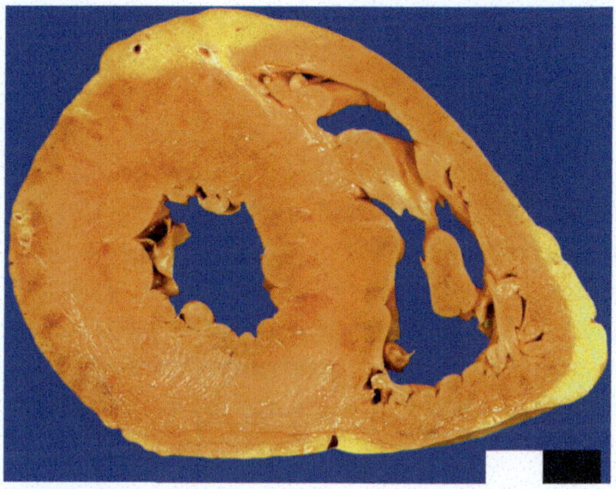

Fig. 11.9 A 26-year-old male who died whilst being arrested by the police as a consequence of excited delirium after cocaine and alcohol abuse. Cardiomegaly (470 g) with concentric LV hypertrophy (20 mm)

11.3.4 Cocaine-Related Myocardial Fibrosis

The high prevalence of myocardial fibrosis encountered in individuals whose deaths are related with cocaine and amphetamine use could highlight the existence of previous ischaemic episodes. However, the healing processes that take place following recurrent contraction band necrosis could also explain the anatomical distribution of the areas of fibrosis reported in cocaine consumers. Underlying silent ischaemia, or myocardial fibrosis as a substrate for re-entry arrhythmias, or both, can increase the risk of arrhythmias. In cases of cardiac hypertrophy secondary to pressure overload, there is also progressive accumulation of collagen fibres in the interstitium, particularly in the LV (Fig. 11.10).

Myocardial fibrosis, either ischaemic or non-ischaemic, has been demonstrated by cardiac MRI in a high percentage of asymptomatic cocaine users.

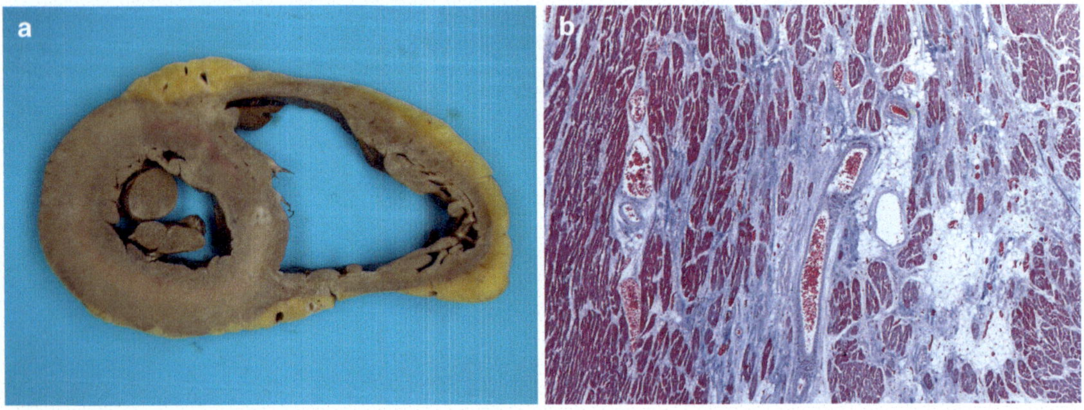

Fig. 11.10 A 31-year-old male, with a history of cocaine use, who died suddenly whilst playing football. Toxicology in blood and urine was negative. (**a**) Heart of 440 g with an area of septal fibrosis caused by an old myocardial ischaemia. Marked dilatation of the RV. (**b**) Replacement fibrosis secondary to chronic ischaemia (Masson's trichrome)

11.3.5 Cocaine-Related Eosinophilic Myocarditis

Eosinophilic myocarditis has been reported in individuals with cocaine in blood, who died from cocaine overdose or homicide. The frequency of myocarditis in these subjects varies substantially, with some works describing eosinophilic myocarditis in around 5 % of autopsy series, while others report this in up to 20 % of cases (Fig. 11.11).

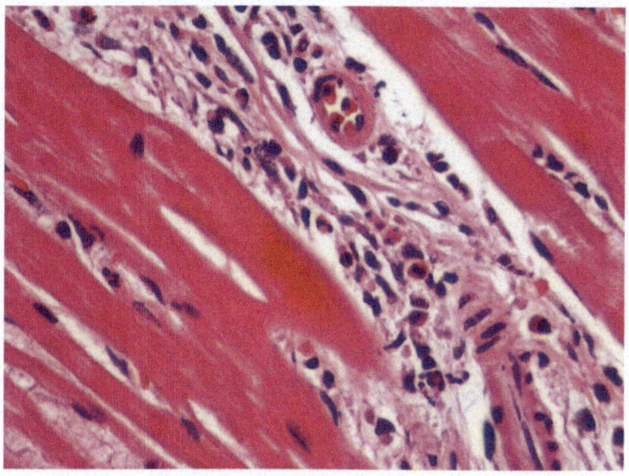

Fig. 11.11 A 26-year-old male who died suddenly after cocaine and alcohol intake. An interstitial inflammatory infiltrate consisting mainly of eosinophils was present in the myocardium

11.4 Synthetic Amphetamines

As a general rule, the cardiovascular effects caused by this heterogeneous group of compounds are quite similar to those of cocaine, with some particularities. Pharmacological actions of these agents are the following:

- Increasing the release and decreasing the reuptake of noradrenaline.
- Inhibiting monoamine oxidase (MAO).
- They are serotonergic agonists.
- Increasing the release of vasopressin.

Based on these mechanisms of action, amphetamine and its derivatives (methylenedioxymethamphetamine –MDMA- or “ecstasy” and methylenedioxyethylamphetamine -MDEA- or “eve”) cause an identical type of cardiotoxicity as cocaine, with increased heart weight, areas of fibrosis and contraction band necrosis. There is also a greater predisposition to suffer accelerated atherosclerosis and intramyocardial small vessel disease, with elevated catecholamine levels as the underlying mechanism (Fig. 11.12).

Methamphetamines have multiple effects on the cardiovascular system, with myocardial infarction, aortic dissection and cardiomyopathy emerging as the most important. The histology of methamphetamine-induced cardiomyopathy is similar to that described in laboratory animals: myocardial remodelling that manifests as concentric LV hypertrophy, atypical nuclei, interstitial and perivascular fibrosis, vacuolization of myocytes and hypertrophy in the medial layer of the intramyocardial small vessels. According to Karch (2011), cardiomyocyte hypertrophy, vacuolization and nuclear atypia constitute the diagnostic criteria in cases of death from cardiac arrest related to methamphetamine abuse.

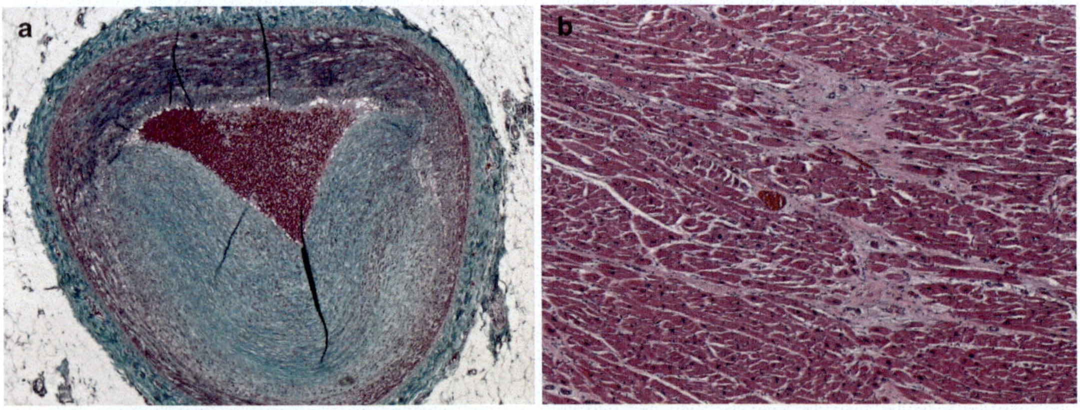

Fig. 11.12 A 39-year-old male who died suddenly. Left ventricular hypertrophy and ischaemic cardiac disease with severe two-vessel (LAD and RCA) coronary disease (**a**) Severe atherosclerosis in LAD coronary artery (Masson’s trichrome). (**b**) Large areas of replacement fibrosis in the anterior wall of the LV. Toxicology revealed the presence of amphetamines in blood, urine and vitreous humour

11.5 Neuroleptics

Patients with schizophrenia have a three-times higher risk of suffering SCD, and this is associated with an unhealthy lifestyle (smoking, unbalanced diet, drug abuse), parasympathetic hypofunction (which causes changes in heart rate or a decrease in baroreflex sensitivity) and pharmacological treatment with antipsychotics (Fig. 11.13). Antipsychotic medications cause metabolic side effects that increase the risk of cardiovascular disease; but above all, the risk of sudden death associated with antipsychotics is due to a prolongation of the QT interval that can lead to ventricular arrhythmias.

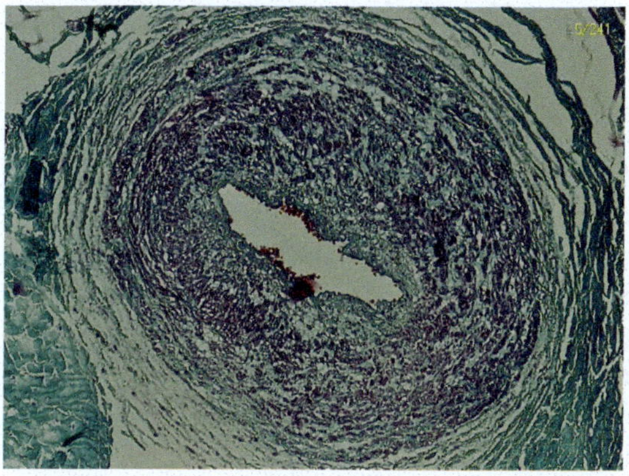

Fig. 11.13 A 29-year-old male diagnosed with schizotypal personality disorder, on medication with Thioridazine, Risperidone and Lorazepam, who died suddenly. Severe stenosis with dysplasia in the AV nodal artery (Masson's trichrome). Toxicology revealed the presence of the drugs prescribed at therapeutic levels

11.6 Chemotherapeutic Agents

Anticancer drugs are responsible for important advances and clinical benefits in the oncology field. However, a wide range of chemotherapeutic agents have been associated with cardiotoxicity, especially with dilated cardiomyopathy and heart failure. Although there is a plethora of antineoplastic drugs with putative cardiotoxicity, the most well known cause of chemotherapy-induced cardiotoxicity due to its incidence and clinical relevance, are anthracyclines, and particularly adriamycin.

Anthracyclines are among the most effective anticancer treatments and have been used since 1960 to combat a broad spectrum of cancer types, both haematological and solid tumours, especially breast cancer. Their main adverse effect is cardiotoxicity (Fig. 11.14), which appears in 10–17 % of the cases, and this cardiotoxicity usually depends on the cumulative dose. Other antineoplastic compounds such as tyrosine kinase inhibitors (Sunitinib) and anti-HER-2 monoclonal antibodies (Trastuzumab) have also been associated with ventricular dysfunction and heart failure, although with much lower incidence. Finally, other agents such as cyclophosphamide, paclitaxel or radiotherapy, despite being used at doses that do not cause significant myocardial damage, can exacerbate or potentiate the damage produced by adriamycin.

The pathophysiology of myocardial damage appears to be caused by the toxicity of the free radicals generated during the metabolism of adriamycin, in combination with the absence of the enzymes and glycoproteins (GP P170) required for its metabolism in the myocardium. These free radicals, especially hydrogen peroxide, which are transported by stable adriamycin molecules are toxic at the cellular level, and in particular for mitochondria. This deleterious effect on the mitochondrial respiratory chain decreases bioenergetic efficiency and ATP production, leading to alterations in the actin-myosin interactions and in the electro-mechanical coupling between cardiomyocytes.

Histological examination shows cellular necrosis and vacuolization of myocytes due to loss of myofibrils ("Adria" cell). Gross examination reveals areas of replacement fibrosis in the myocardium. However, there is no clear clinico-histological correlation and, actually, the toxic actions on the myocardium can appear as soon as the treatment is initiated or later on. Three types of clinical presentation have been described: acute, subacute and chronic. Acute toxicity is the least-common presentation and manifests with altered heart rhythms such as sinus tachycardia or increased supraventricular and ventricular extrasystoles. In some instances myocarditis may ensue, which may be associated with heart failure and even sudden death. The subacute and chronic forms usually manifest as nonspecific episodes of heart failure secondary to dilated cardiomyopathy, which appear between 3 and 5 months after receiving the chemotherapy cycles in the subacute form, and several years after chemotherapy is finished in the chronic one.

The role of the cardiologist is multiple. Thus, the patient's risk of developing clinically-relevant ventricular dysfunction needs to be estimated before starting chemotherapy treatment in order to plan the periodicity schedule for monitoring check-ups, or to reconsider the type of anticancer treatment. For this purpose, it is necessary to evaluate the initial ventricular function, the age of the patient (children and the elderly are considered groups at higher risk of cardiotoxicity), the doses to receive and the coadjuvant treatment, essentially other chemotherapeutical agents that could potentiate the toxic effects.

Ventricular function should be periodically monitored during the chemotherapy cycles, commonly by echocardiography, and by determination of biomarkers with prognostic value: troponins and NT-proBNP. Pharmacological treatment should be administered as soon as structural damage or ventricular dysfunction is detected. Treatment will include drugs commonly used in cases of systolic heart failure: beta-blockers, angiotensin-converting enzyme (ACE) inhibitors, angiotensin II receptor blockers (ARBs) and anti-aldosterone drugs. Diuretics may also be prescribed in congestive states and positive inotropic agents in situations of low cardiac output. From an oncological point of view, several possibilities have been proposed to reduce cardiotoxicity, including a controlled slow release of the chemotherapeutic drugs, the use of liposomal formulations or even the switch to other anticancer agents that tend to be less effective.

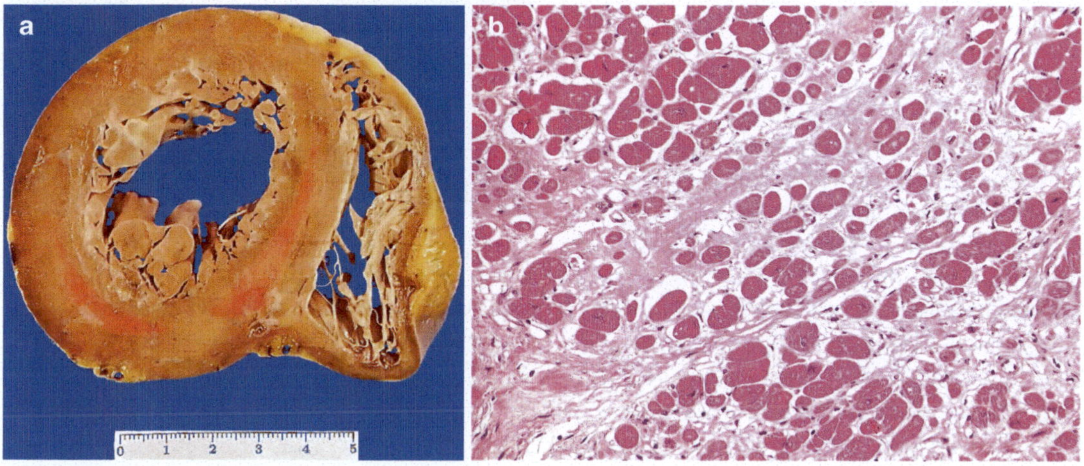

Fig. 11.14 Chronic phase of the cardiotoxic action of anthracyclines used in breast cancer chemotherapy. (**a**) Gross examination reveals diffuse fibrosis mainly subendocardial. (**b**) Microscopic image of the same case where loss of cardiomyocytes and extensive fibrosis can be appreciated

11.7 Carbon Monoxide

Carbon monoxide (CO) poisoning constitutes a common cause of death encountered in forensic pathology, due mainly to home accidents caused by unsafe gas appliances. CO has 250 times-greater affinity for haemoglobin than oxygen, and binds to it forming carboxyhaemoglobin (COHb) that is toxic to cells due to a reduction in the cellular uptake of oxygen, resulting in tissue hypoxia. CO can also bind to mitochondrial cytochrome *c* oxidase, thus interfering directly with cellular respiration. These two mechanisms are considered responsible for the neurological involvement and myocardial damage (Fig. 11.15). Although the cardiotoxic actions of CO have not been fully studied, myocardial damage is considered a direct consequence of a moderate-severe CO poisoning, and could be one of the reasons for the high mortality associated with cardiovascular events that is seen in these patients.

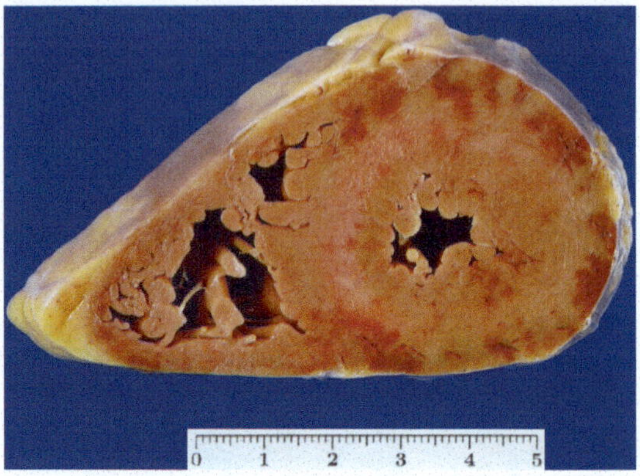

Fig. 11.15 A 49-year-old woman, without known medical history, who died at the family business. Autopsy revealed a generalised red colour in all the viscera. Toxicology of carboxyhaemoglobin levels showed 64 % saturation. Image of the heart where a reddish tabby colour can be appreciated in the subepicardial region

Bibliography

Afonso L, Mohammad T, Thatai D. Crack whips the heart: a review of the cardiovascular toxicity of cocaine. Am J Cardiol. 2007;100:1040–3.

Aquaro GD, Gabutti A, Meini M, Prontera C, et al. Silent myocardial damage in cocaine addicts. Heart. 2011;97:2056–62.

Benzaquen BS, Cohen V, Eisenberg MJ. Effects of cocaine on the coronary arteries. Am Heart J. 2001;142:402–10.

Bertolet BD, Freund G, Martin CA, Perchalski DL, Williams CM, Pepine CJ. Unrecognized left ventricular dysfunction in an apparently healthy alcohol abuse population. Drug Alcohol Depend. 1991;28:113–9.

Capasso JM, Li P, Guideri G. Myocardial mechanical, biochemical and structural alterations induced by chronic ethanol ingestion in rats. Circ Res. 1992;71:346–56.

Chatterjee K, Zhang J, Honbo N, Karliner JS. Doxorubicin cardiomyopathy. Cardiology. 2010;115:155–62.

Cohle SD. Fatal coronary artery intimal hyperplasia due to amphetamine use. Cardiovasc Pathol. 2013;22:e1–3.

Darke S, Kaye S, Duflou J. Cocaine-related fatalities in New South Wales, Australia 1993–2000. Drug Alcohol Depend. 2005;77:107–14.

de la Fuente L, Molist G, Espelt A, Barrio G, Guitart A, Bravo MJ, Brugal MT. Mortality risk factors and excess mortality in a cohort of cocaine users admitted to drug treatment in Spain. J Subst Abuse Treat. 2014;46(2):219–26.

Dettmeyer RB. Histopathology in drug abuse. In: Dettmeyes RB, editor. Forensic histopathology. Fundamentals and perspectives. Berlin/Heidelberg: Springer; 2011. p. 67–94.

Egred M, Davis GK. Cocaine and the heart. Postgrad Med J. 2005;81:568–71.

Fernandez-Sola J, Nicolas JM, Oriola J, Sacanella E, Estruch R, Rubin E, Urbano-Marquez A. Angiotensin-converting enzyme gene polymorphism is associated with vulnerability to alcoholic cardiomyopathy. Ann Intern Med. 2002;137:321–6.

Figueredo VM. Chemical cardiomyopathies: the negative effects of medications and nonprescribed drugs on the heart. Am J Med. 2011;124:480–8.

Galicia M, Nogué S, Burillo-Putze G. Ten years of emergency attendances for cocaine-users in Spain. Med Clin (Barc). 2014. doi:10.1016/j.medcli.2013.11.023 [Epub ahead of print].

Gill JR. The syndrome of excited delirium. Forensic Sci Med Pathol. 2014;10:223–8.

Guzzo-Merello G, Cobo-Marcos M, Gallego-Delgado M, Garcia-Pavia P. Alcoholic cardiomyopathy. World J Cardiology. 2014;6:771–81.

Guzzo-Merello G, Segovia J, Dominguez F, Cobo-Marcos M, Gomez-Bueno M, Avellana P, Millan I, Alonso-Pulpon L, Garcia-Pavia P. Natural history and risk factors in alcoholic cardiomyopathy. JACC Heart Failure. 2015 (in press).

Henning RJ, Li Y. Cocaine produces cardiac hypertrophy by protein kinase C dependent mechanisms. J Cardiovasc Pharmacol Ther. 2003;8:149–60.

Henning RJ, Cuevas J, Ivansicts D, Sanchez A. Cocaine activates calcium/calmodulin kinase II and causes cardiomyocyte hypertrophy. J Cardiovasc Pharmacol. 2006;48:802–13.

Henry CR, Satran D, Lindgren B, Adkinson C, Nicholson CI, Henry TD. Myocardial injury and long-term mortality following moderate to severe carbon monoxide poisoning. JAMA. 2006;295:398–402.

James TN. Small arteries of the heart. Circulation. 1977;56:2–14.

Karch SB. Cocaine cardiovascular toxicity. South Med J. 2005;98:794–9.

Karch SB. Cocaine. In: Karch SB, editor. Pathology of drugs of abuse. 4th ed. Boca Raton: CRC Press; 2009. p. 1–207.

Karch SB. The unique histology of methamphetamine cardiomyopathy: a case report. Forensic Sci Int. 2011;212:e1–4.

Karch SB, Green GS, Young S. Myocardial hypertrophy and coronary artery disease in male cocaine users. J Forensic Sci. 1995;40:591–5.

Karch SB, Stephens B, Ho CH. Relating cocaine blood concentrations to toxicity. An autopsy study of 99 cases. J Forensic Sci. 1998;43:41–5.

Knuepfer MM. Cardiovascular disorders associated with cocaine use: myths and truths. Pharmacol Ther. 2003;97:181–222.

Kolodgie FD, Virmani R, Cornhill JF, Herderick EE, Smialek J. Increase in atherosclerosis and adventitial mast cells in cocaine abusers: an alternative mechanism of cocaine-associated coronary vasospasm and thrombosis. J Am Coll Cardiol. 1991;17:1553–60.

Kolodgie FD, Wilson PS, Mergner WJ, Virmani R. Cocaine-induced increase in the permeability function of human vascular endothelial cell monolayers. Exp Mol Pathol. 1999;66:109–22.

Koponen H, Alaräisänen A, Saari K, Pelkonen O, Huikuri H, Raatikainen MP, Savolainen M, Isohanni M. Schizophrenia and sudden cardiac death. A review. Nord J Psychiatry. 2008;62:342–5.

Lacovoni A, De Mariab R, Gavazzia A. Alcoholic cardiomyopathy. J Cardiovasc Med (Hagerstown). 2010;11:884–92.

Lange R, Hillis N. Mechanism by which cocaine may induce myocardial infarction or ischemia. N Engl J Med. 2001;345:351–8.

Lazarevic AM, Nakatani S, Neskovic AN, et al. Early changes in left ventricular function in chronic asymptomatic alcoholics: relation to the duration of heavy drinking. J Am Coll Cardiol. 2000;35:1599–606.

Lippi G, Rastelli G, Meschi T, Borghi L, Cervellin G. Pathophysiology, clinics, diagnosis and treatment of heart involvement in carbon monoxide poisoning. Clin Biochem. 2012;45:1278–85.

Lucena J, Blanco M, Jurado C, Rico A, Salguero M, Vázquez R, Thiene G, Basso C. Cocaine-related

sudden death: a prospective investigation in southwest Spain. Eur Heart J. 2010;31:318–29.
McKenna CJ, Codd MB, McCann HA, Sugrue DD. Alcohol consumption and idiopathic dilated cardiomyopathy: a case control study. Am Heart J. 1998;135:833–7.
Noord C, Sturkenboom MC, Straus SM, Witteman JC, Stricker BH. Non-cardiovascular drugs that inhibit HERG-encoded potassium channels and risk of sudden cardiac death. Heart. 2011;97:215–20.
Om A, Ellahham S, Vetrovev GW, Guard C, Reese S, Nixon JV. Left ventricular hypertrophy in normotensive cocaine users. Am Heart J. 1993;125:1441–3.
Patel MM, Belson MG, Wright D, Lu H, Heninger M, Miller MA. Methylenedioxymethamphetamine (ecstasy)-related myocardial hypertrophy: an autopsy study. Resuscitation. 2005;66:197–202.
Piano MR. Alcohol and heart failure. J Card Fail. 2002;8:239–46.
Ray WA, Chung CP, Murray KT, Hall K, Stein M. Atypical antipsychotic drugs and the risk of sudden cardiac death. N Engl J Med. 2009;360:225–35.
Robaei D, Grieve SM, Nelson GC, Bhindi R, Figtree GA. Cocaine-induced epicardial coronary artery thrombosis resulting in extensive myocardial injury assessed by cardiac magnetic resonance imaging. Eur Heart J. 2010;31:2446.
Sicouri S, Antzelevitch C. Sudden cardiac death secondary to antidepressant and antipsychotic drugs. Expert Opin Drug Saf. 2008;7:181–94.
Urbano-Marquez A, Estruch R, Navarro-Lopez F, Grau JM, Mont L, Rubin E. The effects of alcoholism on skeletal and cardiac muscle. N Engl J Med. 1989;320:409–15.
Virmani R, Robinowitz M, Smialek SE, Smyth DF. Cardiovascular effects of cocaine: an autopsy study of 40 patients. Am Heart J. 1988;115:1068–76.

Cardiac Tumours

12

Beatriz Aguilera, M. Paz Suárez-Mier,
and Gonzalo Guzzo-Merello

Contents

B. Aguilera (✉) • M.P. Suárez-Mier
Histopathology Service, National Institute of Toxicology and Forensic Sciences,
Las Rozas (Madrid), Spain
e-mail: beatriz.aguileratapia@justicia.es;
mariapaz.suarez@justicia.es

G. Guzzo-Merello
Heart Failure Unit, Cardiology Department, Hospital Universitario de Getafe, Getafe (Madrid), Spain

12.1 Introduction

Tumours found in the heart can be classified, like other tumours, as benign or malignant, and these as primary or metastatic tumours. Diagnosis of cardiac tumours has increased in the last decades mainly due to the widespread use of advanced imaging techniques (CT and CMR). The estimated frequency of primary cardiac tumours ranges from 0.001 to 0.03 %, and 75 % of them are benign. Cardiac myxomas are, by far, the most common primary tumours of the heart. The heart is also a target organ for tumour metastasis from distant organs. According to several studies, metastatic cardiac lesions are found in up to 12 % of the hearts examined in autopsies performed on cancer patients. Both primary and metastatic cardiac tumours have been described among people of all age groups, from foetuses to the elderly. However, some types of cardiac tumours appear to preferentially affect certain age groups (Table 12.1). From a clinical point of view, cardiac tumours should be differentiated from other cardiac masses such as thrombus, which are much more frequent.

J.S. Lucena et al. (eds.), *Clinico-Pathological Atlas of Cardiovascular Diseases*,
DOI 10.1007/978-3-319-11146-9_12, © Springer International Publishing Switzerland 2015

Table 12.1 Classification of primary cardiac tumours based on the age at presentation

Predominantly in children	Predominantly in adults
Rhabdomyoma	Myxoma
Fibroma	Papillary fibroelastoma
Purkinje cell tumour	Lipomatous hypertrophy of the interatrial septum. Lipoma
Teratoma	Hemangioma-Angiosarcoma
Rhabdomyosarcoma	Lymphomas
Myxoma	Paraganglioma
Inflammatory myofibroblastic tumour	Undifferentiated and synovial sarcomas
	Pericardial mesothelioma

Cardiac masses, either tumoural or thrombotic, may cause a variety of clinical presentations. In asymptomatic patients, diagnosis is usually incidental as a result of an imaging test obtained for an unrelated reason. In patients with symptoms, clinical presentation depends on the size of the mass, anatomic location, and friability or ability to embolise. With regards to size and location, these lesions occupy the intracardiac space, and as a consequence the filling of the cardiac chamber where they locate may be compromised. Masses in the left atrium and ventricle may generate clinical signs of pulmonary congestion (dyspnoea, intolerance to supine decubitus), whereas masses in the right heart chambers largely promote signs of venous congestion (oedema or jugular ingurgitation). Malignant tumours that invade the pericardium can provoke episodes of cardiac tamponade.

Independent of their dimensions, if lesions invade the myocardium in depth, clinical presentation will result from different types of arrhythmias: bradycardia due to cardiac blocks, if the lesion alters the integrity of the cardiac conduction system (AV node, His bundle), or atrial or ventricular tachycardias if the lesion damages or irritates the myocardium. In a study by Cina et al. (1996), on a series of 120 cases, the authours reported that AV node tumours, fibromas and atrial myxomas were among the type of tumours in which the first manifestation was mainly sudden death. Finally, depending on the morphological features, the mobility and the disintegration risk, cardiac masses can embolise and provoke symptomatology due to the resultant ischaemia produced in the target organ, which is commonly the brain, although peripheral territories or even coronary arteries can also be involved.

A definite diagnosis can only be completed after histological examination of samples taken at autopsy or biopsy. In clinical practice, biopsies can be obtained by catheterisation in situations where the mass is located in the right chambers, since a possible embolism secondary to the procedure would have less clinical consequences than if the mass was located in the left chambers. Accessibility and mobility also determine whether or not a biopsy is recommended. In those instances where a biopsy is not a safe choice, a surgical intervention should be performed.

Patient prognosis depends on the type of primary tumour, invasiveness in the heart and other organs, and on the possibility of surgical resection. Regarding the latter, imaging techniques such as echocardiography, CT and CMR are of special value. Recent advances in imaging have allowed a better prediction of the different possible diagnoses that, as mentioned previously, can only be certain after histopathological examination.

Echocardiography is the most available imaging technique and the one most commonly used to establish the presence of a myocardial mass. However, it does not accurately differentiate between a thrombus and a benign or malignant tumour. Clinical features help in the diagnosis; for example, atrial fibrillation in the presence of a mass in the atrial appendage, or masses located on a post infarction ventricular aneurysm, are indicative of a thrombus. MRI offers a high spatial resolution and is the best modality for characterising tumour tissue and vascularity. When MRI is not indicated, for example in patients with claustrophobia or carrying a pacemaker or defibrillator, then computed tomography (CT) is an alternative. CT can be used to locate and identify calcifications and to visualise the mass in relation to the coronary arteries. It is generally assumed that the following features suggest malignancy:

location in the right chambers, ill-defined borders, wide implantation base, size greater than 5 cm, involvement of several chambers, mediastinum or great vessels, and associated pericardial effusion.

Diagnostic approaches and differential diagnosis of the distinct cardiac masses, based on the characteristics identified in each of the imaging modalities, are summarised in Table 12.2.

Remarks
Primary cardiac tumours:
75 % of them are benign
Their incidence is increasing
Have been described in all ages, from foetuses to the elderly
Can manifest with obstruction, embolisation and cardiac arrhythmias
Can cause or debut as sudden death
Definite diagnosis is done by histopathology

Table 12.2 Characteristics of cardiac tumours based on imaging techniques

Mass	Echocardiogram	Magnetic resonance	Tomography
Thrombus	Commonly located on ventricular aneurysms, atrial appendage or mechanical prosthesis. Use of contrast agents does not produce an enhancement	Appearance of the lesion may vary in T1–T2, without first-pass or delayed contrast enhancement	Low attenuation with possible attenuation foci
Myxoma	Heterogeneous appearance. Often pediculated. Main differential diagnosis with thrombus. Most common location is in the left atrium, typically in the atrial septum in the region of the *fossa ovalis*. Unlike thrombus myxoma appears enhanced after contrast administration	T1-weighted sequences show isointense or hypointense signals in relation to the myocardium, whereas lesions are hyperenhanced in T2. Contrast enhancement. Occasional heterogeneous appearance may be due to necrotising and haemorrhagic areas	14 % show calcifications
Fibroelastoma	Small and mobile. Generally located in valves or in subvalvular apparatus. Signal intensity is equal to or denser than the myocardium	Difficult to characterise given the small size and mobility. Well-circumscribed. Intermediate signal, similar to endocardium, in T1 and T2. Homogeneous delayed enhancement due to fibroelastic tissue	Low attenuation focal lesion
Lipoma	Homogeneous appearance. Frequently located in septum or intramyocardial	Well-circumscribed. Can be diagnosed due to fatty suppression in the T1 and T2 sequences. Bright T1 sequence with the "black blood" technique. Low intensity signal on T2	Circumscribed lesion with fatty attenuation
Haemangioma	Very variable appearance and location, intramural or cavitaries In general, heterogeneous and hyperechogenic lesions. Cavern images, sometimes multiloculated, with trabeculae and free spaces. It can also appear as globally hypodense	Isointense on T1-weighted images and hyperintense T2 sequences relative to the myocardium. Signal intensity intermediate between blood and myocardium, in the "white blood technique". Heterogeneous, but intense and prolonged delayed enhancement	Calcification foci and significant contrast enhancement

(continued)

Table 12.2 (continued)

Mass	Echocardiogram	Magnetic resonance	Tomography
Fibroma	Usually intramyocardial and causes a thickening which is hard to distinguish from LV hypertrophy. Hyperechogenic with no contractility. Alterations in strain rate parameters	Well circumscribed and of low intensity. Ring that surrounds and compresses the myocardium. Isointense on T1- and hypointense on T2- weighted images, relative to the myocardium. Absence of gadolinium contrast enhancement due to low vascularisation, although sometimes lesions offer delayed homogeneous or heterogeneous enhancement	Calcification foci are present in 50 % of the cases
Rhabdomyoma	Young patients, often associated with tuberous sclerosis in up to 50 % of cases. Intramyocardial location, often in the septum. They are typically multiple, between 0.1 and 9 cm in size, hyperechoic and not calcified	Iso- to hyperintense signals on T1 weighted images and hyperintense on T2 sequences (in contrast to fibroma). Uniform and intense delayed enhancement	
Rhabdomyosarcoma	Arises from the myocardium towards any of the chambers. Often affects or invades the valvular apparatus	Heterogeneous signals on T1 and T2- weighted images. Solid peripheral component hyperintense with contrast enhancement	Low attenuation signal
Angiosarcoma	75 % of these are found in the right atrium and involve the pericardium, tricuspid valve and right ventricle. Blood flow inside the lesion. Pericardial effusion is common	Isointense and hyperintense on T1- and T2-weighted images, respectively. Marked enhancement, blood flow within the neoplasm	Mass of heterogeneous enhancement. Highly attenuated pericardial effusion
Sarcoma with osteo-, myo or fibroblastic differentiation	Predominantly in the left atrial wall that can spread and invade adjacent areas. Young adults	Heterogeneous, infiltrating	Osteosarcoma shows calcification
Lymphoma	Preferentially in the right chambers. Cardial wall thickening. Ventricular dysfunction. Pericardial effusion	Hypointense as seen on T1-weighted images and hyperintense on T2 sequences. Heterogeneous contrast enhancement	Hypodense or isodense relative to the myocardium. Heterogeneous contrast enhancement. Mediastinum involvement

12.2 Cardiac Myxomas

Myxomas are tumours of uncertain histogenesis that appear exclusively in the heart. Cardiac myxomas represent the most common primary cardiac tumour, especially in persons between the ages of 30 and 60 years (mean age at presentation is 50 years), with two thirds of the cases involving women. Around 80–90 % of myxomas are located in the left atrium, arising from the *fossa ovalis*, and around 15 % appear in the right atrium. The majority of myxomas are pedunculated and have a stalk that inserts into the region of the *fossa ovalis* (Fig. 12.1a); this can elongate, and allow the protrusion of the mass through the valvular orifice. Myxomas are heterogeneous in size, and in general range from 1.5 to 7 cm, although in some cases they can be as large as 15 cm in diameter. In 90 % of the cases, there is a single tumour. The morphological characteristics of cardiac myxomas are summarised in Table 12.3.

Clinical presentation is very variable. They can manifest with congestive symptoms due to the obstruction of blood resulting from the presence of a mass within the cardiac chambers. Left atrial myxomas cause pulmonary congestion whereas right atrial myxomas provoke generalised oedemas. In more than one third of myxomas, symptomatology results from embolisation occurring in any systemic tissue, or in the pulmonary territories, dependent upon tumour location. This is especially frequent in myxomas with a papillary and friable surface.

Approximately 20 % of patients present a constitutional syndrome secondary to IL-6 production by tumour cells, with fever, arthralgia, weight loss, anorexia and elevated erythrocyte sedimentation rate. Myxomas smaller than 4 cm are usually asymptomatic and may be discovered incidentally. Myxomas can debut as sudden death. In children, there are reports of sudden death due to myxomas located in the right atrium or tricuspid valve, as a consequence of pulmonary thromboembolism. Although infection of a myxoma is rare, when present, the initial manifestations mimic those of infective endocarditis or vasculitis.

Around 5 % of all cardiac myxomas are associated with Carney complex, which should be suspected in patients with spotty skin pigmentation, atypical location, or multiple or recurrent myxomas. Familial myxomas affect younger people compared with sporadic myxomas, and usually recur in 10 % of the cases. Carney complex is most commonly caused by mutations in the *PRKAR1A* gene, which is a tumour suppressor. In these cases, a familial study is recommended.

Echocardiography, which is the method of choice for diagnosis, reveals the presence of pediculated and mobile masses. Differential diagnosis with atrial thrombus, fibroelastoma or atrial septal aneurysm (described below) should be carried out. However, a definitive diagnosis cannot be established until pathological examination is completed. This is performed by biopsy or after tumour resection, indicated if diagnosis of thrombus is inconclusive.

Table 12.3 Pathology of cardiac myxomas

Macroscopic features:
35 % of them have a papillary, gelatinous and friable surface (tends to embolise) (Fig. 12.1a)
65 % of them are solid, with a smooth surface (clinically they cause heart failure) (Fig. 12.2a)
Usually present haemorrhagic areas that confer a heterogeneous colour, and are surrounded by fibrin (Fig. 12.2b)
The tumour may become calcified, more common in myxomas located in the right atrium (Fig. 12.2b)
Microscopic features:
Tumoral cells are stellate, round or fusiform, dispersed on loose stroma rich in acid mucopolysaccharides
Myxoma cells may be arranged in cords or around the vessels
The stroma contains infiltrates of lymphocytes or plasma cells, haemorrhagic areas with siderophages, collagenised zones and ossification (Fig. 12.3b)
The surface is delineated by grooves of endothelial cells and fibrin deposits, which can embolise with the tumour
The majority of myxomas are diagnosed through H&E staining
Immunohistochemistry: myxoma cells are positive for calretinin (Fig. 12.4) and endothelial markers e.g. CD34 and CD31

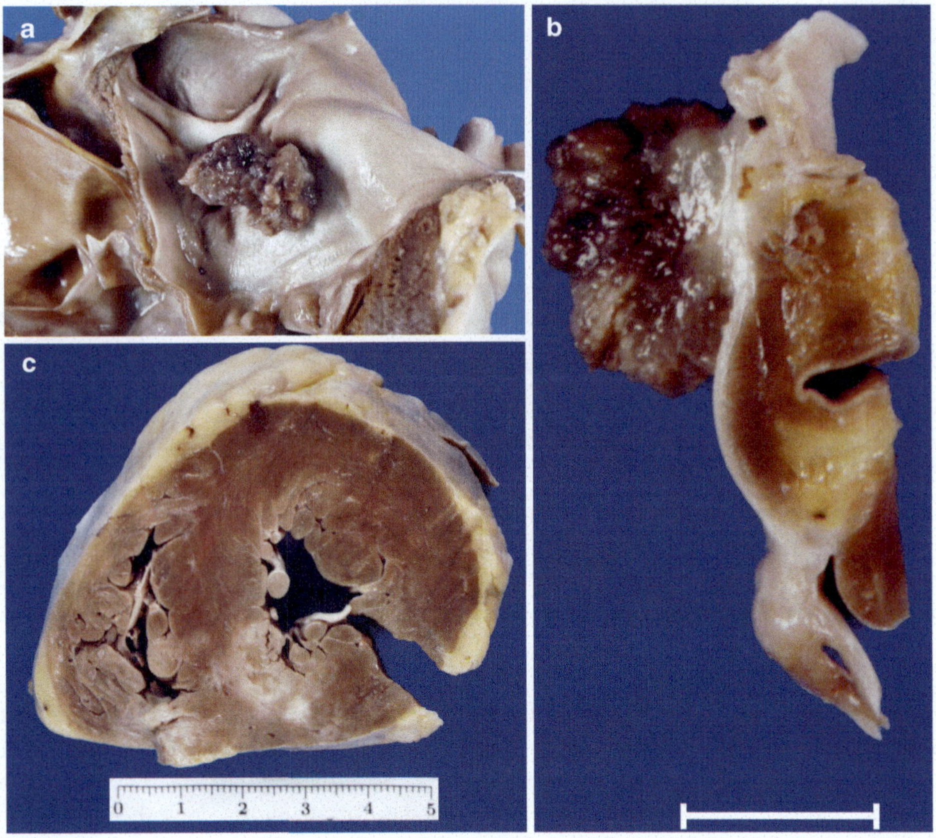

Fig. 12.1 *Myxoma in the left atrium*. A 43-year-old woman with a history of anxiety and depression was feeling unwell and died while being taken to hospital. Heart weight was 314 g. Coronary arteries were permeable. (**a**) A tumour of 1.7 cm, polyploid, with a jelly-like consistency, irregular surface, and friable, was present in the left atrium at the level of the *fossa ovalis*. (**b**) Section shows the tumour implanted on the endocardium, without infiltrating the atrial wall. (**c**) Several whitish and softened areas can be observed in the myocardium; the largest area in the posterior zone of the ventricular septum. (**d**) Microscopically, the tumour consists of a proliferation of cells that are arranged forming cords embedded in an abundant myxomatous stroma, where isolated stellate cells can be observed. (**e**) In samples taken from the softened areas of the myocardium, infarcted areas and the presence of several intramural arteries that contain emboli from the myxoma can be observed. Infarcts show different degrees of evolution, with some areas at the level of the septum with recent myocyte necrosis, suggestive of repeated embolisation episodes

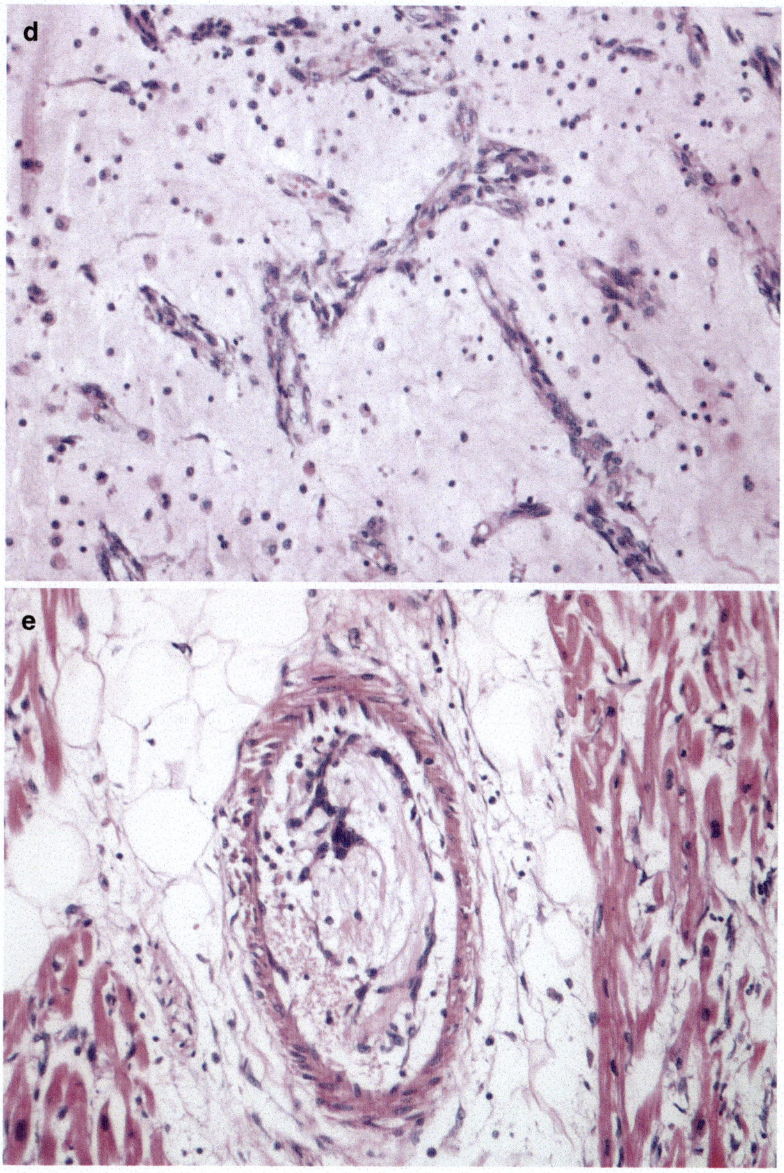

Fig. 12.1 (continued)

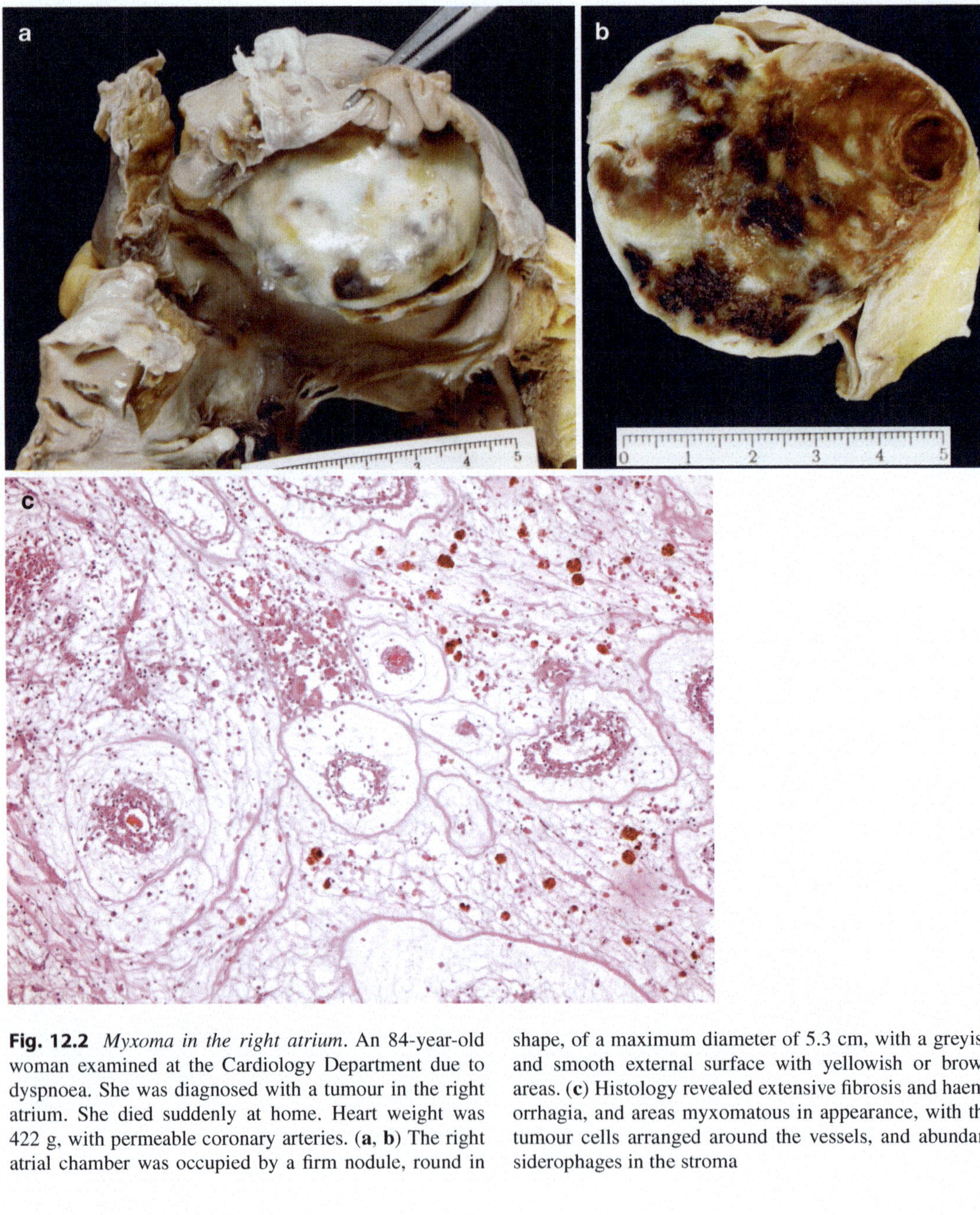

Fig. 12.2 *Myxoma in the right atrium*. An 84-year-old woman examined at the Cardiology Department due to dyspnoea. She was diagnosed with a tumour in the right atrium. She died suddenly at home. Heart weight was 422 g, with permeable coronary arteries. (**a**, **b**) The right atrial chamber was occupied by a firm nodule, round in shape, of a maximum diameter of 5.3 cm, with a greyish and smooth external surface with yellowish or brown areas. (**c**) Histology revealed extensive fibrosis and haemorrhagia, and areas myxomatous in appearance, with the tumour cells arranged around the vessels, and abundant siderophages in the stroma

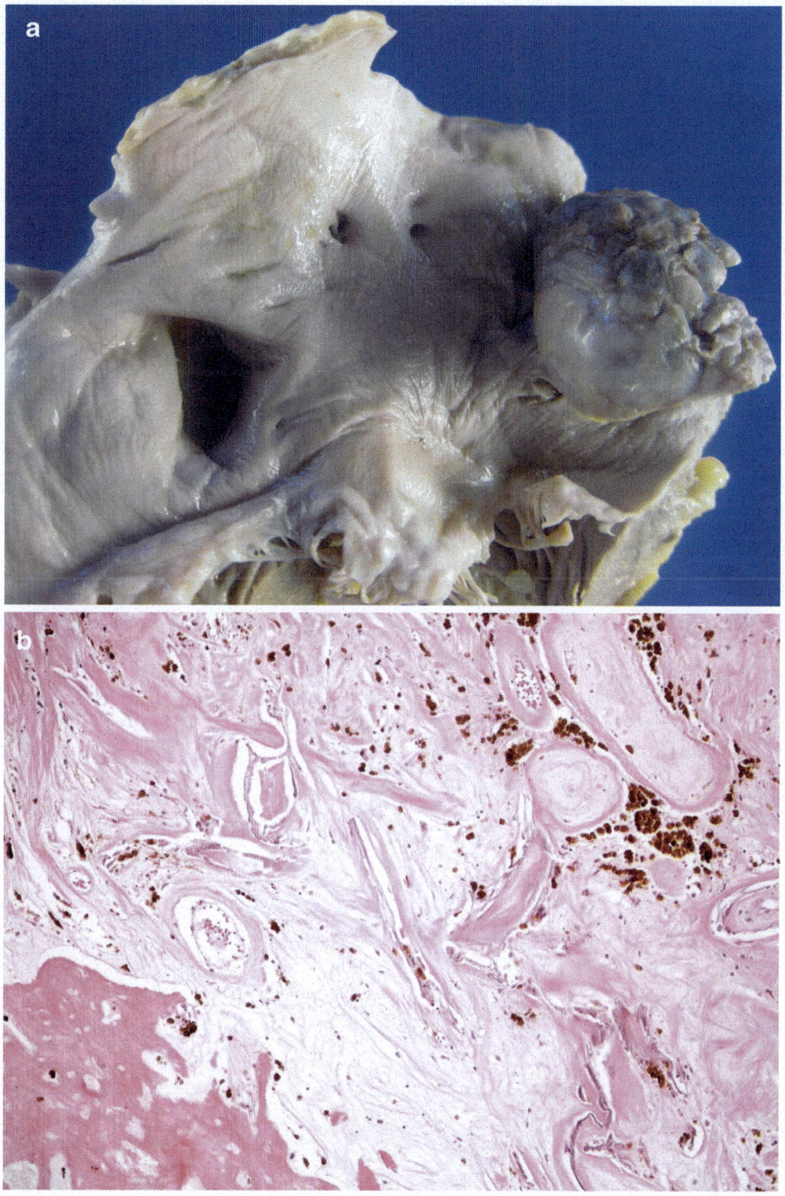

Fig. 12.3 *Myxoma in the left atrium*. A 68-year-old woman who died suddenly at home. Heart weighted 434 g and coronary arteries were permeable. (**a**) The sole autopsy finding was a tumoural node, round in shape, 3 cm in diameter, with a part of its surface smooth and the other lobulated. The tumour had a stalk of 1 cm in length arising from the *fossa ovalis*. Sectioning showed a firm tumour, with calcified areas and zones ochre in colour. (**b**) Histology showed extensive fibrosis, abundant siderophages and a focus of bone metaplasia around myxoma debris, with the typical perivascular arrangement

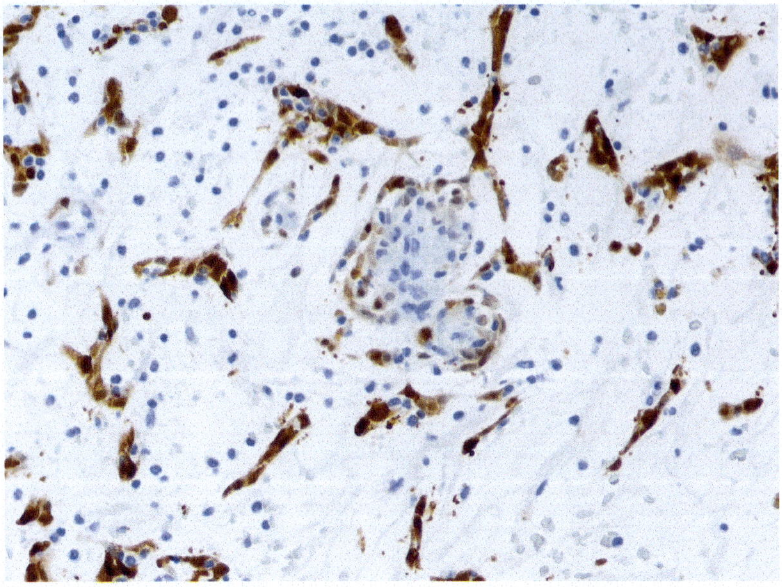

Fig. 12.4 *Myxoma*. Strong positivity for calretinin (Courtesy of Sara Chaves. FJD-Capio)

12.3 Tumours of Cardiac Muscle

Rhabdomyomas and rhabdomyosarcomas occur in the paediatric age group. There is also an extremely rare type of tumour that is formed by mature myocytes, termed hamartoma.

12.3.1 Rhabdomyomas

Cardiac rhabdomyomas represent the most common cardiac tumour in paediatric patients. As occurs with other tumours of the heart, clinical presentation depends on the location, size and number. Tumours may manifest as arrhythmias, blood flow obstruction, heart failure, pericardial effusion and sudden death. Echocardiography is the most effective imaging technique to provide adequate information for diagnosis of rhabdomyomas. Furthermore, given that these types of tumours constitute an important cause of foetal hydrops and intrauterine deaths, echocardiography is also especially important during the gestational period. Tumours appear as intramural or intracavitary masses, well circumscribed and echogenic, which can occur in any location in the heart but are more common in the ventricles.

Cardiac rhabdomyomas can be divided in three groups: sporadic, those that appear in tuberous sclerosis, and those associated with cardiac malformations such as hypoplastic left heart syndrome, tetralogy of Fallot, transposition of great vessels, etc. Most children with rhabdomyoma have tuberous sclerosis. Two altered genes have been confirmed for tuberous sclerosis: *TSC-1* on chromosome 9q34 that encodes hamartin, and *TSC-2* on chromosome 16p13 that encodes tuberin. Both genes are tumour suppressors.

Rhabdomyomas have a natural history of spontaneous regression and hence, regular monitoring by echography is sufficient. Surgical resection is limited to cases of blood flow obstruction, or arrhythmias that do not respond to pharmacological treatment. Morphology is summarised in Table 12.4.

Table 12.4 Pathology of cardiac rhabdomyomas

Macroscopic features:
They most often occur in the ventricles (Fig. 12.5a)
Well-demarcated nodules, generally single, that can become quite large, especially in sporadic cases
Whitish and firm nodules, in contrast to the surrounding myocardium
In patients with tuberous sclerosis, nodules are multiple and can be numerous
Microscopic features:
They are formed by enlarged cells with a monomorphic central nucleus and a clear and vacuolated cytoplasm (Fig. 12.5c)
Some cells adopt a morphology that resembles a spider ("spider cells"), due to a radial distribution of the myofilaments around the nucleus
Cytoplasm is PAS (+) due to abundant intracellular glycogen content
Immunohistochemical studies demonstrate cells positive for myoglobin, desmin, actin and vimentin

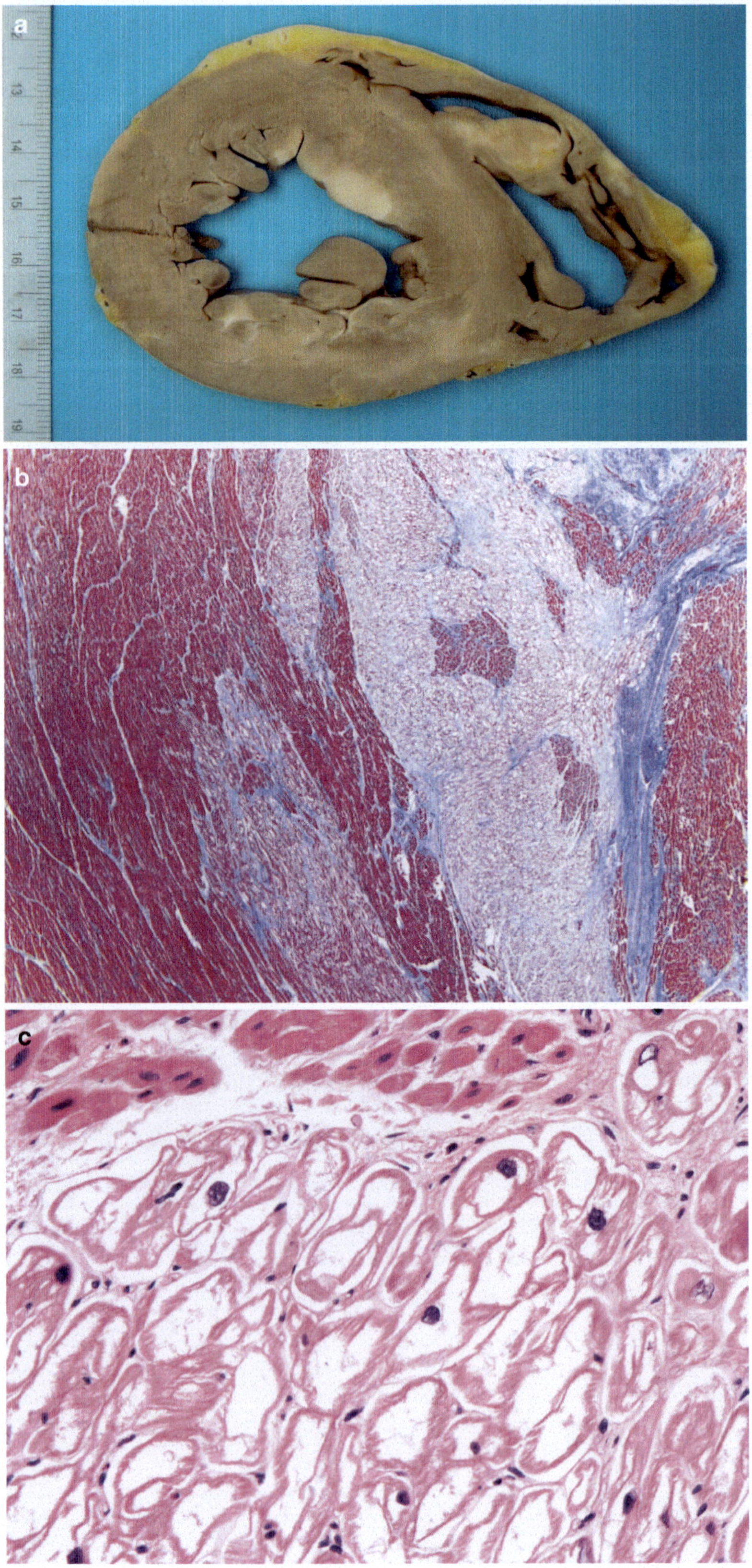

Fig. 12.5 *Rhabdomyoma*. A 15-year-old girl, with a slight developmental delay. She did not show any symptomatology before suffering syncope and cardiac arrest. Heart weight, 310 g. (**a**) Cross-section of the ventricles showed well-circumscribed pale areas mainly in the ventricular septum. (**b**) Microscopic low-power view showing pale appearance of the tumour in contrast to the surrounding myocardium (Masson´s trichrome). (**c**) High magnification of the tumour shows cells with abundant clear cytoplasm, in contrast to the normal myocardium that appears at the top of the picture

12.3.2 Rhabdomyosarcomas

Rhabdomyosarcomas are malignant tumours with a morphology that resembles skeletal muscle. Primary cardiac rhabdomyosarcomas are very rare and are observed primarily in children and young adults, although some cases have also been reported in the elderly. They occur in the atria or the ventricles, but can also arise from the mitral valve (Fig. 12.6a). Due to their large dimensions, they can provoke an obstruction in the blood flow of the heart.

Histologically, cardiac rhabdomyosarcomas are embryonal, and rhabdomyoblasts can be identified by the presence of cytoplasm cross striations or a typical tadpole shape (Fig. 12.6b). These cells are positive for desmin.

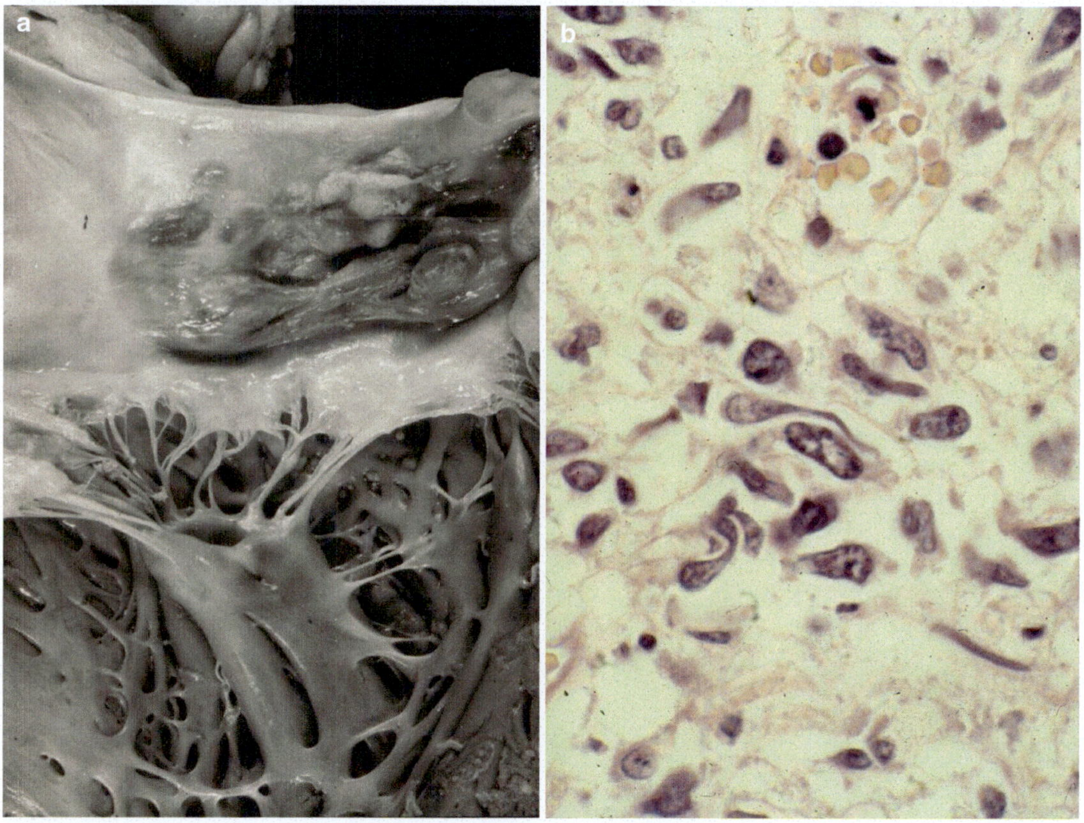

Fig. 12.6 *Rhabdomyosarcoma*. A 36-year-old patient. (**a**) Autopsy reveals an exophytic mass in the posterior wall of the left atrium. (**b**) Photomicrography showing a lesion formed by pleomorphic cells, some of them with the typical tadpole shape and cytoplasm striations (*Courtesy of Dr. Gaetano Thiene and Dr. Cristina Basso. Department of Cardiac, Thoracic and Vascular Sciences. Padua University. Italy*)

12.4 Fibromas

Cardiac fibromas are mesenchymal tumours composed of fibroblasts and collagen. They are discovered in children, often before 1 year of age, teenagers and young adults. There is no sex predominance. Although tumours are well circumscribed macroscopically, histological examination shows undefined borders due to infiltration in the surrounding myocardium. Symptomatology is caused by obstruction in blood flow or interference with valvular function, and can manifest as arrhythmias, syncope, or even sudden death. Total or partial resection is indicated when the tumour grows uncontrollably and/or if the patient is clinically compromised, usually with congestive signs. Heart transplantation is an option when the tumour proves unresectable. Morphology is summarised in Table 12.5.

Table 12.5 Pathology of cardiac fibromas

Macroscopic features:
Whitish and single intramural nodule, well circumscribed but not encapsulated
Located in the ventricular septum (Fig. 12.7) or in the free wall of the LV (Fig. 12.8a)
Tumours have also been reported in the RV and, exceptionally, in the atria
Maximum diameter varies between 1 and 10 cm
Microscopic features:
Monomorphic proliferation of fibroblasts, often with great cellularity (Fig. 12.8b)
More evolved tumours are less cellular, with large amounts of collagen and often focal calcification (Fig. 12.8c)
Poorly defined borders, with the tumour infiltrating the surrounding myocardium (Fig. 12.8d)

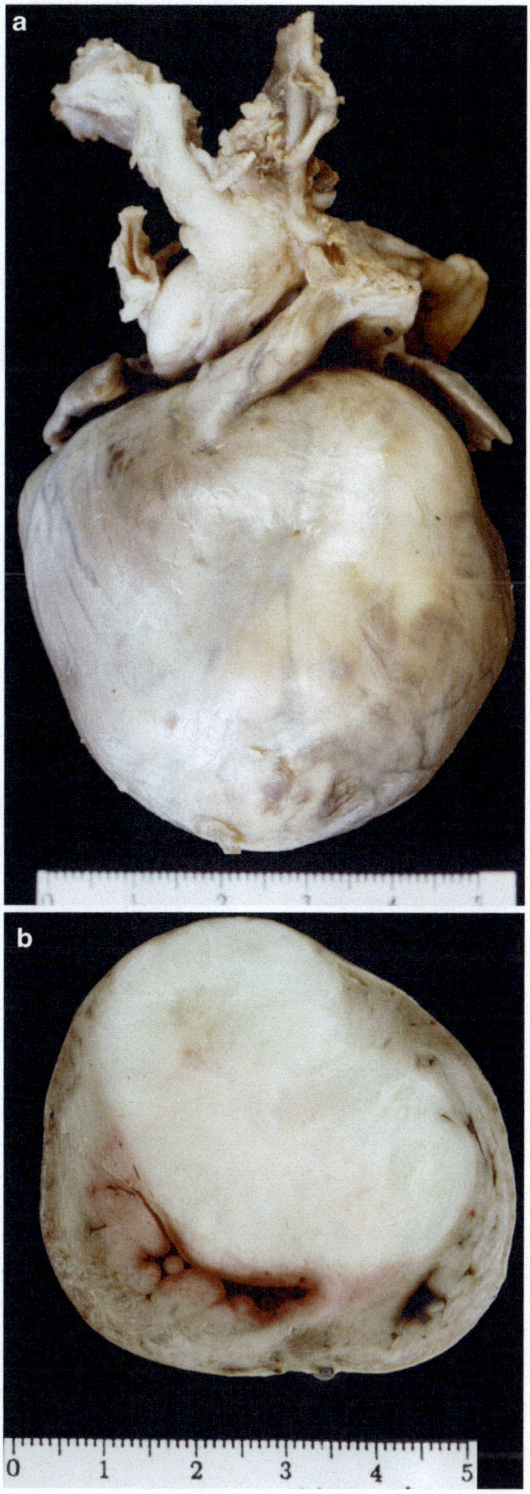

Fig. 12.7 *Fibroma*. A 1-month-old infant. Heart weight of 93 g (normal weight ± 30 g). (**a**) An enlarged and protruded silhouette of the LV can be observed. (**b**) Cross-section of the ventricles shows a whitish rounded mass of 4.5 cm in diameter that occupies the ventricular septum and the anterior wall, which grows and obliterates the ventricular chambers

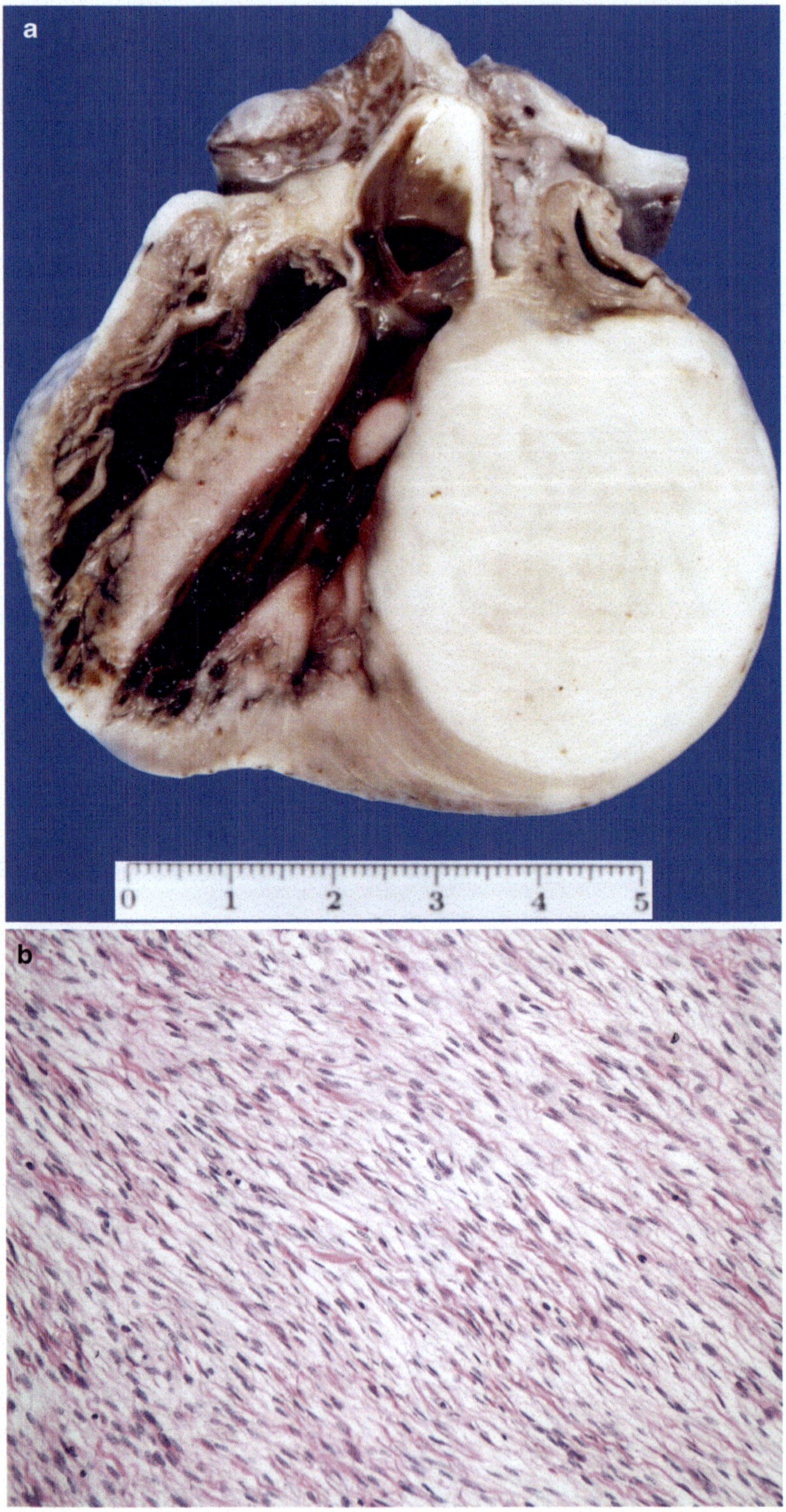

Fig. 12.8 *Fibroma.* A 9-month-old infant with unknown medical history who died suddenly at home. Heart of 108 g and globoid in shape (normal weight for the age ± 43 g). (**a**) Frontal section of the heart showing a well-circumscribed whitish tumour in the lateral wall of the LV. The tumour is firm, whorled, with an antero-posterior diameter of 4.5 cm, and causes obstruction in the LV outflow tract. (**b**) Tumour consists of a monomorphic proliferation of spindle, bland fibroblasts, intermingled with collagen fibres. (**c**) Areas with predominant collagen formation, focal calcification and lymphocytic infiltrate. (**d**) Histology shows an irregular border and infiltration in the surrounding myocardium

Fig. 12.8 (continued)

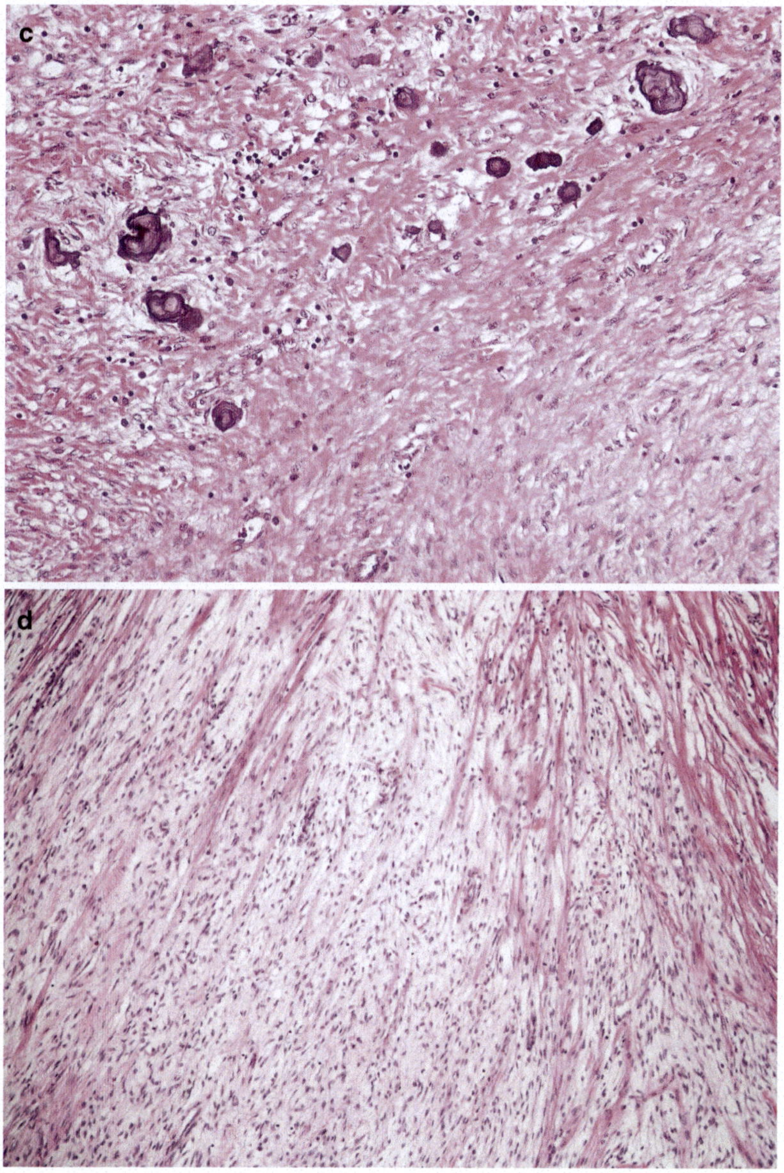

12.5 Papillary Fibroelastoma

Papillary fibroelastoma is the second most frequent tumour of the heart. These types of tumours are proliferations that resemble Lambl's excrescences and are considered hamartomas originating from the endocardium. They are usually mobile due to the small size, the pedunculated morphology and the location, generally on heart valves or subvalvular complexes that are in constant movement. For this reason, the most common clinical manifestations, especially when located on the aortic valve, are the result of myocardial ischaemia produced by intermittent obstruction of the coronary *ostium*, or by embolisation within the coronary artery. Recurrent pulmonary thromboembolism has been described in tumours located in the right chambers. Diagnosis is performed by echocardiography. Surgical resection of the tumour, while keeping the native valve, is indicated in cases of papillary fibroelastomas with great mobility. Morphology is summarised in Table 12.6.

Table 12.6 Pathology of papillary fibroelastoma

Macroscopic features:
Pediculated lesions with papillary fronds (Figs. 12.9 and 12.10)
Tumours are found most commonly on valves, preferentially on the aortic valve (Fig. 12.9a)
Lesions can be multiple
Tumours of up to 3 cm in size have been described, although the majority are less than 1 cm
Microscopic features:
They are branching papillary avascular structures, surrounded by endothelium (Fig. 12.9b)
Papillae with elastic fibres arranged in parallel to the endothelium (Fig. 12.9c)
Tumours may be surrounded by fibrin
Papillae can detach and embolise

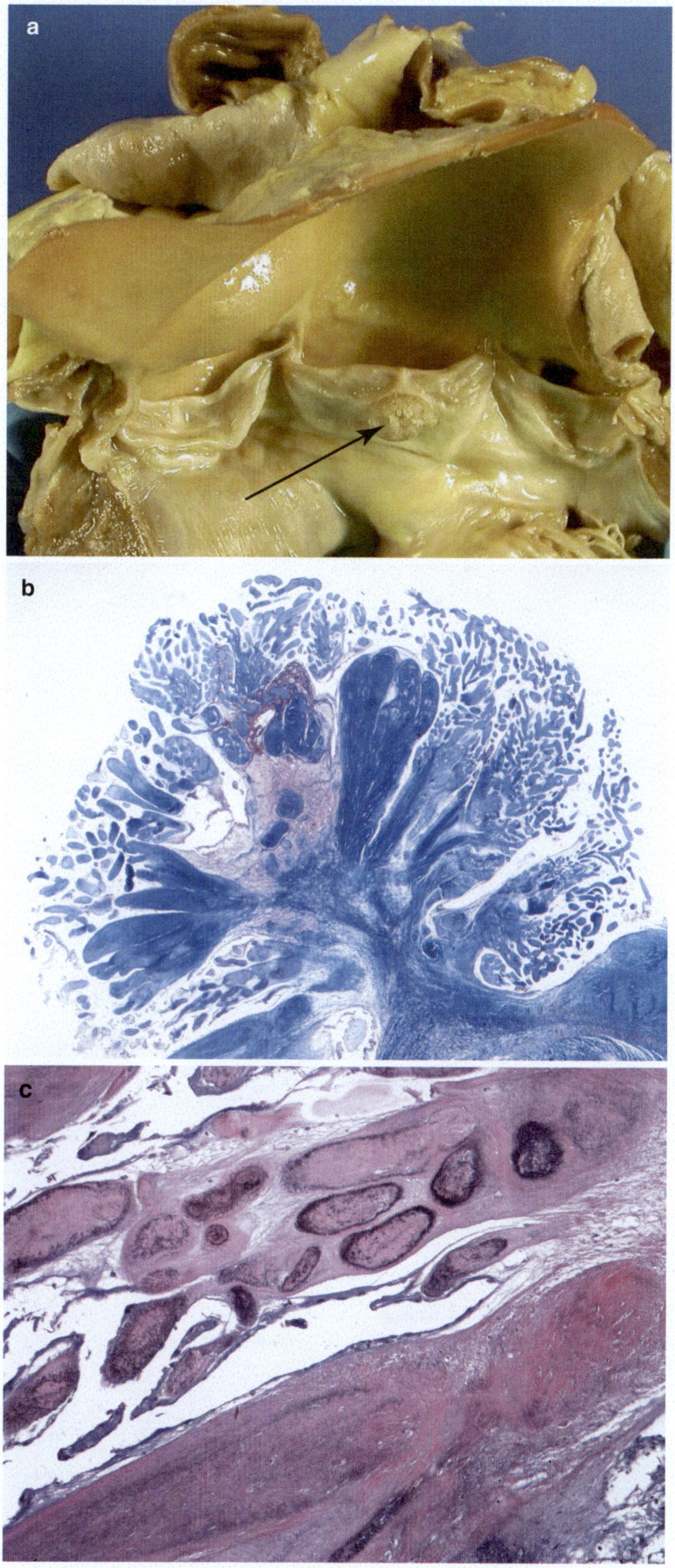

Fig. 12.9 *Papillary fibroelastoma.* Incidental finding in a case of drowning in a 45-year-old male, associated with alcohol consumption. (**a**) A non-pediculated nodule, 1 cm in diameter and with a papillary surface (*arrow*) can be observed in the non-coronary leaflet of the aortic valve, in the region of the Arantius'nodule. (**b**) Low magnification image showing a short stalk from which numerous papillae arise (Masson's trichrome). (**c**) Elastic fibres around the core of the papillae (Weigert's stain for elastic fibres) (*Courtesy of Mª Soledad Sánchez de León. National Institute of Toxicology and Forensic Sciences. Madrid. Spain*)

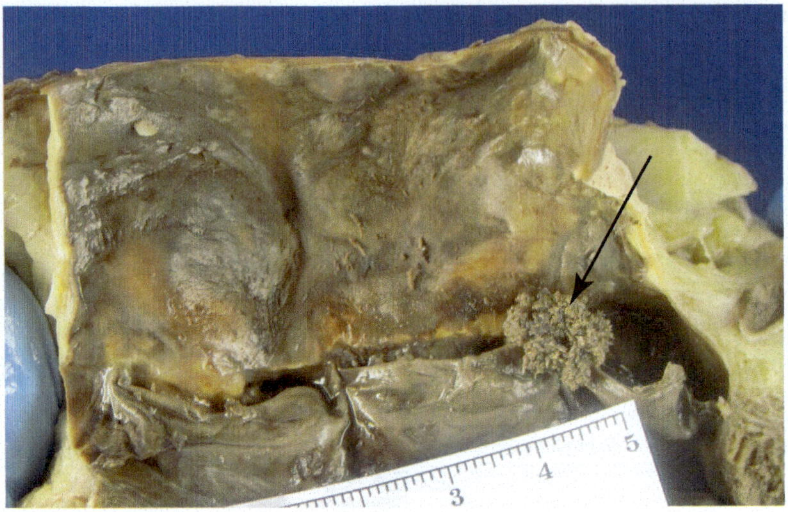

Fig. 12.10 *Papillary fibroelastoma*. Sudden death in a 65-year-old male, with hypertension. There was a mass of 1 cm in diameter in the aortic valve, implanted on the left commissure, with a papillary surface, typical of fibroelastoma (*arrow*) (*Courtesy of Rosario M[a] Mosquera. National Institute of Toxicology and Forensic Sciences. Madrid. Spain*)

12.6 Tumours of Adipose Tissue

12.6.1 Lipomatous Hypertrophy of the Interatrial Septum

Lipomatous hypertrophy of the interatrial septum is a benign cardiac mass rather than a true neoplasm. It is produced by massive and non-encapsulated adipose tissue deposits in the interatrial septum, and can reach a thickness greater than 2 cm (Table 12.7). Fat deposition in the interatrial septum increases with age and is more common in obese people. Atrial septum no thicker than 1.5 cm is considered normal.

Histogenesis of lipomatous hypertrophy of the interatrial septum is unclear. For some authours, it represents a hamartomatous lesion and for others it may be a metabolic disorder. The mean age at presentation is the seventh decade. It is usually asymptomatic and incidentally found at autopsy in overweight and obese people. When symptomatology appears, it can present as supraventricular arrhythmias, ectopic rhythms and compression of the superior vena cava, which may make surgical resection necessary. Table 12.7 describes the pathologic findings.

Table 12.7 Pathology of lipomatous hypertrophy of the interatrial septum

Macroscopic features:
Fat deposits in the interatrial septum thicker than 2 cm (Figs. 12.11a and 12.12)
The deposit bulges towards the atrial chambers
Well-circumscribed
It can extend around the orifice of the superior *vena cava* (Fig. 12.11a)
Microscopic features:
It consists of mature adipocytes intermingled with brown adipose tissue (characterised by a multivacuolated cytoplasm and a small central nucleus), fibrosis and cardiac myocytes, in a variable proportion (Fig. 12.11b)
Entrapped myocytes can be hypertrophic, with large and hyperchromatic nuclei

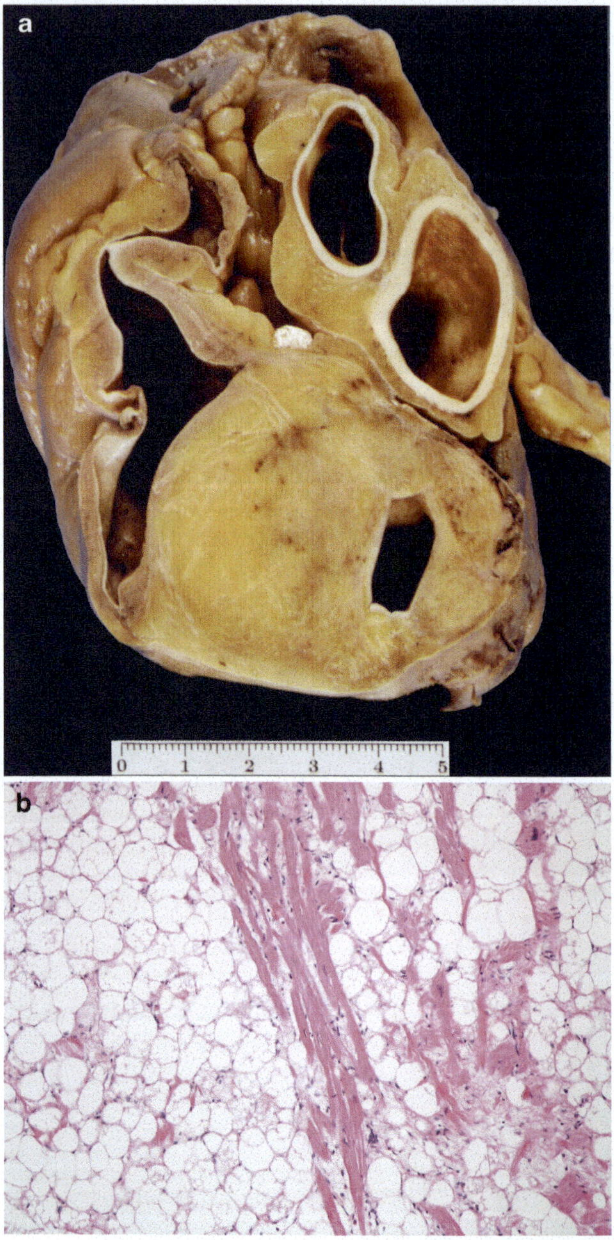

Fig. 12.11 *Lipomatous hypertrophy of the interatrial septum*. A 52-year-old male, with a history of alcohol consumption and depressive disorder who died suddenly. Heart weight of 504 g and severe coronary artery atherosclerosis. Thickness of the interatrial septum is 3.5 cm. (**a**) Cross-section at the level of the atrial roof. Adipose tissue extends around the superior *vena cava*. (**b**) Microscopically, the lesion is formed by mature adipocytes, brown adipose tissue (with a microvacuolated cytoplasm), cardiac myocytes and scant collagen deposition

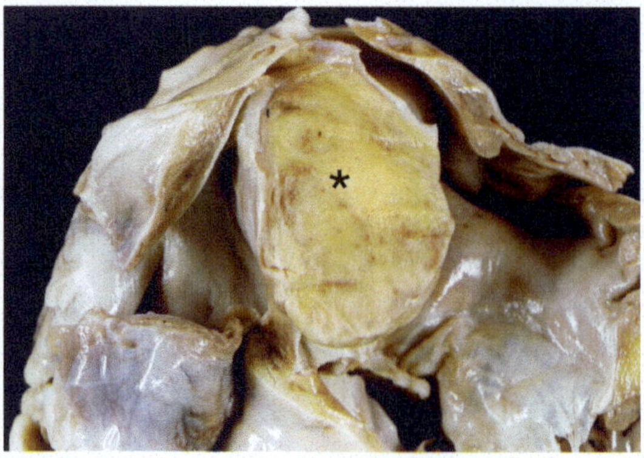

Fig. 12.12 *Lipomatous hypertrophy of the interatrial septum*. A 60-year-old woman who died from thrombosis in the right coronary artery. Longitudinal sectioning of the interatrial septum shows a well-defined nodule, non-encapsulated, formed by adipose tissue. The nodule is 2.5 cm thick, extends between the orifices of the *venae cavae* and protrudes towards the atrial chambers (*)

12.6.2 Lipomas

Cardiac lipomas are benign tumours composed of mature adipocytes. They are extremely rare and account for 3–4.5 % of cardiac tumours, according to different surgical series. In general, they are subepicardial nodules, single and asymptomatic, and in many cases are found incidentally at autopsy. Multiple cardiac lipomas (Fig. 12.13a) are associated with tuberous sclerosis, although some isolated forms have been reported that may cause syncope and sudden death due to the large size or the compression of the coronary arteries. Histologically, cardiac lipomas consist of capillaries, small vessels and mature adipocytes without atypias (Fig. 12.13b). They can be delineated by a fine capsule.

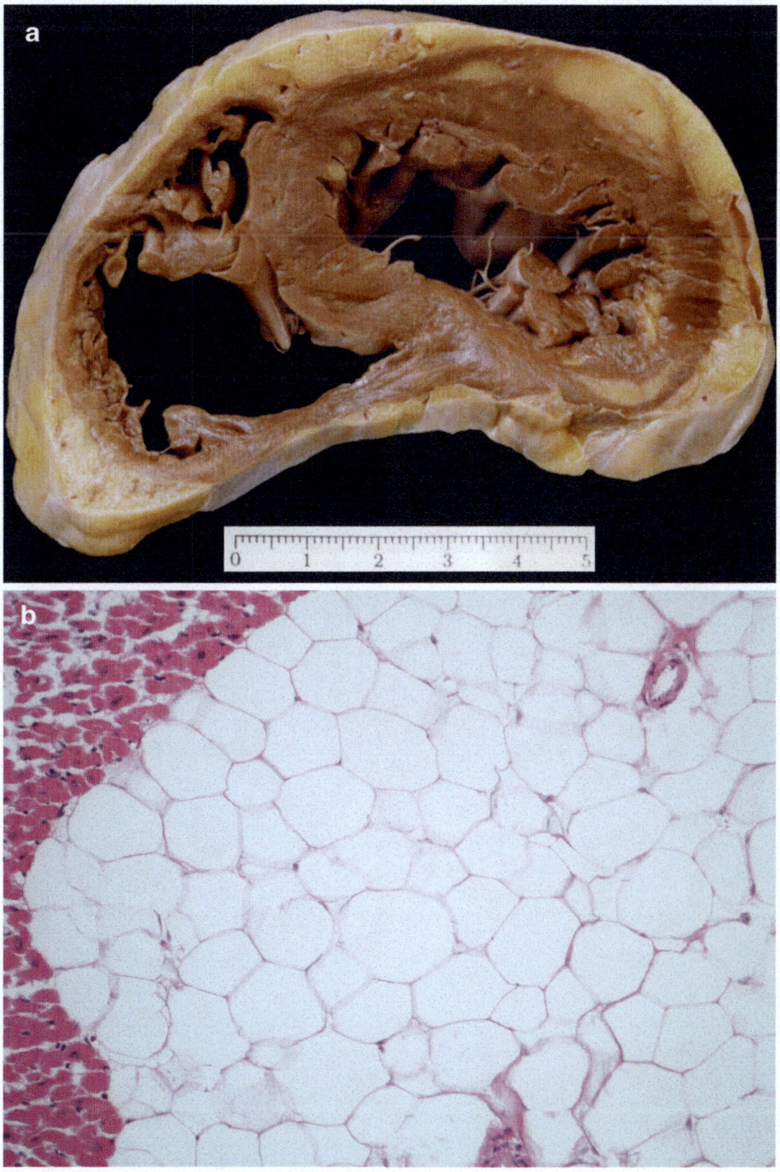

Fig. 12.13 *Multiple lipomas*. A 50-year-old woman, with a history of tuberous sclerosis and epilepsy diagnosed in childhood, on anticomicial treatment. She was found dead in bed. (**a**) Yellow intramyocardial nodules in the LV. The largest nodule is 3 cm and in contact with the subepicardial adipose tissue. (**b**) Nodules consist of mature adipocytes that displace the surrounding myocardium

12.7 Vascular Tumours

Vascular tumours of the heart can be benign (haemangiomas) or malignant (angiosarcomas).

12.7.1 Haemangiomas

Cardiac haemangiomas are very rare tumours. They can occur anywhere in the heart and arise from the pericardium, myocardium or the parietal or valvular endocardium. They are benign tumours composed predominantly of blood vessels. According to morphology, they have been classified in three types: cavernous (composed of vessels with wide lumen) (Fig. 12.14), capillary, and arterio-venous malformations (angiodysplasias) (Fig. 12.15). Clinical presentation varies according to size and location. Patients may present with pericardial effusion, heart failure, heart valve stenosis or sudden death, although most patients are asymptomatic and cardiac haemangiomas are discovered incidentally during medical examination or at autopsy.

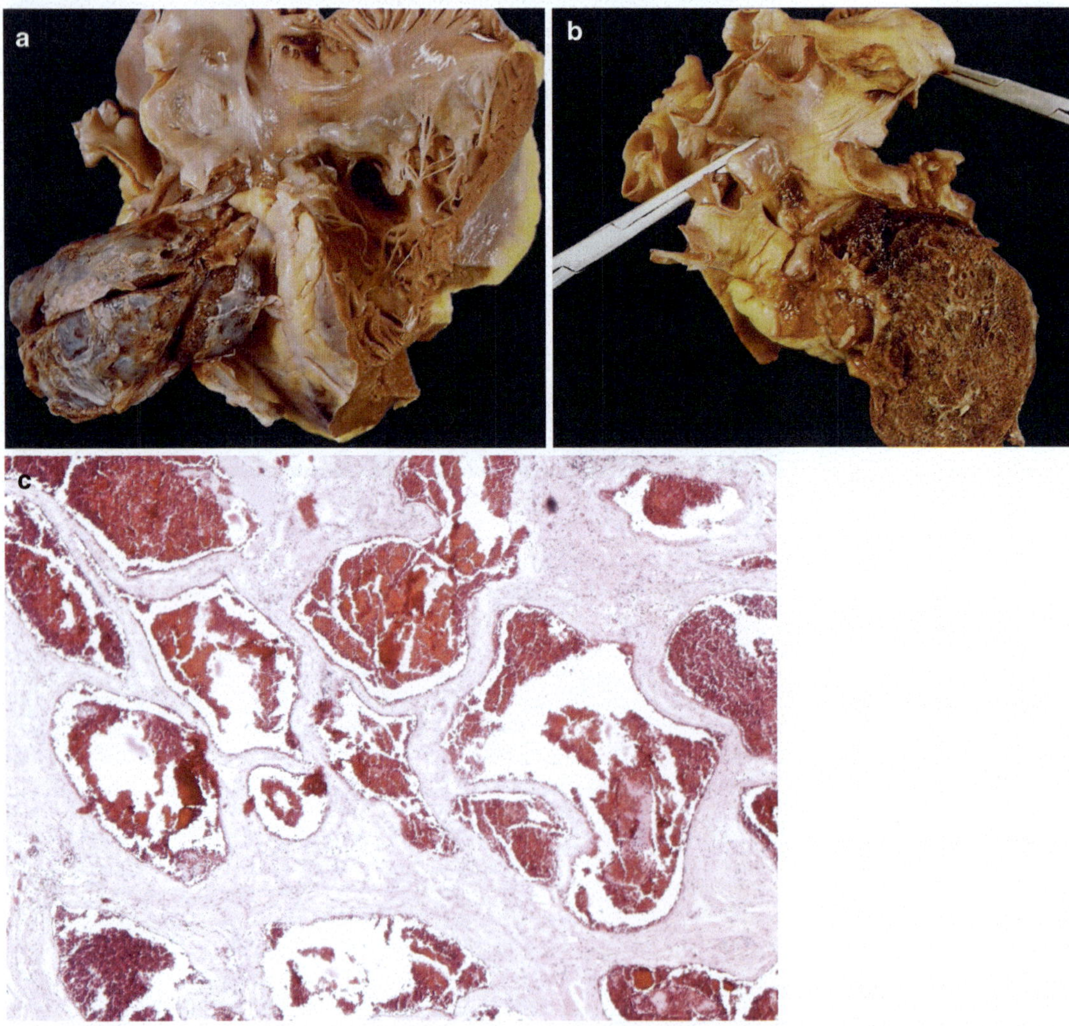

Fig. 12.14 *Cavernous haemangioma*. A 57-year-old male who died in a car accident. Heart weight of 322 g. Permeable coronary arteries. (**a**) There is an ovoid, thin-encapsulated and purple nodule of 7×4×4 cm in the atrioventricular groove, arising from the pericardium. (**b**) When sectioned, the nodule presents a spongy appearance. (**c**) Histology reveals a tumour formed by vessels lined with flattened endothelium, with no atypias and with a large lumen filled with blood. Vessels are separated by collagen tissue, typical of cavernous haemangioma

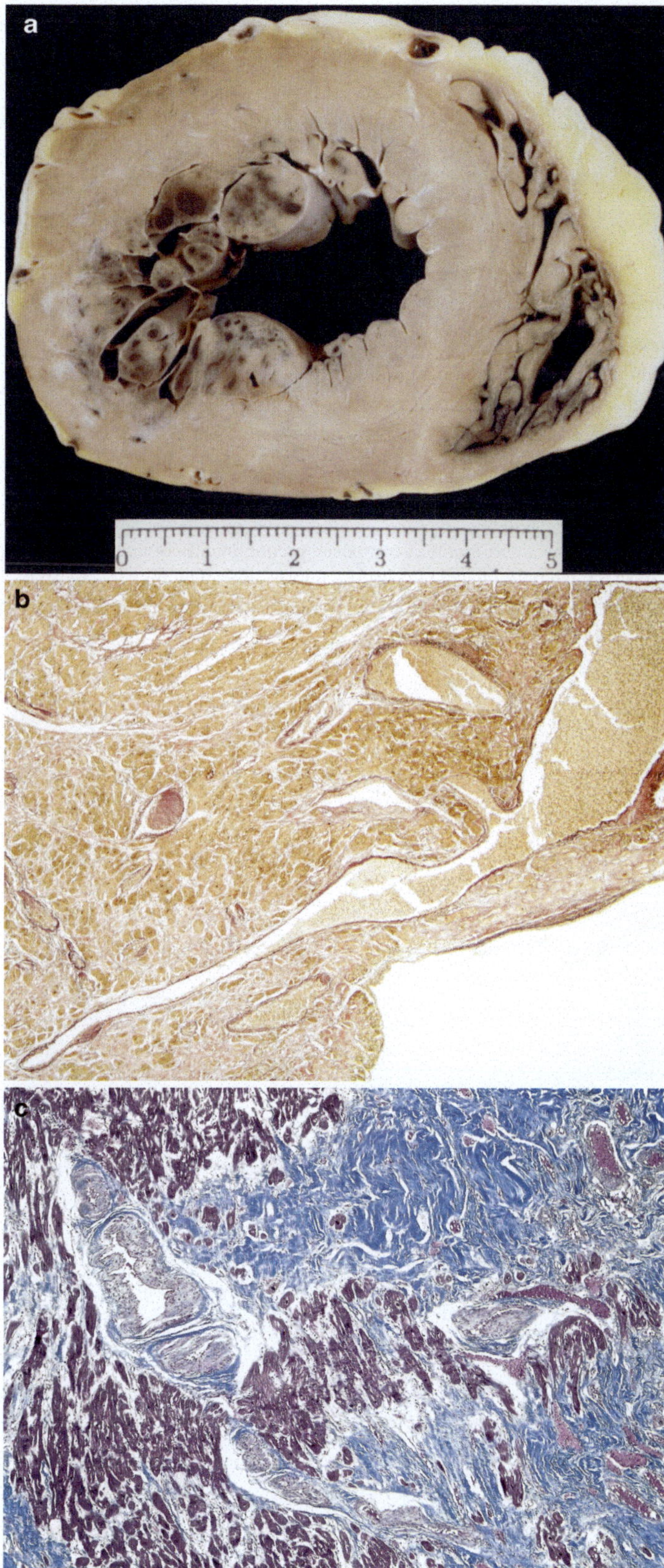

Fig. 12.15 *Arterio-venous malformation*. A 25-year-old woman with a history of interatrial communication that closed spontaneously during childhood. She suffered sudden death with no warning signs. (**a**) Heart weight was 270 g. The only finding was the congestive appearance of the papillary muscles and the subendocardial myocardium in the LV. (**b**) Microscopically, the area corresponds to a proliferation of vessels, with dilated and tortuous lumen, at the level of the trabeculae and the papillary muscles. The thickness of the vessel walls is very heterogeneous and shows a transition from venous to arterial structures, and areas with intimal proliferation (Weigert's stain for elastic fibres). (**c**) In the areas where arterial walls are very thickened and there is luminal reduction, myocardial fibrosis exists (Masson's trichrome)

12.7.2 Angiosarcoma

Angiosarcomas are the most common cardiac sarcomas and show a very aggressive behaviour. They preferentially locate in the right atrium or the pericardium, and occur mainly in adults with a peak incidence in the fourth decade. The most common presenting symptoms are heart failure, chest pain, hemopericardium and pleural effusion caused by pulmonary tumour embolism. Atrial rupture has been described in exceptional cases, leading to cardiac tamponade. Morphology is summarised in Table 12.8.

Table 12.8 Pathology of cardiac angiosarcomas

Macroscopic features:
Multilobulated masses, dark in colour, with areas of haemorrhage and necrosis
The majority involves the right atrium and infiltrates the pericardium
Exceptionally, they locate in the pericardium or the left atrium
Microscopic features:
Well-differentiated angiosarcomas are formed by irregular vascular channels lined by atypical endothelial cells that grow forming papillae
There are some angiosarcomas composed predominantly of anaplastic spindle cells, with extravasated red blood cells in the interstitium and poorly-formed vascular channels (Fig. 12.16a)
Immunohistochemical staining shows tumour endothelial cells positive for Factor VIII, CD31 and CD34 (Fig. 12.16b)

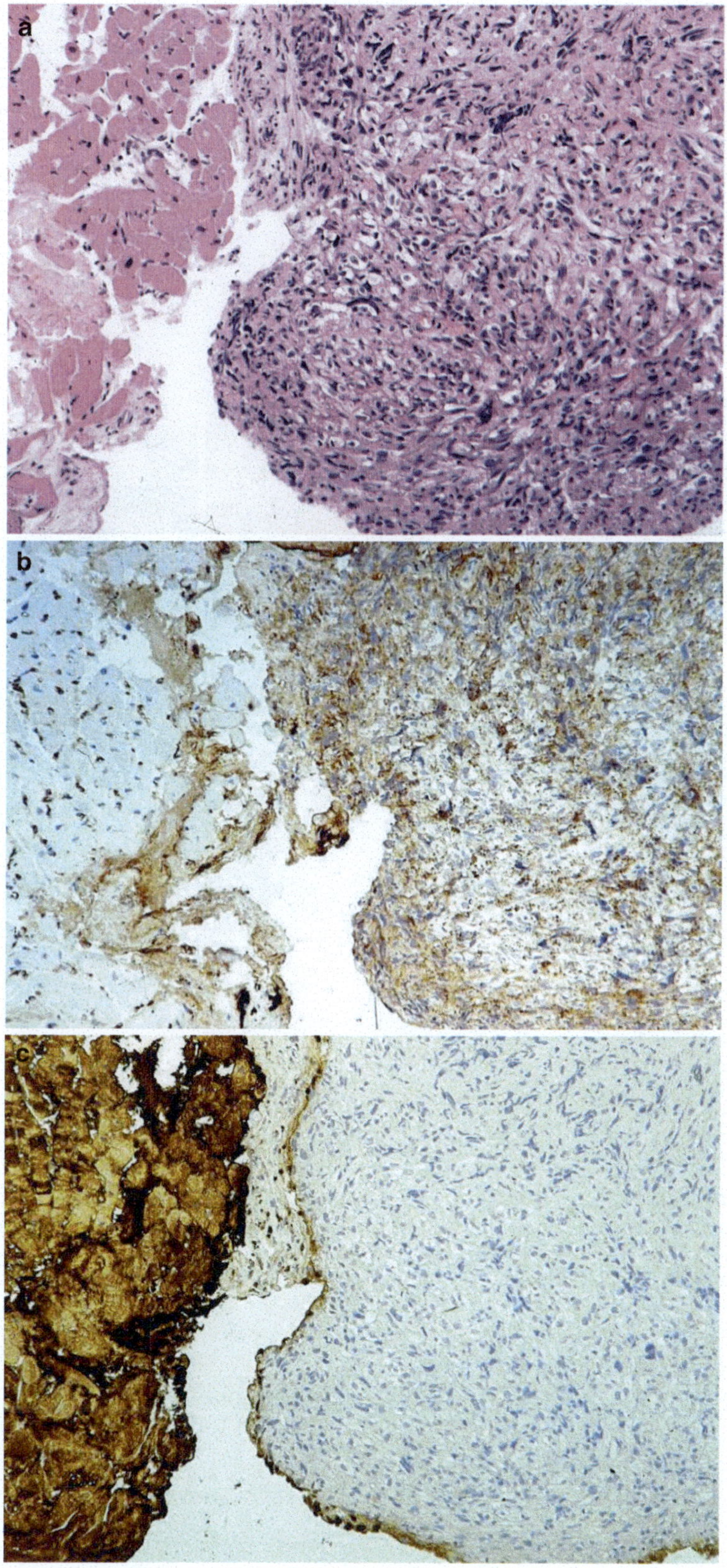

Fig. 12.16 *Angiosarcoma*. A 36-year-old male with dyspnoea and fever. Echocardiography revealed an endocavitary mass of 8 × 6 cm in the right atrium that was spreading to the ventricle. (**a**) Endomyocardial biopsy showed myocardium infiltrated by a proliferation of highly pleomorphic fusiform cells. (**b**) Immunohistochemistry showed positivity for Factor VIII, indicative of the vascular origin of this malignancy. (**c**) Tumour cells were desmin-negative, in contrast to the neighbouring myocardium (*Courtesy of Dr. Gaetano Thiene y Dra. Cristina Basso. Department of Cardiac, Thoracic and Vascular Sciences. University of Padua, Italy*)

12.7.3 Other Vascular Lesions

From a clinical point of view, two types of lesions with endothelial lining exist that, despite not being neoplasms and in general not producing symptoms, have to be included in the differential diagnosis of vascular tumours. These lesions are cardiac varicose veins and hematic cysts of the valves.

12.7.3.1 Cardiac Varicose Veins

Cardiac varicose veins typically locate in the right atrium, in the inferior border of the *fossa ovalis*. These are bulging and purple lesions that stand out on the atrial endothelium (Figs. 12.17a, b). Histologically, they are cavities full of blood, surrounded by endothelium. Thrombosis may occur, and they then appear as whitish firm nodules (Fig. 12.17c).

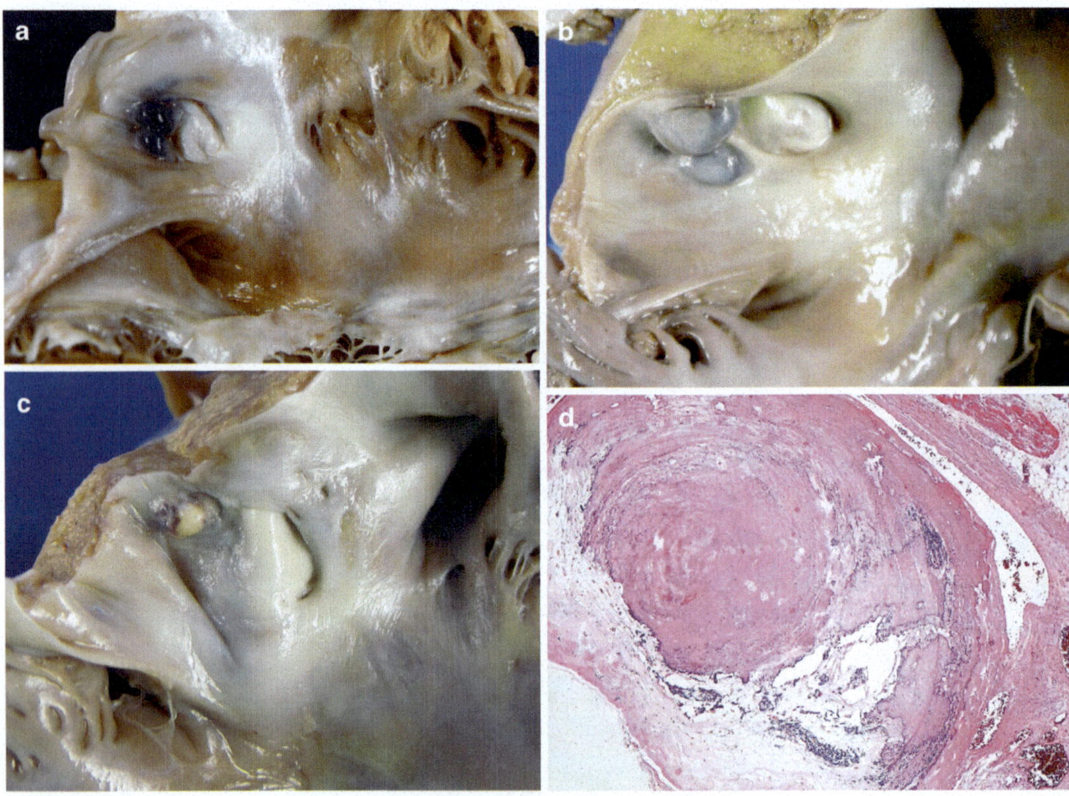

Fig. 12.17 *Varicose veins*. (**a**, **b**) Two soft purple nodules can be appreciated at the inferior border of the *fossa ovalis*. (**c**) Another case showing a whitish and more consistent nodule in the same region. (**d**) Photomicrograph of the nodule shown in (**c**), where an organised and partially-calcified thrombus can be observed in the varicose vein lumen

12.7.3.2 Hematic Cysts

These are congenital cysts that result from endothelial invagination at the level of the line of closure of the atrioventricular valves (Fig. 12.18). In exceptional cases, these cysts involve the semilunar valves. They are diagnosed during the first year of life, and then they involute. These cysts can be single or multiple and are usually a few millimetres in size, although larger sizes have also been reported, causing an obstruction to blood flow. Histologically, hematic cysts are cysts full of blood surrounded by a thin wall lined by endothelium.

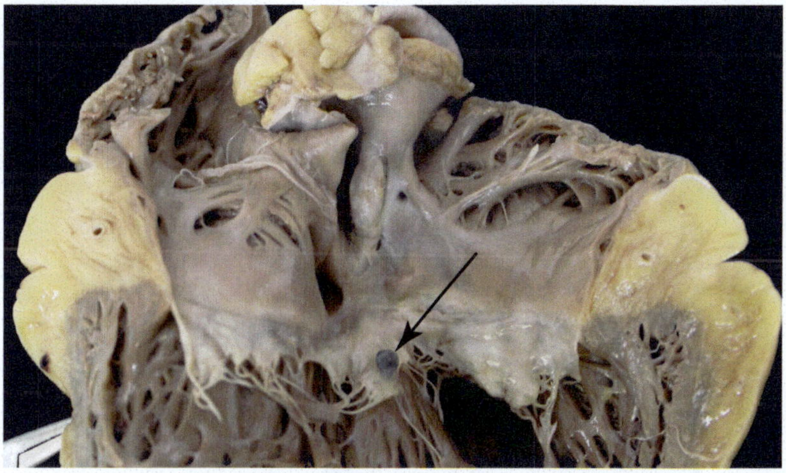

Fig. 12.18 Hematic cyst in the tricuspid valve septal leaflet (*arrow*) (*Courtesy of Trinidad Argente del Castillo, Institute of Legal Medicine. Albacete. Spain*)

12.8 Inflammatory Myofibroblastic Tumour

This tumour type is composed of myofibroblasts and has been defined as a low-grade myofibroblastic sarcoma. In locations other than the heart, it has been described as an inflammatory pseudo-tumour. These tumours generally appear implanted in the mural endocardium and occasionally in the valve leaflets. They are found predominantly in children and teenagers and appear as multiple and grouped polyploid lesions. Histologically, it is composed of myofibroblasts in bundles, with abundant mitotic figures and variable amounts of inflammatory infiltrates (Fig. 12.19).

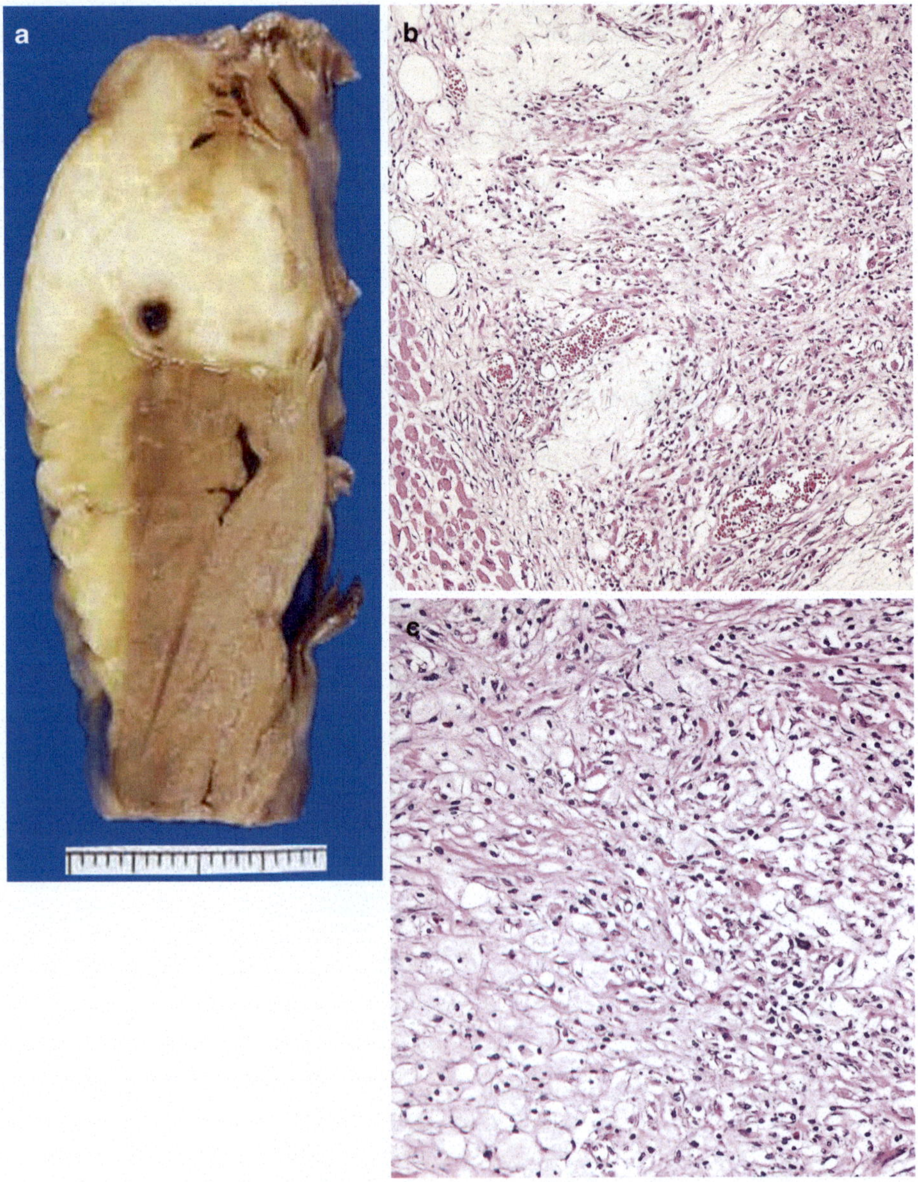

Fig. 12.19 *Inflammatory pseudo-tumour*. A 61-year-old male who died from ischaemic heart disease. (**a**) A white-yellow nodule was found growing around the lateral portion of the RCA. (**b**) The tumour consists of a proliferation of fusiform and inflammatory cells. (**c**) In some areas, numerous histiocytes with foamy cytoplasm, lymphocytes and plasma cells were observed. A similar lesion was found in the renal sinus

12.9 Primary Cardiac Lymphoma

Primary cardiac lymphoma (PCL) is a rare malignancy characterised by lymphomatous infiltration limited to the heart or pericardium. PCL accounts for 1 % of primary cardiac tumours and 0.5 % of extranodal lymphomas. Most PCL cases arise in immunocompetent patients, but the incidence of the tumour and its aggressiveness is increasing among immunocompromised individuals. The preferred locations are the right atrium and the pericardium. Clinical manifestations include precordial pain, pericardial effusion, dyspnoea, and altered heart rhythm with atrioventricular blocks. A cytological study of pericardial effusions usually establishes the diagnosis.

Histologically, PCLs are B-cell non-Hodgkin lymphomas. They can be diffuse or follicle centre cell lymphomas (Fig. 12.20), and in immunocompromised individuals, the predominant subtype is immunoblastic lymphoma. Early diagnosis and adequate treatment allow a relatively favourable prognosis.

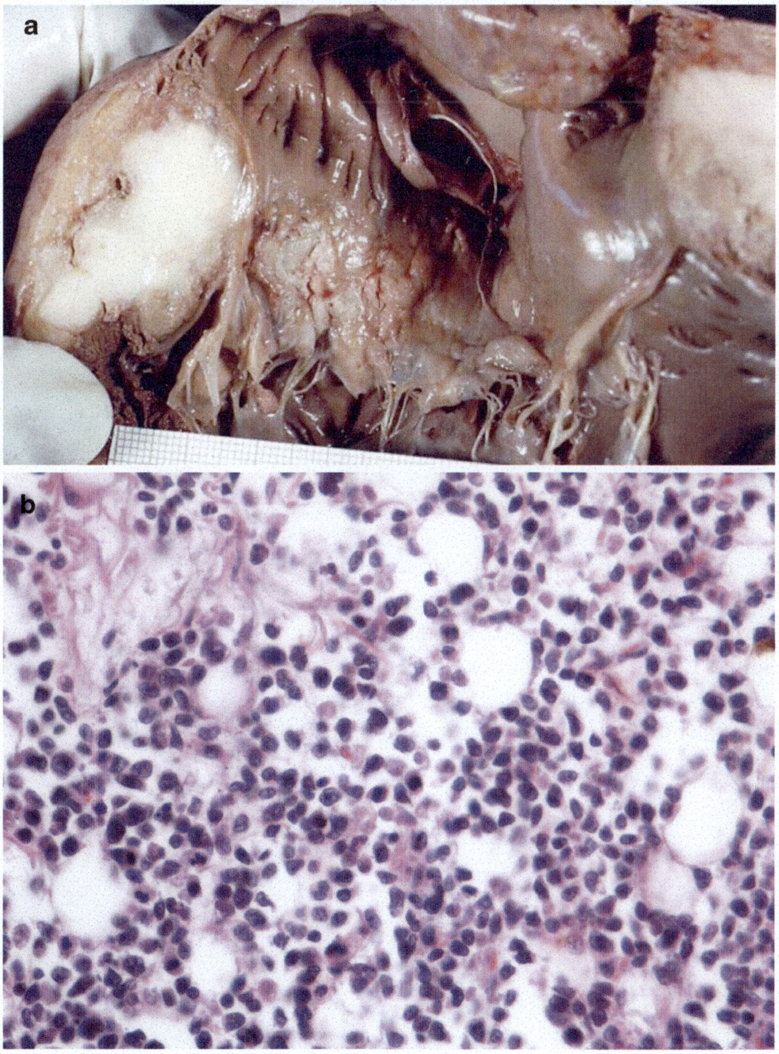

Fig. 12.20 *Cardiac lymphoma*. An 81-year-old male admitted to hospital with febrile syndrome. Multiple negative hemocultures, negative bone marrow biopsy and absence of adenopathies. While in hospital he fell out of a window. Heart weight of 480 g with coronary artery atherosclerosis. (**a**) A whitish and firm tumour mass growing around the RCA, without compromising the arterial lumen can be appreciated in the right atrioventricular groove. (**b**) Microscopically, the tumour was a centroblastic lymphoma

12.10 Secondary or Metastatic Tumours

Any tumour with an ability to spread can metastasise to the heart. Cardiac metastases are 100 times more common than primary heart tumours, with an incidence that ranges from 2.3 to 18.3 % of autopsy series. The majority correspond to metastatic carcinomas since these are much more frequent than sarcomas.

Malignancies spread to the heart by four different means: by direct extension, *via* blood, *via* lymphatics, or by intracavitary extension from the inferior *vena cava* or pulmonary veins. Mediastinal tumours and pleural mesotheliomas spread to the heart by direct extension. Tumours generally spread to the heart *via* lymphatics, most commonly through the epicardial lymphatic vessels; sometimes the spread can obstruct the lymphatic drainage of the heart, facilitating the proliferation of tumour cells in the myocardium. Renal cell carcinomas and, exceptionally hepatocarcinomas, typically propagate through the *vena cava*. Lung carcinomas can grow within the pulmonary veins and reach the left atrium; this type of propagation to the atria may affect the normal function of the atrioventricular valves. Bronchogenic carcinomas can grow and compress one or both pulmonary arteries, causing an increase in the right ventricular intracavitary pressure.

In an individual with a malignant neoplasm, the presence of pericardial effusion, enlargement of the heart, arrhythmias or heart failure should alert the physician to suspect development of cardiac metastases. Generally, the heart is not the only organ involved; it is very rare to find metastases involving only the heart. Among all the malignant neoplasms, melanomas are the tumours that most commonly metastasise to the heart.

Regarding the location of the metastases, any cardiac structure can be compromised, although the pericardium is the structure most predominantly involved (Figs. 12.21 and 12.22). Virtually any primary malignancy may metastasise to the heart. Cardiac metastases are in general asymptomatic and are found incidentally during autopsies; however, in pulmonary or breast carcinomas, cardiac metastases can result in severe pericardial effusion (serohaematic), which is a medical emergency given that it could in turn cause cardiac tamponade. A tumour should be discarded in adults presenting with pericardial effusion, by cytological examination of the pericardial fluid.

On some occasions, gross examination of the tumour infiltration merely reveals a shaggy pericardium without any apparent tumour masses (Fig. 12.23).

Remarks

Tumours reach the heart by four different means:

- By direct extension
- *Via* blood
- *Via* lymphatics
- By intracavitary extension from the inferior *vena cava* or pulmonary veins

Cardiac metastases:

- Are predominantly epicardial and usually asymptomatic
- Serohaematic pericardial effusion is common
- Very rarely, tumour metastases are confined to the heart
- Melanomas, leukaemias and lymphomas are the malignancies that most commonly involve the heart, although the most frequent are metastatic carcinomas, given their greater incidence

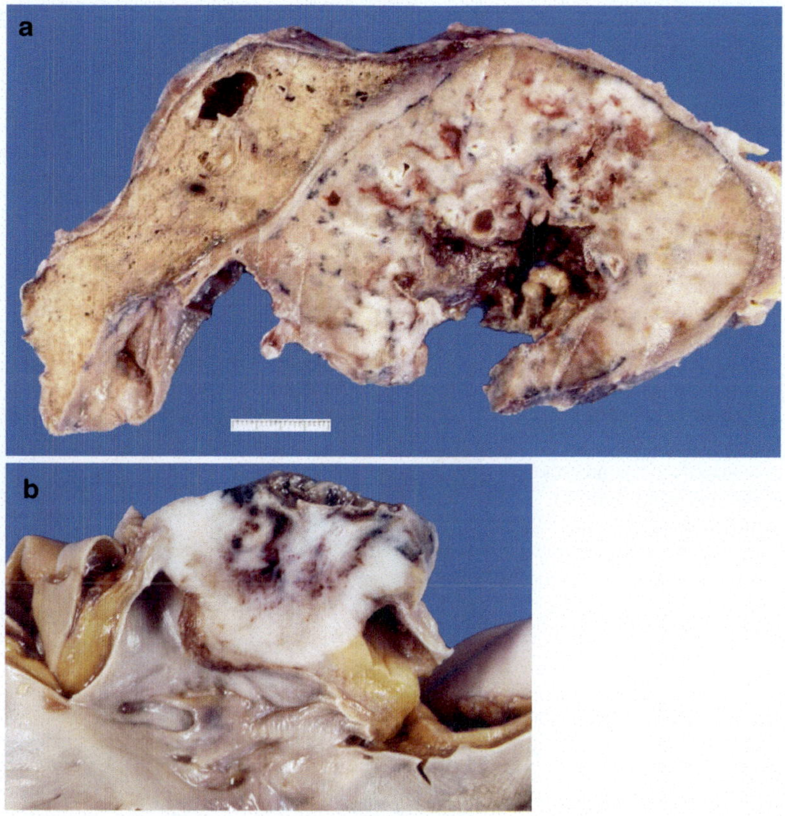

Fig. 12.21 *Metastatic lung carcinoma*. A 58-year-old male, alcoholic and smoker, with massive haemoptysis. (**a**) Autopsy revealed a cavitated carcinoma, with haematic content, in the inferior lobule of the right lung. (**b**) Metastasis in the left atrial surface. Microscopic examination showed a poorly-differentiated squamous carcinoma of the bronchus

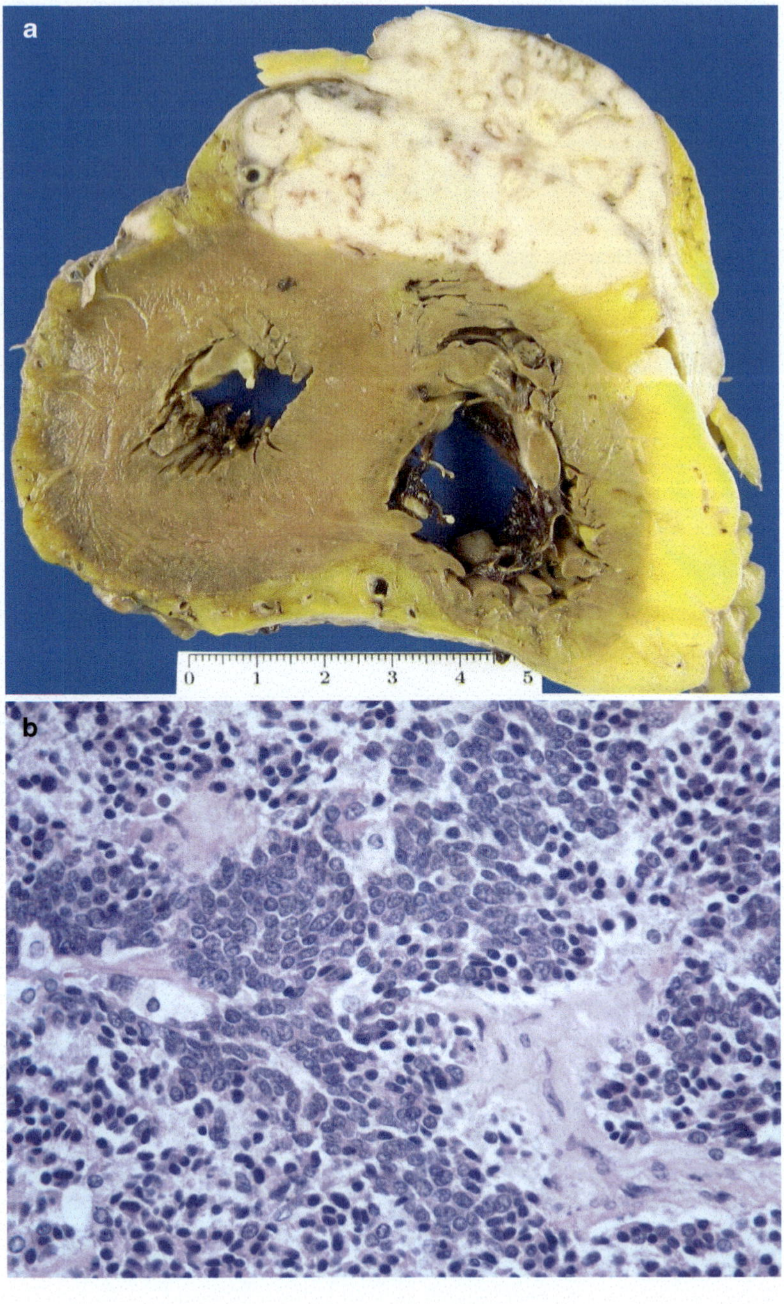

Fig. 12.22 *Neuroendocrine carcinoma infiltration of possible thymic origin*. An 83-year-old woman with massive weight loss, asthenia and anorexia not reported to the doctor. She was found dead in bed. The only finding at autopsy was a mediastinal tumour. (**a**) Whitish tumour in the thymic region that broadly infiltrates the epicardium and reaches the anterior wall myocardium. The tumour was growing around the great arterial trunks. (**b**) The morphology of the tumour corresponds to a neuroendocrine carcinoma; with the tumour cells arranged forming cords and showing nuclear atypia and mitosis

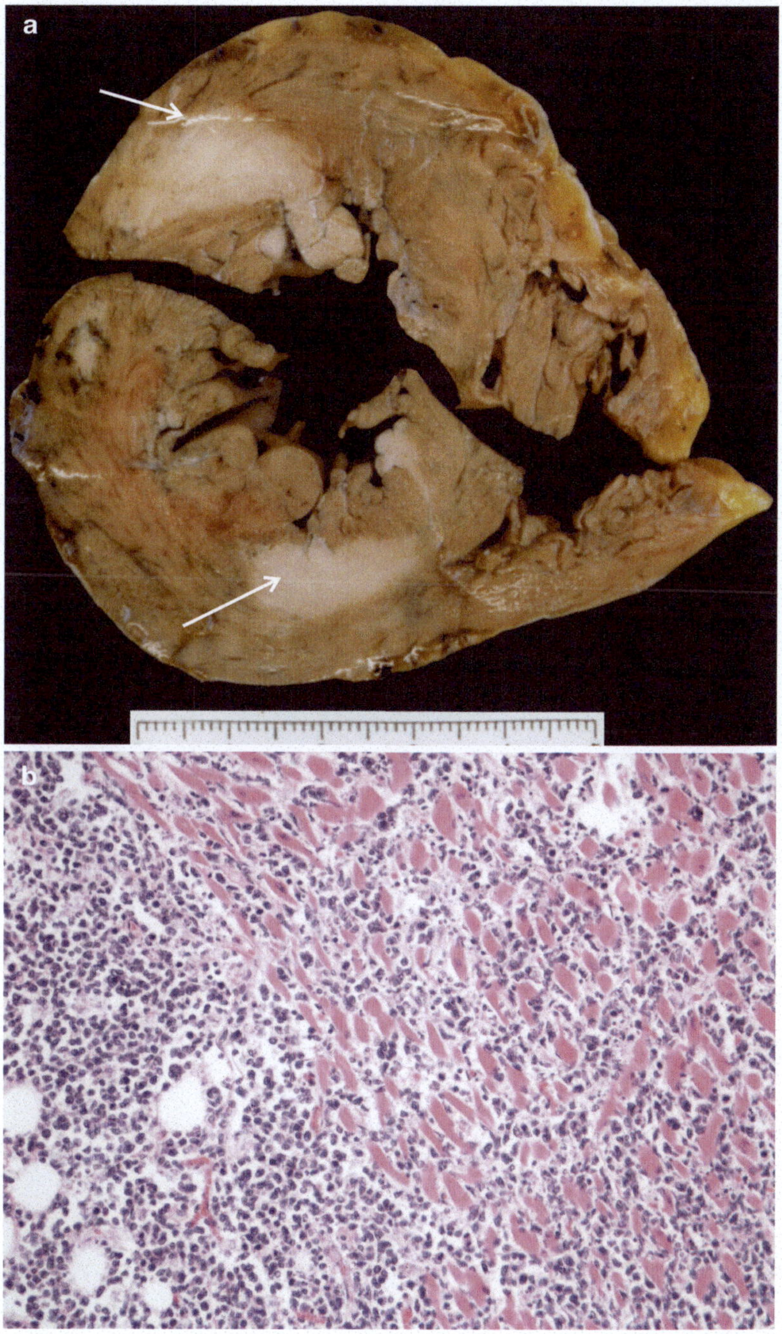

Fig. 12.23 *Metastasis of a non-Hodgkin lymphoma*. A 40-year-old woman who lost 24 kg during the previous months. She died suddenly while admitted to hospital for examination. Autopsy revealed a large tumour in the gastric antrum that had infiltrated the serosa layer. (**a**) Heart weight of 495 g, with multiple whitish nodules in the LV and a single nodule in the interatrial septum. (**b**) Myocardium and epicardial adipose tissue with lymphoid infiltration. (**c**) Immunohistochemistry showed high positivity for the lymphocyte common antigen CD45. (**d**). The lymphoma infiltrates the AV node (*arrows*)

Fig. 12.23 (continued)

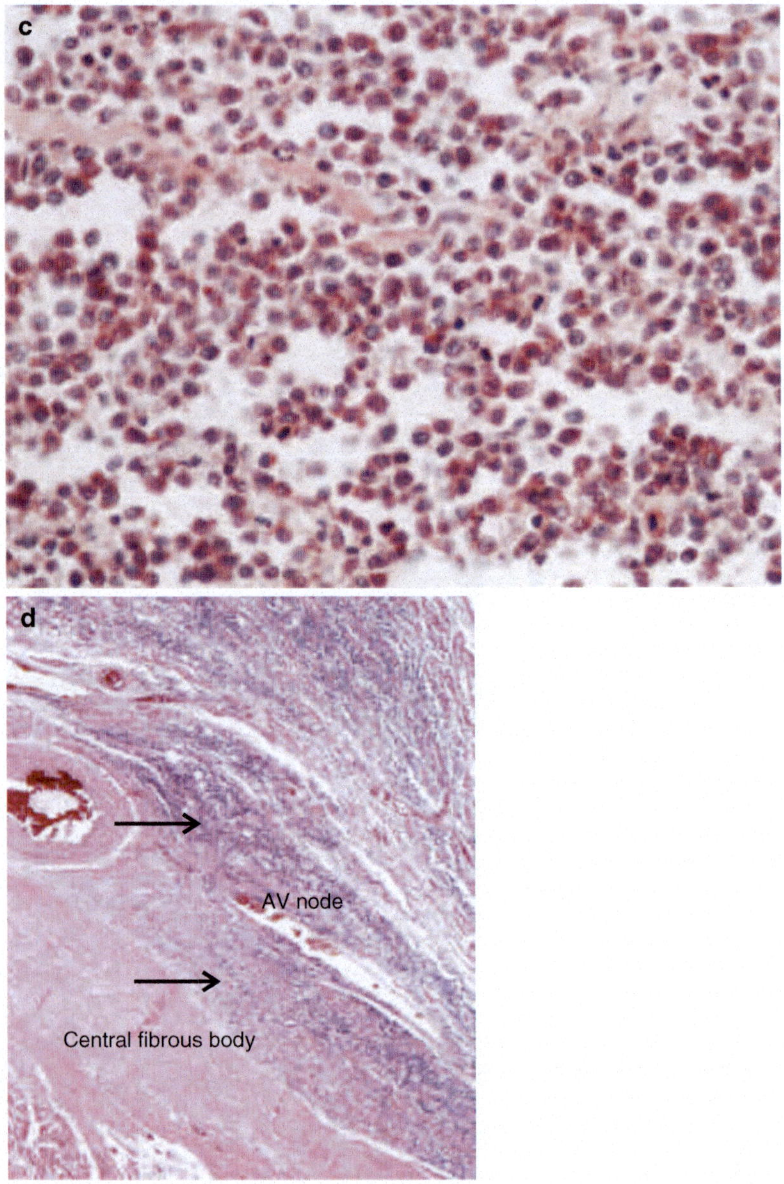

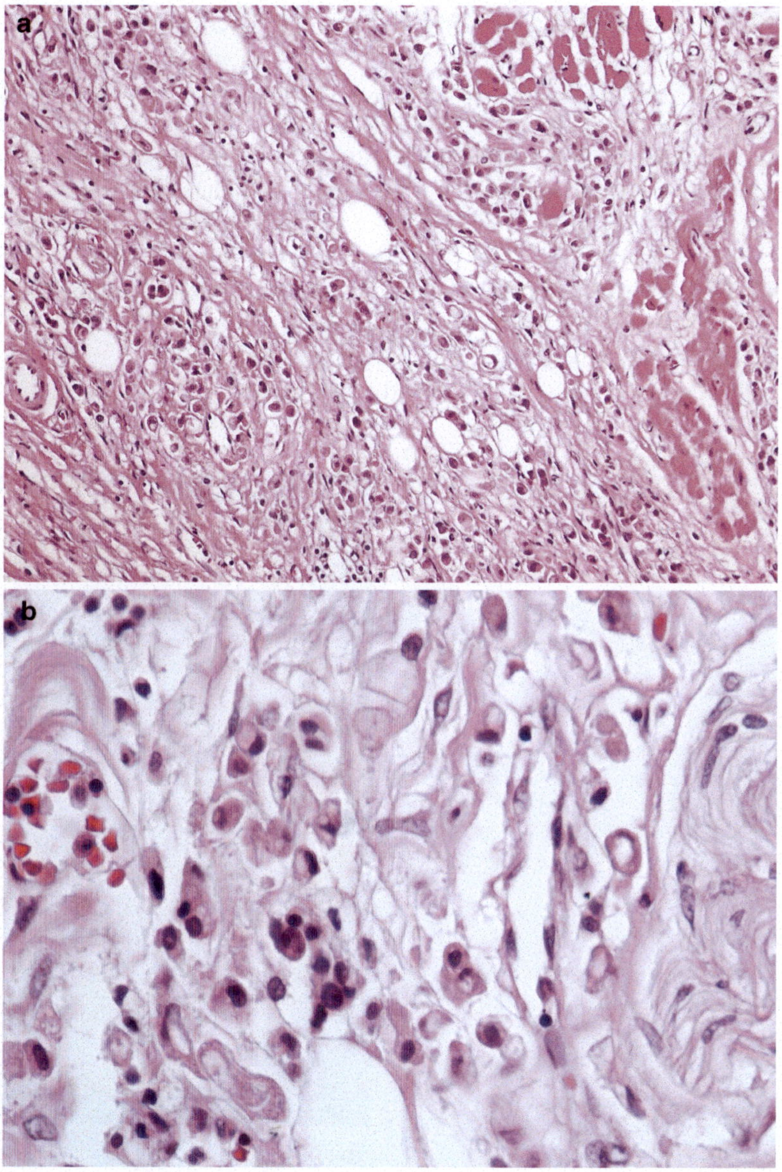

Fig. 12.24 *Metastasis of a gastric carcinoma*. A 63-year-old male with a history of gastrectomy due to linitis plastica-type adenocarcinoma, and an enolic habit. Years later, he was admitted to hospital with cardiac decompensation. Autopsy revealed pericardial effusion. (**a**) Diffuse infiltration of epicardial adipose tissue and myocardial interstitium by atypical cells. (**b**) Higher magnification image showing cells with the signet-ring appearance, typical of linitis plastica

12.11 Tumor-Like Lesions

There are several lesions that produce cardiac masses that are neither neoplasms nor hamartomas, but need to be considered in the differential diagnosis of cardiac tumours.

12.11.1 Hydatid Cyst

Hydatid cyst is produced by the cystic larvae of *Echinococcus granulosus*. Hydatid cysts are commonly located in the liver, followed by the lungs. Only isolated and sporadic cases of cardiac involvement (0.5–2 % of all cases) have been reported. Generally, it appears as a single intramyocardial cyst, predominantly located in the LV (Fig. 12.25). Clinical management requires differential diagnosis with other cystic cardiac tumours. Pathologic study allows accurate diagnosis.

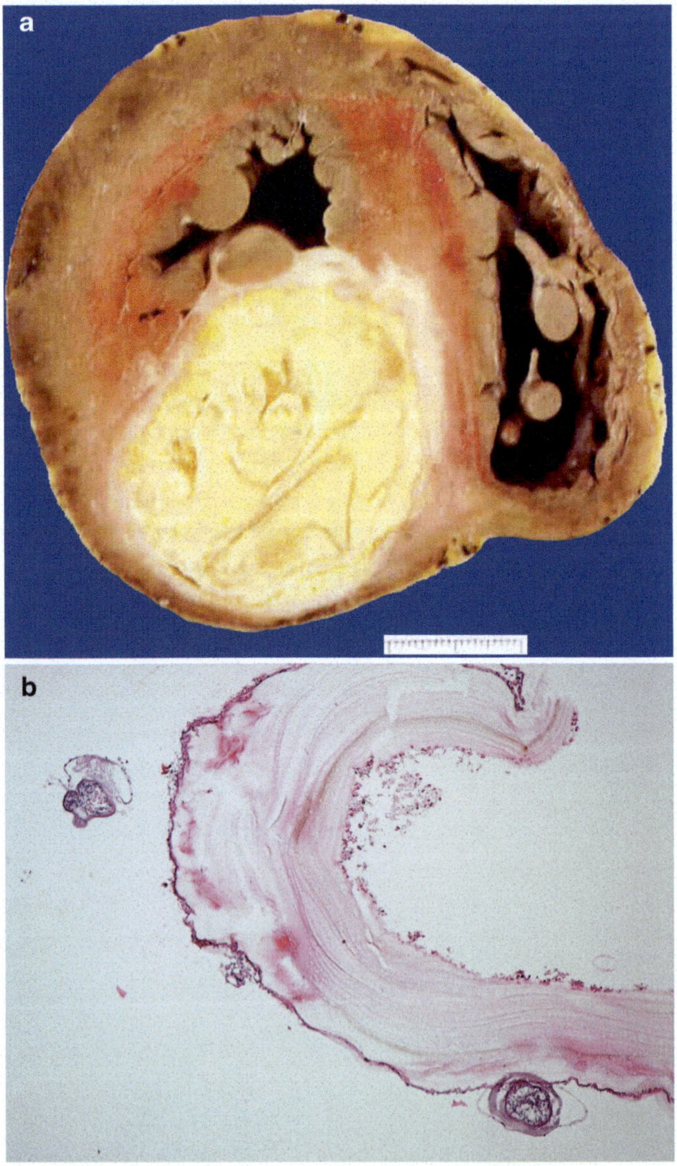

Fig. 12.25 *Hydatid cyst.*
A 25-year-old male who suffered sudden death. Hydatid cysts were found in the liver. Heart weight of 502 g.
(**a**) The posterior wall of the LV was occupied by a cystic mass of 5.5 cm in diameter that contained the typical hydatid membranes. (**b**) Histological image of one of these membranes, showing adhered scolices (*Courtesy of Rosario Mª Mosquera. National Institute of Toxicology and Forensic Sciences. Madrid. Spain*)

12.11.2 Mural Thrombi

Mural thrombi can locate in any one of the four cardiac chambers and are associated with lesions that hamper blood flow. Atrial thrombi can also appear in patients with atrial fibrillation with or without associated valve disease (Fig. 12.26). They have also been described in association with coagulopathies, antiphospholipid syndrome and disseminated lupus erythematosus. Thrombi can detach and embolise.

Microscopically, mural thrombi are composed of fibrin, platelets and entrapped blood cells. Starting from the base of the thrombi, cells start arranging and forming new vessels, which obliges a differential diagnosis with haemangiomas (Fig. 12.26b). Focal calcification may occur. It is probable that the clinical entity known as *cardiac calcified amorphous tumour* (*CAT*), characterised by cardiac masses with intracavitary growth, represent diffuse calcified thrombi.

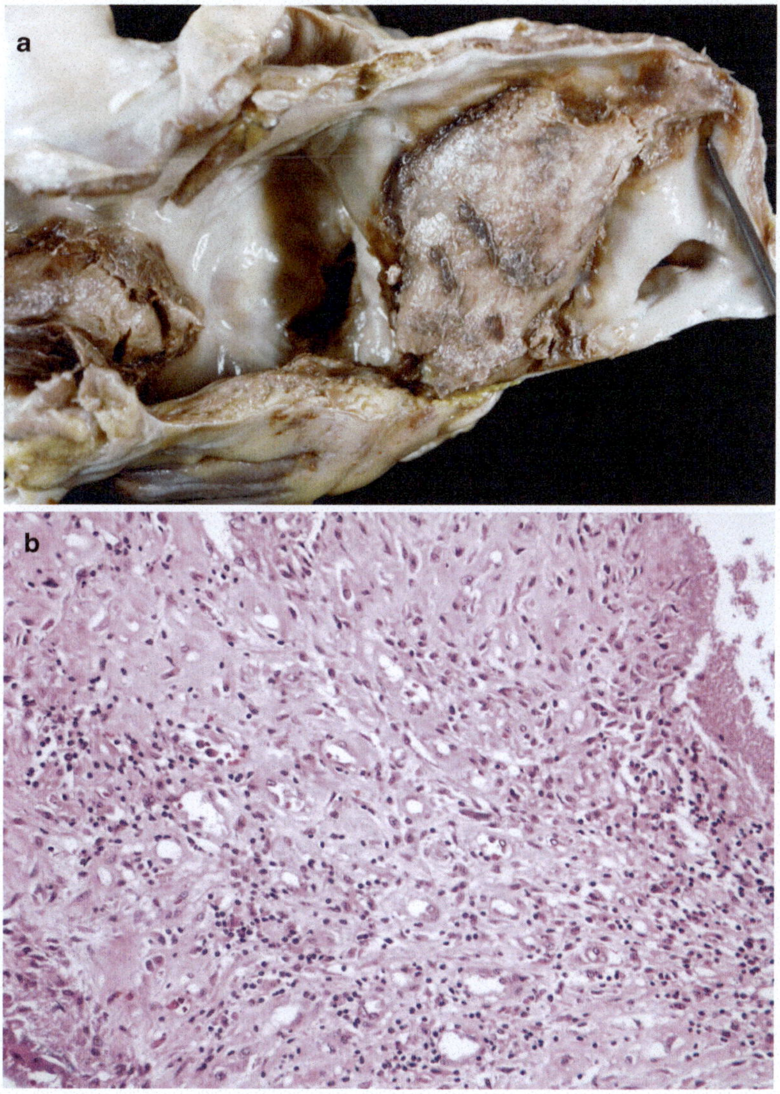

Fig. 12.26 *Mural thrombus*. An 85-year-old woman with mitral stenosis. (**a**) Dilatation of the left atrium and large thrombi adhered to the endocardium in the anterior and posterior walls. (**b**) Organised thrombus with neovascularisation

12.11.3 Atrial Septal Aneurysm

Atrial septal aneurysm (ASA) is a localised saccular deformity of the atrial septum in the region of the *foramen ovale*, characterised by its free movement towards one or both atria. Diagnosis is established when the plane extends more than 5 mm in relation to the atrial septal plane (Fig. 12.27). ASA has been reported incidentally in patients with supraventricular arrhythmias or stroke, and can be associated with *Patent foramen ovale*. Differential diagnosis should include myxomas of the left atrium, since they also arise from the *fossa ovalis*.

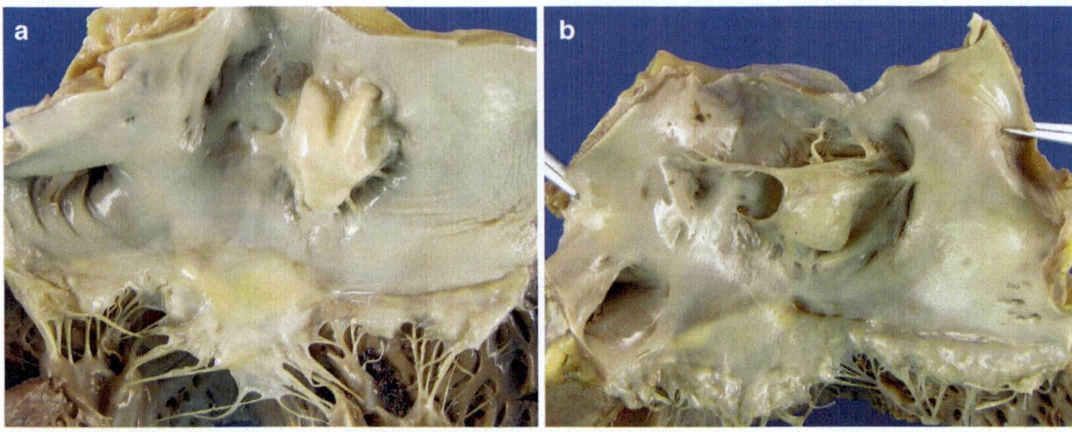

Fig. 12.27 *Foramen ovale aneurysm*. (**a**) Autopsy finding in a 78-year-old male who died from ischaemic heart disease. (**b**) A 46-year-old male who died from myocardial infarction. In this case, the aneurysm was associated with a *Patent foramen ovale*. In both cases, the aneurysm protrudes towards the left atrium

Bibliography

Aguilera B, Suárez-Mier MP, Argente T. Cardiac arteriovenous malformation causing sudden death. Cardiovasc Pathol. 2004;13:296–8.

Aguilera B, Suárez-Mier MP, García García R. Muerte súbita por embolia cerebral de mixoma de la aurícula izquierda. Cuad Med Forense. 2011;17:149–53.

Barreiro M, Renilla A, Jimenez JM, Martin M, Al Musa T, Garcia L, Barriales V. Primary cardiac tumors: 32 years of experience from a Spanish tertiary surgical center. Cardiovasc Pathol. 2013;22:424–7.

Basso C, Valente M, Thiene G. Tumori del cuore. Monografie di Cardiologia. Società Italiana di Cardiologia, 2005. Novate-Milan: Arti Grafiche Color Black.

Basso C, Valente M, Poletti A, Casarotto D, Thiene G. Surgical pathology of primary cardiac and pericardial tumors. Eur J Cardiothorac Surg. 1997;12:730–7.

Basso C, Valente M, Thiene G, editors. Cardiac tumor pathology, Current clinical pathology series. New York: Humana Press; 2013.

Burke A, Tavora F. Heart tumours. In: Practical cardiovascular pathology: an atlas. Philadelphia: Wolters Kluwer Health/Lippincott Williams & Wilkins; 2011. p. 373–415.

Burke A, Virmani R, editors. Atlas of tumor pathology: tumours of the heart and great vessels. 3rd series, fasc 16. Washington, DC: Armed Force Institute of Pathology; 1996.

Burke A, Jeudy Jr J, Virmani R. Cardiac tumours: an update. Heart. 2008;94:117–23.

Bussani R, De-Giorgio F, Abbate A, Sivestri F. Cardiac metastases. J Clin Pathol. 2007;60:27–34.

Centella T, Oliva E, García Andrade L, Lamas MJ, Epeldegui A. [A patient with cardiac angiosarcoma who survived for four years. Case report and literature review]. Rev Esp Cardiol. 2005;58:310–2.

Cina SJ, Smialek JE, Burke A. Primary cardiac tumors causing sudden death: a review of the literature. Am J Forensic Med Pathol. 1996;17:271–81.

Dell'Amore A, Lanzanova G, Silenzi A, Lamarra M. Hamartoma of mature cardiac myocytes: case report and review of literature. Heart Lung Circ. 2011; 20(5):336–40.

Gómez-Aldaraví Gutiérrez R, Otero Coto E, Chorro Gascó FJ, Muñoz Gil J, Losada Casares A, López MV. Cardiac cysts. A case of isolated cardiac hydatidosis. Rev Esp Cardiol. 1999;52(5):355–8.

Hayek E, Hughes M, Speakman E, Miller H, Stocker P. Cardiac paraganglioma presenting with acute myocardial infarction and stroke. Ann Thorac Surg. 2007;83:1882–4.

Janion M, Sielski J, Ciuraszkiewicz K. Sepsis complicating giant cardiac myxoma. Am J Emerg Med. 2008; 26(3):387.e3-4.

Krous H, Chadwick A, Isaacs H. Tumors associated with sudden infant and childhood death. Pediatr Dev Pathol. 2005;8:20–5.

Li L, Burke A, He J, Chang L, Zielke R, Fowler D. Sudden unexpected death due to inflammatory myofibroblastic tumor of the heart: a case report and review of the literature. Int J Legal Med. 2011;125:81–5.

Manfroi W, Vieira SR, Saadi EK, Saadi J, Alboim C. Multiple recurrences of cardiac myxomas with acute tumoral pulmonary embolism. Arq Bras Cardiol. 2001;77:161–6.

Mas JL, Arquizan C, Lamy C, Zuber M, Cabanes L, Derumeaux G, et al. Recurrent cerebrovascular events associated with patent foramen ovale, atrial septal aneurysm, or both. N Engl J Med. 2001;345:1740–6.

Mongal LS, Salat R, Anis A, Esrig BC, Oz M, Klapholz M, et al. Enormous right atrial hemangioma in a asymptomatic patient: a case report and literature review. Echocardiography. 2009;26(8):973–6.

Nascimento AF, Winter GL, Pinkus GS. Primary cardiac lymphoma: clinical, histologic, immunophenotypic, and genotypic features of 5 cases of a rare disorder. Am J Surg Pathol. 2007;31(9):1344–50.

Padalino MA, Basso C, Milanesi O, Vida VL, Moreolo GS, Thiene G, et al. Surgically treated primary cardiac tumors in early infancy and childhood. J Thorac Cardiovasc Surg. 2005;129(6):1358–63.

Patel J, Sheppard MN. Pathological study of primary cardiac and pericardial tumours in a specialist UK Centre: surgical and autopsy series. Cardiovasc Pathol. 2010;19(6):343–52.

Piazza N, Chughtai T, Toledano K, Sampalis J, Liao C, Morin JF. Primary cardiac tumours: eighteen years of surgical experience on 21 patients. Can J Cardiol. 2004;20:1443–8.

Poletti A, Cocco P, Valente M, Fassoli G, Chioin R, Thiene G. In vivo diagnosis of cardiac angiosarcoma by endomyocardial biopsy. Cardiovasc Pathol. 1993;2:89–91.

Reynen K, Köckeritz U, Strasser RH. Metastases to the heart. Ann Oncol. 2004;15(3):375–81.

Reynolds C, Tazelaar HD, Edwards WD. Calcified amorphous tumor of the heart (cardiac CAT). Hum Pathol. 1997;28(5):601–6.

Roberts WC. Neoplasms involving the heart, their simulators and adverse consequences of their therapy. Proc (Bayl Univ Med Cent). 2001;14(4):358–76.

Sancho Pellicer A, Labrador Fuster J. [Atrial septal aneurysms. Presentation of three cases]. Rev Esp Cardiol. 2000;56:1129–31.

Selkane C, Amahzoune B, Chavanis N, Raisky O, Robin J, Ninet J, et al. Changing management of cardiac myxoma based on a series of 40 cases with long-term follow-up. Ann Thorac Surg. 2003;76(6):1935–8.

Shapiro LM. Cardiac tumours: diagnosis and management. Heart. 2001;85(2):218–22.

Strecker T, Rösch J, Weyand M, Agaimy A. Primary and metastatic cardiac tumors: imaging characteristics, surgical treatment, and histopathological spectrum: a 10-year-experience at a German heart center. Cardiovasc Pathol. 2012;21:436–43.

Swartz MF, Lutz C, Chandan V, Landas S, Fink G. Atrial myxomas: pathological types, tumor location, and presenting symptoms. J Card Surg. 2006;21(4):435–40.

Val-Bernal JF, Acebo E, Gómez-Román JJ, Garijo MF. Anticipated diagnosis of left atrial myxoma following histological investigation of limb embolectomy specimens: a report of two cases. Pathol Int. 2003;53(7):489–94.

Index

Page numbers followed by 'f' denote figures and those followed by 't' denote tables.

J.S. Lucena et al. (eds.), *Clinico-Pathological Atlas of Cardiovascular Diseases*, DOI 10.1007/978-3-319-11146-9, © Springer International Publishing Switzerland 2015